J. Fries

# Quality Improvement Through Planned Experimentation

# Quality Improvement Through Planned Experimentation

Ronald D. Moen

Thomas W. Nolan

Lloyd P. Provost

Second Edition

McGraw-Hill

New York San Francisco Washington, D.C. Auckland Bogotá
Caracas Lisbon London Madrid Mexico City Milan
Montreal New Delhi San Juan Singapore
Sydney Tokyo Toronto

**Library of Congress Cataloging-in-Publication Data**

Moen, Ronald D.
Quality improvement through planned experimentation / Ronald D. Moen, Thomas W. Nolan, Lloyd P. Provost.—2nd ed.
p. cm.
Includes bibliographical references and index.
ISBN 0-07-913781-4 (alk. paper)
1. Quality control—Statistical methods. 2. Process control—Statistical methods. 3. Experimental design. I. Nolan, Thomas W. II. Provost, Lloyd P. III. Title.
TS156.M6223 1998
658.5'62'015195—dc21 98-8290
CIP

McGraw-Hill

A Division of The McGraw-Hill Companies



The first edition was published by McGraw-Hill in 1991 with the title *Improving Quality Through Planned Experimentation*.



3 4 5 6 7 8 9 10 FGR/FGR 0 5 4 3 2 1

P/N 043952-4
PART OF
ISBN 0-07-913781-4

*The sponsoring editor for this book was Robert Esposito, the editing supervisor was David E. Fogarty, and the production supervisor was Tina Cameron. It was set in Century Schoolbook by Ron Painter of McGraw-Hill's Professional Book Group composition unit.*

*Printed and bound by Quebecor World/Fairfield*

This book is printed on recycled, acid-free paper containing a minimum of 50% recycled, de-inked fiber.

*To the Memory of W. Edwards Deming,*
*1900–1993*

# Contents

# Foreword

This book by Ronald D. Moen, Thomas W. Nolan, and Lloyd P. Provost breaks new ground in the problem of prediction based on data from comparisons of two or more methods or treatments, tests of materials, and experiments.

Why does anyone make a comparison of two methods, two treatments, two processes, or two materials? Why does anyone carry out a test or an experiment? The answer is to predict—to predict whether one of the methods or materials tested will in the future, under a specified range of conditions, perform better than the other one.

Prediction is the problem, whether we are talking about applied science, research and development, engineering, or management in industry, education, or government. The question is, What do the data tell us? How do they help us to predict?

Unfortunately, the statistical methods in textbooks and in the classroom do not tell the student that the problem in the use of data is prediction. What the student learns is how to calculate a variety of tests ($t$-test, $F$-test, chi-square, goodness of fit, etc.) in order to announce that the difference between the two methods or treatments is either significant or not significant. Unfortunately, such calculations are a mere formality. Significance or the lack of it provides no degree of belief—high, moderate, or low—about prediction of performance in the future, which is the only reason to carry out the comparison, test, or experiment in the first place.

Any symmetric function of a set of numbers almost always throws away a large portion of the information in the data. Thus, interchange of any two numbers in the calculation of the mean of a set of numbers, their variance, or their fourth moment does not change the mean, variance, or fourth moment. A statistical test is a symmetric function of the data.

In contrast, interchange of two points in a plot of points may make a big difference in the message that the data are trying to convey for prediction.

The plot of points conserves the information derived from the comparison or experiment. It is for this reason that the methods taught in this book are a major contribution to statistical methods as an aid to engineers, as well as to those in industry, education, or government who are trying to understand the meaning of figures derived from comparisons or experiments. The authors are to be commended for their contributions to statistical methods.

W. EDWARDS DEMING
*Washington, July 14, 1990*

# Preface

This book is about planned experimentation to make improvements. We believe that statistical methods of planned experimentation are powerful aids to managers, designers, engineers, scientists, and technicians. But these methods have been applied in only a small fraction of the circumstances in which they would have been useful. The aim of this book is to provide a system of planned experimentation in such a way that there will be a substantial increase in the number of people who will use these methods.

Our approach to accomplish this aim contained several components. We continually strove to increase our understanding of the needs of managers, engineers, and others with regard to methods of experimentation. We studied and integrated the theory and methods of others with our own ideas and experiences. We were especially influenced by studying the works of W. Edwards Deming, George Box, Stuart Hunter, William Hunter, and Genichi Taguchi. Finally we developed a system of experimentation that met the essential needs of experimenters but required a lower level of mathematical and statistical sophistication than was previously necessary. We learned a great deal from Deming's papers on analytic studies (studies to improve a product or process in the future), including

- Prediction as the aim of an analytic study
- The importance of conducting analytic studies over a wide range of conditions
- The limitations of commonly used statistical methods such as analysis of variance to address the important sources of uncertainty in analytic studies
- The importance of graphical methods

The chapters on factorial and fractional factorial designs in *Statistics for Experimenters* by Box, Hunter, and Hunter (Wiley, 1978) provided the foundations for our chapters on these subjects. Especially useful to us was their description of factorial designs at two levels as a link of paired comparisons and their system of fractional factorial designs.

Taguchi's contributions to the application of planned experimentation to the design of product and processes have been integrated throughout the book, but especially in Chapter 9. We are in agreement with Taguchi that most experimenters are in need of a small number of design matrices (called orthogonal arrays by Taguchi) that cover most of their applications with only minor adaptations. Our approach to fractional factorial designs blends that approach with the important concepts of confounding included in the book by Box, Hunter, and Hunter.

The spirit of our approach to analysis of data from experiments owes much to Dr. Tukey's methods of exploratory data analysis. We have blended these ideas with Deming's counsel that confirmation of the results of exploratory analysis comes primarily from prediction rather than from the use of formal statistical methods, such as confidence intervals. Satisfactory prediction of the results of future studies conducted over a wide range of conditions is the means to increase the degree of belief that the results provide a basis for action.

In 1984 we were asked by Dr. Deming to write a book on planned experimentation from the viewpoint of analytic studies described in several of his papers and books. As notes on earlier drafts became available, they were used in almost one hundred seminars and were improved based on observation of their usefulness to the participants both during and after the seminars.

There are several aspects of this book that, when combined, make it different from others currently available. The book is written to be compatible with Deming's viewpoint of analytic studies. Also, we have presented planned experimentation as a system in the context of a Model for Improvement, which is based on the Shewhart/Deming PDSA cycle. This system includes the integration of methods of statistical design of experiments with statistical process control. As part of the system we have emphasized the sequential use of experiments and provided guidance on how to build the sequence.

We have included guidance on many of the practical aspects of planning experiments and have provided a form so that these aspects are addressed during the planning phase. Most of the examples in the book are based on our experiences. We have included examples that contain some of the problems often encountered in actual experimentation in a manufacturing plant or research facility. We offer suggestions on what to do in those situations. Another distinctive attribute of the book is the almost exclusive use of graphical methods for analysis of data from experiments.

Since we are aiming at a relatively broad audience, we have sometimes substituted methods that could be learned and used by this broad audience in place of the more traditional methods. Our feedback from experimenters convinced us of the importance of the use of original engineering units in interpreting data from experiments. We have not included designs that would require development of a mathematical model for analysis. That is why designs for mixtures or formulations and central composite designs are discussed only briefly. These are extremely useful designs, but the model-based approach necessary for their analysis would not have been compatible with approaches used throughout the rest of the book. In the case of central composite designs, we have substituted factorial or fractional factorial designs.

We have not included fairly common statistical methods such as standard errors, confidence intervals, and analysis of variance. We recognize that experimenters need to distinguish between variation that is a result of planned changes in the factors and variation that results from other sources, e.g., measurement. Based on the principles of analytic studies in Chapter 3 and the concept of common and special causes described in Appendix A, we have chosen to use graphical methods to help experimenters ascertain how much the planned changes in the factors are contributing to the variation in the data.

## Second Edition

We have included the following changes in this second edition of the book:

A chapter on testing changes (Chapter 2) develops the use of the PDSA cycle to run tests of change. This chapter comes before the introduction of the language and structure of experimental design and can be used by anyone to improve the way he or she runs tests. The importance of studying data over time is emphasized in this chapter, and graphical methods of analysis are introduced.

An appendix on evaluating measurement systems (Appendix B) has been added to make this important aspect of improving systems readily available. Understanding and improving the process for measurement play an important role in the improvement of both processes and products. The chapter on control charts in the first edition has been moved to Appendix A. The material in both of these appendixes is thus available when you use this book in teaching or as a reference guide to plan experiments.

The previous chapters on analytic studies and principles for designing analytic studies have been merged in a new Chapter 3, to give a single, complete theory and approach to designing and analyzing analytic studies.

Other emphases and examples have been added to better illustrate the important role of planned experimentation in all types of environments. For example, a discussion of experimenting on large systems has been included in Chapter 8. Application of planned experimentation to new product design is expanded in Chapter 9.

In addition, an easy-to-use software package to design and analyze experiments has been added. This software complements the book with a built-in planning form, all graphical plots, and data files by page number for the examples used throughout the book.

We thank the users of the first edition for their encouraging feedback and suggestions for improvement.

## Acknowledgments

We wish to thank Dr. W. Edwards Deming for his encouragement and his work on the distinction between enumerative and analytic studies. We appreciate the comments and suggestions of our associates Jerry Langley and Kevin Nolan. Most importantly, we wish to thank the experimenters whom we had the opportunity to work with and learn from. We appreciate their patience with us as we used earlier, sometimes very rough, drafts of this book.

*Ronald D. Moen*
*Thomas W. Nolan*
*Lloyd P. Provost*

Chapter

# 1

# Improvement of Quality

## 1.1 Introduction

Global competitive pressures are causing organizations to find ways to better meet the needs of their customers, to reduce costs, and to increase productivity. Continuous improvement of quality of products and services has become a necessary and integral part of the business strategy of organizations.

Since improvement of quality is predicated on change, making effective changes in how businesses are run has become a matter of survival. The rate and extent of improvement to products, processes, and systems are directly related to the nature of the changes that are developed and implemented. One way to develop a change is to examine the current system for flaws, problems, and opportunities for improvement. Alternatively, designing a new system without recourse to the way things are currently done is another way to develop a change.

Although improvement requires change, not all changes are improvement. What changes should we make? How will we know that the change is an improvement? To ensure that changes result in improvement requires experimentation.

Optimizing resources when making changes requires engineers, people in research and development, marketing researchers, or managers to answer questions like the following:

How is the best supplier selected for a new process?

How is the current service approach changed to achieve higher customer satisfaction?

How are the best concepts or features in a new product selected for meeting customer needs from the many that are in contention?

How are the vital few parameters for design selected from the hundreds of choices?

How is a new product designed to work under the wide range of conditions that will be encountered during actual production and use by the customer?

How are the best operating conditions chosen for a manufacturing process from the hundreds of choices?

Answers to these questions cannot be found without some form of experimentation. The most common practice is to test one concept, feature, parameter, or condition at a time. This practice consumes resources and may not lead to the best solution. This book provides strategies and methods of experimentation to answer these questions rapidly with minimal resources. Running two or three experiments with multiple concepts, features, parameters, or conditions can cut the learning time significantly.

### Purpose of this book

The purpose of this book is to provide the philosophy, principles, and methodologies to plan and conduct experiments to test whether changes to product, process, or system are improvements. Study of the proposed methods of planned experimentation will help people to learn about the many factors that impact the performance of the product, process, or system and to use this knowledge to make changes to improve quality. This planned approach maximizes learning relative to the resources expended.

This book uses a model for a sequential building of knowledge called the *Model for Improvement.* This model is designed to increase knowledge that in turn leads to improvement of the product, process, or system. The model is used throughout the book as a framework for many different types of studies.

The concepts behind improvement of quality and the Model for Improvement are developed in the next two sections of this chapter. Chapter 2 looks at testing a change. Chapter 3 lists the tools and properties of a good experiment and defines the concepts behind the design and interpretation of analytic studies.

With these foundations for experimentation in place, Chaps. 4 through 8 present the methods of planned experimentation. Chapter 9 is devoted to the application of methods to design a new product or service. Two case studies in Chap. 10 show how the model and the methods work together as a system of experimentation. Appendix A looks at the role of control charts in planned experimentation. Appendix B looks at evaluating the measurement process. Appendix C provides forms and design matrices for designing a planned experiment.

Chapters 3 through 10 can be used to teach the fundamentals to plan and conduct experiments. It may also be used as a reference in running specific experiments using the designs given in Chaps. 4 through 9. Anyone planning to test a single change should read Chap. 2. All examples in the chapters are real-world applications from our own experiences. The exercises at the end of each chapter are designed to reinforce the concepts.

## 1.2 Improvement of Quality

This section lays a foundation for experimentation by identifying the basic activities for change and the role of predictions.

### Basic activities for change

Deming (1994) views the organization as a system that includes the goal of improvement of quality in every stage from receipt of incoming materials to the consumer, as well as redesign of products and services for the future. All functions and activities are directed at a common purpose. Deming illustrated production as a system by means of a flow diagram. This diagram, first used by Dr. Deming in 1950, is reproduced in Fig. 1.1.

The flow diagram starts with ideas about a possible product or service. What might the customer need? This prediction is the 0th stage

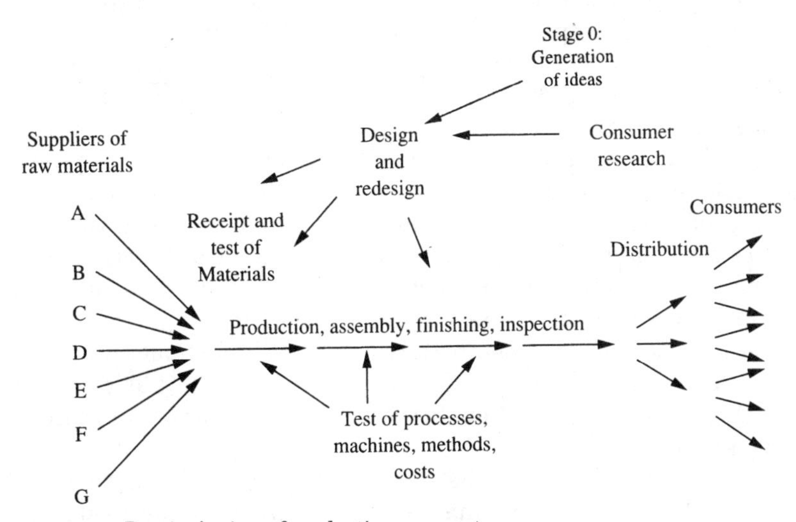


**Figure 1.1** Deming's view of production as a system.

and leads to the design of a product or service. Will the market be sufficient to keep us in business? Continuation through the cycle, including observations on the use of the product in the hands of the customer, leads to redesign—new prediction. The cycle goes on and on, design and redesign.

Improvement of quality begins with identifying the future needs of the customer through consumer research. At the design and redesign phase, products or services are designed to better meet those needs. Processes are designed to produce the product or service. The matching of products and services to a need is ongoing.

The five approaches to change follow from Fig. 1.1:

1. Design of a new product or service
2. Redesign of an existing product or service
3. Design of a new process
4. Redesign of an existing process
5. Improvement of the system as a whole

These five activities may be carried out within various functions of the organization. Efforts must be coordinated and focused on a common purpose. Cooperation between departments is needed so that people in research, design, sales, and production can work as a team in any of the basic activities. The needs of the customer must guide work in all these functions.

A planned experiment is a change made to a system (product or process) in a controlled manner for the purposes of learning.

A planned experiment can be a relatively small change to a local work process, e.g., a supervisor changing the frequency of preventive maintenance to determine whether breakdowns of equipment are reduced or a nurse changing the method of keeping track of a patient's drugs in a hospital to see whether errors are reduced.

A planned experiment could involve a very complex set of changes. For example, many factors such as amount and type of catalyst, temperature profile, pressure, and percentages of ingredients in the mix will affect the quality of a product made in a chemical reaction. The engineer may desire the settings for each of the factors that optimize the output of the process. Marketing managers may be interested in the set of factors and circumstances which attract customers to their products. They may choose to devise an experiment that changes many of these factors in a thoughtful way to learn how to optimize their marketing plan.

Planned experiments may be part of a very high-risk decision process. A national landscape maintenance company is interested in using a new chemical on the lawns of the large office complexes that it maintains. The proper use of the chemical has the potential to simultaneously give a better appearance to the lawns and slow the rate of growth of the grass. This could result in large gains in quality and productivity. The benefits are huge and so are the risks. The fertilizer improperly applied could ruin the lawns. A sound experiment to determine the correct amount of fertilizer and the proper frequency and timing of application for the many variations of soil and weather conditions is sorely needed.

Similar high-risk decisions are made everyday by executives deciding whether to move a new manufacturing process from pilot plant to full scale or to move a new product from prototype to production. Sound experiments would help lessen the risks by providing a sounder basis for making the predictions inherent in these situations.

These examples illustrate the strategic importance of effective experimentation. A company wishing to spread continuous improvement throughout its organization needs employees who are capable of designing and running simple experiments as part of daily work. A company whose success depends on introducing new products quickly must have at least some employees who are a predictable source of new ideas. However, it also must have some employees capable of designing relatively sophisticated experiments to increase the pace at which the products and manufacturing processes that result from these ideas are optimized. Effective experimentation is a core competency for any business whose success relies on high-risk decisions based on predictions of future performance or circumstances.

The greatest leverage for improvement of quality will come during the design of the new product or service and the design of a new process. The potential for improvement during these phases is many times greater than that for stages downstream. The uncertainty of quality improvement at these phases is increased since the results of tests must be extrapolated to predict how the product or process will perform in the future. There are many approaches to design. Planned experiments run during the design phases can reduce product or service development time, prevent future quality problems, and lower costs.

### Prediction

The primary reason to carry out an experiment is to provide a basis for action on the product, process, or system to improve its performance in the future. Interpreting the results of an experiment is prediction—that a change in a product, process, or system will lead to improvement in the future.

The prediction is a statement made in advance of what the state of a product, process, or system will be after the change is made. The prediction will usually be stated in terms of particular characteristics, measures, or outcomes. For example, "I predict that if I ride my bicycle to work, I can get there more quickly than by car." Here the proposed change is to ride a bicycle instead of drive a car to work. The focus of the prediction is on the time required to complete the process of going to work.

The formulation of a scientific basis for prediction has its beginnings with W. A. Shewhart in his study of the nature of stable systems. Shewhart (1931) said, "A phenomenon will be said to be controlled when, through the use of past experience, we can predict, at least within limits, how the phenomenon may be expected to vary in the future." Shewhart (1939) added

> There are three important components of knowledge: a) the data of experience in which the process of knowing begins, b) the prediction in terms of data that one would expect to get if he were to perform certain experiments in the future, and c) the degree of belief in the prediction based on the original data or some summary thereof as evidence.

In a letter to W. Allen Wallis concerning the development of statistical courses to aid in the war efforts, W. Edwards Deming (1942) stated, "The only useful function of a statistician is to make predictions, and thus provide a basis for action. The theory of estimation does not tell a person what action to recommend."

Deming (1950) expanded the idea of improving performance of a product or service in the future by differentiating enumerative and analytic studies. In an enumerative study, action is taken on the universe. In an analytic study, action will be taken on the causal system to improve performance of a product or a process in the future. Deming states that most problems in industry are analytic. This book is concerned with the design and analysis of analytic studies. (See Chap. 3 for more details on analytic studies.)

## 1.3 Model for Improvement

The Model for Improvement (Langley et al., 1996) shown in Fig. 1.2 is a framework for improvement that is widely applicable and easy to learn and use. The model is a reflection of the fact that we rationally choose which actions to take and not to take, based on existing knowledge and in the interest of accomplishing the chosen objectives. As planned actions are taken and effects studied, the body of useful knowledge grows in a way that enhances the effectiveness of the actions.

The model becomes a flexible framework for developing and testing changes and, if appropriate, the use of application-specific tools and

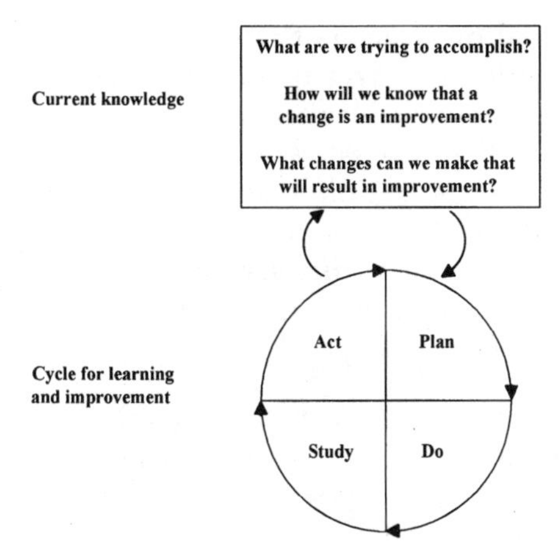


**Figure 1.2** The Model for Improvement.

methodology. The model is made up of two components: current knowledge and the plan-do-study-act (or PDSA) cycle for learning and improvement. A description of each component follows.

## Current knowledge—The three questions

The three questions that form the first component of the model are:

1. What are we trying to accomplish?
2. How will we know that a change is an improvement?
3. What changes can we make that will result in improvement?

These questions define the endpoint. Any effort to improve something would result in answers to these questions.

The answer to the first question provides an aim for the improvement effort that will guide and keep the effort focused. For example, suppose a team is formed "to eliminate quality problems." Customers' expectations are not being met with the current product. Inspection of product has historically been the approach to quality problems. Quality is obtained at high cost.

The team's answer to the first question is to "redesign the process that produces the product." The production process has a history of rework, mistakes, delays, and snags.

The answer to the second question provides the foundation for the learning that is fundamental to improvement. An experiment to test a change is run. Is the result an improvement? Criteria or measures for improvement need to be identified. Multiple measures are almost always required, to provide a balance among competing interests and to help to ensure that the system as a whole is improved.

The team's answer to the second question is that in "a capable process, all product meets specification for three quality characteristics." These measures are the response variables for testing any specific changes to the production process.

What quality characteristics are needed? A difficulty in improving quality lies in translating the needs of the customer to measurable characteristics. Garvin (1987) proposed eight dimensions of quality. The following list is an expansion of Garvin's eight dimensions:

| | |
|---|---|
| Performance | Primary operating characteristics |
| Features | Secondary operating characteristics, added touches |
| Time | Time waiting in line, time from concept to production of a new product, time to complete a service |
| Reliability | Extent of failure-free operation |
| Durability | Amount of use until replacement is preferable to repair |
| Uniformity | Low variation among repeated outcomes of a process |
| Consistency | Match with documentation, advertising, deadlines, or industry standards |
| Serviceability | Resolution of problems and complaints |
| Aesthetics | Characteristics that relate to the senses |
| Personal interface | Characteristics such as punctuality, courtesy, and professionalism |
| Harmlessness | Characteristics relating to safety, health, or the environment |
| Perceived quality | Indirect measures or inferences about one or more of the dimensions; reputation |
| Flexibility | Willingness to adapt, customize, or accommodate change |
| Usability | Relating to logical and natural use; ergonomics |

A product or service can rank high on one dimension and low on another. An understanding of the relationships of selected characteristics is basic to any improvement effort.

Translating the needs of the customer to measurable characteristics is facilitated by a diagram called a *quality characteristic diagram.* This diagram is illustrated in Fig. 1.3. The needs of the customer are listed at the top. The first column defines quality at a primary level in the language of the customer. Examples are: easy to service, easy to close, does not rattle, reliable, comfortable, or right size. The primary quality characteristics may be broken into several subcategories (secondary and tertiary), until a measurable quality characteristic is obtained. A review of the dimensions of quality is useful to uncover all the important quality characteristics.

| Needs of the customer: | | |
|---|---|---|
| | Quality characteristic[1] | |
| Primary[2] | Secondary[3] | Tertiary[3] # |
| | | |

[1] Do not include design factors in this list. (*Test*: You should *not* be able to set the levels of these quality characteristics.)
[2] Express in the language of the customer.
[3] To add more detail, subdivide into two or more quality characteristics.

**Figure 1.3** Quality characteristic diagram.

Examples of quality characteristics for different types of processes are given in Fig. 1.4. These quality characteristics are usually performance measures or results of the output of a process. (*Note:* See Chap. 9 for more on the use of quality characteristic diagrams in new product design.)

There are other sources for measures. Customer measures involve either identifying the needs and expectations of the customer to design new products or services or capturing the level of customer satisfaction for the current products or services.

**Marketing Sales/Service**

- Time to process a customer request
- Error in filling out dealer orders
- Overdue accounts
- Customer complaints
- Wrong counts
- Customer satisfaction
- Sales performance
- Slow/missed deliveries

**Engineering**

- Time to process engineering change
- Number of engineering design changes
- Failure time of product
- Change requests
- Shortage of parts

**Manufacturing**

- Downtime
- Laboratory precision
- Repair time
- Physical dimensions
- Quality outgoing
- Viscosity of batch process
- Amount of scrap
- Amount of rework
- Level of inventory
- Cost of inspection
- Employee suggestions

**Administrative**

- Time to process reports
- Errors in accounts receivable
- Cost of inspection
- Incoming calls
- Computer downtime
- Errors in purchase orders
- Idle time of cars
- Telephone usage
- Waiting time
- Transit times
- Time filling orders
- Amount of supplies
- Clerical errors
- Cost of warranty

**Management**

- Number of accidents
- Time lost by accidents
- Absenteeism
- Turnover of people
- Appraisal of people
- Training and educating people
- Percent of overtime
- Wasted worker hours due to the system
- Variance from budget
- Cost of health care

**Figure 1.4** Examples of quality characteristics.

In-process measures monitor the task and activities that produce a given result. They are usually upstream from the performance measures and are more diagnostic in nature. All three types of measures are illustrated in the supplier-customer model given in Fig. 1.5.

A common problem arises in the attempt to define the performance of a system by a single measure or a few measures concentrated in one dimension. Given one measure of success, almost any group can be successful in the short term by optimizing that measure at the expense of the other important measures. A family of measures of the system should be developed to serve as both an indicator of the pre-

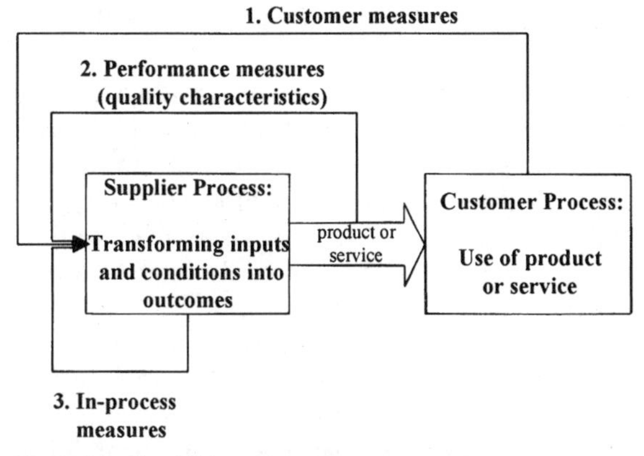


**Figure 1.5** Supplier-customer measures.

sent performance of the system and a predictor of how the system will perform in the future and to understand improvement as changes are made.

Test methods must be developed that operationally define the concepts in these measures. For physical characteristics, chemists or metallurgists may be needed. For a survey, a psychologist may be needed. Once a measurement process is operationally defined, its quality must be assessed. High quality means there is a clear "window" in which products, processes, or systems are observed. A hazy view will result in actions that may not produce the intended results. (See App. B for evaluating measurement processes.)

These measures play the role of response variables in a planned experiment. They represent the outcome of an experiment as one or more factors are changed to observe the impact on the response variable. (See Chap. 3 for more on definitions of terms associated with planned experimentation.)

It is possible to run a planned experiment without measures of the response variable. The outcomes are observed with respect to changes in the factors under study. Two three-factor experiments come to mind: scuff in main bearings and wrinkles in fabric. In both experiments the product was observed to select the most important factors.

The third question of the model was, What changes can we make that will result in improvement? This question requires some ideas for a change. Sometimes the answer is obvious. The knowledge to support a specific change already exists. Sometimes the answer is not obvious. How are these ideas generated (stage 0 in Fig. 1.1)?

Ideas for a change often come from knowledge about the current system. Sometimes ideas for a change can come from a new technology.

Some important ideas throughout history came about through chance, accident, or mistake.

Planned experimentation will play an important role in developing, testing, and implementing these changes. For example, a team working on eliminating a production problem may run an experiment on the various types of materials or different suppliers of the same material. The team may work on optimizing the machine time by changing six different factors on the machines. Or the team could test different maintenance frequencies of the machines.

Suppose the team wishes to improve the production system as a whole. Larger improvements can be made by putting processes in the context of the system in which they are embedded. To develop effective changes, the team needs to understand the nature of the relationships among the processes (interactions) that make up the activity to be improved. In a system, not only the parts but also the relationships among the parts become the opportunities for improvement.

If no new ideas for a change are being generated, moving to a more general concept level may be useful. A concept helps generate new ideas. Many of these new ideas relate back to the aim or objective. This approach helps separate the current ideas from the concepts to which they are attached. See Provost and Sproul (1996) for more on creative thinking tools as an approach to innovation and creativity.

Langley et al. (1996) recommend 70 *change concepts* for developing specific ideas for change that result in improvement. These change concepts are listed in such categories as eliminate waste, improve work flow, optimize inventory, change the work environment, enhance the producer-customer relationship, manage time, manage variation, design systems to avoid mistakes, and focus on the product or service. The change concepts are not specific enough to use directly. They must be applied to specific situations and turned into ideas for making changes.

Another approach to generating new ideas is by the theory of inventive problem solving, the translation for the Russian words *Teoriya Resheniya Izobretatelskikh Zadatch* (TRIZ). This methodology was originally formulated by Genrich S. Altshuller, an employee in the patent department of the Soviet Navy and a mechanical engineer. Through his research of worldwide patent and technology files, he identified 40 principles that had been used over and over to resolve technical contradictions. See Clarke (1996) and Altshuller (1996) for more on TRIZ.

Innovation often fails on the application end. Often, there is a fine line between success and failure. What new ideas should be used to develop changes? Screening studies from the fractional factorial designs of Chap. 6 or the nested designs of Chap. 7 are used to focus on the ideas with the greatest potential.

Once ideas for a change are developed, they need to be tested. Chapter 2 looks at the approach to designing, running, and analyzing experiments to test changes.

The focus on the three questions accelerates the building of knowledge by emphasizing a framework for learning, the use of data, and the design of effective tests by use of the *plan-do-study-act* (*PDSA*) cycle.

### Cycle for learning and improvement —The PDSA cycle

The second component of the Model for Improvement is the PDSA cycle. The PDSA cycle is an adaptation of the scientific method. Its application will enhance learning about the product, process, or system. Variations of this cycle have been called the *Shewhart cycle,* the *Deming cycle,* and the *plan-do-check-act,* or *PDCA, cycle.* Deming (1994) calls the cycle "the Shewhart cycle for learning and improvement: the PDSA cycle."

The PDSA cycle is used primarily to test and implement changes. Will the change result in improved performance of the product, process, or system in the future? What additional knowledge is necessary to take action?

The PDSA cycle is a vehicle for learning. A deduction (prediction) based on some theory is made, observation is taken (data collection), a comparison is made of the data to the predicted consequences, and a modification of the theory (learning) is made when the consequences and the data fail to agree.

Knowledge becomes useful when it results in action. Many times a person has a preconceived notion of the course of action and searches for data to support the action. No learning takes place; hence, improvements in quality may not result.

The PDSA cycle has four steps. A description of each of these steps and how it is used follows.

**Step 1: Plan.** The planning step of the PDSA cycle starts with stating the specific objective for the cycle. Some examples of objectives of a PDSA cycle are the following:

- Conduct a survey to understand customer needs.
- Do a Pareto analysis to set priorities.
- Develop control charts to study the stability of the process.
- Conduct an experiment to study the cause-and-effect relationships in the process.
- Conduct a test to evaluate changes to a product or process.

- Run an experiment to choose among competing concepts or features for a new product or service.

Develop questions raised by the specific objective of the cycle. The set of questions to be answered by the data will be necessary input to the plan. Predict the answers to these questions by use of current knowledge. Do others agree with these predictions? Develop a plan to carry out the cycle (who, what, where, when). Include methods for collecting data and analysis of the data.

The use of data from measures plays an important role in learning. *Data* are defined as the documentation of an observation or a measurement. What information is contained in the patterns of variation in data? How is this information used to guide actions for improvement? Plotting data over time maximizes the learning from any data you collect and allows the information to unfold as it happens. Is the process dominated by common or special causes? (See App. A for a discussion of common and special causes.)

**Step 2: Do.** The second step begins by carrying out the plan developed in the previous step. The plan could be to test a change aimed at improvement. Observations made in carrying out the plan should be documented. Identify the things observed that were not part of the plan. Evaluate the data for changes over time. Document what went wrong during the collection of data. This step includes control of the quality of the data being obtained.

**Step 3: Study.** Once the data are obtained, they are analyzed by methods considered during the planning step. Study the results. Current knowledge is modified if the data contradict certain beliefs about the process. If the data confirm the existing knowledge about the product or process, there will be an increased degree of belief that the current knowledge provides sufficient basis for action.

Compare the analysis of the data to current knowledge. Do the results of this cycle agree with the predictions made in the planning step? Under what conditions could the conclusions from this cycle differ? What are the implications of the unplanned observations and problems during the collection of the data? Synthesis as well as analysis is needed. What has been learned?

Summarize the new knowledge gained in this cycle. Revise the current knowledge to reflect this new information. Will this new knowledge apply elsewhere?

**Step 4: Act.** Based on the results of step 3, decide whether to make a change to the product, process, or system. The decision may be to go through the cycle without making a change or to plan a new cycle. Questions to consider include these:

- Has the appropriate action or change been developed or selected?
- Have the changes been tested on a small scale?
- Will the actions or changes improve performance in the future?

Assign responsibilities for implementing and evaluating the changes in the product or process. List forces in the organization that will help or hinder the changes. Identify the organizations and people affected by the changes. Communicate and implement the changes. What should be the objective of the next cycle? Figure 1.6 is a work sheet to aid in documenting the effort.

Because the PDSA cycle is so simple, many people say they use it. Some questions for the users of the PDSA cycle might include these:

- Is the planning based on theory? Is the theory stated?
- Are the predictions made prior to data collection?
- Are multiple cycles run to enhance the iterative learning process to develop, test, and implement changes?
- Is there documentation of what was learned?
- Does the learning provide a basis for action?

### Sequential experimentation*

By repeated use of the PDSA cycle, knowledge of the product or process is increased sequentially. Degree of belief in the prediction is increased as knowledge of the product or process is increased. Figure 1.7 illustrates the iterative nature of this building of knowledge through the multiple cycles to develop, test, and implement changes that will result in improvement.

Small-scale rapid cycles are recommended as an effective trial-and-learning approach to making changes. For systemwide changes, multiple PDSA cycles are usually run simultaneously. Figure 1.8 lists a typical sequence of cycles that a team might perform. The order of the cycles will differ depending on the situation.

George Box (1997) presents learning as a deductive-inductive iterative process (PDSA cycles) and tells the story of the Wright brothers who conducted hundreds of experiments, each of which led to the next

---

*Dr. Deming's choice for the title of this book was "Sequential Experimentation for Quality."

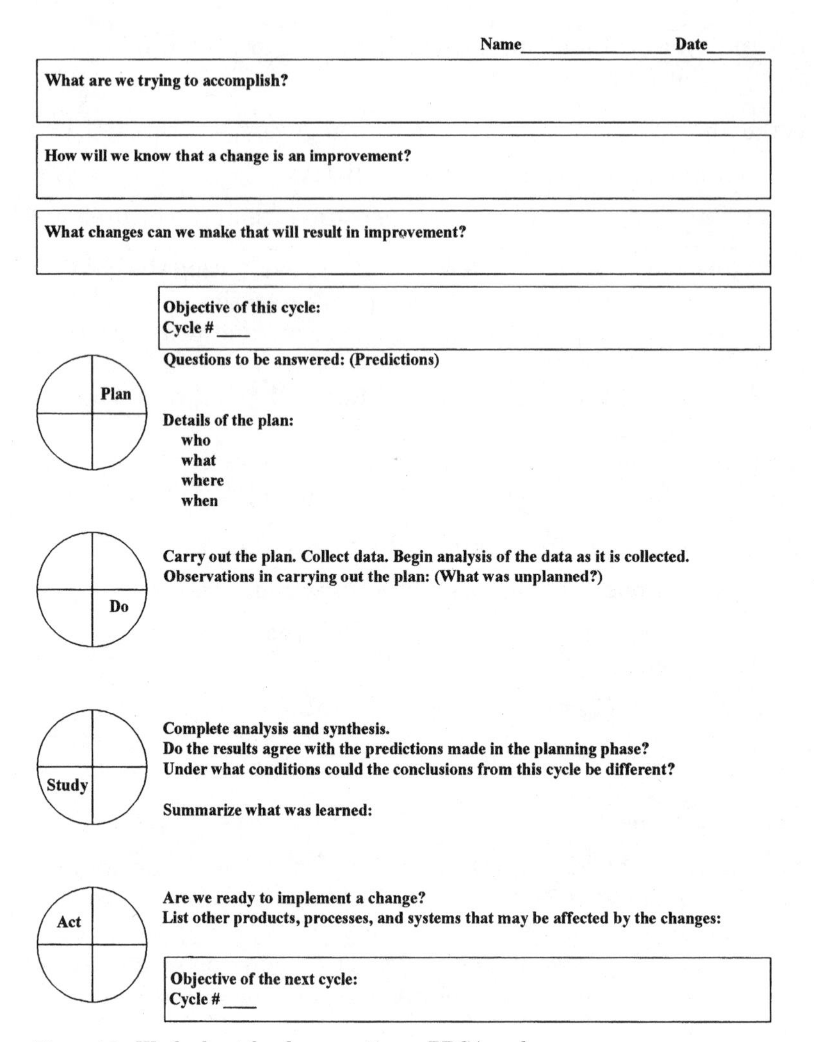
Name______________ Date______

| What are we trying to accomplish? |
| --- |
| |

| How will we know that a change is an improvement? |
| --- |
| |

| What changes can we make that will result in improvement? |
| --- |
| |



Objective of this cycle:
Cycle # ____

Questions to be answered: (Predictions)

Details of the plan:
who
what
where
when

Carry out the plan. Collect data. Begin analysis of the data as it is collected.
Observations in carrying out the plan: (What was unplanned?)

Complete analysis and synthesis.
Do the results agree with the predictions made in the planning phase?
Under what conditions could the conclusions from this cycle be different?

Summarize what was learned:

Are we ready to implement a change?
List other products, processes, and systems that may be affected by the changes:

Objective of the next cycle:
Cycle # ____

**Figure 1.6** Work sheet for documenting a PDSA cycle.

(from kites to gliders to powered aircraft). During these experiments the Wright brothers discovered several of the current theories about flight were wrong.

Planned experimentation utilizes multiple PDSA cycles to learn about a product, process, or system and uses this knowledge to make improvements. Does the theory need to be modified? Has the knowledge increased so that a prediction of the results of future experiments can be made? What will be the impact downstream, at a later stage in the process? Should a new condition be tested with another PDSA cycle? As knowledge is increased, the focus moves from developing a change with

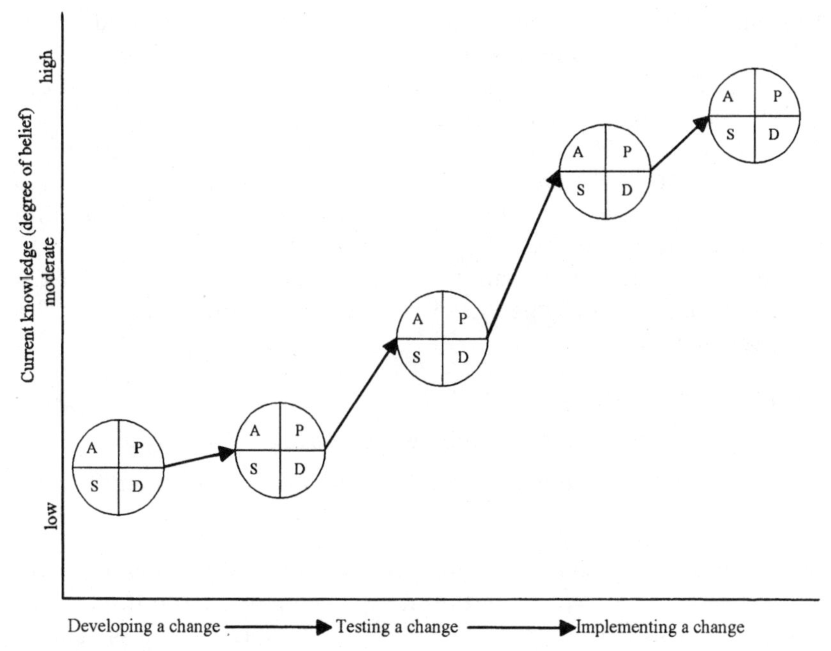


**Figure 1.7** Sequential building of knowledge.

| Improve a process | Improve a product |
|---|---|
| *Cycle 1*. Standardize the process. | *Cycle 1*. Survey to determine quality-characteristics. |
| *Cycle 2*. Study the measurement system | *Cycle 2*. Test different concepts. |
| *Cycle 3*. Identify and remove special causes. | *Cycle 3*. Run a planned experiment on prototypes. |
| *Cycle 4*. Remove dominant common causes. | *Cycle 4*. Pilot-test the new product. |
| *Cycle 5*. Monitor process. | *Cycle 5*. Survey to determine the acceptance of the new product. |

**Figure 1.8** Potential PDSA cycles.

a screening experiment (Chap. 6) to making sure the change is an improvement by use of confirmatory experiments (Chap. 4).

## 1.4 Summary

Improvement of quality of a product or service is accomplished through five approaches: design or redesign of products, design or redesign of processes, or improvement of the system that produces the products and services. The starting point is to view the organization as a system. Products and services from the system must match the needs of the customer.

Making predictions is inherent in the strategy. These predictions and the decisions based on them can be high-risk, but also high in potential benefit. Effective sequential experimentation can reduce the risks while increasing the benefits.

Any system must have a generation of new ideas. This system must have employees who have the capability of designing relatively sophisticated experiments to increase the pace at which products and services result from these ideas.

The Model for Improvement with three questions and the PDSA cycle provides a "knowledge-based road map" for improvement. The three questions are

1. What are we trying to accomplish?
2. How will we know that a change is an improvement?
3. What changes can we make that will result in improvement?

A planned experiment is a change made to a system in a controlled manner for the purpose of learning. The use of the Model for Improvement and sequential experimentation to develop, test, and implement changes will result in improvements to products and services. Applying this strategy to the five activities will provide "organizational learning," not just a learning organization. The result will be every person working on improvement of quality to enhance customer satisfaction and an organization with products and services that will be able to compete in the international marketplace.

## References


Altshuller, G. (1996): *And Suddenly the Inventor Appeared—TRIZ, the Theory of Inventive Problem Solving,* Technical Innovation Center, Inc., Worcester, Mass.

Box, George (1997): "Scientific Method: The Generation of Knowledge and Quality," *Quality Progress,* vol. 30, no. 1, pp. 47–50, January.

Clarke, Dana W. (1996): "Enhancing the Value of the Correlation Matrix through Utilization of the Theory of Inventive Problem Solving, TRIZ," *QFD Transactions,* QFD Institute, Ann Arbor, Mich., pp. 323–354.

Deming, W. Edwards (1942): Letter dated April 24 to W. Allen Wallis of Stanford University. Available for view at The Deming Collection, Library of Congress, Washington, D.C.

Deming, W. Edwards (1950): *Some Theory of Sampling,* John Wiley, New York (reprinted by Dover Publishing, 1960).

Deming, W. Edwards (1994): *The New Economics for Industry, Government, Education,* 2d ed., Massachusetts Institute of Technology, Center for Advanced Engineering Study, Cambridge, Mass.

Garvin, David A. (1987): "Competing on the Eight Dimensions of Quality," *Harvard Business Review,* vol. 87, no. 6, pp. 101–109, November.

Langley, G. J., K. M. Nolan, T. W. Nolan, C. L. Norman, and L. P. Provost (1996): *The Improvement Guide,* Jossey-Bass, San Francisco.

Provost, L. P., and R. M. Sproul (1996): "Creativity and Improvement: A Vital Link," *Quality Progress,* vol 29, no. 8, August.

Shewhart, W. A. (1931): *The Economic Control of Quality of Manufactured Product,* American Society for Quality Control, Milwaukee, Wis. (reprinted 1980).
Shewhart, W. A. (1939): *Statistical Method from the Viewpoint of Quality Control,* The Graduate School, USDA, Washington, D.C.

## Exercises

**1.1** List major processes in your organization. Identify suppliers and inputs for each process. Identify outcomes and customers for each process. List key stages in each process. What are the quality characteristics for selected inputs and outcomes of each process? What are some important in-process measures? How is customer feedback obtained?

**1.2** Identify some objectives and methods that might be useful in the PDSA cycle.

**1.3** Give two specific examples of each of the five basic activities for improvement.

**1.4** Identify (1) in-process measures, (2) quality characteristics, and (3) customer feedback for: (*a*) engineering process, (*b*) manufacturing process, (*c*) managing process, and (*d*) teaching process.

**1.5** Discuss how the PDSA cycle may have aided your past experimental investigations. Can you think of an experimental investigation that was not iterative?

**1.6** List some examples of changes that have been carried out in your organization. Were they successful? Were data used? Were predictions made prior to the collection of data? How might planned experimentation have been useful in developing or testing these changes?

Chapter

# 2

# Testing a Change

## 2.1 Introduction

When a change is developed, how should the potential improvement from the change be evaluated before implementation of the change? A common approach is to collect data before and after the change and to use the difference between the two sets of data to describe the potential improvement. In this chapter, we show why this approach is often inadequate, and we demonstrate the utility of annotated run charts to understand the effect of the change. The discussion addresses test of change at a level that anyone can use to improve the way to run tests. Later chapters develop the methods of planned experimentation for more complex testing situations.

This chapter describes how to use the PDSA cycle to test changes. Three basic principles of testing a change are described:

1. Test on a small scale and build knowledge sequentially.
2. Collect data over time.
3. Include a wide range of conditions in the sequence of tests.

Following these principles will result in conducting tests in which the effects observed can be clearly tied to the change of interest. Before we discuss these principles, we present some background information on testing changes.

## 2.2 Prediction and Degree of Belief

Chapter 1 introduced the idea of a change as a prediction. When planning to test a change, a team or individual is making a prediction that the change will be beneficial in some way in the future. When one is considering making predictions about a change, it is important to recognize that the conditions in the past, during the test, and in the future will be different. A very limited set of conditions will be present during the test. Circumstances unforeseen or not present at the time of the test will arise in the future. Will the change still result in an improvement under these new, future conditions?

The process of developing and stating a prediction prior to making a change forces the improvement team or individual to assess the theory or theories that support this change. This statement also puts team members in a position to learn as they test changes.

How well are the changes being considered understood? How precisely can the effect of making a change to a process be predicted? The concept of *degree of belief* introduced in Chap. 1 provides a way to think about and assess the depth of an improvement team or individual's knowledge about a change. Knowledge about the change is knowledge based on the specific subject matter on which the change is based, as well as knowledge about the environment in which the change will be implemented. Extrapolation of the test results to the future is the primary source of uncertainty when a change is tested. Determining whether the change resulted in improvement during the test is important, but this determination is usually much less difficult than consideration of the effect of the change in the future.

Satisfactory prediction of the results of tests conducted over a wide range of conditions is the means for team members to increase their degree of belief that they are ready to implement a change. Thus, degree of belief is increased as tests of changes are conducted and predictions begin to agree with the results of the tests. The connection between knowledge of the subject matter from which the change is developed and analysis of the data from a test of the change is essential to effective improvement. The PDSA cycle should be used as the framework in which to carry out such tests.

## 2.3 Using the PDSA Cycle to Test a Change

The PDSA cycle, introduced in Chap. 1, provides the framework for testing changes. The four steps in the cycle consist of planning the details of the test and making predictions about the outcomes (*plan*), conducting the test and collecting data (*do*), comparing the predictions to the results of the test (*study*), and taking action based on the new knowledge (*act*). Figure 2.1 describes each step of the PDSA cycle for testing a change.

Development of a good plan for a test is critical to its success. A simple form to help develop the plan for a test cycle is shown in Fig. 2.2. The plan begins with a statement of the specific objective of the cycle.

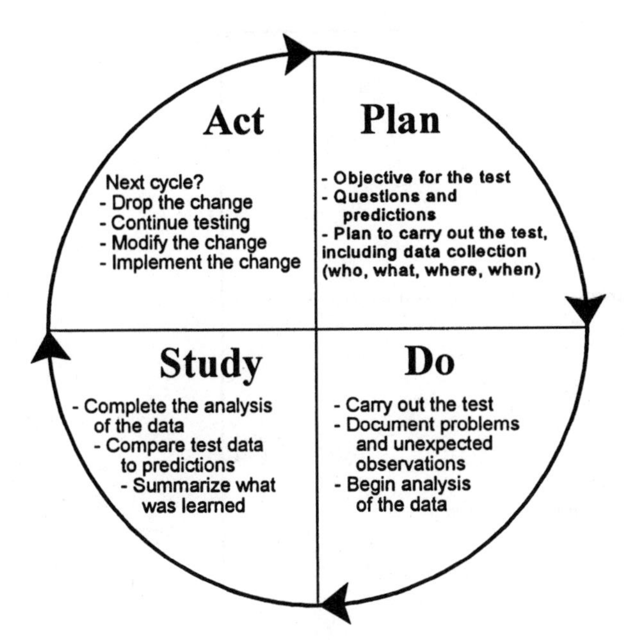


**Figure 2.1** Using the cycle to test a change.

Cycles to test a change will have varying objectives depending on the current degree of belief. Some objectives of test cycles might be to

- Increase the degree of belief that the change will result in an improvement
- Decide which of several proposed changes will lead to the desired improvement
- Evaluate how much improvement can be expected if the change is implemented
- Decide if the proposed change will work in the actual environment of interest
- Evaluate cost implications and possible side effects of the change
- Give individuals a chance to experience the change to minimize resistance upon implementation
- Decide which combinations of changes will have the desired effects on the measures

The objective of the cycle, which is developed in the *plan* step, will clarify the specific focus of testing the change. After the objective is agreed upon, the specific questions to be answered in the cycle should be stated by the team (or individual) involved in the improvement effort. These questions provide the format for stating the team's predic-

1. Objective of the Test

2. Key Background Information

3. Change Being Tested

4. Questions to Be Answered and Predictions about the Change

5. Measure(s)

6. Design of the Test

   - Scale of the test and the risks involved
   - Type of study
   - Method of data analysis
   - How will a range of conditions be included?
   - Has randomization been considered?

7. Who, When, Where for the Test

**Figure 2.2** Form for planning a test.

tions about the change(s) that will be tested in the cycle. The predictions should be stated in such a way that the results of the tests conducted in the cycle can be compared to the predictions.

Next, a plan for the test is developed. The plan should answer these questions:

- Who will schedule and conduct the test?
- Exactly what changes will be made and what results measured?
- Where will the test be conducted?
- When will the test be done?

Some hints for planning useful cycles for testing changes include these:

- Think a couple of cycles ahead of the initial test (future tests, implementation).
- Scale down the size and decrease the time required for the initial test.

- Do not require buy-in or consensus as a prerequisite for the test; recruit volunteers.
- Use temporary supports to make the change feasible during the test.
- Be innovative to make the test feasible.

In the *do* step of the cycle, the test is performed and the data collected are analyzed. The information obtained during this step should prepare for the effective study. What if the test of a change is not successful? There are a number of possible reasons:

1. The change was not properly executed.
2. The support processes required to make the change successful were not adequate.
3. The change was executed successfully, but the results were not favorable.

Information should be obtained during the *do* step to clearly differentiate which of these situations occurred.

The third step, *study*, brings together the predictions made in the *plan* step and the results of the test conducted in the *do* step. This synthesis is achieved by comparing the results of the data analysis to the predictions made for the specific questions asked during the planning step. This is where learning occurs and degree of belief about the change is increased. If the results of the test match the predictions, the team's degree of belief about its knowledge is increased. If the predictions do not match the data, there is an opportunity to advance their knowledge through understanding why the prediction was not accurate. Also, things that were learned about the change that will be useful in implementation can be documented.

In the *act* step of the cycle for testing a change, the team (or individual) must decide on the next course of action:

- Is further testing needed to increase the team's degree of belief about the change?
- Do alternative changes need to be tested?
- Is it important to learn about other implications (e.g., costs) of the change?
- Is the team ready to implement the change on a full-scale basis?
- Should the team modify the proposed change or develop an alternative change?
- Should the proposed change be dropped from consideration?

A change should be abandoned only when the study indicates that the change was executed properly, but the improvement was less than expected. The decision on a specific course of action will lead to developing the next PDSA cycle. The use of multiple cycles allows knowledge to increase as the team progresses from testing to implementing a change. It allows for risk to be minimized. As degree of belief that the change will be successful is increased, the scale of the test can be increased. Suppose a change is developed in a manufacturing process. In the first cycle, people with knowledge of the subject might review it. Then, in the second cycle, it could be tried in a pilot plant. In the third cycle, the change might be tested on one line in the production area. The change might then be revised and tested in a fourth cycle. If the learning from the first four cycles increases degree of belief to a high level that the change will result in improvement, then all or part of the change could be implemented in full-scale production. The collection and analysis of data in each of these cycles are essential to the learning process.

Based on the results of a test, a change or some part of a change could be implemented as is, the change could be modified and retested, or it could be abandoned. Figure 2.3 illustrates changes in degree of belief as a team or individual uses cycles to go from the development of a change to testing and implementing it. Cycles for implementing a change differ from test cycles in a number of ways:

- Support processes need to be developed to support the change as it is implemented.
- Failures are not expected when the change is implemented.

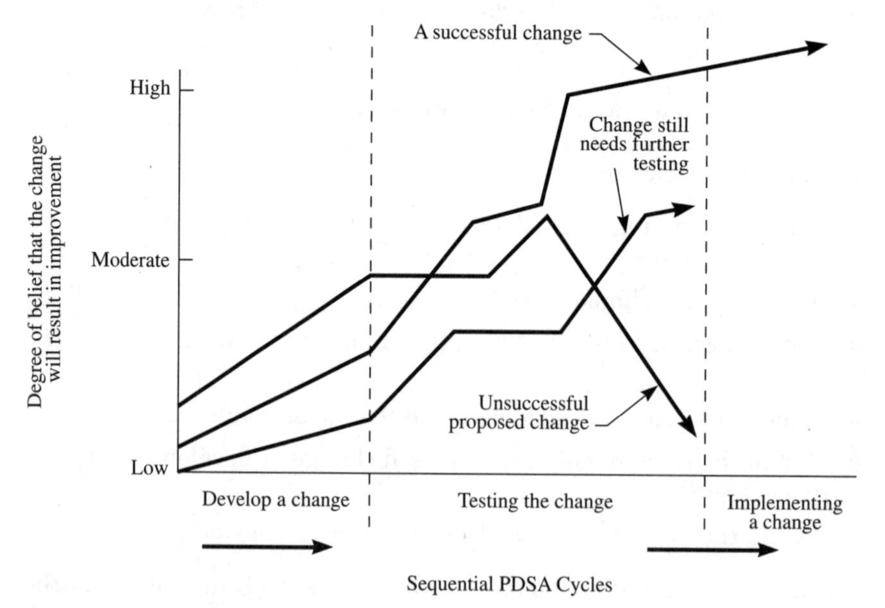


**Figure 2.3** Degree of belief when making changes to improve.

- Increased resistance to the change can be expected as it impacts more people.
- Cycles for implementing a change take longer than rapid test cycles.

**Example 2.1** The concept of increasing degree of belief through cycles is illustrated in the following example. A teacher, Tim, was considering ways to improve the learning environment in his classroom. Using a traditional classroom approach, he was considered an effective teacher by his colleagues and former students. He had recently become concerned about the attitudes of many of his students toward learning and the increasing dropout rate in his school. Tim had read of an experience in another school, in which the students were asked to work in teams and were given more control of the direction of the curriculum in the classroom. In this other school, the teachers had defined the basic requirements and goals for the class, and then they played the role of resource and facilitator to the teams of students. Improvements were reported in test scores at the end of the year as well as in other measures, such as absenteeism and required disciplinary actions. Should he consider implementing this change in his classes next year?

Initially, Tim had a very low degree of belief that this change would result in an improvement in his classes. He felt the students did not know how to work together, that they would not focus on the important subject matter, and that he would lose control of the learning environment. He also felt that some individual students would just rely on their teammates and not actively participate in the class. He had been successful for years with his traditional approach to teaching, and he found it hard to believe that giving this freedom to students would lead to improvement. Thus, Tim's degree of belief that the change would lead to an improvement was very low.

Rather than implement his change, first he decided to learn more about the changes. He wrote to one of the teachers at the school that had reported success with this approach. He received more information that described the theory behind the changes and why they had worked. In a follow-up call, he learned more about how the changes were initiated and some specific examples of the role that the teachers played as a resource to the teams. His degree of belief about the potential for the change was increased by this information, but he still was not ready to risk implementing the change in his classes.

Tim's next plan involved a visit to the school where the change had been implemented. He spent time in three different classrooms where the new teaching methods were being used. He also had follow-up conversations with each of the teachers. He learned that two of the teachers felt very strongly about the impact of the changes, but the third teacher was still questioning the approach. From this experience, his degree of belief that the new methods could improve classroom learning was greatly increased. But would it be effective in his classrooms? Although his degree of belief had increased, he was still not ready to predict that the change would be successful in his classes.

An opportunity to test this type of change arose in an environment away from his school. Tim was asked to lead a Sunday school class for a month while the regular teacher was away. He decided to test the new teaching methods during his month with this class. His degree of belief that he could make these methods work was not very high, but he felt that any failure would not have important consequences. He asked the regular teacher to assess the current attitude of the class and asked her if she would assess it again when she returned to teach the class. At his first class, he described "his" approach to teaching to the class, gave the students goals for the next month, put students in teams, and asked them to go to work. He noted attendance each week and compared the data to previous attendance. At the end of the month, he asked the students for feedback on his sessions. He also called the regular teacher a week later and got her assessment of student morale. The results of this test were generally favorable. But Tim was concerned that he did not really "teach" that much in the classes. Although the students were actively working toward the goals for the month, they did not need his help very often. His degree of belief that the new teaching methods would improve the learning environment in the classroom was now high, but he was not sure he could personally adjust to the changes.

Since a new semester was beginning, Tim decided to test the changes in one of his classes. He discussed his plans with the school principal and began his test the first day of the semester. He kept track of test scores, absenteeism, and discipline in each of his classes. He also noted how he felt about his teaching experience in each of his classes each week. At the end of the semester, he was enthusiastic about the results in the test class and felt as if he had learned to "teach" with these changes. Tim's degree of belief that these changes to teaching would lead to improvements was very high. He decided to implement the changes in all his classes the next semester. His knowledge about the changes and how he could personally work with the changed environment had greatly increased. He felt he could accurately predict results of making the changes in his other classes next semester.

Note that in this example, the teacher's degree of belief increased as he conducted tests of the changes and evaluated how well the predictions agreed with the results of the tests. How could you quantify degree of belief? Unlike a probability, confidence level, or statistical significance level, degree of belief is a concept, not a calculated value. The belief is about a prediction, not a past occurrence, and there is not a proven theory to make quantitative statements about the future.

Dr. Daniel Johnson of the National Bureau of Standards (1973) described a chemistry study in which the authors were asked to quantify their confidence in the reported values from their experiment. They provided the following assessment:

> We think our reported value is good to 1 part in 10,000. We are willing to bet our own money at even odds that it is correct to 2 parts in 10,000. Furthermore, if by any chance our value is shown to be in error by more than 1 part in 1000, we are prepared to eat the apparatus and drink the ammonia.

This statement of degree of belief certainly communicates their convictions about the prediction.

## 2.4 Designing Tests of Change

Planned experimentation, as described in Chaps. 3 through 10, is a collection of approaches and methods to help increase the rate of learning about improvements to systems, processes, and products. An experiment is a change to a process or product and an observation of the effect of that change on one or more measures of interest. The methods of planned experimentation are appropriate for understanding the important causes of variation in a process and evaluating changes to the process.

What is the minimum amount of information from a test or an experiment that can provide an adequate degree of belief that the change, when implemented, is likely to lead to an improvement? One of the primary issues is whether the effects observed can be clearly tied to the change of interest. The answer to this question depends on what the team already knows about the change, i.e., its current degree of belief that the change will lead to improvement. If the change of interest has a history of lots of success in similar environments, then a simple verification of results in the environment of interest may be all that is needed. But if the change has not been previously used in similar environments, then more sophisticated tests may be required.

One of the most common designs for a test of change is the *before-and-after test* (or *pretest, posttest*) design. In this design for a test, a change is made, and the circumstances after the change are compared to the circumstances before the change. The collection of data before the change provides the historical experience that is the basis of the comparison. Is this type of design useful for testing changes?

Figure 2.4 shows the results of such a before-and-after test. Data were collected on week 4, the change was made after week 7, and then data were collected again on week 11. The reduction in cycle time from 8 h to 3 h was considered very significant for the process of interest. The top graph in Fig. 2.4 shows a summary of the test data. Does this test, summarized in this way, provide an adequate degree of belief that the change, when implemented, will lead to an improvement? Are there other feasible explanations of the reduction in cycle time after the change was introduced?

The run chart at the bottom of Fig. 2.4 (case 1) shows one possible scenario that could have yielded the results observed in the test. The run chart shows results for cycle times for weeks 1 to 14 (three weeks before the change was made until three weeks after the second test observation was made). The run chart in case 1 confirms the conclusion that the change did result in meaningful improvement.

Figure 2.5 shows run charts for five other possible scenarios that offer alternative explanations of the test results. In each case a run

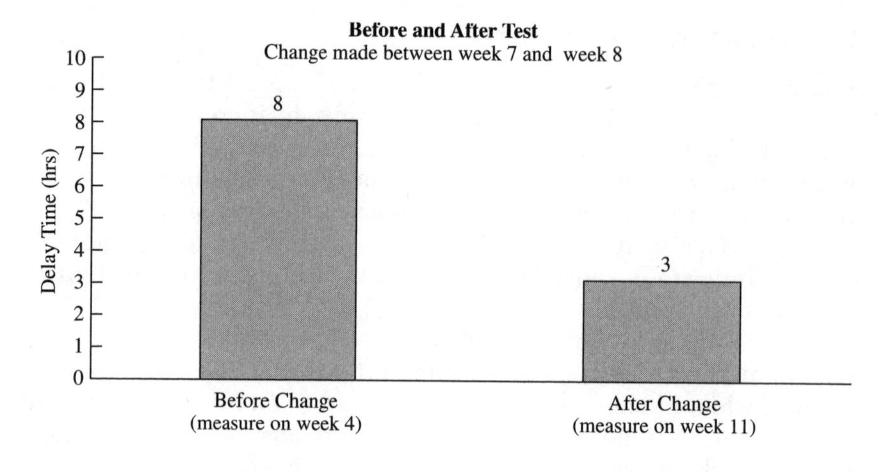


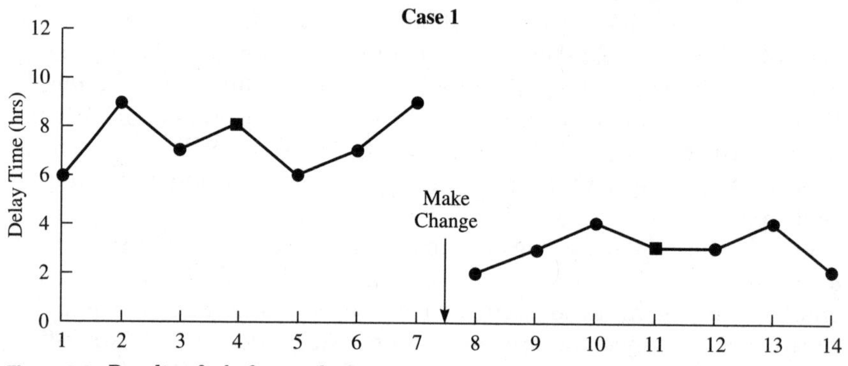


**Figure 2.4** Results of a before-and-after test.

chart of cycle time for weeks 1 to 14 is shown. The test results for week 4 (cycle time of 8) and week 11 (cycle time of 3) are the same for all cases.

In case 2 there is no obvious improvement after the change is made. The measures made during the test are typical results from a process that has a lot of week-to-week variation. The conclusion from study of the run chart is that the change did not have any obvious impact on the cycle time.

In case 3 it appears that the process has been steadily improving over the 14-week period. The rate of improvement did not change when the change was introduced. Although the cycle time for the process has certainly improved, there is no evidence that the change made any contribution to the steady improvement in the process over the 14 weeks.

In case 4 an initial improvement is observed after the change is

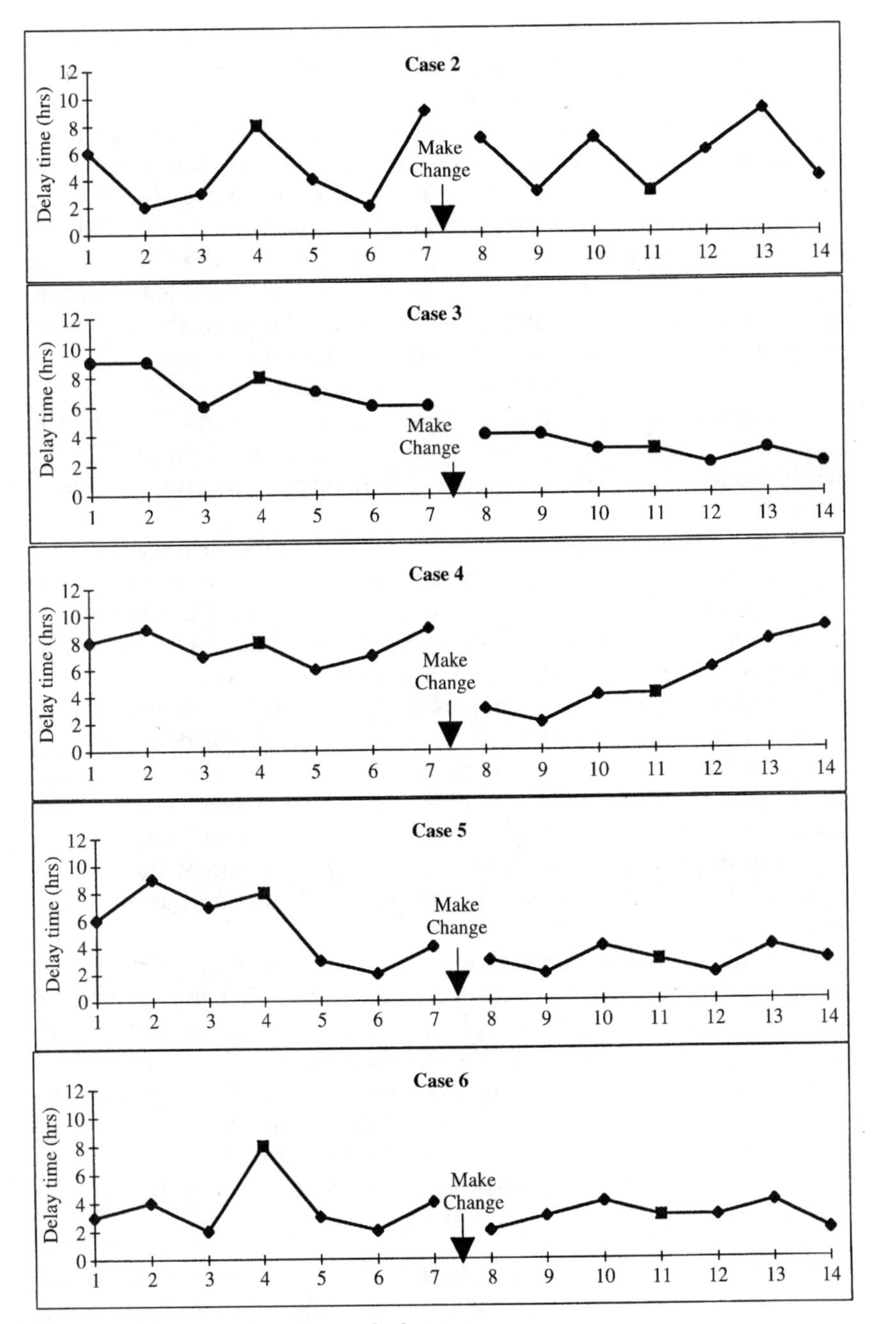


**Figure 2.5** Run charts for before-and-after tests.

made, but in the last three weeks the process seems to have returned to its prechange level of cycle time. The results may be due to the *Hawthorne effect.* The Hawthorne effect is named after some tests on productivity conducted at the Western Electric Hawthorne plant in the 1920s. Whenever changes were made in the work environment, initial improvements were observed. But performance quickly returned to normal levels after workers became used to the change. This effect is similar to a *placebo effect.* Because of the focus on the process, an initial improvement is observed due to increased attention paid to the measures of interest. Later, when focus on the change is lessened, the cycle times revert to the original process levels.

In case 5, an improvement in the process cycle times has occurred, but it appears that the improvement occurred in week 5 before the change was made after week 7. The improvement in cycle time should be attributed to some other phenomenon, not the change of interest.

In case 6 the process appears to be stable, except for a special case that occurred in week 4 when the pretest results were obtained. The unusually high result on week 4 made it appear as if the typical result on week 11 were an improvement. Once again, there is no evidence that the change contributed to any improvement.

From this example, it is obvious why the simple before-and-after test is often not rigorous enough to increase degree of belief about a change. This design should be used only when the change has had previous testing and has proved effective in a wide variety of environments. Then the test is being done to confirm performance in the specific environment of interest or to measure side effects of the change. What can be done to increase the rigor of the before-and-after test in other situations?

Often, the simplest alternative is to conduct the test over a longer period, both before and after the change is made. Plotting the test results on run charts like those in Figs. 2.4 and 2.5 usually will provide convincing evidence of the effect of the change.

But even with a run chart over a long time, it is always possible that some other cause, which occurred at nearly the same time as the change of interest was made, could be responsible for the observed effects. Some other ways to increase the rigor of the test include these:

1. Remove the change and see whether the process returns to its initial levels. This can be repeated as many times as necessary (make the change, observe results over time, remove the change) until an adequate degree of belief is obtained.
2. Add some type of control group to the study. The same measures

will be made for the control group as for the environment where the change is made. A run chart for the control group can then be compared to the environment undergoing the change.

3. Use the methods of planned experimentation described in the remaining chapters of this book.

There is an important difference between conducting a planned experiment to test a change and other types of studies to assess the impact of a change. The issue has to do with the *experimental units* in the test. An experimental unit is the smallest division of material in a study such that any two units may receive different *changes* (combinations, levels, intensity, etc.) applied to them. A week is the experimental unit in the previous example. Other typical experimental units in studies are individual people, one pound of material, a batch, a classroom, a 10 $ft^2$ plot of land, etc. (see Chap. 3 for more on experimental units). In a planned experiment, the individual or team designing the test decides which experimental unit will have certain changes made to it. In other types of designs, those conducting the study do not have the option of making that decision.

In analyzing the results from tests not following the principles of planned experimentation, one must make assumptions about how the specific conditions came to be associated with particular experimental units. The relevance and importance of these assumptions to the test will determine how much the degree of belief changes as a result of the study. Both planned experiments and the simpler test designs described earlier are useful in testing changes.

As an example of how both types of studies could be used to test a change, a team wanted to test a new method of measuring patients' temperatures in a hospital. The new method, which was advertised to be quicker than and just as accurate as the current method used in the hospital, had been suggested for use by one of the administrators. The team could approach this test as a planned experiment or as a simple observational study:

| | |
|---|---|
| Observational study | Collect data from two other hospitals that are using the new method. Compare results (time and accuracy) to similar data from the team's hospital using the current method. |
| Planned experiment | Select ten patients in the hospital for the experiment. Randomly assign five of the patients to be tested with the current method and five of the patients to be tested with the new method. Collect data for both groups of patients and compare the results. |

The important difference between the observational study and the planned experiment is that, in the observational study, the team did not decide which patients would be tested by each method. Other considerations and factors made that determination. Why is this an important issue? Unless the assignment of the change or level of change is made to the experimental units by the study team, there are always possible alternatives to what actually caused the results of the test. In the hospital example, faster measurements in the other hospitals could be attributed to many more factors than the change that was being tested. Thus there can be questions about the validity of the study which affect the degree of belief that the change will lead to an improvement.

When you are making a decision about an observational study or a more formal planned experiment, the following issues should be considered:

1. Are historical data available to provide a basis of comparison? If so, take advantage of information in the first test cycle.
2. How likely are threats of misinterpretation of the results because some external event is present on or about the same time the change is made? A historically stable process makes this threat less likely.
3. Can data continue to be collected over a long period after the change is made, mitigating the Hawthorne effect or misinterpretation of external events?
4. How large an improvement is expected? Large improvements are much easier to isolate with simple tests.
5. Are there lots of different conditions that need to be included in the test? Planned experimentation techniques make this feasible.

In considering the design of the test, a run chart that includes multiple measurements before and after the change is made provides minimal complexity and excellent protection from misinterpreting the results of the test.

The next section discusses some basic principles that should be applied regardless of the sophistication of the design of the test cycle. These principles will be developed further in Chap. 3.

## 2.5 Principles for Testing a Change

After consideration of the basic design that will determine whether the effects observed can be clearly tied to the change of interest, the primary source of uncertainty involved in testing a change is the uncertainty of extrapolation of the results of the test to the future. Will the results of the test be obtained when the change is implemented? In addition to the uncertainty of extrapolation, another difficulty in testing a change is that the change may not result in improvement and may even make the situation worse. Three basic principles to help overcome these difficulties were introduced at the beginning of the chapter:

- Test on a small scale and build knowledge sequentially.
- Collect data over time.
- Include a wide range of conditions in the sequence of tests.

Each of these principles is discussed below and illustrated in the examples in the remainder of the chapter.

#### Test on a small scale and build knowledge sequentially

What is the most rapid way to proceed from a new idea to an improvement obtained from implementing a specific change based on the idea? One approach is to spend time thinking about all the possible options, ramifications, and implementation issues before proceeding with a test of the idea. Another approach is to very quickly run a test of the idea. Experience has shown that this latter approach will lead to much more rapid improvement in an organization.

It is also important to minimize the negative impact that can result from a change that does not result in improvement. When a team or individual makes a change to improve a product or process, the prediction is that the change will in fact result in improvement. Predictions come from a team's current knowledge about the change. The more complete the current knowledge, the better the prediction will be. The degree of belief that the current knowledge provides sufficient basis for action is directly related to the ability to predict future performance of the process or results of future studies.

Knowledge is built by an iterative process of developing a theory, making predictions based on the theory, testing the predictions with data, improving the theory based on the results, making predictions based on the revised theory, and so forth. Tests of change are designed to answer questions that come from a combination of theory about the subject matter and conclusions from analysis of data from past studies. When studies are designed for testing a change, the planned tests should match this sequential nature of building knowledge.

Table 2.1 summarizes the appropriate scale of the test for different situations. Planning one large cycle to attempt to get all the answers should always be avoided. Testing the change during full-scale implementation should be considered only when

**TABLE 2.1 Deciding the Scale of a Test**

| Consequences of a failed test | Degree of belief in success | |
|---|---|---|
| | Low | High |
| Minor | Medium-scale tests | One cycle to implement the change |
| Major | Very small-scale tests | Small- to medium-scale tests |

- The team has a high degree of belief that the change will result in improvement.
- The risk is small (losses from a failed test are not significant).
- The team cannot find a way to test the change on a small scale.

During the design of studies to test a change, those responsible for developing the change should continually be asking themselves how they can reduce the risks of the test and still gain some knowledge. The following are some ways to design a test on a small scale:

- Simulate the change (physical or computer simulation).
- Have others who are knowledgeable about the change review and comment on its feasibility.
- Test the new product or process on members of the team who developed the change before it is introduced to others.
- Incorporate redundancy in the test by making the change side by side with the existing process or product.
- Conduct the test in one facility or office in the organization, or with one customer.
- Conduct the test over a short period (one hour or one shift) .
- Test the change on a small group of volunteers.
- Develop a plan to simulate the change in some way.

**Collect data over time**

There will be variation in the measures of quality of the product or process due to causes and conditions that are unrelated to the change that is being tested. The effect of the change must be distinguished from the unrelated variation in the process. Data must be analyzed over time to determine whether a system is stable after a change has been made. As discussed in Chap. 3, a stable system provides a rational basis for predicting that the change will result in improvement in the future. Effective studies to test a change will include data, plotted on a run chart, before as well as after the change is made.

If all possible data are collected and used in the study, is it still important to look at the data over time? For example, if a team collects data from all patients one month before the change and all patients one month after the change, is there any useful information in a run chart? Figure 2.6 shows the results of a test of change in two different hospitals. In hospital A, the run chart supports the improvement indicated in the bar graph. But in hospital B, where the bar graph is identical to that for hospital A, there is no evidence that the change had any impact on the cycle time. Sampling is not the issue; looking at data over time is always important to protect from misinterpretation of the results of a test.

The use of replication of tests is the primary way to build time into a study. For example, if a meaningful test of the change can be conducted in 4 h, replicating the test 2 times per day over a 2-week period will result in 28 replications of the test. Combining data collected during these 28 tests with similar data collected during the 2 weeks prior to making the change will provide results that give the team a high degree of belief about the effect of the change. For some changes, replication of the change at different times of the year may be important. For example, 100 replications of a test in the summer may not increase the degree of belief about the effect of the change in the winter.

Incorporating time into the study will be a more important consideration than the number of experimental units incorporated in the study. There is almost always more information in a sample selected over a long time than in a larger one collected over a relatively short period. Including data spread over time is also a convenient way to include a range of conditions in the study, the last principle for testing a change.

### Include a wide range of conditions in the sequence of tests

Making a change to a process, system, or product for the purpose of improvement involves making a prediction that the change will be beneficial in some way in the future. The conditions in the future will be different from the conditions of the test. Circumstances unforeseen or not present at the time of the test will arise. Will the change still be an improvement under these new conditions? How can a team increase its degree of belief that the change will be effective in the future? Including a wide and varied set of conditions in the test is the best way to increase the degree of belief the change will result in improvement in the future.

Too often, tests of changes are not conducted over a broad range of conditions. Some reasons given for limiting the conditions include limited resources, time constraints, difficulty in analysis of the data, lack

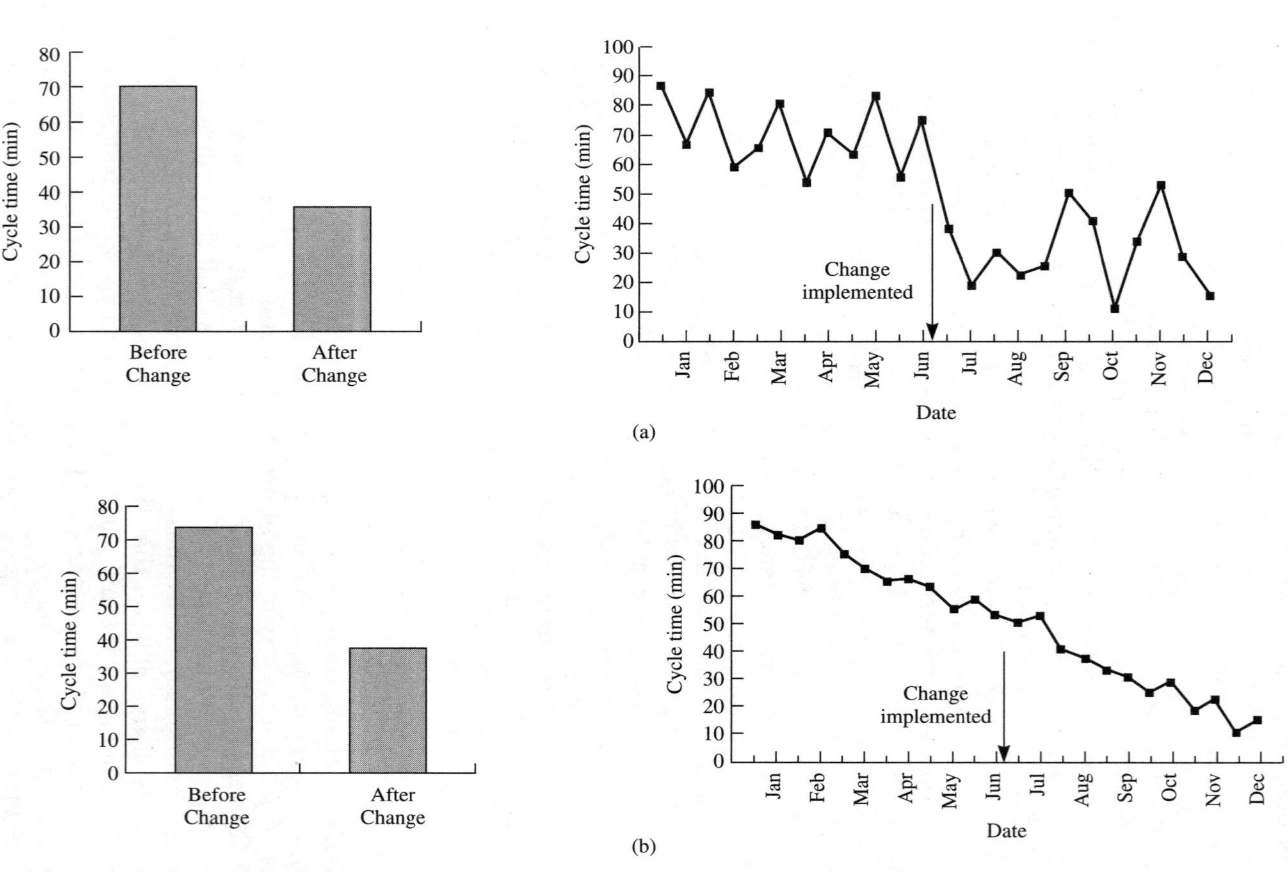


**Figure 2.6** Run charts versus before-and-after summaries. Improvement in cycle time for (*a*) hospital A and (*b*) hospital B.

of knowledge of how to efficiently include different conditions, and too many possible conditions to consider. Degree of belief in the results of a study is increased as the same conclusions are drawn for a variety of test conditions. If a particular supplier's material proves the best over different environmental conditions, on different days, one feels much easier in using the results to select a supplier than if the study had been run on one day under constant environmental conditions. The experimenters might also consider running the study using material from more than one machine setup or batch. Maybe the material will be made on more than one shift or in several plants. Incorporating some of or all these conditions in the study will increase the degree of belief in the results if similar conclusions are seen for all conditions.

The selection of experimental units for a test provides one opportunity to consider a wide range of conditions. It is rare in a test of a change that a random selection of conditions or experimental units is preferred to a selection made by a subject matter expert (called a *judgment sample*).

Consider a study to choose one of four different types of instruments for use by operators in the plant to replace the current instrument. The objective of the cycle is to choose an instrument for future use that will provide the best precision when used by any operator. Resources are sufficient to allow 7 of the 30 operators presently working in the plant to participate in the study. How should the seven operators be chosen?

The seven could be randomly selected, and the magnitude of measurement variation under the conditions of the study could be estimated. But it is important to remember that the conditions of the study will not be seen again. There will be new operators working under new conditions, training may change, etc. A judgment selection of operators will provide a higher degree of belief concerning the performance of the instruments in the future. One option is to choose operators who have the greatest experience and some with the least experience. If an instrument performs best when used by experienced and inexperienced operators, the degree of belief that a good choice of instrument has been made will usually be greater than if the instrument performed best when used by seven randomly selected operators.

The concept of testing over a wide range of conditions can be summarized by saying that the degree of belief is increased as results are repeated in tests conducted under a variety of conditions that might occur in the future. In most tests of a change, conditions are a more important consideration than sample size. A change may have the predicted effect in the short term, but most tests also have the aim of determining whether the effect will persist in the future. Since conditions will naturally change over time, more information is usually ob-

**TABLE 2.2 Guidelines for Sample Size to Test a Change**

| Number of points | Situation |
|---|---|
| Less than 10 | Expensive tests, expensive prototypes, or long periods between available data points. Large effects are anticipated. |
| 15–50 | Usually sufficient to discern patterns indicating improvements that are moderate or large. |
| 50–100 | The effect of the change is expected to be small relative to the variation in the system. |
| Over 100 | The change is intended to affect a rare event. |

tained from a sample selected over a long period than from a larger one collected over a short period. (Note that one cycle or several cycles might be used to provide an appropriate period of time.) The time required to increase the degree of belief that the change will result in improvement and persist is a matter of judgment.

The importance of time in determining whether a change is an improvement sets a context for determining sample size for the test. The sample size needs to be adequate to detect patterns that indicate improvement. Table 2.2 contains some guidelines for selecting the sample size.

Usually 15 to 30 data points on a run chart will be sufficient to recognize patterns indicating improvement. Sometimes as many as 50 points might be necessary if no historical data are available to establish a baseline before the change is made. From 50 to 100 points would be necessary only when the effect of the change was anticipated to be small relative to the variation in the system. Examples are situations in which the variation in the measurements themselves are large or when the variation among people—students, patients, or customers—can mask the effect of the change.

More than 100 points might be needed if the change were intended to affect a rare event, such as the side effect of a new drug or a serious but rarely occurring defect. In these situations, it is a good idea to consult a statistician to help in the design of the test.

## 2.6 Analysis of Data from Tests of Changes

The specific approach to analysis of data from a test cycle will differ depending on the design of the study that is conducted. Knowledge of the subject matter is important in studying the results of tests of a change. Run charts should be used to analyze data to keep everyone involved in the analysis. The following elements will be common to the analysis of all tests of change:

1. Plot the data in the order in which the tests were conducted. This

is an important means of identifying the presence of trends and unusual results in the data.

2. Rearrange this plot to study other potential sources of variation that were included in the study design, but not directly related to change under study. Examples of such variables are batches of raw material, measurements, operators, and environmental conditions.
3. Use graphical displays to assess how much of the variation in the data can be explained by the change(s) of interest. These displays might include using different symbols to identify the change or ordering the test results to highlight data before and after the change.
4. Summarize the results of the study with appropriate graphical displays.

Some examples of cycles to test changes follow to illustrate these guidelines for analysis.

Figure 2.7 contains a run chart that was used to analyze the results from a test cycle conducted by an improvement team in an order center. The test was designed to determine whether a new approach to receiving orders would be an improvement over the existing procedure. One of the measures used in the evaluation was the percentage of orders that required rework. *Rework* was defined as any changes to the order after initial receipt from the customer. In the first test cycle, rework was measured for 20 days from orders taken by three members of the team. Then the three team members began using the new procedure. Rework was measured for 20 days after the initiation of the new procedure. Collecting data over time allows us to see whether patterns in the data indicating improvement coincide with the time of the change. In the run chart in Fig. 2.7, the reduction in percentage of rework coincides with the use of the new procedure.

Since the basis of comparison in this before-and-after test is historical experience, the test is vulnerable to misinterpretation if a special cause

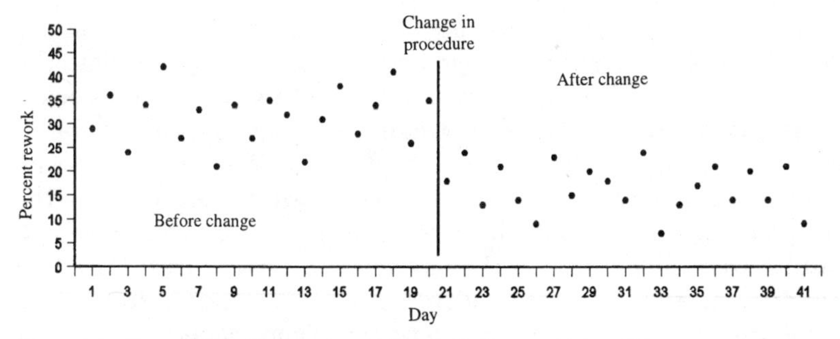


**Figure 2.7** Run chart of percentage of rework from a test cycle.

unrelated to the change occurs at or about the same time as the change is made. Were there other changes in the order center or in the business that occurred at the same time as the change in procedure was introduced? It is up to those conducting the test to make the judgment that the effect seen is due to the change being tested. If there are no obvious external events and the system has been stable in the past, there is a rational basis for this judgment. The members of the improvement team believed that the reduction in the percentage of rework was the result of the new procedure. They decided to plan another cycle to test the use of the procedure with all the order center personnel.

In the example illustrated in Fig. 2.7, it was possible to collect a sequence of data during the test. In some situations, it might only be practical to collect data once before and once after the change. For example, scores on a pretest and posttest might be used to evaluate the effect of new audiovisuals in a history class; or a group of patients undergoing rehabilitation might be asked to evaluate the amount of pain they feel before and after a new exercise program. In such situations, displaying the data on a frequency plot is a useful way to do the analysis. But, as described earlier, the conclusions of the test are vulnerable to other explanations that must be considered when the test results are studied.

In a simultaneous comparison test (commonly called a *paired-comparison study*), two or more alternatives are compared at the same time, in the same space, or under other similar conditions. By comparing alternatives in such a way, the effect of external events on the different alternatives can be studied during the test. Therefore, a simultaneous comparison test can help rule out alternate explanations for the improvement seen.

Figure 2.8 shows the results of a test cycle with a simultaneous comparison test. A team in a hotel management company was trying to reduce the time to clean a hotel room. The data were collected during a simultaneous comparison test of a new cleaning procedure. Based on benchmarking experiences and some simulations of the new procedure, the improvement team's degree of belief was high that the new procedure would result in improvement.

The first run chart displays the data in time order for both the new and the old procedures. Different symbols are used to differentiate the procedures. The second graph shows a reordering of the time order chart to group the results of the procedure. A review of the run charts reveals the impact of the change and the effect of any external events on both the current and the new methods. The results were best for the new procedure during the last 4 days of the test. Note that on the fourth day, results were high for both the old and new procedures. Based on the results of this test, the team's degree of belief was increased that the new procedure would result in reduced time to clean rooms.

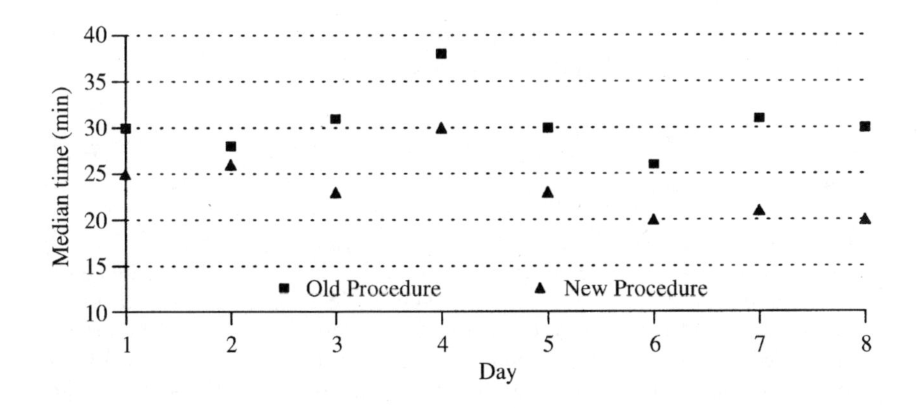


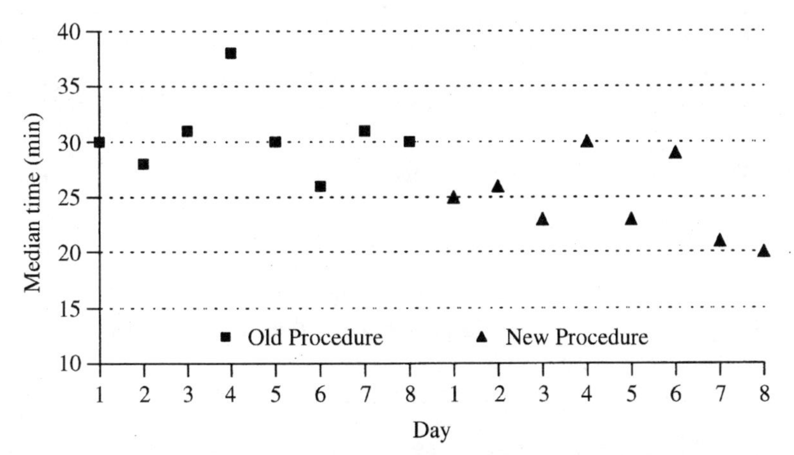


**Figure 2.8** Run charts for simultaneous comparison test cycle. Median time to clean hotel rooms.

After reviewing the run charts in this example, one might question whether the two groups differ not because of the cleaning procedure used but because of some preexisting difference. One way to overcome this possible alternate explanation for the results of the test is to use random assignment. As discussed earlier, random assignment plays a key role in planned experimentation. Randomization is discussed in Chap. 3.

A simultaneous comparison test can be used to rule out alternate explanations for the improvement seen during a test. The design of such a test can also be used to include a wide range of conditions. If the change is an improvement over broad conditions, one's degree of belief that the change will be an improvement in the future is increased. To make the results of the test easier to interpret, it is wise to include the different conditions in a systematic way.

**Example 2.2** A sales manager was interested in reducing the time required to close a sale. Currently it took up to 3 weeks to finalize an agreement after a proposal was made. The sales manager had developed a new approach that potentially could cut the wait time in half. She had run a test cycle with one customer and her most experienced salesperson. The results of this cycle were favorable. Based on the results of her test, the sales manager believed that the new approach could significantly reduce the time to close a sale. She was interested now in learning whether the procedure would prove successful when used in different sales offices, by different salespersons, and with different customers. She decided to run another cycle to consider a wide range of conditions in her test. She planned to set up two groups with extreme conditions to determine whether the new sales approach would result in improvement in both groups. She selected two sales offices—one small and one large; two salespersons—one with 2 years' experience and one with 10 years'; and she selected two customers—a new customer and a customer who had done business with her company for a number of years.

The two groups the sales manager planned for the test cycle were as follows:

| *Group 1* | *Group 2* |
|---|---|
| Small sales office | Large sales office |
| Salesperson with 2 years' experience | Salesperson with 10 years' experience |
| New customers | Long-term customers |

The sales manager purposely set up two groups that had very different conditions. If the new approach resulted in improvement in both groups during the test, her degree of belief would be high that the change would result in improvement if used in the future. To conduct the test, she ran a simultaneous comparison study in both groups. Once the customers were selected for a particular group, they were randomly assigned to receive the new or old procedure. The results of the test are shown in Fig. 2.9; a separate run chart was prepared for each group.

The run charts show that the new approach resulted in improvement in both the groups. Although the closing time was generally longer in group 1, the new procedure still resulted in decreased times. The sales manager's degree of belief was increased that the new procedure would result in improvement in the future under the different conditions in which it might be applied.

In this type of design for a test cycle, there is always the possibility that the change will show improvement in some planned groups but not in others. If this does occur, further cycles should be used to better understand the relationship between the change being tested and the different conditions in the groups.

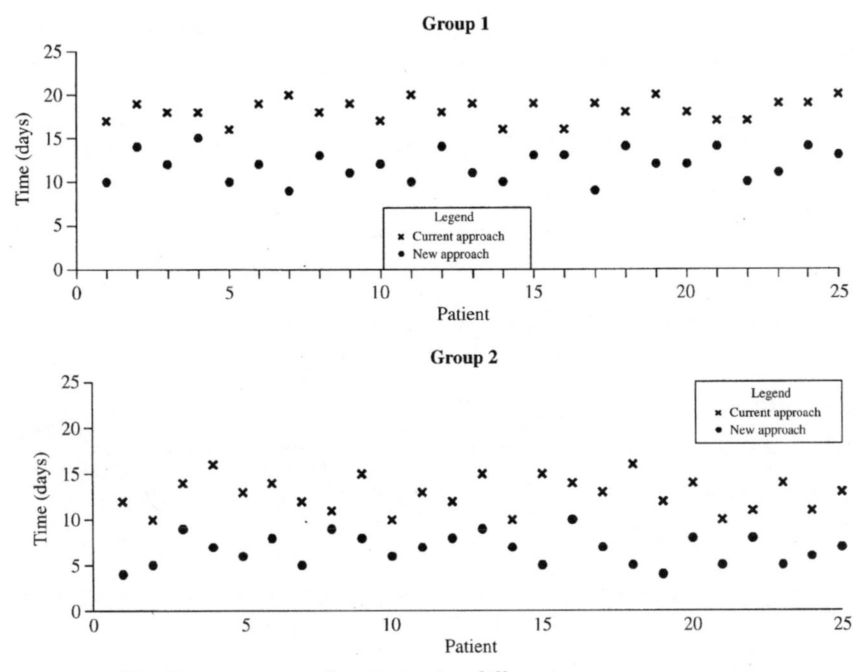


**Figure 2.9** Simultaneous comparison test using different groups.

This approach to forming groups of background variables for the test is called *planned grouping.* The method is discussed in detail in Chap. 3.

## 2.7 Summary

This chapter described how to use the PDSA cycle to test changes. The focus was on answering the third fundamental question of the model: What changes can we make that will result in improvement? The objective in designing cycles is to conduct tests in which the effects observed can be clearly tied to the change of interest.

These are some of the key ideas about testing change in the chapter:

1. Making a change to a system, process, or product for the purpose of improvement involves making a prediction.
2. The concept of degree of belief is used to describe one's conviction that the change will lead to improvement in the future. Degree of belief can be increased by testing.
3. Conduct a sequence of rapid PDSA test cycles to accelerate the rate of improvement.
4. Knowledge of the subject matter is important in studying the re-

sults of tests of a change. Run charts should be used to analyze data to keep everyone involved in the analysis.

Three basic principles of testing a change are described in this chapter to help teams and individuals increase their rate of learning about changes:

- Test on a small scale and build knowledge sequentially.
- Collect data over time.
- Include a wide range of conditions in the sequence of tests.

## References

Johnson, Daniel (1973): National Bureau of Standards, personal communication.
Judd, Eliot R., Charles M. Smith, and Louise H. Kidder (1991): *Research Methods in Social Relations,* 6th ed., Holt, Rinehart, and Winston, Inc. Orlando, Fla.
Langley, G., K. Nolan, T. Nolan, C. Norman, and L. Provost (1996): *The Improvement Guide: A Practical Approach to Enhancing Organizational Performance,* Jossey-Bass Publishers, San Francisco.
*Nuclear Instruments and Methods* (1973): "Round-Table Discussion on Statement of Data and Errors," vol. 112, pp. 391–395.

## Exercises

**2.1** Why is it important to state predictions about a change prior to running a test? Can you think of examples of testing where you would have no basis on which to make a prediction?

**2.2** Describe the concept of degree of belief in your own words. Write down your assessment about your degree of belief about a change you would like to make.

**2.3** Think about a recent change in your organization. How was the change tested and implemented? Design a series of cycles which could have been used to test the change. List the objective of each cycle.

**2.4** Describe a before-and-after test cycle for a change with which you are familiar. Sketch the run chart for this test cycle.

**2.5** Describe a simultaneous comparison test cycle for a change with which you are familiar. Sketch the run chart for this test cycle.

Chapter

# 3

# Principles for Design and Analysis of Planned Experiments

The first two chapters discussed the Model for Improvement, the role of planned experimentation, and testing a change in making improvements. This chapter lays the foundation for the more formal approach to experimentation covered in the remainder of this book. The chapter starts by defining the language peculiar to experimental design, the different types of planned experiments, and the principles and tools needed in planned experimentation. The last section discusses the approach to the analysis of planned experiments that will be used throughout the book. The remaining chapters will discuss the specifics in using the principles and tools in the construction of experimental designs and the analysis of data from these designs.

## 3.1 Definitions

The following important terms associated with planned experimentation are defined for future use in the book:

**response variable** A variable observed or measured in an experiment, sometimes called a **dependent variable.** The response variable is the outcome of an experiment and is often a quality characteristic or a measure of performance of a product, process, or system. An experiment will have one or more response variables.

**factor** Sometimes called an **independent variable** or **causal variable,** a variable that is deliberately varied or changed in a controlled manner in an experiment to observe its impact on the response variable. The factor can be either qualitative (e.g., machine A, B, or C) or quantitative (for example, a temperature of 90, 100, or 110 degrees).

**background variable** Sometimes called a **noise variable** or **blocking variable,** a variable that potentially can affect a response variable in an experiment but is not of interest as a factor. The objective of the study will differentiate factors and background variables. Typical background variables are lot, time, operator, cavity within a mold, and instrument. Background variables can be controlled in a study by holding the variable constant, by the use of blocks (to be defined), or by measuring the background variable and accounting for the effect in the analysis of the data.

**nuisance variable** An unknown variable that can affect a response variable in an experiment, sometimes called a **lurking variable** or an **extraneous variable.** A nuisance variable is a background variable that is unknown at the time the experiment is planned. A nuisance variable will appear in an experiment as noise (either common cause or special cause of variation). The impact of nuisance variables can be minimized by randomization and diagnostic analysis of the response data.

**experimental unit** The smallest division of material in an experiment such that any two units may receive different combinations of factors. Examples of experimental units are parts, batches, one pound of material, an individual person, or a 10 $ft^2$ plot of ground.

**blocks** Groups of experimental units treated similarly in an experimental design. Blocks are usually defined by background variables. The variation of a response variable within a block is expected to be less than the variation within the entire experiment. For example, experimental units produced and tested at one time (a block defined by time) might be expected to vary less than experimental units produced at other time periods.

**level** A given value or specific setting of a quantitative factor or a specific option of a qualitative factor that is included in the experiment. The levels of a factor selected for study in the experiment may be fixed at certain values of interest, or they could consist of a random selection from many possible values.

**effect** The change in a response variable that occurs as a factor or background variable is changed from one level to another. The effect must be further described in terms of the context in which it is used (a linear effect, an interaction effect, etc.).

The distinction between response variables, factors, and background variables is especially important in planning an experiment. The factors and background variables can be thought of as causes, and the response variables as effects. The objective of the experiment should clarify the distinction between response variables, factors, and background variables.

## 3.2 Types of Planned Experiments

An experiment consists of a series of tests of a system carried out by changing levels of factors and background variables and an observation of the effect of that change on one or more response variables. The primary reason to carry out an experiment is to provide a basis for action on the system.

Deming (1950, 1975) classified studies into two types depending on the type of action that will be taken:

> An *enumerative study* is one in which action will be taken on the universe.
>
> An *analytic study* is one in which action will be taken on a cause system to improve performance of a product, process, or system in the future.

The universe in an enumerative study is the entire group of items (e.g., people, materials, invoices) possessing certain properties of interest. The universe is accessed by the frame, a list of identifiable, tangible units, some or all of which belong to the universe and any number of which may be sampled and studied. The aim in an enumerative study is estimation—estimation about some aspect of the universe. Action will be taken on the universe based on this estimate through sampling the frame.

An example of an enumerative study is the U.S. Census, which is carried out every 10 years. The number of Representatives in Congress from an area depends on the number of inhabitants in an area, as counted by the last census. Determination of a fair price to pay for an inventory is another enumerative study. For a thorough review of enumerative studies see Deming (1960).

The aim of an analytic study is prediction—prediction that one of several alternatives will be superior to the others in the future. The choice may encompass different concepts for a product, different materials, or different conditions for operating a process. As part of the analysis of the data from an analytic study, it is generally useful to compare estimates of the performance of the alternatives under the conditions of the study. However, it is important that the experimenters not lose sight of the fact that the ultimate aim is prediction of performance in the future.

| Aspects of a study | Type of study<br>Enumerative | <br>Analytic |
|---|---|---|
| Aim | Estimation | Prediction |
| Method of access | Frame and sample | Models of product or process |
| Major source of uncertainty | Sampling error | Extrapolation to the future |
| Major source of uncertainty quantifiable? | Yes | No |
| Environment of the study | Static | Dynamic |
| Role of the statistician | Assess important effects | Support subject matter expert |
| Role of the subject matter expert | Define the universe, approve the frame | Identify variables, levels; assess conditions in the future; assess degree of belief |

**Figure 3.1** Important aspects of enumerative and analytic studies.

A study of material from three suppliers to decide which supplier should be given a contract is an example of an analytic study. The study is done to determine the supplier whose material will be most advantageous for the plant to use in the future. Use of a control chart to study a process to bring it into a state of statistical control is an analytic study. An accelerated life test would be an analytic study. Another example of an analytic study is a Presidential election survey 6 months before the election. An exit poll survey is an example of an enumerative study.

A summary of the important aspects of enumerative and analytic studies is presented in Fig. 3.1. In an analytic study, the focus is on the cause system. There is no identifiable universe, as there is in an enumerative study, and, therefore, no frame. Since the purpose of an analytic study is to improve performance in the future, the outcomes of interest have yet to be produced.

In the case of a new product, the production process itself does not exist; studies are performed on laboratory or pilot versions of the process or prototypes of a product. Moreover, the potential customers live in a dynamic environment such that a decision today may be quite different a year or two from now. All studies done by marketing, engineering, or manufacturing during product development are analytic studies.

Deming's dichotomy for classifying problems into enumerative and

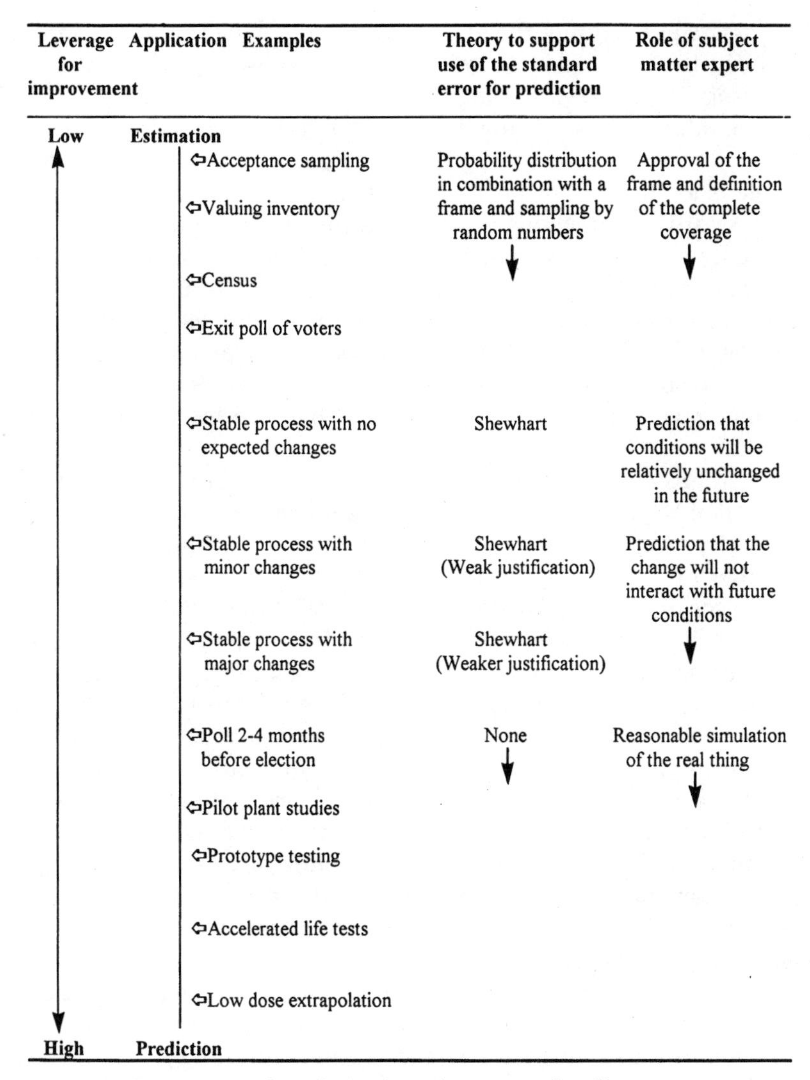


**Figure 3.2** Estimation and prediction in various types of studies.

analytic, based on the type of action that will be taken, can be expanded into a continuum based on the degree of prediction in the inferences drawn from the study. Figure 3.2 illustrates the continuum between purely enumerative studies and strongly analytic studies.

Acceptance sampling, valuing inventory, sampling approximations to censuses, and exit polls of voters are all essentially purely enumerative studies and fall at one end of the spectrum. At that end, the theory of probability models and sampling supports the use of the standard error for estimation. The subject matter expert has a relatively small role here, restricted to defining the universe and approving the frame. Most importantly, the knowledge we gain here may be useful, but it gives us little leverage for improving things in the future.

In the middle of the spectrum, we consider stable processes. If probabilistic predictions are to be appropriate here, the subject matter expert's role is to ensure that future conditions will remain relatively unchanged, or if they do change, to ensure that these changes will not substantially affect the process. Prediction in this case is supported by the theory behind the Shewhart control chart, which hinges on the assumption of a common-cause system. The more this assumption appears to be violated, the weaker the justification for prediction. Yet the knowledge available from these methods give us greater leverage for identifying ways to improve the process by changing the cause system at work.

At the other end of the spectrum, in situations of pure prediction—such as a poll taken several months before an election, pilot studies of a prototype product design, accelerated life tests, or low-dose extrapolation—there is absolutely no theory to support the use of the standard error for prediction. These are analytic studies in which if predictions are to be reasonable, the subject matter expert has to ensure that the studies, conducted as they are at a given time, are reasonable representations of the actual situations to which the results will be applied. Here the subject matter expert's knowledge is critical, and traditional statistical methods provide no rational quantification of the accuracy of the prediction. Note that these are precisely the situations with the highest leverage for improvement, where there is the greatest opportunity to influence future outcomes by making useful changes.

Methods used to conduct and analyze analytic studies must be suited to this dynamic environment. Because of the effect of changing conditions, the primary source of uncertainty in an analytic study lies in identifying which variables will have the greatest influence on future outcomes of the product or process. There is *no* statistical theory that allows a quantification of the magnitude of this uncertainty.

Experiments on processes or products are very common in industrial environments. These experiments are carried out to predict whether one of the methods or materials tested will perform better in the future. How do the data from a planned experiment help us to predict? This question is important for government and education as well as industry. Statistical methods for the design of analytic studies and analysis of data from these studies will be presented in the remainder of this book.

Chapter 2 looked at single experiments to test a change. Other types of experiments range from very informal investigations to well-planned studies involving a series of specific experiments. Table 3.1 lists some types of experiments ordered by increasing level of formality.

**TABLE 3.1 Types of Experiments**

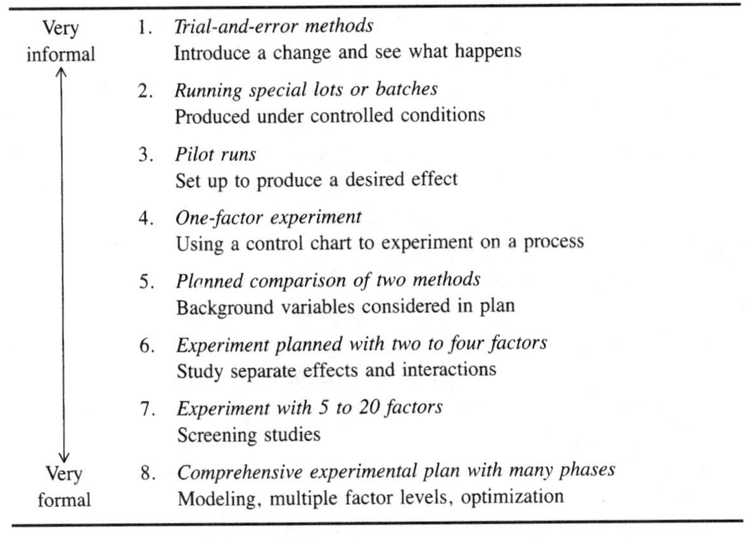

| Scale | No. | Type of experiment |
|---|---|---|
| Very informal ↑ | 1. | *Trial-and-error methods*<br>Introduce a change and see what happens |
| | 2. | *Running special lots or batches*<br>Produced under controlled conditions |
| | 3. | *Pilot runs*<br>Set up to produce a desired effect |
| | 4. | *One-factor experiment*<br>Using a control chart to experiment on a process |
| | 5. | *Plnnned comparison of two methods*<br>Background variables considered in plan |
| | 6. | *Experiment planned with two to four factors*<br>Study separate effects and interactions |
| | 7. | *Experiment with 5 to 20 factors*<br>Screening studies |
| ↓ Very formal | 8. | *Comprehensive experimental plan with many phases*<br>Modeling, multiple factor levels, optimization |

SOURCE: Adapted from Western Electric Co. (1956).

## 3.3 Principles for Designing Analytic Studies

R. A. Fisher (1947) first expounded principles for sound experimentation, and many others have extended these concepts during the past 50 years. The following five principles are applicable to analytic studies:

1. *Well-defined objective:* The objective of the experiment must be clearly defined, preferably with a statement of planned action based on the results.
2. *Sequential approach:* Experimentation should proceed sequentially, with knowledge gained in previous experiments used to design new experiments.
3. *Partitioning variation:* The experiment should allow the variation in the response variables to be clearly partitioned into components due to factors, due to background variables, and due to nuisance variables.
4. *Degree of belief:* Conclusions from the experiment should be drawn with an adequate degree of belief. The study should be conducted under a range of conditions to increase the degree of belief that the results will hold in the future.
5. *Simplicity of execution:* The design should be as simple as possible, while still satisfying the first four principles.

### Objective

A careful statement of the objective will allow efficient allocation of resources for designing the experiment. If the experiment is part of the plan for an improvement cycle, the objective of the cycle should be stated in such a way that it provides guidance to those designing the experiment. The objective should clarify whether the experiment involves screening a large number of variables to find the most important ones, studying in depth a few variables, or confirming the results of past studies under new conditions.

All interested parties should contribute to the objective before other work on the experiment is begun. For example, an experiment to be designed and conducted in research is often of interest to those in manufacturing, engineering, marketing, or management. Discussion of the objective with people in these groups can often result in a study that is more generally useful.

A statement of the results required for taking action should be considered in the objective when appropriate. Examples of objectives for experiments might be to identify factors that can be used to improve yield by 5 percent, determine if the machines differ by more than 10 $lb/in^2$, and identify the three factors that have the greatest effect on the variation of the parts.

Another consideration in the statement of the objective is to make clear the predictive nature of the experiment and identify the various courses of action that could be followed based on the results of the study. The objective should be helpful in identifying the response variables and appropriate factors for the study.

### Sequential approach

The sequential nature of learning should be considered in planning experiments. The Model for Improvement (see Chap. 1) stresses the iterative nature of development of knowledge of the product or process. Many iterations of the cycle to develop and test changes will include the design of an experiment. Screening studies are used for developing a change for improvement. Screening studies are discussed in Chap. 6, where fractional factorial designs to examine five or more factors in an initial study are presented. Chapter 7 presents nested designs that can be used to focus on the areas of the process with the greatest potential for improvement.

As knowledge is gained, experiments will be repeated using new levels of factors previously studied and some new factors. If it is desired to study in depth the relationships among the factors, two to four factors can be studied in a factorial design (Chap. 5). A fractional

**TABLE 3.2 Strategy for Experimentation**

| Current knowledge | Types of experiments |
|---|---|
| Low knowledge | Fractional factorials (screening studies) |
| Moderate knowledge | Fractional factorials (study interactions)<br>Factorial studies (new levels, new factors) |
| High knowledge | Confirmatory |

factorial design can be used for five factors. As new theories are developed, experiments will be confirmatory in order to increase the degree of belief from previous experiments (see Table 3.2).

To begin an analytic study, the current knowledge about the relevant products and processes is documented by use of models such as flowcharts and cause-and-effect diagrams. These models provide information on variables and factor levels.

The iterative nature of the PDSA cycle is an important feature for analytic studies. Knowledge of the system is increased sequentially as experiments are completed. Experiments are designed to answer questions based on a combination of theory from experts in the subject matter and conclusions from analysis of data from past studies. The more complete the current knowledge, the better the prediction of the answers to the questions. Confidence that the current knowledge provides sufficient basis for action is directly related to the ability to predict future performance of the process or results of future studies. Experiments performed to increase the degree of belief, by testing the results of past studies under different conditions, are referred to as *confirmatory studies.*

### Partitioning of variation

Determination of important factors and estimation of the effects of these factors on the response variables are important in any experiment. These decisions and estimates should not be confounded by background or nuisance variables. The factors chosen for the experiment are usually those that the experimenter believes will have the greatest effect on the response variable. In many experiments, the variation due to background or nuisance variables will be as great as or greater than the variation due to the factors chosen. To help determine whether the most important factors have been studied, the experimental design must allow the variation in the response variable to be partitioned into components due to factors, due to background variables, and due to nuisance variables.

### Degree of belief

Most experiments are carried out to determine whether a change will result in better performance in the future. The wider the range of conditions included in the experiment, the more generally applicable will be the conclusions from the experiment. The degree of belief in the validity of the conclusions is increased by running the experiment using different machines, different operators, different days, different times of the year, different batches of raw materials, and so on.

The range of conditions selected for the study will ultimately determine the degree of belief in the actions taken as a result of the experiment. The expert in the subject matter must determine what is an "adequate" degree of belief for taking action. This determination will depend on the magnitude of the change in the system being considered and the degree of extrapolation necessary. Changes in existing manufacturing processes may require a completely different degree of belief than changes in the process of development of a new product.

### Testing over a wide range of conditions

Too often, analytic studies are not conducted over a broad range of conditions. The reasons include tight budgets, time constraints, difficulty in analysis of the data, lack of knowledge of how to efficiently include them, and too many possible conditions to consider. Fortunately, there are methods that help to eliminate or reduce these difficulties in planning experiments. The use of blocks allows the grouping of experimental runs so that a wide variety of conditions can be included in the study and still allow comparisons of different alternatives under uniform conditions.

The degree of belief in the results of an analytic study is increased as the same conclusions are drawn under a wide range of test conditions. How wide a range is sufficient? How close to field conditions do research conditions have to be? These are questions that only an expert in the subject matter should attempt to answer. They cannot be answered by statistical theory; however, the theory of statistical design of experiments provides tools to increase the degree of belief in the results and hence to increase the expert's potential to make improvements.

### Simplicity of execution

The simplicity of the experiment should be considered one of the most important properties of a planned study to improve a system. Simplicity is important in the design, conduct, and analysis of a planned experiment. An experimental design should be as simple as possible while still satisfying the other properties of a well-planned experiment. Simplicity allows all interested parties to be involved in all aspects of the study.

Simplicity also allows the experimenter the flexibility to adjust for changes that are often required during the conduct of the study.

Simplicity requires that all the practical aspects of conducting an experiment be considered. Some important aspects include the degree of difficulty in changing levels of a factor, the ability to control background variables, and the ability to measure important response variables.

## 3.4 Tools for Experimentation

The five principles discussed in the previous section should always be considered in planning an experiment. Subject matter knowledge is needed to effectively practice these principles when planning an experiment. These subject matter experts may be managers, operators, technicians, scientists, engineers, physicians, etc.

R. A. Fisher (1935) described the use of four tools to help ensure that an experiment follows these principles.

1. *Experimental pattern:* the arrangement of factor levels and experimental units in the design
2. *Planned grouping:* blocking of experimental units
3. *Randomization:* the objective assignment of combinations of factor levels to experimental units
4. *Replication:* repetition of experiments, experimental units, measurements, treatments, and other components as part of the planned experiment

Table 3.3 summarizes how these four tools can be used. The use of each is discussed in the following sections.

### Experimental pattern

The experimental pattern is the schedule for conducting the experiment. Each test to be conducted or measurement to be made is identi-

**TABLE 3.3 Using Experimental Tools to Attain the Properties of a Good Experiment**

| | Tool | | | |
|---|---|---|---|---|
| Property | Experimental pattern | Planned grouping | Randomization | Replication |
| Well-defined objective | X | | | |
| Sequential approach | X | X | | X |
| Partitioning variation | X | X | X | X |
| Degree of belief | X | X | | X |
| Simplicity of execution | X | | X | |

fied in the pattern. The experimental pattern is sometimes called the *test plan.* The pattern may include consideration of any of the other tools (planned grouping, randomization, and replication).

The selection of an appropriate experimental pattern is the primary tool for ensuring the attainment of the objectives of the experiment. The pattern will identify the factor combinations and factor levels to be included in the study. The experimental pattern selected will also aid in attaining the other properties of a good experiment:

- The change in degree of belief resulting from the study will be limited by the pattern.
- The proper pattern will allow estimation of important effects.
- The properly selected experimental pattern will simplify the analysis of data from the study.
- The costs and other resources required by a study are controlled through the experimental pattern.

There are many types of experimental patterns. The names of the particular patterns are often used to describe the type of experiment being conducted. One of the most common patterns used in experiments for improvement is the *factorial design* (Chap. 5). In a factorial design, tests are arranged in a pattern such that multiple factors are studied with each test. Figure 3.3 is an example of a factorial pattern.

Another common experimental pattern is the nested or hierarchical design (Chap. 7). In a nested design, levels of different factors are studied within a given level of another factor. Figure 3.4 is an example of a nested design.

| | | *Pressure 1* | | *Pressure 2* | |
|---|---|---|---|---|---|
| | | *Temp. 1* | *Temp. 2* | *Temp. 1* | *Temp. 2* |
| Batch 1 | Load 1 | #6 | #2 | #8 | #15 |
| | Load 2 | #11 | #13 | #4 | #7 |
| Batch 2 | Load 1 | #10 | #9 | #5 | #12 |
| | Load 2 | #14 | #1 | #16 | #3 |

Four-factor experiment (Factors: pressure, temperature, batch, and load)
Each factor at two levels (pressure 1, pressure 2, etc.)
Test numbers ( #1 — #16 ) shown in a randomized order

**Figure 3.3** A factorial experimental pattern.

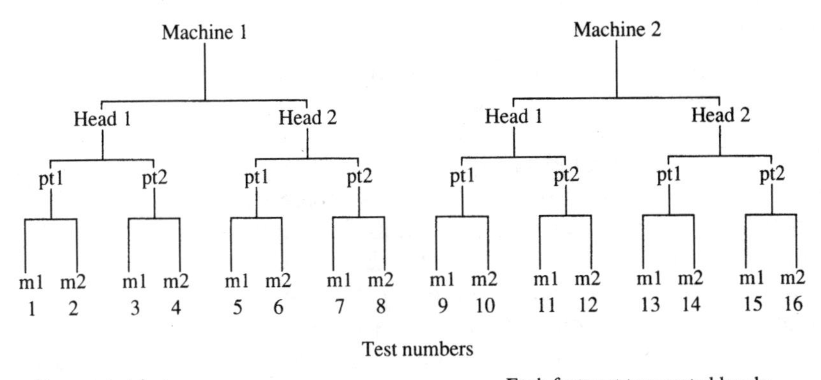


**Figure 3.4** A nested experimental pattern.

Incomplete designs have patterns that include only a subset of a full factorial or nested experimental pattern. These designs include the fractional factorial design and the unbalanced or partially balanced nested design.

Composite designs are another class of experimental pattern. These patterns are either combinations of factorial and nested designs or one of the basic experimental patterns augmented with additional tests. Designs for experiments involving mixtures are another important class of patterns. Much of the discussion in Chaps. 4 through 8 will deal with the selection and construction of experimental patterns appropriate for different situations.

### Planned grouping

Planned grouping is an important tool for addressing background variables. There are two important decisions to make concerning background variables:

- How to control the background variables so that the effects of the factors are not distorted by them
- How to use the background variables to establish a wide range of conditions for the study to increase the degree of belief or to aid in designing a robust product or process

There are three basic methods of controlling background variables:

1. Hold them constant in the study.
2. Measure them and adjust for their effects in data analysis.
3. Use planned grouping to set up blocks.

The concept of planned grouping takes advantage of naturally occurring homogeneous experimental units in the conduct of an experiment. When planned grouping is used in developing the experimental pattern, the design is often called a *block design*.

Examples of common background variables used to form blocks are time of day, operator, machine, shift, batch of raw materials, agricultural plot, cavity in a mold, test stand, and season. In studying tire wear for cars during normal use, an individual car is a natural blocking variable. The amount and type of driving for different cars could vary considerably. Different brands (or factor combinations) of tires could be tested on any particular car to minimize the car-to-car variation when comparing the brands. The natural block size is four, so four different brands could be "grouped" on each car. In studying the wear of soles of shoes, a natural block size of two would correspond to the two feet of each individual wearing the shoes.

Planned grouping plays a very important role in establishing the degree of belief for the results of the study. Should the study be conducted on one machine in the plant or on multiple machines? Including different machines will add variation but will also increase the degree of belief that the results can be extrapolated to all machines of that type in the plant. By treating the machines as blocks, the experiment can include multiple machines to increase the degree of belief in the results through analysis of interactions between the machines and the factors in the study.

Often, naturally occurring blocks are formed by a combination of background variables. Learning about an important block effect would require additional study to determine the specific variables that contribute to the effect. Such a background variable has been called a *chunk-type factor* by Ott (1975). For example, if parts for a study were selected from two different lots of available materials, the block effect from the two different lots might be caused by different vendors, different types of raw materials, different preprocessing equipment, or different storage conditions. The selection of different lots as a chunk variable for the study would potentially allow the effects of any of these factors (plus others unknown) to be exposed in the study. This concept is extremely important in most planned experiments. Other common chunk-type block variables are time and location.

Grouping important background variables into blocks is an important aspect of designing an experiment. If the sizes of parts can possibly alter the effect of certain factors in the study, how can we incorporate part size in the experiment? One way is to form two groups of parts based on their measured sizes. One group would be parts at the low end of the size specification, and the other group would be parts at the high end of the specification. Combinations of factors would be tested using parts from each of these planned groups or blocks.

What if there are five background variables (e.g., room temperature, operator, machine, material supplier, and line speed) that need to be considered in the study? Blocks formed by all combinations of these variables may result in too large a study. Two or three blocks could be developed by grouping extreme conditions for each of the background variables. Block 1 could comprise high temperature, operator A, machine 8, supplier X, and low line speed; and block 2 could include low temperature, operator B, machine 1, supplier Z, and high line speed. These blocks would be designed to represent extreme conditions based on the judgment of the expert in the process. If the effect of the block (or the block-factor interaction) were found to be large, the next experiment could be used to determine which background variable or combination of background variables caused the effect.

Figure 3.5 summarizes a one-factor design that creates a chunk variable (four blocks) to incorporate background variables in the design.

Some types of experimental patterns are named for the planned grouping pattern used to accommodate background variables. A paired-comparison experiment requires a single factor at two levels and one background variable (which could be a chunk-type variable). Two experimental units must be available for each grouping of the background variable. Factor level differences are evaluated within each block. Examples of background variables appropriate for a paired-comparison design are identical twins, front fenders on a car, and two adjacent plots of land.

The randomized block design extends the approach of the paired-comparison design to more than two factor levels, still using only one background variable (for defining the blocks). Each factor level combination must occur in each block an equal number of times. An incomplete block design relaxes the requirement that each combination occur an equal number of times in each block.

The split-plot design occurs when the different factors or combinations of factors are not assigned to experimental units (possibly within a block) in a random manner (see the following section on randomization). The subplots or subgroups resulting from the assignment become chunk-type blocks.

Another approach to accommodating background variables is the outer array suggested by Taguchi and Wu (1979). This concept is discussed further in Chap. 9.

Objective: Run an experiment to compare three material suppliers. Each of the three suppliers will submit four prototypes.

A. Identify background variables in the plant that could affect the response variables of interest:

| Background variable | Levels |
|---|---|
| Machine | #7, #4 |
| Operator | Joe, Susan, George |
| Gage | G-102, G-322 |
| Saw blade | 20 blades available |
| Time (day-to-day) | Many different days possible. |

B. Create four blocks with widely varying conditions based on these background variables:

| | Block 1 | Block 2 | Block 3 | Block 4 |
|---|---|---|---|---|
| Machine | #7 | #4 | #7 | #4 |
| Operator | Joe | Susan | George | Joe |
| Gage | G-102 | G-322 | G-102 | G-322 |
| Saw blade | Blade 1 | Blade 2 | Blade 3 | Blade 4 |
| Time | Day 1 | Day 2 | Day 3 | Day 4 |

C. Evaluate one prototype from each supplier (A, B, C) in each block (random order within each block):

| Test | Block 1 | Block 2 | Block 3 | Block 4 |
|---|---|---|---|---|
| 1 | B | B | C | A |
| 2 | A | C | A | C |
| 3 | C | A | B | B |

D. Analyze supplier differences within each block; evaluate consistency of differences across blocks.

**Figure 3.5** Example of planned grouping of background variables using chunk-type blocks.

Planned grouping is a key method for an analytic study. Instead of holding conditions constant in a study, degree of belief can be increased by varying conditions between blocks while simultaneously holding them constant within blocks.

## Randomization

Randomization is the use of random numbers to determine the assignment of factor combinations to experimental units or to the order of performing some aspect of the study. Whereas the experimental pattern addresses the variables identified as factors in the experiment, and planned grouping is used to accommodate background variables, randomization is a tool that addresses the nuisance variables. Nuisance variables are process variables that affect the response variables but that have not been identified by the experimenter. It is common in conducting studies that many variables that can affect the response variables are not known to the persons conducting the experiment. Typical nuisance variables are environmental effects such as temperature or humidity, drifts in measurement equipment, batch-to-batch variation in raw materials, machine warm-up effects, imprecision of measurement instruments, and position effects in a chamber.

Randomization helps prevent the variation due to nuisance variables from being confused with the variation due to the factors or due to the background variables. Randomization is often compared to insurance: You only need it when a problem arises (i.e., a big effect of a nuisance variable). If important unknown nuisance variables are present during an experiment, use of randomization can allow their effects to be separated from the effects of factors or background variables. Diagnostic analyses to uncover nuisance variables are also facilitated by randomization. Unlike insurance, when randomization is not done, the experimenter may not be aware of the impact of nuisance variables and might draw incorrect conclusions.

Randomization requires a formal procedure and not just haphazard selection or ordering. Random number or random permutation tables should be used for this purpose. Appendix C contains examples of random number tables and instructions for their use. Mechanical devices that simulate a random process can also be used. Examples of such procedures are flipping a coin, rolling a pair of dice, and pulling numbers (after mixing) from a container.

Randomization is often required at several different levels or phases of an experiment. For example, an experiment to evaluate several factors in an industrial process could require random selection of raw materials stored in a warehouse, random ordering of the combinations of the factors to be evaluated, and random ordering of the tests conducted to obtain the response data on the product during the experiment.

In most situations it is not desirable to completely randomize the conduct of an experiment. Randomization should be restricted as much as possible based on knowledge of background variables. Planned grouping, as previously discussed, is the most common restriction put on randomization. When blocking is employed, the randomization is done within each block rather than over all tests in the design. For example, an experiment might require 20 parts, five from each of four cavities of a mold. Randomization would be used to assign combinations of factors to the five parts within each mold.

In other cases, a random order for the conduct of a test could be prohibitively expensive. For example, changing the levels of one particular factor might require a machine to be shut down for a day, whereas changing all the other factors could be done in a few minutes. Extra care is required in analysis and interpretation in these situations. Some examples will be discussed in later chapters.

The decision of what type and level of randomization to use should be based on the particulars of the experiment. The advantages of randomization should be weighed against the additional resources required to randomize. Situations in which randomization is particularly important include

- Experiments conducted when the important response variables have not been brought into a state of statistical control
- Experiments that will be conducted by many different operators and technicians
- Experiments in which the variation due to nuisance variables is expected to be large relative to the magnitude of effects of important factors
- Formal experiments in which results must be evaluated by others (such as customers or senior staff) for action to be taken

After restricting for important background variables, randomize in all remaining situations unless the constraints to randomization have been objectively considered and found to be prohibitive. Randomization procedures for different types of experimental patterns will be presented in later chapters of this book.

How should experimental units be selected? In an analytic study, there is no universe from which to draw a sample. However, in designing an analytic study there are decisions to make concerning the conditions under which the product, process, or system will be run during the study and the outcomes that will be measured for each set of conditions. Deming (1975) makes the point that all analytic studies are conducted on judgment samples. The judgment of the expert in the subject matter determines the conditions to be studied and the measurements to be taken for each set of conditions. It is rare in an analytic study that a random selection of conditions or outcomes is preferred to a judgment selection. The opposite is true in an enumerative study (see Fig. 3.6).

| *Type of study* | *Method of selection* | |
|---|---|---|
| | *Random* | *Judgment* |
| Enumerative | Good | Bad |
| Analytic | Fair | Good |

**Figure 3.6** Methods of selection of units for enumerative and analytic studies.

It is often impractical to measure all the outcomes of the process for each set of conditions. However, a random selection is usually not the best method for selecting which outcomes to measure.

It is sometimes useful to combine units selected randomly and by judgment. Consider a study of a molding process in which different operating conditions for the process are studied. If the mold has 100 cavities, it may be impractical to study the parts produced in all cavities under each set of conditions. A portion of the experimental units could be selected based on the judgment of an engineer. The engineer may suggest some cavities to be included—possibly those in the corners of the mold, some that are fairly worn out, some near the point of injection, and others far away. Remaining cavities could be selected randomly.

Initially, it is usually best to select conditions or experimental units at the extremes (within practical constraints). Alternatives can be compared at these extremes and further experiments performed as needed. Incorporating temporal spread into the study will be a more important consideration than the total size of the sample.

### Replication

Replication refers to repeating particular aspects of an experiment. Replication plays an important role in attaining many of the properties of a good experiment. It is the primary tool for studying stability of effects and for increasing the degree of belief in the results. There are many different types of replication, including

- Repeated measurements of experimental units
- Multiple experimental units for each combination of factors
- Partial replication of the experimental pattern
- Complete replication of the experimental pattern

Each of these types of replication will require different interpretations in the analysis of the results of an experiment. For example, for comparing differences between factor combinations, replications that include all sources of variation typical of a run of either combination are desirable.

Replication plays the primary role in providing a measure of the magnitude of variation in the experiment due to nuisance variables. Replication also aids in minimizing the impact of nuisance variables on factor effects by (1) possibly enabling nuisance variables to be averaged out and (2) allowing the study of interactions between factors and experimental conditions. Replication, in conjunction with planned grouping, can be used to expand the study to a wide range of conditions.

Since the amount of replication in an experiment directly affects the resources (time, budget, materials, etc.) required for the study, replication is often established by constraints on these resources. Since experiments should be run sequentially in studies for improvement, the amount of replication for any one experiment is not critical. Usually it is desirable to obtain 5 to 10 comparisons for the different levels of each factor for the initial stages of the study. Another guideline is that the initial experiment should consume about 25 percent or less of the resources allocated for the study. This will allow the completion of at least four improvement cycles within the budgetary constraints on the study.

The most important consideration in an analytic study is not the number of experimental units per level of a factor, but the breadth of conditions under which the comparisons can be made. Replication over similar conditions provides little increase in the degree of belief. For example, in a high-speed manufacturing operation, a study based on 5 to 10 experimental units selected from different days' production and made from different material lots would be much preferred to a study based on 50 consecutively produced units.

## 3.5 Form for Documentation of a Planned Experiment

Figure 3.7 shows a form that can be used to document the planning of experiments. The form is helpful in communicating the experimental plan and in documenting the considerations given to the various tools of experimentation discussed in this chapter. The form can serve as a tool in the planning phase of the improvement cycle. In many studies the form will serve as a summary of more extensive documentation and background information. Copies of this and other forms used in this book are included in App. C.

### Objective

The first item on the form is a summary statement of the objective of the experiment. In the statement of the objective, consideration should be given to

- Response variables and factors of interest
- Level of knowledge about the process under study
- Actions to be taken as a result of the study
- Analytic nature of the study
- Statement of the results required in order to take action

1. **Objective:**

2. **Background information:**

3. **Experimental variables:**

| A. | Response variables | Measurement technique |
|---|---|---|
| 1. | | |
| 2. | | |
| 3. | | |

| B. | Factors under study | Levels |
|---|---|---|
| 1. | | |
| 2. | | |
| 3. | | |
| 4. | | |
| 5. | | |
| 6. | | |
| 7. | | |

| C. | Background variables | Method of control |
|---|---|---|
| 1. | | |
| 2. | | |
| 3. | | |

4. **Replication:**

5. **Methods of randomization:**

6. **Design Matrix:** (attach copy)

7. **Data collection forms:** (attach copies)

8. **Planned methods of statistical analysis:**

9. **Estimated cost, schedule, and other resource considerations:**

**Figure 3.7** Form for documentation of a planned experiment.

## Background information

The second item on the form is background information. In analytic studies, it is important to proceed sequentially, building on knowledge gained previously from theory or from other improvement cycles. This *current knowledge* should be used to design the current study. The experiment being considered should be put in context with other studies in previous improvement cycles.

Examples of useful background information are summaries of results of previous experiments on the process under study, control charts from the process for the response variables in the study, laboratory prototype experiments, studies of similar products or processes, and studies or control charts of the measurement processes to be used in the experiment.

As part of the summary of the current knowledge of the process, a prediction of the outcome of the study is useful. This prediction can be used during the analysis of the data to help determine whether the study confirms previous beliefs or suggests a need to reconsider those beliefs.

#### Variables in the study

The response variables, factors, and background variables are documented next. The objective should make clear whether a particular process variable will be a response variable, a factor, or a background variable. In choosing levels for the factors, the levels should be set far enough apart that their effects will be large relative to the variation caused by the nuisance variables. However, the levels should not be so far apart that the following types of trouble could develop:

- Conditions that make the experimental run unsafe
- Conditions that cause substantial disruption of a manufacturing facility
- Important nonlinearities or discontinuities hidden between the levels
- Substantially different cause-and-effect mechanisms under the different conditions in the experiment

The method of control (hold constant, measure, or planned grouping in blocks) for each background variable identified should be included.

Based on the information in the first three areas of the form, an experimental pattern can be developed and the amount and type of replication determined. Any randomization or restrictions on the randomization should be documented, including

- Assignment of treatments to experimental units
- Determination of the order of running the study
- Determination of the order of testing

Another important part of the documentation of the experiment is the form for recording the data. Careful consideration should be given to developing these forms. The form should be designed for simplicity of recording, not for analysis. The randomized experimental pattern should be built into the form. The form should also include space for recording any significant events that happened during the study, including things that did not go according to plan. A carefully designed form for recording the data will help communicate the intentions of the planners of the experiment to those running the study.

A brief summary of the planned methods of analysis should be documented next. Finally, some information on cost, schedule, and other necessary resources required by the experiment should be given. An example of a completed form is presented in Fig. 3.8.

## 3.6 Analysis of Data from Analytic Studies

In an enumerative study, the existence of a distribution for the characteristic of interest is ensured by the existence of a frame. Summary statistics such as a mean and standard deviation can be used to estimate parameters of the distribution. These estimates will have a quantifiable measure of uncertainty if the sample from the frame is chosen using a random number table. This type of analysis is usually an important step in accomplishing the aim of an enumerative study.

In an analytic study the aim is prediction. A distribution useful for even short-term (days or weeks) prediction may not exist for any characteristics of a new product. The standard error of a statistic or the standard deviation does not address the most important source of uncertainty in an analytic study—factors outside the conditions of the study that will change in the future.

Running an experiment involves making changes to the system in a thoughtful way. When a well-planned experiment is run on a stable system, it is usually straightforward to determine what effects the changes in the factors have had on the response. When a system is unstable, the planning and the analysis become more difficult. An experimenter must be able to separate the effects of changes in the factors from the variation induced by special causes.

One should be cautious about running factorial experiments on an unstable system. The preferred strategy is to identify and remove special causes using control charts, standardization of procedures, or other methods before running such experiments. However, factorial experiments can be used to improve unstable systems, but doing so effectively requires a high level of skill of the experimenters.

Often in analytic studies, we are interested in long-term (months or years) predictions or predictions extending far beyond the conditions of the study. Although stability is important in increasing the degree of belief in an analytic study, stable processes in the past do not guar-

1. **Objective:**
Maximize the profit of a movie theater chain (from *Forbes* magazine on March 11, 1996)

2. **Background information:**
Increased competition has resulted in a downturn in ticket sales.

3. **Experimental variables:**

| A. Response variables | Measurement technique |
|---|---|
| 1. Profit | revenue - expenses (for a 2-week period) |

| B. Factors under study | Levels | |
|---|---|---|
| 1. Ticket price | current | increase |
| 2. Advertising | current | bigger ads |
| 3. Popcorn | current price | free |

| C. Background variables | Method of control |
|---|---|
| 1. Day of the week | identify days for each test |
| 2. Movie showing | held constant |
| 3. Competitors' movies | recorded |
| 4. Weather | recorded |

4. **Replication:**
Replicated one time

5. **Methods of randomization:**
The order of the factor combinations was randomized using a random number table.

6. **Design Matrix:** (attach copy)
$2^{3-1}$

7. **Data collection forms:** (attach copies)
Not shown

8. **Planned methods of statistical analysis:**
Analyze the effects of the three factors using the design matrix, and collapse to a full factorial design if one of the factors is unimportant.

9. **Estimated cost, schedule, and other resource considerations:**
Plan to run the experiment over a two-month time period. Advertising budget is increased 10%.

**Figure 3.8** Example of a completed form (three-factor example).

antee a constant cause system in the future. Responsibility for extrapolation still rests with the experts in the subject matter.

Also, in many circumstances the conditions under which the results will be used are very different from the conditions of the study. This is often the case for work in product development, where a study is commonly conducted on prototypes to learn about how the products will function years later in the field. (See more on the use of planned experimentation in new product design in Chap. 9.) The same can be said of laboratory tests on a chemical process intended to foster some insight on how the full-scale process should be designed. In these and many other situations, the experimental plan is the most important statistical tool for increasing the degree of belief. Plans that allow results to be compared over a wide range of conditions are especially useful.

### Basic principles for analysis

Three basic principles underlie the methods of analysis for analytic studies:

1. The analysis of data, the interpretation of the results, and the actions that are taken as a result of the study will be closely connected to the current knowledge of experts in the relevant subject matter.
2. The conditions of the study will be different from the conditions under which the results will be used. An assessment of the magnitude of this difference and its impact by experts in the subject matter should be an integral part of the interpretation of the results of the experiment.
3. Methods for the analysis of data will be almost exclusively graphical, with minimum aggregation of the data before graphical display. The aim of the graphical display will be to visually partition the data among the sources of variation present in the study.

The first principle relates to the important role that knowledge of the subject matter plays in the analysis and interpretation of data from analytic studies. The PDSA cycle is used to build the knowledge necessary for improvement. The first step in the cycle is the plan. During the planning phase, predictions of how the results of the study will turn out are made by the experimenters. This is a means of bringing their current knowledge into focus before the data are collected for the study.

In the do step of the PDSA cycle, the study is performed and the analysis of the data is begun. The third step, study, ties together the current knowledge and the analysis of the study. This is done by comparing the results of the analysis of the data to the predictions made during the planning phase.

The degree of belief that the study forms a basis for action is directly related to how well the predictions agree with the results of the study. This connection between knowledge of the subject matter and analysis of the data is essential for the proper conduct of an analytic study.

The second principle relates to the difference between the conditions under which an analytic study is run and the conditions under which the results will be used. In some studies, the conditions are not too different. An example of a small difference is a study in which minor changes are made to a stable manufacturing process. Pilot plant studies and medical experimentation on animals to develop drugs or procedures to eventually be used for humans are examples of substantial differences in conditions.

Differences between the conditions of the study and the conditions of use can have substantial effects on the accuracy of the prediction. In most analytic studies, some aspects of the product or the process are changed, and the effects of these changes are observed. Estimation of these effects is an important part of the analysis. However, it should not be assumed that the factors produce the same effects under all conditions. The factors changed in the study may interact with one or more background conditions, such as environmental conditions, time of day, or purity of materials. This interaction can cause the effect of a factor to differ substantially under different conditions.

A wide range of conditions will allow the experimenter to assess the presence of such interactions and the implications for interpretation of results. Graphical methods of analysis will be used to determine the stability of effects over different conditions. During all phases of an analytic study, the experimenters should have firmly in mind the answer to the question, How will the results of this study be used?

The third principle relates to the synthesis of theory and the data presented with the aid of graphical methods. In analytic studies, it is essential to

- Include knowledge of the subject matter in the analysis.
- Include those who have process knowledge—operators, technicians, engineers, clerical personnel, scientists, and managers.
- Apply this subject matter knowledge creatively to uncover new insights about the relationship between variables in the process.

These three principles provide the basis for analysis of all experiments in this book. The specific approach to analysis will differ depending on the type of experiment conducted, but the following elements will be common to the analysis of all experiments:

1. Plot the data in the order in which the tests were conducted. This is an important means of identifying the presence of special causes of variation in the data.

2. Rearrange this plot to study other sources of variation that were included in the study design but are not directly related to the aim of the study. Examples of such variables are batches of raw material, measurement, operators, and environmental conditions.

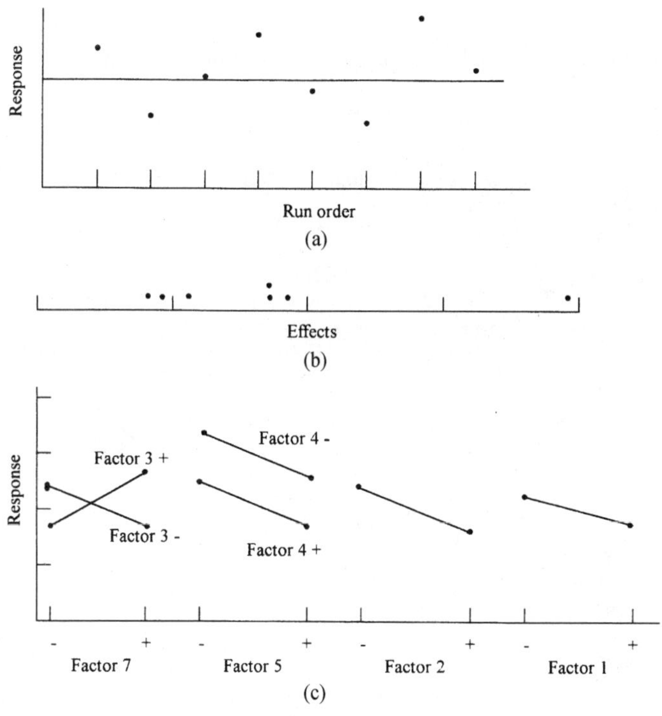


**Figure 3.9** Examples of typical graphical displays. (*a*) Run chart; (*b*) dot diagram; (*c*) response plot.

3. Use graphical displays to assess how much of the variation in the data can be explained by factors that were deliberately changed. These displays will differ depending on the type of experiment run.
4. Summarize the results of the study with graphical displays.

Examples of typical graphical displays that will be used for analysis in this book are given in Fig. 3.9.

## 3.7 Summary

This chapter has laid a foundation for all the planned experiments discussed in the remaining chapters. This foundation is based on Deming's concept of analytic studies. Planned experiments to improve product, process, or system are analytic studies, where the aim is prediction. This requires that the experiment use the principles and tools presented in Secs. 3.3 and 3.4. The planning form given in Sec. 3.5 is a sort of checklist for these principles and tools. The analysis of data from an analytic study will rely heavily on subject matter knowledge and simple graphical methods.

## References

Deming, W. Edwards (1950): *Some Theory of Sampling,* Wiley, New York. (Reprinted by Dover Publishing, 1960.)
Deming, W. Edwards (1960): *Sample Design in Business Research,* Wiley, New York.
Deming, W. Edwards (1975): "On Probability as a Basis for Action," *The American Statistician,* vol. 29, no. 4, pp. 146–152.
Fisher, R. A. (1935): *Design of Experiments,* Hafner Publishing Company, New York (first published in 1935, 8th edition in 1966).
Fisher, R. A. (1947): *The Design of Experiments,* 4th ed., Oliver and Boyd, Edinburgh, Scotland.
Ott, Ellis R. (1975): *Process Quality Control,* McGraw-Hill, New York, chap. 4.
Taguchi, Genichi, and Yu-In Wu (1979): *Introduction to Off-Line Quality Control,* Central Japan Quality Control Association, Nagaya, Japan, chap. 5.
Western Electric Co. (1956): *Statistical Quality Control Handbook,* Indianapolis, Ind., pp. 76–77.

## Exercises

**3.1** Pick a process or product you are familiar with. To plan a study of that process or product, list examples of each of the following:

(*a*) Response variables
(*b*) Factors
(*c*) Background variables
(*d*) Nuisance variables
(*e*) Experimental units
(*f*) Blocks

**3.2** How does one distinguish a factor from a background variable in a study?

**3.3** What is the relationship between degree of belief and the conditions studied in an experiment?

**3.4** What are the four tools for planned experimentation? Give an example of the specific application of each of the tools in a study.

**3.5** The manager of a large office wants to compare two different types of computer terminals for future use in the office. The terminal vendors have made three of each type of terminal available for one week on a trial basis. List potential background variables in the office that should be considered in a study of the two types of terminals. Develop three chunk variables to be used as blocks to incorporate these background variables in the study.

**3.6** Consider a study in which you have recently been involved. What was the aim of the study? Was the study an enumerative or an analytic study? For any analytic study mentioned, assess the difference between the conditions under which the study was conducted and the conditions under which the results will be used.

**3.7** Planned experimentation is one of the most important statistical tools for analytic studies. Why is planned experimentation not as useful for enumerative studies?

**3.8** How would you define *experimental error* for analytic studies?

**3.9** The 14 members of the accounting department will participate in a study. Each person will collect data for a one-week period. Since the study will take more than 3 months to complete, nuisance variables could affect the outcome. Use each of the three random number tables in App. C to develop an order for participation in the study. (*Hint:* To use the 1 to 8 random permutations, divide the participants into two blocks of 7 each.)

**3.10** Management has identified some important areas for improvement in a continuous production unit, including yields, costs, efficiencies, scrap, and rework. A number of process changes have been proposed to address these areas, including new mixing procedures, alternative flow control, and changes in the catalyst operation (catalyst type, amount of charge, frequency of catalyst charge). The previous studies of these alternatives have not yielded clear results because of different operating procedures between the shifts. Also, the ambient weather conditions can have a big effect on the performance of the process. Plan an experiment to improve this process.

Complete an experimental design planning form for the study (it is not necessary to complete the design matrix or data collection form for the study). What other important information needs to be determined before you can plan the study? Use your imagination to fill in these details.

Chapter

# 4

# Experiments with One Factor

This chapter discusses planned experimentation when there is only one factor of interest. The single-factor experiment forms the basis for the multifactor experiments presented in later chapters. The principles and procedures introduced in this chapter for developing experimental designs and analyzing data from the experiments will be used in the remainder of the book.

## 4.1 General Approach to One-Factor Experiments

Chapter 2 discussed experiments for testing a change. This one-factor experiment had two levels: before and after the change. The rigor of the test was increased by replication (removing the change as many times as necessary until an adequate degree of belief is obtained). Testing a change is a special case of experiments with one factor.

In an experiment with only one factor, one or more response variables are measured or observed for different levels of the factor. The levels of the factor may be either qualitative (e.g., machine A, B, or C) or quantitative (e.g., temperature of 60, 80, or 100 degrees). One replication of the experiment requires the factor to be tested one time at each level. Additional replications can be done to satisfy the other properties of a good experiment. Each level of the factor is assigned to an experimental unit by a random process. The rationale of how the factor and levels are defined is important in any good design.

In an analytic study involving one factor, background variables should always be considered. In the simplest experimental pattern, important background variables are measured or are held constant. In block designs the background variables are handled by grouping the experimental units into blocks. Important block designs for one-

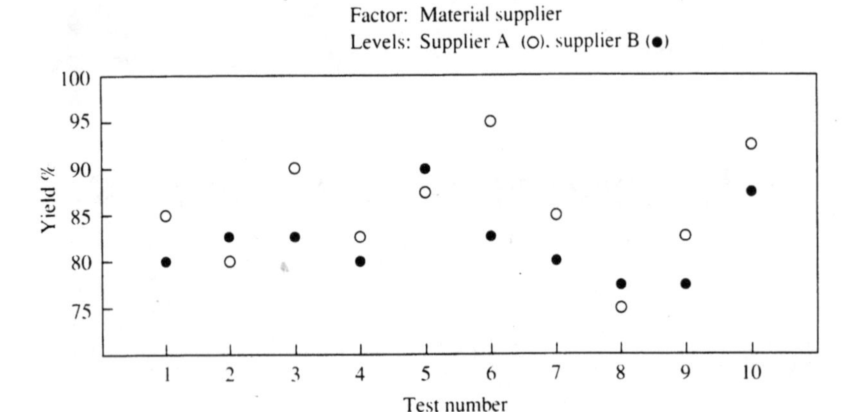


**Figure 4.1** Example of a run chart for a one-factor experiment.

factor experiments are the paired-comparison design, the randomized block design, and the balanced incomplete block design. Multiple background variables are considered by using the concept of chunk variables, introduced in Chap. 3. In these block designs, the factor levels are assigned to the experimental units within each block. The last three sections of this chapter discuss block designs.

The primary tool for analyzing data from a one-factor experiment is the run chart with the factor levels identified. The run chart can also be stratified by factor level. Figure 4.1 is an example of a run chart for a one-factor experiment with two levels of the factor included in the study. The usual steps in the analysis of data from a one-factor design are as follows:

1. Plot a run chart or control chart of the data with the factor levels identified.
2. Reorder the run chart according to factor level.
3. Remove the effects of the background variables, and plot the adjusted data by factor level.

Explanation of these tools will be given and clarified in the examples presented in this chapter.

## 4.2 Using the Control Chart for a One-Factor Experiment

Probably the most common use for the one-factor experiment in activities to improve quality is the active use of the control chart. The

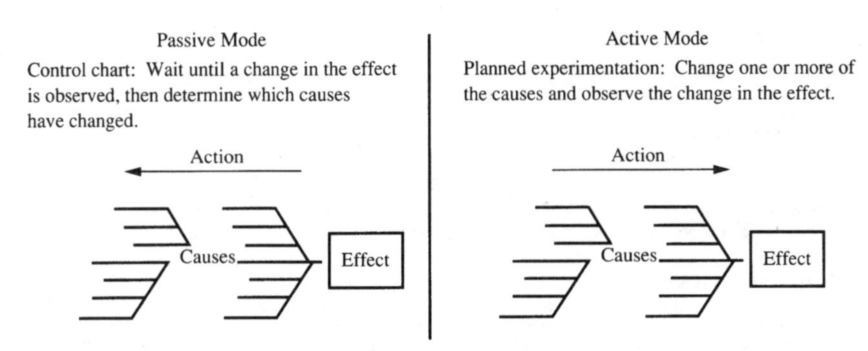


**Figure 4.2** Cause-and-effect diagram with a control chart.

cause-and-effect diagram (Fig. 4.2) is useful in comparing the reactive (passive) and active uses of the control chart.

When a control chart is used in the passive mode, action begins after the effect (i.e., a special cause) has occurred. At that time, a search is done to determine which of the causes or combination of causes produced the effect. Sometimes the search is not successful because of the large number of potential causes (both known and unknown) for the effect.

When the control chart is used in the active mode, changes in the process (the cause variables) are made, and then the effect of these changes on the response variable being plotted on the control chart is observed. When only one cause (or factor) is changed, this can be considered a single-factor experimental design. The degree of planning that takes place distinguishes a one-factor experiment from common fiddling with variables in the process. Example 4.1 illustrates the use of a control chart in the active mode.

**Example 4.1** The $X$-bar and $R$ control charts for a critical dimension (measured in thousandths of an inch) had been maintained for 6 months. The charts had remained in statistical control until recently, when changes in the average occurred for a short period when tools were replaced. A cause-and-effect diagram was prepared to identify the process variables that may have affected the variation in this dimension. One of the variables, clamp pressure, had not been previously studied by any of the operators or engineers on the improvement team. The team decided to study this factor to quantify its effect on the variation in the dimension.

Figure 4.3 shows the form for documentation of a planned experiment developed for this effort. The control chart for the day prior to the study and the 3 days of the study is shown in Fig. 4.4. No special causes that could not be attributed to the clamp pressure changes were noted during the experiment. Increasing the clamp pressure appeared to reduce the variation in the dimension. The averages of the ranges were computed for each day of the experiment and are summarized in Fig. 4.5.

1. **Objective:**
Determine the effect of clamp pressure on the variation of the critical dimension.

2. **Background information:**
The variation of the process (range chart) has remained in control for the past six weeks. Previously the variation was reduced by modifying the gage used for the measurement. Currently the variability of the gage is less than 5% of the total variation. No prior testing of clamp pressure has been done. The gage on the clamp pressure adjustment reads from 1 to 1000 pounds, and the current setting is 30 pounds.

3. **Experimental variables:**

A.

| Response variables | Measurement technique |
|---|---|
| 1. Dimension variation | Range of four measurements (in thousandths of an inch) using the modified gage |

B.

| Factors under study | Levels |
|---|---|
| 1. Clamp pressure | 20, 40 and 80 pounds |

C.

| Background variables | Method of control |
|---|---|
| 1. Thirty-five other variables that might affect the range had been identified on the cause-and-effect diagram | None of these variables will be changed during the study |

4. **Replication:**
The clamp pressure will be changed at the beginning of the shift and remain at each level for the eight-hour shift. Subgroups will be selected once an hour during the shift for each level. With three levels, the experiment can be completed in three days.

5. **Methods of randomization:**
The order for setting the three levels was determined by selecting the numbers 1 (low level), 2, and 3 (high level) from a table of random numbers.

6. **Design matrix:**

| | Day 2 | Day 3 | Day 4 |
|---|---|---|---|
| Level | 80 | 20 | 40 |

7. **Data collection forms:**
Data will be recorded and plotted on the current control chart.

8. **Planned methods of statistical analysis:**
Interpretation of the current control chart

9. **Estimated cost, schedule, and other resource considerations:**
Maintenance will be required to set the clamp pressure each day.

**Figure 4.3** Documentation form for Example 4.1.

Based on this experiment, it was decided to set the clamp pressure at 80 lb. The control chart and maintenance records will be evaluated during the next week to determine the impact of this change. If the expected reduction in variation is observed, another study will be planned to evaluate the clamp pressure at settings above 80 lb.

This study is typical of a one-factor experiment by use of a control chart. Some comments on the design of the study:

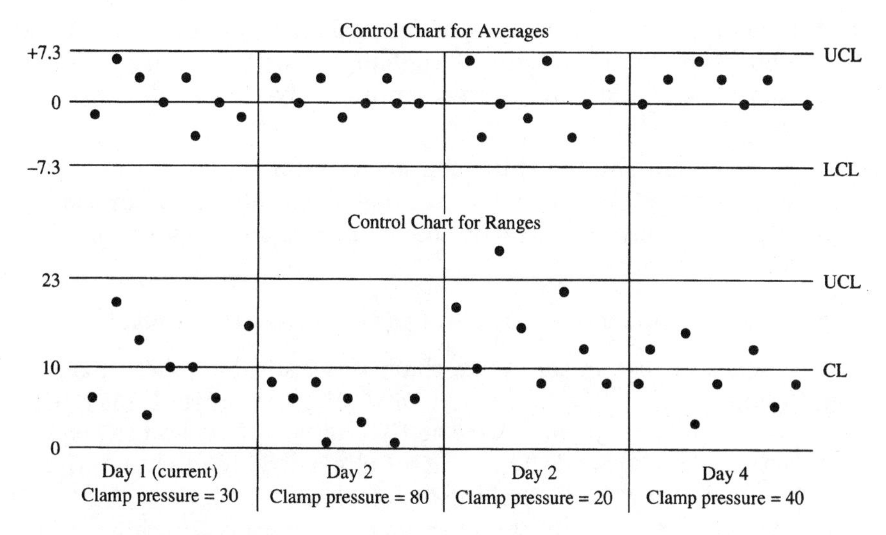


**Summary of control charts:**

Day 1 (prior to experiment): both average and range charts are in control; control limits are based on data from previous month.

Day 2 (clamp pressure changed to 80 lb ): The average remained in control; all of the eight ranges were below the centerline, indicating a reduction in the variation. Since no other changes in the casual variables were detected, the reduction is probably due to the increased clamp pressure. The average of the eight ranges was 5.0 (thousandths of an inch).

Day 3 (clamp pressure changed to 20 lb ): The variation increased with the range on the third hour above the UCL. This increase is probably due to the low clamp pressure. Average range = 16.0.

Day 4 (clamp pressure changed to 40 lb ): Both charts are in control. Average range = 8.0.

**Figure 4.4** Control chart for Example 4.1.

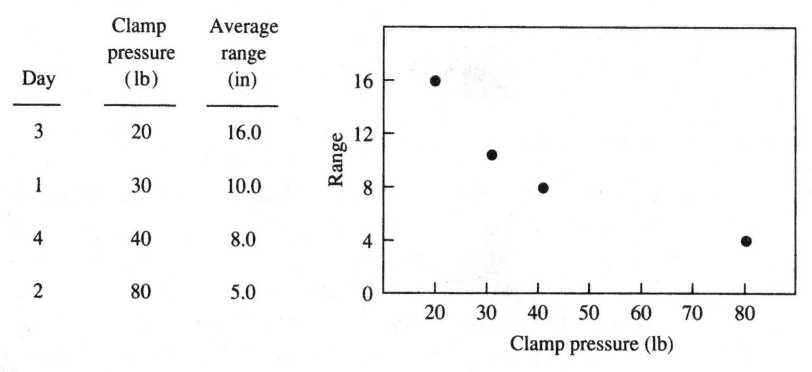

| Day | Clamp pressure (lb) | Average range (in) |
|---|---|---|
| 3 | 20 | 16.0 |
| 1 | 30 | 10.0 |
| 4 | 40 | 8.0 |
| 2 | 80 | 5.0 |



**Figure 4.5** Summary of results of experiment for Example 4.1.

- Each clamp pressure level was evaluated on only one day (shift), so the effect of clamp pressure is confounded with days (i.e., the observed effect could be clamp pressure or day-to-day differences or both).
- Variation from shift to shift has not been observed for the past 6 weeks, so it was reasonable to include this confounding in the design of the study, as frequent changing of clamp pressure is difficult.

Some comments on the analysis of the data from the study:

1. A special cause in the range chart was indicated twice (a run of eight points on day 2, a point beyond the upper control limit (UCL) and two out of three points near the UCL on day 3) when the level of the factor was changed. The control chart provided the analysis required to detect this effect of the factor, clamp pressure.
2. The summary plot of the average range versus the clamp pressure is important since the levels of the factor are quantitative and can be ordered. The degree of belief that greater clamp pressure will reduce the variation is increased by the consistency in this plot.
3. The next cycle is a confirmation of the expected reduced variation with the clamp pressure at 80 lb. Another cycle may be planned to study the effects of increasing the clamp pressure even more.

There are two important differences between this one-factor experiment and the typical use of the control chart:

1. The active nature of the study. Experimenters made changes in the factor rather than wait for changes in the response variable.
2. The amount of planning that was done prior to making the changes in the factor levels. The form in Fig. 4.3 was the guide used for the planning.

See App. A for more on control charts.

## 4.3 Example of a One-Factor Design

Almost all experiments for analytic studies will include background variables, as discussed in Chap. 3. The following example of a one-factor experiment is given to illustrate the importance of run charts in planned experiments. These charts should be developed while the study is in progress to take advantage of information as it is obtained. The run chart will be used to do an initial partitioning of the variation in the response variables into variation due to the factors, variation due to background variables, and variation due to nuisance variables (either common or special causes). The examples in the remainder of the chapter cover the introduction of background variables into one-factor experiments.

**Example 4.2** Alternative control methods were being evaluated for a continuous chemical process. These included the current control methods, an approach using a series of manually operated control valves, and an automated flow-metered system. A study was outlined to evaluate the alternatives. Changing the control methods required that the unit be shut down for a significant time. The key measures of process performance that could be affected by the control methods were the sulfate content, percentage of water, and amount of a critical ingredient X. The target levels for these variables were ≤1 percent sulfate, ≤3 percent water, and ingredient X = 15 percent.

Figure 4.6 shows the documentation form summarizing the planned study. Table 4.1 lists the data obtained from the study.

The first step in the analysis is a run chart for each of the response variables. Figure 4.7 contains run charts for the three response variables in this study. The following conclusions were made from these plots:

*Percent water:* The variation in the data can almost all be attributed to a linear trend. The identification of the nuisance variable causing this trend is the most important next step. The effect of the factor is small relative to this nuisance variable effect.

*Percent sulfate:* The majority of the variation is due to nuisance variables. The process appears stable throughout the study. No factor effects or special causes are obvious.

*Ingredient X:* The factor, control methodology, has a major effect on ingredient X. Two special causes (July 10 and 15) are also important. Other nuisance variables contribute a small amount of variation. The current control method resulted in a proportion slightly above the target of 15 percent. The valve showed lower results, and the meter had higher results. The special causes resulted in one of the valve measures being high and one of the meter measures being low.

These conclusions are summarized in Table 4.2. The follow-up studies for this example should focus on the special causes observed in the study. A decision on choice of control methodology may depend on the outcome of this further investigation.

In this design, the factor levels were not randomly assigned to the experimental units (days). If important special causes due to nuisance variables are present in the data, as in this example, the lack of randomization can limit learning about the factors in the study. The run charts play a critical role in detecting these special causes by distinguishing their effects from those due to one of the factor levels.

1. **Objective:**
Compare alternative control techniques in a continuous process for the following quality characteristics: percent sulfate, percent water, and percent ingredient X.

2. **Background information:**
The current method was recently upgraded, and a number of other special causes have been occurring, so little background information is available. Tests are not usually conducted at the point of interest in this study, but the variation of the characteristics of interest is thought to be small, based on sampling and analysis.

3. **Experimental variables:**

A.

| Response variables | Measurement technique | Target level |
|---|---|---|
| 1. sulfate (%) | Titration | ≤1.0% |
| 2. water (%) | Titration | ≤3.0% |
| 3. Ingredient X (%) | Gas chromatograph | 15% ± 3% |

B.

| Factors under study | Levels |
|---|---|
| 1. Control method | Current, valve, meter |

C.

| Background variables | Method of control |
|---|---|
| 1. None identified | |

4. **Replication:**
A decision on methodology is needed within about three weeks, so tests will be conducted on each method for 5 days (a total of 15 days).

5. **Methods of randomization:**
Because of the extreme expense required to change the control method, tests cannot be randomized. The current method will be tested for five days, then the valve for five days, then the meter for five days.

6. **Design matrix:** (attach copy)
(See Table 4.1.)

7. **Data collection forms:**
Normal laboratory worksheets with "special study" noted at top.

8. **Planned methods of statistical analysis:**
Run charts of data; possibly run charts of adjusted data and statistical summaries of response variables if appropriate.

9. **Estimated cost, schedule, and other resource considerations:**
The study can be completed during a 15-day period with one composite sample per day tested for each parameter.

**Figure 4.6** Documentation form for Example 4.2.

Sometimes the special causes are not obvious by inspection of the run charts. The trends or unusual data values may be hidden by the effects of the factors in the study. Additional analysis may be useful in cases where this might be expected. Each data point in the study can be adjusted by subtracting the factor effect. This analysis is equivalent

TABLE 4.1 **Test Results for Example 4.2**

| Control method | Date | Water, % | Sulfate, % | Ingredient X, % |
|---|---|---|---|---|
| Current | 7/2 | 2.2 | 0.5 | 15.7 |
| | 7/3 | 2.0 | 0.2 | 17.2 |
| | 7/4 | 2.4 | 0.6 | 15.1 |
| | 7/5 | 2.7 | 0.3 | 16.5 |
| | 7/6 | 2.9 | 0.4 | 14.8 |
| Valve | 7/7 | 3.3 | 0.2 | 11.8 |
| | 7/8 | 3.2 | 0.5 | 12.6 |
| | 7/9 | 3.4 | 0.4 | 13.4 |
| | 7/10 | 3.6 | 0.3 | 19.4 |
| | 7/11 | 3.5 | 0.5 | 12.3 |
| Meter | 7/12 | 3.7 | 0.4 | 18.8 |
| | 7/13 | 3.8 | 0.5 | 17.6 |
| | 7/14 | 4.0 | 0.3 | 19.0 |
| | 7/15 | 4.2 | 0.5 | 13.0 |
| | 7/16 | 4.1 | 0.6 | 18.2 |

to *residual analysis* (Box et al., 1978, p. 183) except that the overall mean is added to the residual to return to the original engineering units. For this example, the adjusted data values are computed by subtracting the average value for each factor level from the data points obtained with that factor level, and then adding the overall average. Table 4.3 shows the adjusted data calculated for this example. Figure 4.8 shows run charts of these adjusted data.

The special cause (trend) in the water data is no longer obvious in the adjusted data run chart. Since the factor levels were not randomly assigned to the days, the trend is somewhat confounded with the factor effect. Thus, when the factor effect is removed in adjusting the data, some of the trend effect is also eliminated. If the factor levels had been randomly assigned, the trend effect of the special cause would still be obvious in the adjusted data run chart.

The adjusted data for sulfate are very similar to the original data. This is because the factor had no important effect on the data.

The two special causes in the ingredient X data stand out more clearly in the adjusted data run chart than in the original data plot. If it is questionable whether these two points should be treated as special causes, a control chart for individuals could be developed for the adjusted data to aid in making this determination.

Run charts of the adjusted data do not need to be developed for every study. If the interpretation of the run chart of the original data is not clear, calculating the adjusted data and plotting them in run order may be useful.

There were no background variables identified in this study. Back-

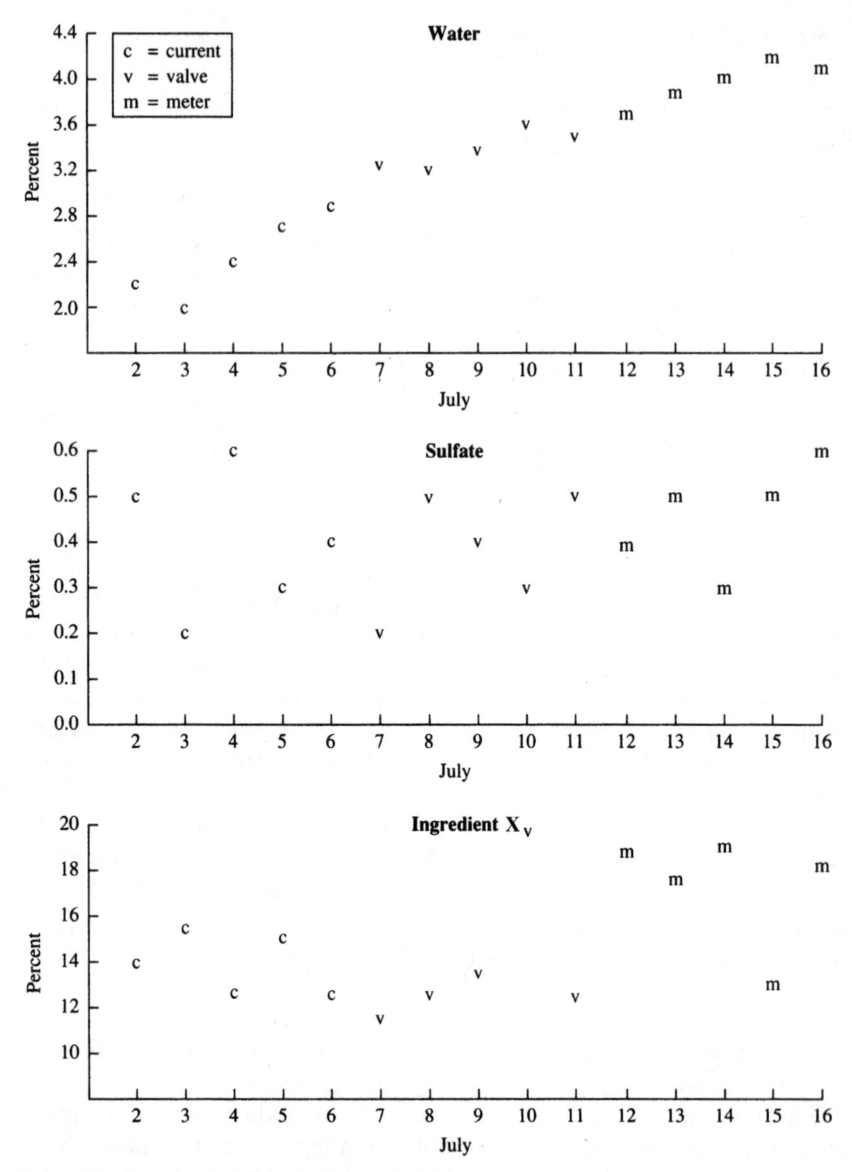


**Figure 4.7** Run chart of data for Example 4.2.

ground variables should be included in analytic studies to evaluate the factors of interest at widely varying conditions in the process. This is important for increasing the degree of belief in the conclusions of the study. The technique of blocking can be used to introduce one or more background variables in an experiment. The next three sections discuss background variables in analytic studies.

**TABLE 4.2 Summary of Important Effects in Example 4.2**

| | Response variable | | |
|---|---|---|---|
| Source of variation | Percent water | Percent sulfate | Percent ingredient X |
| Factor | Small | Small | Important |
| Nuisance variable: | | | |
| Common causes | Small | Important | Small |
| Special causes | Important | None | Important |

**TABLE 4.3 Example 4.2 Data Adjusted for Factor Levels**

| | | Water, % | | Sulfate, % | | Ingredient X, % | |
|---|---|---|---|---|---|---|---|
| Control method | Date | Test result | Adjusted value | Test result | Adjusted value | Test result | Adjusted value |
| Current | 7/2 | 2.2 | 3.03 | 0.5 | 0.51 | 15.7 | 15.53 |
| | 7/3 | 2.0 | 2.83 | 0.2 | 0.21 | 17.2 | 17.03 |
| | 7/4 | 2.4 | 3.23 | 0.6 | 0.61 | 15.1 | 14.93 |
| | 7/5 | 2.7 | 3.53 | 0.3 | 0.31 | 16.5 | 16.33 |
| | 7/6 | 2.9 | 3.73 | 0.4 | 0.41 | 14.8 | 14.63 |
| | | $\overline{X} = 2.44$ | | $\overline{X} = 0.40$ | | $\overline{X} = 15.86$ | |
| Valve | 7/7 | 3.3 | 3.17 | 0.2 | 0.23 | 11.8 | 13.59 |
| | 7/8 | 3.2 | 3.07 | 0.5 | 0.53 | 12.6 | 14.39 |
| | 7/9 | 3.4 | 3.27 | 0.4 | 0.43 | 13.4 | 15.19 |
| | 7/10 | 3.6 | 3.47 | 0.3 | 0.33 | 19.4 | 21.19 |
| | 7/11 | 3.5 | 3.37 | 0.5 | 0.53 | 12.3 | 14.09 |
| | | $\overline{X} = 3.40$ | | $\overline{X} = 0.38$ | | $\overline{X} = 13.90$ | |
| Meter | 7/12 | 3.7 | 3.01 | 0.4 | 0.35 | 18.8 | 17.17 |
| | 7/13 | 3.8 | 3.11 | 0.5 | 0.45 | 17.6 | 15.97 |
| | 7/14 | 4.0 | 3.31 | 0.3 | 0.25 | 19.0 | 17.37 |
| | 7/15 | 4.2 | 3.51 | 0.5 | 0.45 | 13.0 | 11.37 |
| | 7/16 | 4.1 | 3.41 | 0.6 | 0.55 | 18.2 | 16.57 |
| | | $\overline{X} = 3.96$ | | $\overline{X} = 0.46$ | | $\overline{X} = 17.32$ | |
| | | $\overline{\overline{X}} = 3.27$ | | $\overline{\overline{X}} = 0.41$ | | $\overline{\overline{X}} = 15.69$ | |

Adjusted value $= X - \overline{X} + \overline{\overline{X}}$, where $X$ is the data value, $\overline{X}$ is the average for each level of the factor (for each response variable), and $\overline{\overline{X}}$ is the overall average. *Example* (7/2, water): Adjusted value $= 2.2 - 2.44 + 3.27 = 3.03$.

## 4.4 Paired-Comparison Experiments

The simplest type of blocking design is the paired-comparison experiment. This type of design requires a single factor at two levels and one or more background variables. The background variables must be such that two experimental units can be obtained from each grouping of background variables. An example is a study to compare the wear

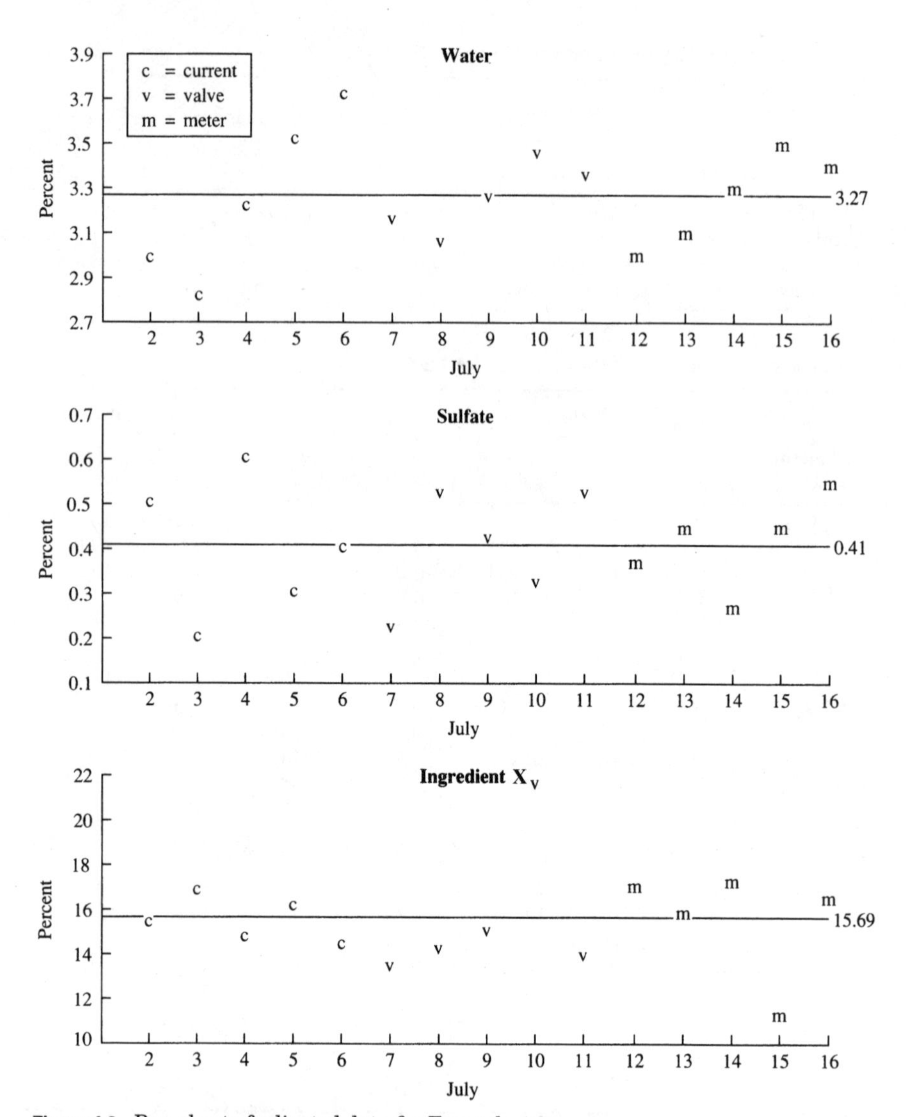


**Figure 4.8** Run chart of adjusted data for Example 4.2.

properties of two brands of shoes by having different people wear the shoes for one month and then measuring the wear. An important background variable here would be the variation in use of the shoes by the different people in the experiment. A paired-comparison design would assign one shoe of each brand to each person in the experiment. The difference in wear between the two brands could then be evaluated in terms of each person.

Other examples of pairs of units appropriate for a paired-comparison experiment are identical twins, front fenders of a car, two adjacent plots of land, two groupings of raw material from the same lot, and two time periods on the same day. The following example illustrates the design and analysis of a paired-comparison experiment.

**Example 4.3** One of the operators in the machining section has suggested an alternative setup procedure designed to increase tool life. The current procedure results in a typical tool life of one-half of a shift. The operator designed an experiment to investigate this suggestion. The single factor to be studied was the setup procedure. The factor was at two levels—the current procedure and the suggested alternative. There was one response variable—the wear rate of the tool.

An important background variable is the quality of the blanks of steel prior to machining. The blanks are purchased in lots representing a single heat at the foundry, with each lot large enough for one shift's production in the machining section. The wear rate of the tool has varied significantly for different lots in the past.

A paired-comparison design was used for the experiment. Figure 4.9 shows the documentation form for the study. Each shift (and thus each lot of blanks) was designated a block. The two setup procedures were paired on each shift, and each was used for a 4-hour period. The assignment of a particular setup method to the first or second 4-hour period was made by flipping a coin. The experiment was run for 10 days. Four parts were selected every 15 min, and a critical dimension was measured. Control charts for the average and range were used to summarize the 16 subgroups for each 4-hour run. A straight line was fit to the 16 averages, and the slope of this line was used as a measure of tool wear. The results in Table 4.4 were obtained.

The first step in the analysis of a paired-comparison experiment is the preparation and study of a run chart of the response data. A run chart of the wear rates in Table 4.4 is shown in Fig. 4.10. The variation in the tool wear rates can be partitioned by studying the run chart:

1. The largest component of the variation is due to the background variables (day and steel lot).
2. The effects of the factor and nuisance variables are somewhat difficult to discern because of the variation caused by the background variable. The wear rates for the new procedure are consistently less than those for the old procedure.

The next step in the analysis is to reorder the run chart to group the levels of the factor together. This chart, also shown in Fig. 4.10, focuses the analysis on the factor of interest. The wear rates for the new setup procedure tend to be less than those for the old procedure, but the data are still dominated by the effect of the background variables.

In a paired-comparison design, it is possible to remove the effect of the background variable but still evaluate the factor of interest at each

1. **Objective:**
Investigate the effect of an alternative tool setup procedure on tool life. A 10% reduction (0.0004 inch per hour) in wear rate would warrant changing to the new procedure.

2. **Background information:**
Using the current procedure, tool life has averaged about one-half of a shift (wear rate of 0.004 inch per hour). The particular lot of steel significantly affects wear; the standard deviation calculated from the range of tools used on the same batch was 0.00025 inch per hour.

3. **Experimental variables:**

| A. Response variables | Measurement technique |
|---|---|
| 1. Wear rate | Slope of straight line fit to data for each tool tested |

| B. Factors under study | Levels |
|---|---|
| 1. Setup procedure | Current (O), new (N) |

| C. Background variables | Method of control |
|---|---|
| 1. Steel lot | Blocking on steel lot |
| 2. Day | Blocking on day |

4. **Replication:**
Ten days were available for the study; this would result in 10 comparisons of the two setup procedures.

5. **Methods of randomization:**
Flip a coin to determine setup procedure for first four-hour period (heads: current procedure; tails: new procedure).

6. **Design matrix:** (attach copy)
(See Table 4.4.)

7. **Data collection forms:** (attach copy)
Use standard $X$-bar and $R$ charts for basic data. Read slope from $X$-bar chart.

8. **Planned methods of statistical analysis:**
Run charts of individual wear rates; run chart with setup types grouped; run chart of adjusted wear rates.

9. **Estimated cost, schedule, and other resource considerations:**
Study can be conducted during production runs during next 10 day shifts of operation.

**Figure 4.9** Documentation form for Example 4.3.

level of the background variable. This is easily done by computing the average of the response variable for each block and then subtracting the appropriate block average from the original data. The overall average of all the data is then added back to keep the data in the original units. These calculations are shown in Table 4.4. The adjusted wear rates contain the effect of the factor of interest (setup procedure), but the effect of the background variables (day and steel lot) has been removed. A run chart of the adjusted wear rates is shown in Fig. 4.11.

The run chart of the adjusted wear rates clearly shows the differ-

**TABLE 4.4 Results of Paired-Comparison Experiment (Example 4.3)**

Wear Rates (in × $10^{-3}$/h) for Each Tool on Each Day

| Day (lot) | First 4 h Setup | First 4 h Tool wear | Second 4 h Setup | Second 4 h Tool wear | Average for day |
|---|---|---|---|---|---|
| 1 | New | 2.0 | Old | 3.0 | 2.50 |
| 2 | New | 4.2 | Old | 5.3 | 4.75 |
| 3 | Old | 1.8 | New | 1.1 | 1.45 |
| 4 | New | 3.6 | Old | 4.4 | 4.00 |
| 5 | Old | 4.1 | New | 3.0 | 3.55 |
| 6 | Old | 2.8 | New | 1.9 | 2.35 |
| 7 | Old | 3.1 | New | 2.0 | 2.55 |
| 8 | New | 2.6 | Old | 4.0 | 3.30 |
| 9 | New | 3.2 | Old | 3.9 | 3.55 |
| 10 | New | 4.0 | Old | 5.2 | 4.60 |
| | | | | Overall average $\bar{\bar{X}}$ = | 3.26 |

Data Adjusted for Block (Day and Steel Lot) Effects

| Day (lot) | Old setup: Wear | − | Daily average | + | $\bar{\bar{X}}$ | = | Adjusted wear | New setup: Wear | − | Daily average | + | $\bar{\bar{X}}$ | = | Adjusted wear |
|---|---|---|---|---|---|---|---|---|---|---|---|---|---|---|
| 1 | 3.0 | − | 2.50 | + | 3.26 | = | 3.76 | 2.0 | − | 2.50 | + | 3.26 | = | 2.76 |
| 2 | 5.3 | − | 4.75 | + | 3.26 | = | 3.81 | 4.2 | − | 4.75 | + | 3.26 | = | 2.71 |
| 3 | 1.8 | − | 1.45 | + | 3.26 | = | 3.61 | 1.1 | − | 1.45 | + | 3.26 | = | 2.91 |
| 4 | 4.4 | − | 4.00 | + | 3.26 | = | 3.66 | 3.6 | − | 4.00 | + | 3.26 | = | 2.86 |
| 5 | 4.1 | − | 3.55 | + | 3.26 | = | 3.81 | 3.0 | − | 3.55 | + | 3.26 | = | 2.71 |
| 6 | 2.8 | − | 2.35 | + | 3.26 | = | 3.71 | 1.9 | − | 2.35 | + | 3.26 | = | 2.81 |
| 7 | 3.1 | − | 2.55 | + | 3.26 | = | 3.81 | 2.0 | − | 2.55 | + | 3.26 | = | 2.71 |
| 8 | 4.0 | − | 3.30 | + | 3.26 | = | 3.96 | 2.6 | − | 3.30 | + | 3.26 | = | 2.56 |
| 9 | 3.9 | − | 3.55 | + | 3.26 | = | 3.61 | 3.2 | − | 3.55 | + | 3.26 | = | 2.91 |
| 10 | 5.2 | − | 4.60 | + | 3.26 | = | 3.86 | 4.0 | − | 4.60 | + | 3.26 | = | 2.66 |
| | Average, old setups | | | | | = | 3.76 | Average, new setups | | | | | = | 2.76 |

ence between the two setup procedures. The adjusted wear rates for the old procedure are consistently around 3.76, whereas the adjusted wear rates for the new procedure are consistently around 2.76—about 1.0 lower (all units are $10^{-3}$/hour). On each of the 10 days (and thus for each lot of steel), the new procedure had a lower wear rate than the old procedure.

The contribution to the variation of nuisance variables can also be seen in Fig. 4.11. The variation of the 10 tests within each setup procedure can be attributed to nuisance variables. This contribution to the variation is small relative to the contributions of the background variable and the factor.

Since the new procedure outperformed the old procedure in each of the 10 comparisons, the process engineer involved in the study would

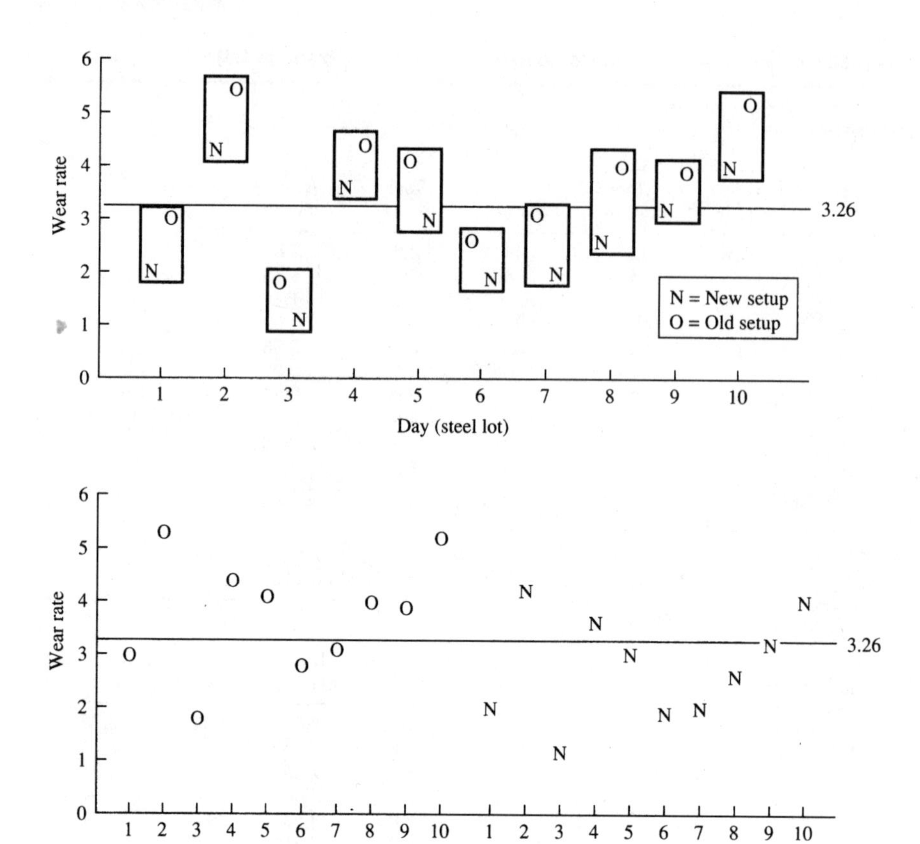


**Figure 4.10** Run charts of individual wear rates (Example 4.3).

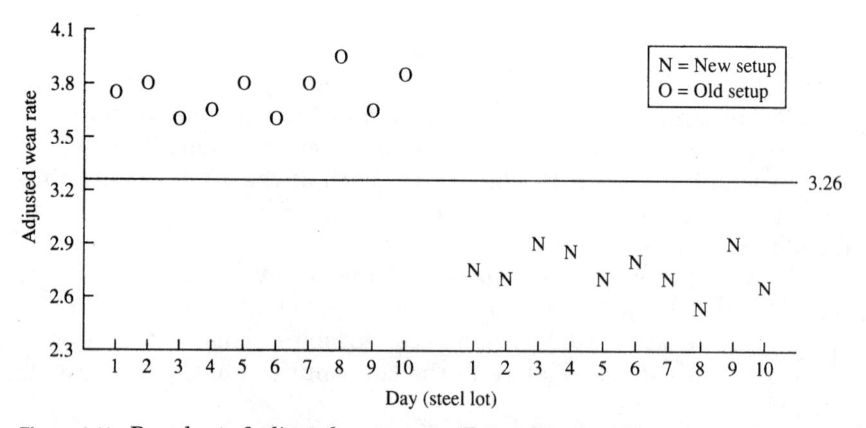


**Figure 4.11** Run chart of adjusted wear rates (Example 4.3).

have a high degree of belief that the new setup procedure would result in lower wear rates in the future. The engineer should assess the magnitude of the lot-to-lot variation in the study relative to previous experience. The wear rates in the study ranged from 1.8 to 5.3 for the old procedure, so a wide range of conditions have been included. After the new setup procedure is implemented, the next improvement cycle should focus on this lot-to-lot variation.

This example is similar to most paired-comparison experiments for an analytic study. By using this design, it is possible to include background variables that have large effects on the response variable. This allows the comparisons of the factor of interest to be made at a wide range of conditions that might be expected in the future. The steps in the analysis of a paired-comparison design are as follows:

1. Plot a run chart of the original data in order of test or measurement.
2. Reorder the run chart by grouping the levels of the factor.
3. If the effect of the background variable(s) is large, adjust the data to remove the effect. This is done by computing the average of each data pair (each block), subtracting that average from each of the two data points, and then adding the overall average to return the data to the original units.
4. Plot a run chart of the adjusted data, grouped by factor level.

By adjusting the data using the block averages, the effect of the background variables is prevented from interfering with the evaluation of the factor. The levels of the factors can be evaluated for stability over the wide range of conditions using the run chart of data adjusted for the block averages.

## 4.5 Randomized Block Designs

The randomized block design extends the approach of the paired-comparison design to more than two levels of the factor. Traditionally, the randomized block design has been used for one background variable, but through the concept of chunk variables (see Fig. 3.5) any number of background variables can be incorporated into the design. The number of factor levels is limited only by the block "size," since each level must occur an equal number of times (at least once) in each block. The blocks of experimental units are defined by the background variables. The factor levels are assigned in a random order within each block, with a different randomization for each block.

The concept of chunk variables arranged in blocks is very important in analytic studies. The degree of belief concerning the effect of a factor can be greatly increased by varying many of the background variables and evaluating the factor under these widely varying conditions. By organizing these background variables in blocks and then evaluating each level of the factor in each block, the comparisons of the factor levels can be made under uniform conditions. The following study is an example of a randomized block design for a one-factor experiment.

**Example 4.4** A program was undertaken to improve the yield of a batch chemical process. Control charts had been continually out of control due to variability of the feedstock. Four batches of the chemical could be made from each tank car of feedstock at a rate of about one batch per day. A major project was conducted with the feedstock supplier to identify the causes that affected yield from shipment to shipment.

The plant chemist made a suggestion to improve yield through changes in one of the catalysts used in the batch process. Three alternative catalysts were identified, each of which could replace the current catalyst. The production manager asked that a test be run during the next month to evaluate the potential of the alternative catalysts. A documentation form for the study is shown in Fig. 4.12.

A randomized block design was used for the study. The single factor to be studied was the catalyst, and the factor had four levels (catalysts X, Y, and Z and the current catalyst C). An important background variable was the feedstock shipment, so the experimental units were grouped into blocks of four production units from each feedstock shipment. Since the study had to be completed within a month, six feedstock shipments were incorporated into the experimental design. Thus, the experiment would require 24 days to complete.

Each of the four catalyst types was assigned to one of the four batches scheduled for each of the six feedstock shipments in a random order (using a table of random permutations). Yield was calculated for each batch by a material balance method. The results of the study are summarized in Table 4.5.

Figure 4.13 shows run charts of the yields from each individual batch. From the first run chart, the effect of the feedstock shipment (the blocks) is obvious. Differences between the catalyst types can be observed in the second run chart, where the batches of a particular catalyst type are grouped together. The highest yield in each block was for the batch using catalyst type Y. Catalysts X and Z had yields close to that of the current catalyst in each block.

A clear picture of the effect of catalyst type on yield is difficult from the run charts of yield because of the effect of the background variable (feedstock shipment). The effect of the background variable can be removed from the yields by adjusting the data. The approach is similar to the adjusting procedure used with the paired-comparison design in the previous section: The average of each block is computed, the block average is subtracted from each individual yield, and the overall average is then added. This calculation removes the effect of the background variable but leaves the adjusted data in the original units (percent yield, in

1. **Objective:**
Choose catalyst (there are three new alternatives available) to give maximum yield. Any improvement over the current catalyst is worthwhile; a practical minimum difference to change type of catalyst is 0.5%.

2. **Background information:**
The new chemical process is still undergoing major improvements. Feedstock variability has a big effect on yield and is being studied by a joint team including the supplier. Yields during the last 30 batches have ranged from 75% to 95%, with big swings for each new feedstock shipment.

3. **Experimental variables:**

| A. Response variables | Measurement technique |
|---|---|
| 1. Yield (%) | Material balance |

| B. Factors under study | Levels |
|---|---|
| 1. Catalyst | New (X, Y, Z), current (C) |

| C. Background variables | Method of control |
|---|---|
| 1. Feedstock shipment | Block defined by each shipment |

4. **Replication:**
There is enough supply of the experimental catalysts to make six batches from each. The study should be completed within a month. Six replications (blocks) of each of the four catalyst types are planned.

5. **Methods of randomization:**
Table of random permutations of 4 was used to assign the four catalyst types to the four batches made from each feedstock shipment.

6. **Design matrix:** (attach copy)
(See Table 4.5.)

7. **Data collection forms:** (attach copy)
Batch cards include material balance calculation of yield.

8. **Planned methods of statistical analysis:**
Run charts of yields; run chart of yields adjusted for background variables.

9. **Estimated cost, schedule, and other resource considerations:**
Initial study can be completed during next 24 days of operation; possible losses or gains in yield from the different catalysts.

**Figure 4.12** Documentation form for Example 4.4.

this case). The yields adjusted by using this procedure are shown in Table 4.5. The adjusted yields are shown on the third run chart in Fig. 4.13, with the data grouped by catalyst type.

The three run charts in Fig. 4.13 can be used to partition the variation in the yields:

1. The first two plots (run chart of the yields ordered by time and by catalyst type) indicate that most of the variation is attributable to the background variable, feedstock shipment. The average yields for the five shipments in the study ranged from 80 to 91.5 percent.

**TABLE 4.5 Randomized Block Design to Evaluate Four Catalysts (Example 4.4)**

Results of Study: Percent Yield (Random Order)

| Catalyst | Block (feedstock shipment) | | | | | |
|---|---|---|---|---|---|---|
| | 1 | 2 | 3 | 4 | 5 | 6 |
| X | 87(4) | 79(1) | 82(2) | 89(4) | 83(1) | 78(2) |
| Y | 93(1) | 84(4) | 89(4) | 96(2) | 86(3) | 87(1) |
| Z | 88(3) | 80(2) | 84(1) | 91(3) | 83(2) | 82(3) |
| C | 88(2) | 77(3) | 83(3) | 90(1) | 82(4) | 79(4) |
| Block average | 89.0 | 80.0 | 84.5 | 91.5 | 83.5 | 81.5 |

Overall average = 85.0

Yield Adjusted for Background Variable (Feedstock Shipment)
(Yield − Block average + Overall average)

| Catalyst | Block (feedstock shipment) | | | | | | Average |
|---|---|---|---|---|---|---|---|
| | 1 | 2 | 3 | 4 | 5 | 6 | |
| X | 83.0 | 84.0 | 82.5 | 82.5 | 84.5 | 81.5 | 83.0 |
| Y | 89.0 | 89.0 | 89.5 | 89.5 | 87.5 | 90.5 | 89.2 |
| Z | 84.0 | 85.0 | 84.5 | 84.5 | 84.5 | 85.5 | 84.7 |
| C | 84.0 | 82.0 | 83.5 | 83.5 | 83.5 | 82.5 | 83.2 |
| | | | | | | Overall average | 85.0 |

2. The third chart (a run chart of the adjusted yields) indicates that the factor, catalyst type, is the next biggest contributor to the variation in the yields. The difference in yield within a feedstock shipment ranges up to 9 percent. The yield for catalyst Y consistently was 6 percent higher than the current catalyst. Catalyst Z had a higher yield than the current catalyst in five of the six shipments, averaging 1.5 percent higher. Catalyst X was similar to the current catalyst.

3. The effect of nuisance variables is best seen from the third chart in Fig. 4.13. The adjusted yields for each of the three alternative catalysts are clustered tightly, with a range of about 3 percent. There are no indications of special causes (trends or individual points) in any of the plots.

From the combination of the data from this study with his previous laboratory experiments with the catalysts, the plant chemist had a high degree of belief that the use of catalyst Y would result in higher yields than continuing use of the current catalyst. A critical result was the consistency of factor effect (catalyst differences) across all the blocks. But even with catalyst Y, the batch yields were as low as 84 percent. The next step in this study would be to begin using catalyst Y, monitor the yields with a control chart, and focus future experiments on the feedstock variation.

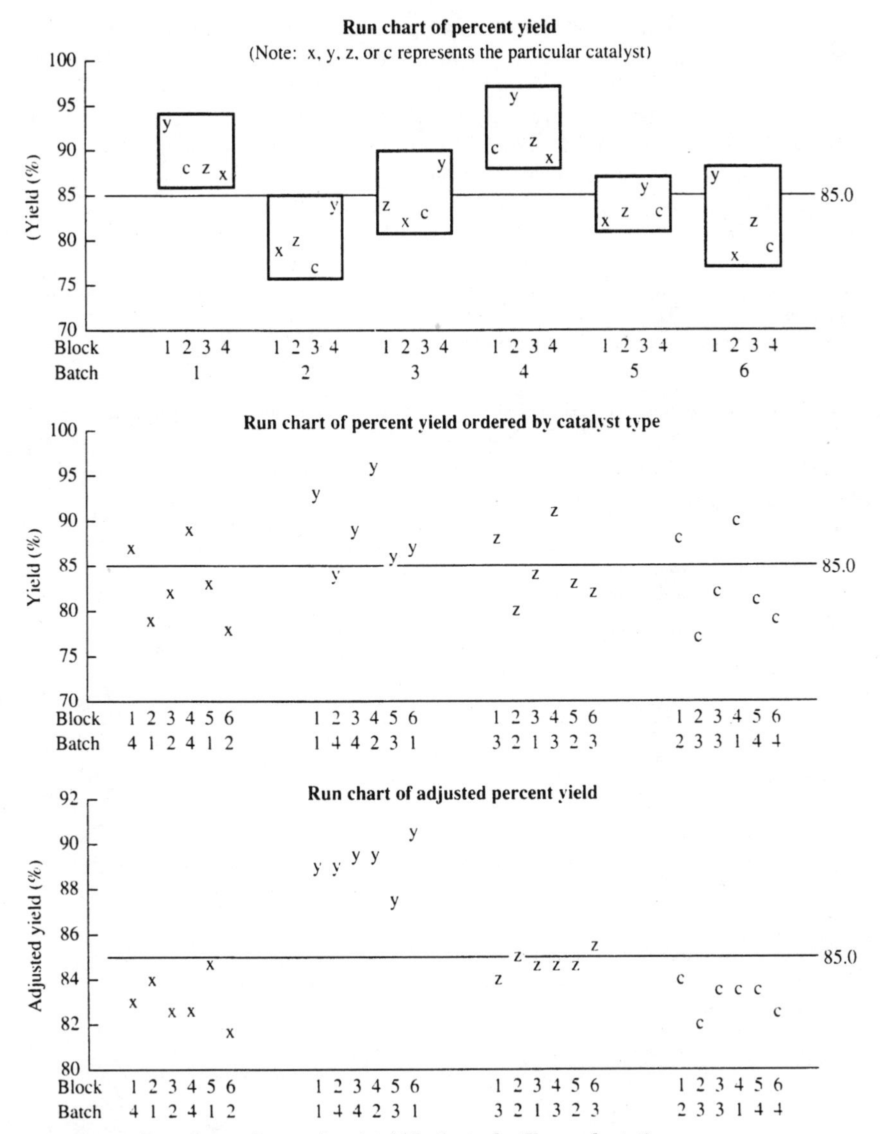


**Figure 4.13** Run charts for randomized block study (Example 4.4).

The next example is another study based on a randomized block design. In this study, a chunk variable is used to define the block. The example also illustrates an interaction between a factor and a background variable.

**Example 4.5** Loose labels on cans had been a problem on and off since the new labeling machine was installed. The application rate of the glue was thought to be the control mechanism for eliminating the problem of loose labels. The label operator would turn up the application rate when loose labels were observed. If runs of glue were observed on the labeled cans, the supervisor would reduce the application rate. During the last month, 7 percent of the cans required rework because of loose labels. The production manager asked the quality team in the label operation to conduct a study to determine the best application rate.

Figure 4.14 shows the documentation form for this study. The application

---

1. **Objective:**
   Determine an application rate for glue to minimize loose labels without causing runs or smears.

2. **Background information:**
   About 7% of the labels during the last month have come loose. The operators have changed the application rate a number of times (from 10 to 20 oz/hour). The rate has been turned down when runs occurred on the cans.

3. **Experimental variables:**

A.

| Response variables | Measurement technique |
|---|---|
| 1. Percent cans with loose labels | 100% visual inspection |
| 2. Percent cans with smears or runs | 100% visual inspection |

B.

| Factors under study | Levels |
|---|---|
| 1. Application rate | 10, 12, 14, 16, 18 oz/hour |

C.

| Background variables | Method of control |
|---|---|
| 1. Humidity in plant | Create two blocks made up of extreme conditions for each background variable (two chunks) |
| 2. Temperature in plant | |
| 3. Label machine speed | |
| 4. Can size | |
| 5. Batch of labels | |
| 6. Batch of cans | |
| 7. Label machine pressure setting | |

4. **Replication:**
   Each of the five tests will be run for one hour (1,000 cans). Cans will be run for one-half hour in between tests to let the new factor level reach stability. The tests will then be repeated on another day (the second chunk block).

5. **Methods of randomization:**
   Table of random permutations was used to determine the order of the five tests on each day.

6. **Design matrix:**
   (See Table 4.6 for summary.)

7. **Data collection forms:**
   Operator checksheets used to record occurrence of labels and runs.

8. **Planned methods of statistical analysis:**
   Run charts of percent labels loose and runs; response plot for each response variable and block.

9. **Estimated cost, schedule, and other resource considerations:**
   Test can be conducted during production runs. Inspector needed for two days to monitor test and coordinate data collection. Possible higher rework costs due to loose labels and runs.

---

**Figure 4.14** Documentation form for Example 4.5.

**TABLE 4.6 Design Matrix and Test Results for Example 4.5**

Development of Two Chunk Blocks

| Background variable | Block 1 | Block 2 |
|---|---|---|
| Plant humidity | Low | High |
| Plant temperature | Cool | Warm |
| Label machine speed | Low (800/h) | High (1200/h) |
| Can size | Small | Large |
| Label batch | Batch 803 | Batch 615 |
| Can batch | Batch 4120 | Batch 4125 |
| Pressure setting | Low | High |

Design Matrix and Test Results

| Application rate, oz/h | Block | Run order | Percent of labels loose | Percent of cans with runs |
|---|---|---|---|---|
| 10 | 1 | 3 | 9.0 | 0.0 |
| 12 | 1 | 1 | 5.4 | 0.2 |
| 14 | 1 | 4 | 1.0 | 0.4 |
| 16 | 1 | 5 | 0.9 | 1.5 |
| 18 | 1 | 2 | 1.0 | 3.0 |
| 10 | 2 | 5 | 10.3 | 0.0 |
| 12 | 2 | 1 | 8.0 | 0.1 |
| 14 | 2 | 3 | 6.7 | 0.4 |
| 16 | 2 | 2 | 7.0 | 1.3 |
| 18 | 2 | 4 | 6.8 | 2.8 |

rate would be studied at five levels (10, 12, 14, 16, and 18 oz/hour). The team felt this range covered the application rates that had been tried in the past. Because of the erratic results that had been experienced, the team decided to test each of these levels under different conditions that occurred in the plant. Seven background variables that could affect the labeling were identified, and two chunk-type blocks were defined for running the test. Table 4.6 shows the design matrix and the test results.

Figure 4.15 shows a run chart for the two response variables. During the study, the percent of loose labels ranged from 1 to 10. The percent of cans with runs ranged from 0 to 3. No outliers or other special causes are apparent on the run charts. Plots of the responses adjusted for the block effect could be prepared, but since application rate is a continuous variable, response plots better display the effect of the factor. In a response plot, results for a response variable are plotted on the vertical axis and the factor levels on the horizontal axis.

Figure 4.16 shows response plots for each of the response variables in this study. The results for each block can be compared for stability by using these plots. The response plot for loose labels indicates a different effect of the application rate between the two blocks. For the first block, the percent of loose labels decreased as application rate in-

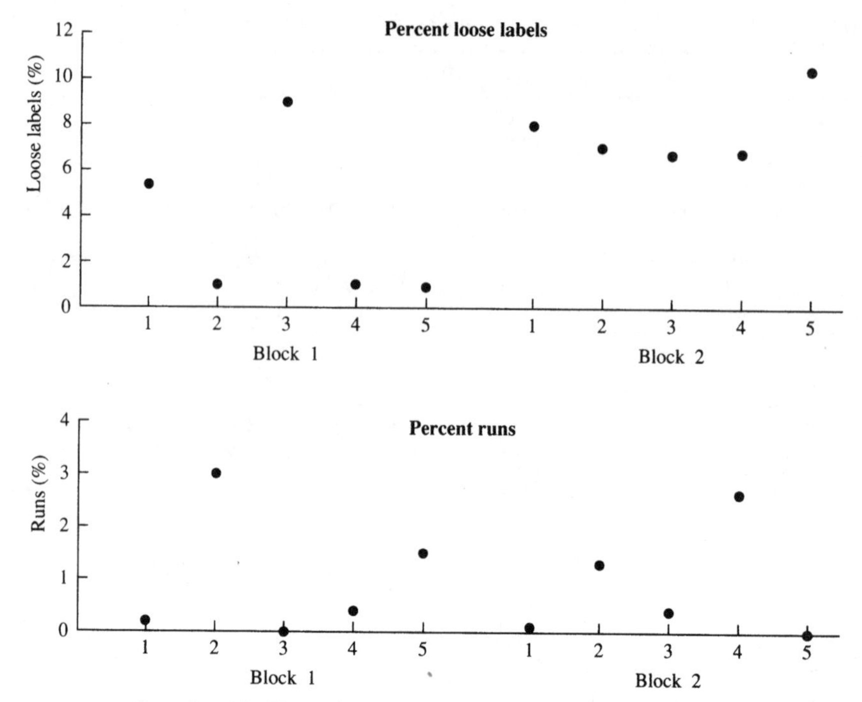


**Figure 4.15** Run chart for Example 4.5.

creased. Results at 14 oz/hour and above were near 1 percent loose labels. For the other block, the percent of loose labels decreased as the application rate changed from 10 to 12 oz/hour, but they remained steady at about 7 percent for the remaining levels. This indicates that the effect of the application rate is dependent on some of the background variables in the plant. Increasing the application rate above 12 oz/hour does not reduce the loose labels under the conditions in block 2.

The response plots for runs are almost identical for the two blocks. In both cases, the presence of runs was less than 0.5 percent until the application rate was increased above 14 oz/hour. So, after testing under widely varying conditions, there is a high degree of belief that the runs can be kept below 0.5 percent if the application rate is maintained at 14 oz/hour or below.

The next cycle in this study should focus on the conditions in block 2 that mitigated the effect of the glue application rate on the percent of loose labels. Under the conditions in block 1, an application rate of 14 oz/hour would keep both loose labels and runs at less than 1 percent. One approach would be to develop a control chart for the loose labels with the background variables noted on the chart. Another would be to set the application rate at 14 oz/hour and study the background variables by using a factorial experimental design (see Chap. 5).

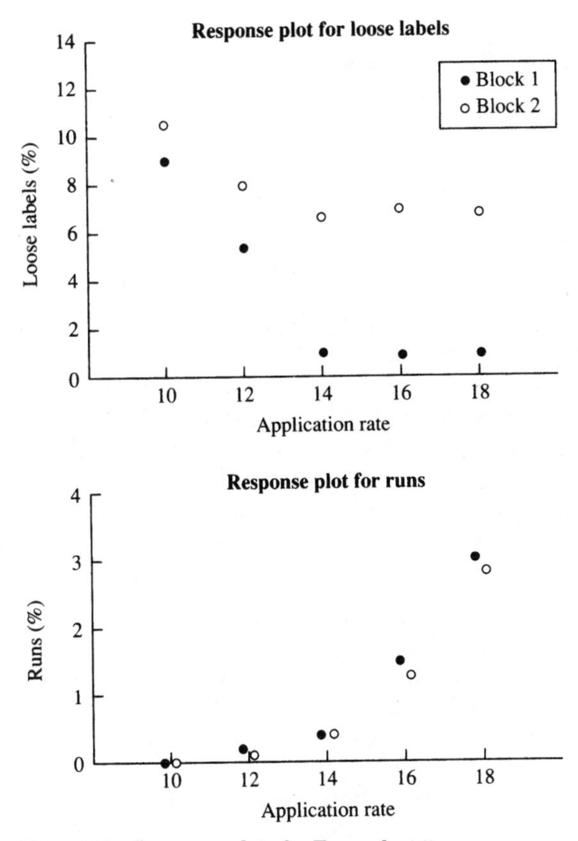


**Figure 4.16** Response plots for Example 4.5.

This example illustrates the importance of stability of factor effects in an analytic study. The past confusion in understanding the effect of application rate on the percent of loose labels was probably due to the interactive effect of one or more of the background variables. If this study had been conducted with the background variables held constant, the results would not have provided a basis for changing the process to improve performance under future plant conditions.

## 4.6 Incomplete Block Designs

Suppose that in Example 4.4 only three batches could be made from each feedstock shipment. How should the four catalyst types be tested? An incomplete block design can be used for this situation. An incomplete block design is an experimental pattern that groups experimental units in blocks (to accommodate one or more background variables), where the number of experimental units in each block is less than the number of factor levels.

A special class of such designs is the *balanced incomplete block design,* which has the following characteristics:

1. Each block is the same size (has the same number of experimental units).
2. Each factor level occurs the same number of times in the design.
3. Any two factor levels occur together in the same block an equal number of times.

Table 4.7 shows two different ways of displaying the experimental pattern for an incomplete block design. Table 4.8 gives some examples of balanced incomplete block designs for experiments with three, four, five, and six levels of the factor or groups to be tested. The design is characterized by the number of levels of the factor, the size of the block, the number of blocks required to achieve the balance, and the amount of replication of each level of the factor in the design. Balanced designs do not exist for some combinations of block size and number of levels.

The concept of a balanced design is not critical to the analysis or interpretation of incomplete block designs presented here. In general, the greater the balance in the design, the easier the interpretation and the higher the degree of belief in the results. Cochran and Cox (1957) give a good presentation of partially balanced incomplete block designs.

The analysis of data from incomplete block designs is similar to that

**TABLE 4.7 Experimental Pattern for Incomplete Block Design**

Five factor levels (A, B, C, D, E)
10 blocks
Block size = 3

| Experimental pattern display | | | | | | Alternative experimental pattern display | | | |
|---|---|---|---|---|---|---|---|---|---|
| | Factor level | | | | | | | | |
| Block | A | B | C | D | E | Block | Factor level | | |
| 1 | X | X | X | | | 1 | A | B | C |
| 2 | X | X | | X | | 2 | A | B | D |
| 3 | X | X | | | X | 3 | A | B | E |
| 4 | X | | X | X | | 4 | A | C | D |
| 5 | X | | X | | X | 5 | A | C | E |
| 6 | X | | | X | X | 6 | A | D | E |
| 7 | | X | X | X | | 7 | B | C | D |
| 8 | | X | X | | X | 8 | B | C | E |
| 9 | | X | | X | X | 9 | B | D | E |
| 10 | | | X | X | X | 10 | C | D | E |

**TABLE 4.8 Some Balanced Incomplete Block Designs**

| | | | | Block number | | | | | | | | | |
|---|---|---|---|---|---|---|---|---|---|---|---|---|---|
| $t$ | $b$ | $k$ | $r$ | 1 | 2 | 3 | 4 | 5 | 6 | 7 | 8 | 9 | 10 |
| 3 | 2 | 3 | 2 | A | A | B | | | | | | | |
| | | | | B | C | C | | | | | | | |
| 4 | 2 | 6 | 3 | A | A | A | B | B | C | | | | |
| | | | | B | C | D | C | D | D | | | | |
| 4 | 3 | 4 | 3 | A | A | A | B | | | | | | |
| | | | | B | B | C | C | | | | | | |
| | | | | C | D | D | D | | | | | | |
| 5 | 2 | 10 | 4 | A | A | A | A | B | B | B | C | C | D |
| | | | | B | C | D | E | C | D | E | D | E | E |
| 5 | 3 | 10 | 6 | A | A | A | A | A | A | B | B | B | C |
| | | | | B | B | B | C | C | D | C | C | D | D |
| | | | | C | D | E | D | E | E | D | E | E | E |
| 5 | 4 | 5 | 4 | A | A | A | A | B | | | | | |
| | | | | B | B | C | B | C | | | | | |
| | | | | C | C | D | D | D | | | | | |
| | | | | D | E | E | E | E | | | | | |
| 6 | 3 | 10 | 5 | A | A | A | A | A | B | B | B | C | D |
| | | | | B | B | C | C | D | C | C | D | E | E |
| | | | | E | F | D | F | E | D | E | F | F | F |
| 6 | 4 | 15 | 10 | A | A | A | A | A | A | A | A | | |
| | | | | B | B | B | B | B | B | C | C | | |
| | | | | C | C | C | D | D | E | D | D | | |
| | | | | D | E | F | E | F | F | E | F | | |
| | | | | A | A | B | B | B | B | C | | | |
| | | | | C | D | C | C | C | D | D | | | |
| | | | | E | E | D | D | E | E | E | | | |
| | | | | F | F | E | F | F | F | F | | | |
| 6 | 5 | 6 | 5 | A | A | A | A | A | B | | | | |
| | | | | B | B | B | B | C | C | | | | |
| | | | | C | C | C | D | D | D | | | | |
| | | | | D | D | E | E | E | E | | | | |
| | | | | E | F | F | F | F | F | | | | |

$t$ = Number of factor levels or combinations
$b$ = Block size (number of experimental units per block)
$k$ = Number of blocks required for balanced design
$r$ = Number of replications of each factor level required
A, B, C, D, E, and F represent factor levels or factor combinations.

for the complete block design. A run chart of the data is prepared first. Next, the data are adjusted to remove the effect of the background variable. The adjusted data are then analyzed as in a single-factor design.

The following example illustrates the design and analysis of an incomplete block design.

**Example 4.6** An evaluation of gaging instruments was proposed in order to select the particular instrument with the best repeatability for future plant use. (Repeatability is measured by the standard deviation of repeated measurements of the same part.) The gages were used by over 100 different operators in the plant, and the level of skill among different operators was known to have an impact on repeatability.

Six different instruments were available for testing. The test protocol for an instrument, established by the quality control department, requires each operator to measure four different parts seven times each. These data are evaluated for stability, and then a pooled standard deviation is calculated and used as a measure of repeatability. This protocol requires about 1 hour per instrument. Since the instrument selected is for general plant use, it is desirable to use a number of different operators in the study. The production manager said that operators could be made available for the study for up to 3 hours each.

This study is a single-factor (instrument) experiment with an important background variable (operator). The response variable is the calculated standard deviation from each test. There are six levels of the factor (instruments A, B, C, D, E, and F), but the blocks defined by the background variable are only of size 3 (maximum 3 hours per operator). Since a randomized block design cannot be used, a balanced incomplete block design is considered. From Table 4.8, such a design is available for six factor levels and a block size of three experimental units. The experimental pattern requires that each instrument be tested 5 times ($r = 5$) and that 10 operators be used in the study ($k = 10$).

Figure 4.17 shows the completed documentation form for this study. Table 4.9 shows the experimental pattern and the response variable for the completed experiment.

The response variable (the standard deviation of the repeat measurements) is plotted in run order in Fig. 4.18. No special causes are obvious in this plot. Some differences among blocks (operators) are present, but there is a wide variation of results for some of the operators. Operator 10 appears to have more variation than the other operators. The second run chart in Fig. 4.18 shows the data (the standard deviations) grouped by instrument type. Instrument C appears to have good precision (a low standard deviation) each time it appears. The differences among operators make it difficult to evaluate the other instruments.

The next step in the analysis is to remove the effect of the operators so that the variation in the factor (instruments) can be studied. This is

1. **Objective:**
Determine which of six available gaging instruments has the best repeatability.

2. **Background information:**
Operators in the plant have widely different abilities in using the gaging instruments. Each operator can test three instruments. The standard deviation of the current instrument is about 1.0 unit.

3. **Experimental variables:**

| A. | Response variables | Measurement technique |
|---|---|---|
| 1. | Standard deviation of repeated readings | Calculated using QC department protocol |

| B. | Factors under study | Levels |
|---|---|---|
| 1. | Gaging instruments | Instruments A, B, C, D, E, and F (letters randomly assigned to six different brands) |

| C. | Background variables | Method of control |
|---|---|---|
| 1. | Operators | 10 operators selected by the production manager, each operator considered a block |

4. **Replication:**
Each operator can test three instruments, 10 operators are selected for the study.

5. **Methods of randomization:**
Operators assigned numbers 1–10 in order of particapation in the study. Random permutation table used to assign operator to particular block of instruments. Test order for each operator randomized using permutation table.

6. **Design matrix:**
Balanced incomplete block design with 10 blocks, three tests per block. (See Table 4. 9.)

7. **Data collection forms:**
QC gage test forms used during study.

8. **Planned methods of statistical analysis:**
Run charts of original and adjusted data.

9. **Estimated cost, schedule, and other resource considerations:**
Schedule developed to complete study in one day, with 10 operators each contributing three hours.

**Figure 4.17** Documentation form for Example 4.6.

done in a similar manner to the randomized block design, by subtracting the average for each block (operator) from the original data and then adding the overall average. Since the experimental pattern is balanced (each instrument is evaluated with every other instrument by the same operator twice in the study), evaluation of the adjusted data to compare instruments is valid. Table 4.9 shows the adjusted standard deviations. Figure 4.19 shows the adjusted data plotted, grouped by instrument type.

From Fig. 4.19 it is much clearer that instrument C performed better in the study than the other instruments. Instrument E was consistently high, whereas the effectiveness of instrument D depended on

**TABLE 4.9 Incomplete Block Design for Example 4.6**

Design Matrix and Results of Study:
Standard Deviation of Repeat Measurements
(Run Order within the Block in Parentheses)

| | Instrument | | | | | | |
|---|---|---|---|---|---|---|---|
| Operator | A | B | C | D | E | F | Operator average |
| 1 | 1.1(2) | | 0.7(1) | | | 0.9(3) | 0.90 |
| 2 | | 1.3(3) | | 1.4(1) | | 1.2(2) | 1.30 |
| 3 | 0.6(1) | 0.5(3) | | | 0.9(2) | | 0.67 |
| 4 | | 0.9(1) | 0.4(3) | 1.2(2) | | | 0.83 |
| 5 | | | | 0.8(3) | 1.6(1) | 1.2(2) | 1.20 |
| 6 | 1.6(2) | 1.4(3) | | | | 1.5(1) | 1.50 |
| 7 | 1.0(2) | | | 0.8(1) | 1.5(3) | | 1.10 |
| 8 | | | 0.3(1) | | 1.0(2) | 0.6(3) | 0.63 |
| 9 | | 1.2(2) | 0.6(1) | | 1.4(3) | | 1.07 |
| 10 | 1.8(3) | | 1.2(1) | 1.7(2) | | | 1.57 |
| Average | 1.22 | 1.06 | 0.64 | 1.18 | 1.28 | 1.08 | $\bar{\bar{X}} = 1.077$ |

Adjusted Standard Deviations
(Standard Deviation − Operator Average + Overall Average)

| | Instrument | | | | | | |
|---|---|---|---|---|---|---|---|
| Operator | A | B | C | D | E | F | |
| 1 | 1.28 | | 0.88 | | | 1.08 | |
| 2 | | 1.08 | | 1.18 | | 0.98 | |
| 3 | 1.01 | 0.91 | | | 1.31 | | |
| 4 | | 1.15 | 0.65 | 1.45 | | | |
| 5 | | | | 0.68 | 1.48 | 1.08 | |
| 6 | 1.18 | 0.98 | | | | 1.08 | |
| 7 | 0.98 | | | 0.78 | 1.48 | | |
| 8 | | | 0.75 | | 1.45 | 1.05 | |
| 9 | | 1.21 | 0.61 | | 1.41 | | |
| 10 | 1.31 | | 0.71 | 1.21 | | | |
| Adjusted average | 1.152 | 1.066 | 0.720 | 1.060 | 1.426 | 1.054 | $\bar{\bar{X}} = 1.080$ |

the operator. Instrument C should be considered for selection. A possible next cycle would be to have the five operators who did not test this instrument in the study try out instrument C. Obtain comments from these operators on ease of use, general applicability in the plant, and other features before a decision is made to purchase instrument C.

Future study should focus on the variability of the operators in using the instruments. The differences in precision among the different operators were at least as great as differences among the instruments in the study. Perhaps a flowchart of the measurement process and a short training session would be appropriate. Operator 10 may require special help.

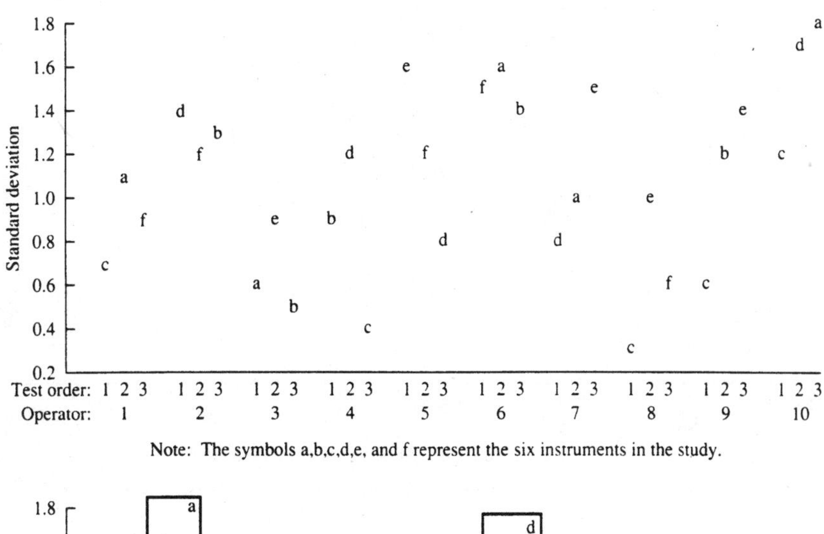


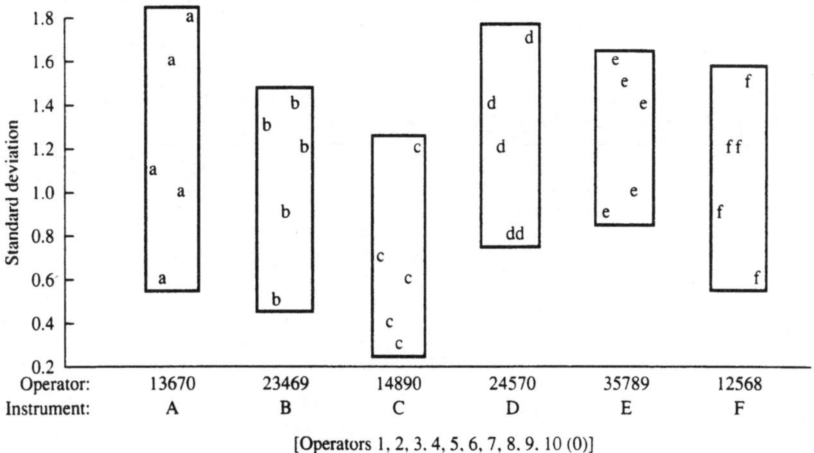


**Figure 4.18** Run charts for Example 4.6.

The analysis used in this example required that a balanced incomplete block design be used. If an unbalanced block design had been used, the data adjustment procedure would not have been valid. With an unbalanced design, there would be an unequal number of differences among the instruments. Paired differences could be used to remove the block effect for unbalanced designs. The response for each instrument could be subtracted from the response of every other instrument in the block to obtain the differences. The results of these calculations for the data in this example are shown in Table 4.10. For example, both operators 3 and 6 tested instrument A and instrument B. The two differences in column B, row A are calculated by using the results from operators 3 and 6. For operator 3, the standard deviation for instrument A was 0.6, and the standard deviation for B was 0.5. Thus, the difference B − A

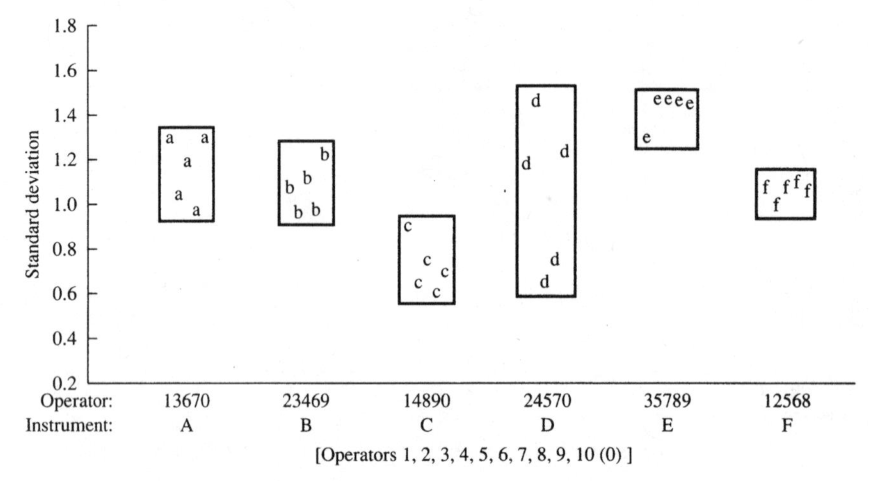


**Figure 4.19** Adjusted standard deviations for each instrument type in Example 4.6.

**TABLE 4.10 Using Differences between Instruments Tested by the Same Operator (Column−Row) to Remove Block Effect (Example 4.6)**

| | A | B | C | D | E | F |
|---|---|---|---|---|---|---|
| A | | −0.1 | −0.4 | −0.2 | 0.3 | −0.2 |
| | | −0.2 | −0.6 | −0.1 | 0.5 | −0.1 |
| B | 0.1 | | −0.5 | 0.1 | 0.4 | −0.1 |
| | 0.2 | | −0.6 | −0.3 | 0.2 | 0.1 |
| C | 0.4 | 0.5 | | 0.8 | 0.7 | 0.2 |
| | 0.6 | 0.6 | | 0.5 | 0.8 | 0.3 |
| D | 0.2 | −0.1 | −0.8 | | 0.4 | −0.2 |
| | 0.1 | 0.3 | −0.5 | | 0.7 | −0.4 |
| E | −0.3 | −0.4 | −0.7 | −0.4 | | −0.8 |
| | −0.5 | −0.2 | −0.8 | −0.7 | | −0.4 |
| F | 0.2 | 0.1 | −0.2 | 0.2 | 0.8 | |
| | 0.1 | −0.1 | −0.3 | 0.4 | 0.4 | |
| Average | 0.21 | 0.04 | −0.54 | −0.03 | 0.52 | −0.16 |

(0.5 − 0.6 = −0.1) is determined. Similarly, for operator 6, the B − A difference (1.4 − 1.6 = −0.2) is calculated.

A plot of these differences grouped by instrument is shown in Fig. 4.20. The effect of the factor, instrument, and the stability of the effect can be studied from this plot of differences. An instrument with average precision will tend to have differences centered on 0.0. An instrument with poor precision will have differences greater than zero, and an instrument with good precision will have negative differences. All the differences for instrument C are negative, indicating that instrument C had better precision than all the other instruments each time it was tested. Instrument E had all positive differences, indicating

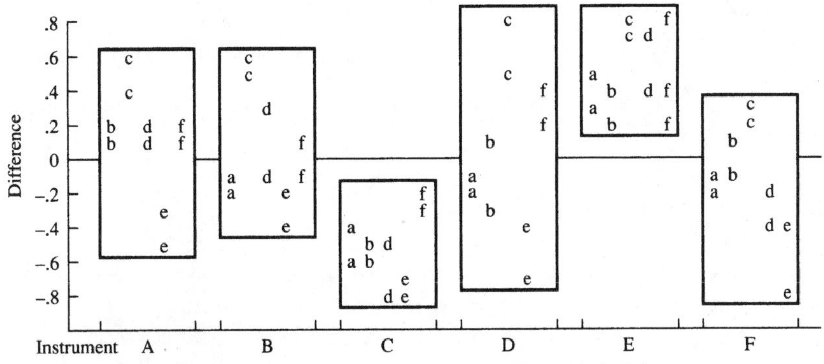


**Figure 4.20** Plot of differences by instrument type (Example 4.6).

that it performed poorly (large standard deviation) in all cases. As for the other instruments, B and D were average, A tended toward poor precision, and F tended to have good precision.

Incomplete block designs can be used for experiments with more than one factor. Chunk-type blocks can be developed to include multiple background variables. Depending on the size of the blocks created, either a randomized block design or an incomplete block design will be appropriate for studying the factors.

## 4.7 Summary

A number of important concepts were illustrated in the examples in this chapter:

- The use of the planning form to consider and document the tools for experimentation introduced in Chap. 3
- The use of the run chart as a key tool for the analysis of data from analytic studies
- The use of blocking to incorporate background variables
- The use of grouping and stratification on the run chart to highlight the factor levels under different conditions defined by the background variables
- Adjustment of data to remove the effect of background variables to facilitate comparison of the factor levels
- The importance of stability in interpreting the data from an analytic study

Each of these concepts will be expanded upon in the following chapters, when more than one factor is incorporated into a study.

## References

Box, G., W. Hunter, and J. S. Hunter (1978): *Statistics for Experimenters,* Wiley, New York, chap. 7.

Cochran, W. G., and G. M. Cox (1957): *Experimental Design,* Wiley, New York, chaps. 9 and 11.

## Exercises

**4.1** Describe how a control chart can be used to conduct a one-factor experiment on a process. List the key steps involved in conducting such a study. How is this different from using a control chart to control a process?

**4.2** Design an experiment with one factor for a process you are familiar with. Complete a planning documentation form, including objective, background information, experimental variables, replication, randomization, design matrix, data collection form, planned analysis, and resource requirements.

**4.3** Describe some situations that would make a paired-comparison design appropriate for studying one factor.

**4.4** Why are randomized block designs important in analytic studies? Describe how blocking could be used in a study to evaluate three alternative teaching methods in a school.

**4.5** The production manager wants to evaluate four new types of drill bits for consideration as the plant standard. There are three machines that could use the drill bits, each with a different operator and each run on two shifts. Historically, there has been variability of drill bit performance among machines and operators.

What type of design is appropriate for this study? Describe the experimental pattern, blocking, replication, and randomization for the study.

**4.6** In the one-factor example on control methods for a chemical process (Example 4.2), two other response variables were also measured. The results for purity (percent) and a contaminant (ppm) were the following:

| Date: | 2 | 3 | 4 | 5 | 6 | 7 | 8 | 9 | 10 | 11 | 12 | 13 | 14 | 15 | 16 |
|---|---|---|---|---|---|---|---|---|---|---|---|---|---|---|---|
| Factor: | c | c | c | c | c | v | v | v | v | v | m | m | m | m | m |
| Purity: | 72 | 69 | 67 | 67 | 64 | 62 | 63 | 66 | 68 | 67 | 70 | 70 | 73 | 74 | 76 |
| Contaminant: | 33 | 28 | 30 | 26 | 31 | 43 | 31 | 48 | 44 | 46 | 35 | 31 | 33 | 29 | 32 |

(*a*) Prepare run charts for these data. Partition the variability of the two response variables among the factor and the nuisance variables. What conclusions can be made about the factor for each response variable?

(*b*) Calculate adjusted data for each response variable, and plot the adjusted data on a run chart. What additional information can be learned from a study of the adjusted data?

**4.7** During the paired-comparison experiment evaluating tool wear (Example 4.3), parts were selected from each lot to evaluate the variability of a specified dimension. The quality control manager was concerned that the new setup procedure might affect the variability of the initial parts produced. A subgroup of five parts was selected during the first half-hour from each lot for each of the setup procedures. The dimension (in thousandths of an inch from nominal) was measured for each part, and the range of the five parts was calculated. The following results (range, in thousandths of an inch) were obtained:

| Lot: | 1 | 2 | 3 | 4 | 5 | 6 | 7 | 8 | 9 | 10 |
|---|---|---|---|---|---|---|---|---|---|---|
| New setup: | 5.3 | 0.2 | 2.1 | 4.6 | 3.0 | 1.2 | 5.8 | 4.0 | 8.4 | 3.0 |
| Old setup: | 6.4 | 1.3 | 2.0 | 6.2 | 3.8 | 3.0 | 8.5 | 5.8 | 8.0 | 5.3 |

Prepare run charts of the ranges and the adjusted ranges. Partition the variation in the ranges among the factor, background variables, and nuisance variables.

**4.8** In the study to choose a catalyst (Example 4.4), another response variable, purity, was evaluated for each of the catalyst types. A composite sample was taken from each batch produced and analyzed for purity (percent). The following data were obtained:

| | Purity of sample for each feedstock shipment | | | | | |
|---|---|---|---|---|---|---|
| Catalyst | 1 | 2 | 3 | 4 | 5 | 6 |
| X | 99.1 | 99.8 | 99.5 | 98.4 | 99.6 | 99.7 |
| Y | 98.0 | 99.2 | 98.3 | 97.5 | 98.4 | 98.9 |
| Z | 97.0 | 98.3 | 97.9 | 96.8 | 97.7 | 98.3 |
| C | 97.9 | 99.1 | 98.6 | 97.2 | 98.5 | 99.0 |

(*a*) Prepare run charts for the purity data (see Example 4.4 for run order of catalyst within each shipment).

(*b*) Calculate the purity adjusted for feedstock shipment, and prepare a run chart for the adjusted values.

(*c*) What are the conclusions from this study for purity? Summarize the importance of the factor, background variable, and nuisance variables. Which catalyst would be expected to give the highest purity in future shipments? How strong is the degree of belief in this conclusion?

(*d*) Combine the conclusions from Example 4.4 on yield and the results on purity to develop recommendations for future catalyst use. The cost of each catalyst ($/lb) is: current, $12; catalyst X, $10; catalyst Y, $16; and catalyst Z, $12. What additional information is needed to develop a cost-effective recommendation for catalyst use?

**4.9** The accounts payable quality improvement team was studying the process of paying invoices. Team members wanted to improve the efficiency and accuracy of payments. During the development of flowcharts, the team found that there was variability in the process among the three clerks and at different time periods. In their attempts to standardize the process, four alternative procedures were developed. The team decided to study the four alternatives and choose the process with the highest efficiency (number of invoices paid) and best accuracy (lowest number of errors) as the standard.

The team was concerned that the evaluation could be affected by variation in the payment process from week to week. This variation included number of invoices received, end-of-month payments, and other peculiarities related to different time periods.

A balanced incomplete block design was chosen for the test. A block was defined by the week in which the test was done. There were four levels of the factor, process alternative, to study. Each of the clerks could use a different alternative during a time period. The design with four factor levels, a block size of three, and four blocks was selected (see Table 4.8). The three process alternatives within a week were randomly assigned to one of the clerks. The experimental pattern was replicated by running the study for 8 weeks. The following data were obtained.

| | Clerk 1 | | | Clerk 2 | | | Clerk 3 | | |
|---|---|---|---|---|---|---|---|---|---|
| Week | Process | Number paid | Number of errors | Process | Number paid | Number of errors | Process | Number paid | Number of errors |
| 1 | C | 32 | 9 | B | 38 | 5 | A | 30 | 10 |
| 2 | A | 42 | 6 | B | 51 | 5 | D | 57 | 12 |
| 3 | D | 43 | 7 | A | 37 | 9 | C | 34 | 3 |
| 4 | B | 70 | 9 | D | 74 | 20 | C | 55 | 8 |
| 5 | B | 41 | 6 | C | 31 | 5 | A | 26 | 15 |
| 6 | B | 51 | 3 | A | 43 | 8 | D | 60 | 10 |
| 7 | A | 35 | 10 | D | 45 | 9 | C | 33 | 4 |
| 8 | C | 60 | 9 | D | 69 | 14 | B | 68 | 13 |

(*a*) Analyze each of the response variables for this study. Prepare run charts, calculate adjusted values, and construct run charts of the differences. Label the factor level (the process alternative) on the run charts. Group the adjusted values by factor level for evaluation of each process alternative.

(*b*) Which alternative process can be expected to have the highest efficiency? Which alternative will give the fewest errors? What kind of study should be completed next by the team?

Chapter

# 5

# Experiments with More than One Factor

In this chapter, designs to study the effects of multiple factors on a response variable are considered. These designs will provide the foundation to study and improve complex processes and products. Because of the complexity of most processes, several factors are usually studied in an experiment. A common approach to experimentation when there is more than one factor is to change one factor at a time. Experimenters cite as a reason in support of this approach that if more than one factor is changed, the experimenter will not be able to determine which factor was responsible for the change in the response.

There are two major deficiencies in studying one factor at a time. The first is that there are often interactions between the factors under study. An interaction means that the effect a factor has on the response may depend on the levels of some other factors. Figure 5.1 contains response plots (described in Chap. 3) illustrating various degrees of interaction. In Fig. 5.1*a,* the change in the response as factor 1 is changed is the same regardless of whether factor 2 (denoted by $F_2$) is at the low or high level (i.e., the slopes of the line are the same). This plot indicates that the two factors independently affect the response, i.e., there is no interaction between the factors. In Figs. 5.1*b, c,* and *d,* the effect of changing factor 1 from a low to a high level differs depending on the level of factor 2. In all three cases, since the slopes of the two lines in the response plots are not the same, an interaction effect exists between the two factors.

The second deficiency in studying one factor at a time is inefficiency. As each factor is changed in turn, the data previously collected to study other factors are set aside, and new data are collected. Each set of data supplies information on only one factor.

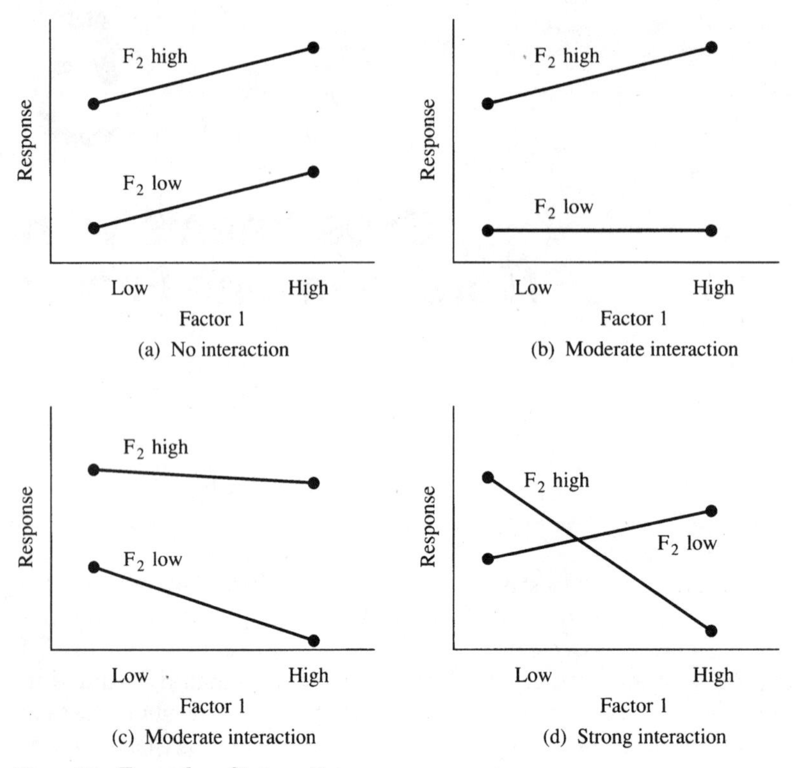


**Figure 5.1** Examples of interactions.

Factorial designs provide an effective alternative to studying one factor at a time. Factorial designs allow the study of interactions between factors. The structure of factorial designs allows all the data from the experiment to be used to study each factor.

## 5.1 Introduction to Factorial Designs

To set up a factorial design, the investigator determines the factors to be studied and the levels for each. A full factorial design consists of all possible combinations of the factors and levels. For example, if three factors are to be studied, the first at two levels, the second at three levels, and the third at five levels, a $2 \times 3 \times 5$ factorial design is used. This design requires 30 ( $= 2 \times 3 \times 5$) tests for one replication of the experimental pattern.

In this chapter, designs are considered for experiments in which each factor is studied at only two levels. This class of designs will be referred to as $2^k$ factorial designs. For example, a $2^3$ factorial design would require $2 \times 2 \times 2 = 8$ tests. Factorial designs with more than two levels will be considered in Chap. 8.

Reasons for emphasizing two-level ($2^k$) factorial designs include these:

- They are easy to use, and the data analysis can be performed by graphical methods. This allows all interested parties to participate in the experimentation.
- Relatively few runs are required. A factorial design for four factors at two levels each requires 16 runs. If each of the factors is studied at three levels, 81 runs are required. For four levels, 256 runs are required.
- The $2^k$ designs have been found to meet the majority of the experimental needs of those engaged in improvement efforts.
- The $2^k$ factorial designs are easy to use in a sequence of studies, so that even complex systems with many variables can be studied in depth by using these relatively simple designs.
- When a large number of factors are studied, fractions of the $2^k$ designs can be used to keep the experiment at a reasonable size (these fractional factorial designs will be discussed in Chap. 6).

The combinations of factors and levels that make up a factorial design (i.e., the design matrix) can be displayed in various ways. When factors are studied at two levels, a common convention is to designate the low level of the factor as $-$ (minus) and the high level as $+$ (plus). When the factor is qualitative, such as type of material, the $-$ and $+$ labels can be assigned arbitrarily. Using this convention, Fig. 5.2 shows examples of displays of $2^2$, $2^3$, and $2^4$ factorial designs. The format in Fig. 5.2 is a simple listing of the combinations of factors in a form often called the *design matrix.* This display of a factorial design is used primarily for documentation of the tests to be made in the experiment and as a basic form for the collection of data. Also, as described later in this section, this format is useful for estimating the effects of the factors.

Figure 5.3 contains a second way to display a factorial design called a *tabular display.* Each of the small squares (called *cells*) in the table corresponds to a specific set of combinations of the factors. The display of the data obtained from the experiment in such a table aids direct visual analysis of the responses from a factorial experiment. For example, consider the tabular display for a $2^3$ design in Fig. 5.3*b*. By comparing the data in the four cells on the left side of the table to the data in the four cells on the right side, the effect of factor 3 can be studied.

Figure 5.4 depicts the third form of displaying factorial designs, called a *geometric display.* Each corner of the square or cube corre-

| Test | Factor 1 | Factor 2 |
|---|---|---|
| 1 | – | – |
| 2 | + | – |
| 3 | – | + |
| 4 | + | + |

(a) $2^2$ design

| Test | Factor 1 | Factor 2 | Factor 3 |
|---|---|---|---|
| 1 | – | – | – |
| 2 | + | – | – |
| 3 | – | + | – |
| 4 | + | + | – |
| 5 | – | – | + |
| 6 | + | – | + |
| 7 | – | + | + |
| 8 | + | + | + |

(b) $2^3$ design

| Test | Factor 1 | Factor 2 | Factor 3 | Factor 4 |
|---|---|---|---|---|
| 1 | – | – | – | – |
| 2 | + | – | – | – |
| 3 | – | + | – | – |
| 4 | + | + | – | – |
| 5 | – | – | + | – |
| 6 | + | – | + | – |
| 7 | – | + | + | – |
| 8 | + | + | + | – |
| 9 | – | – | – | + |
| 10 | + | – | – | + |
| 11 | – | + | – | + |
| 12 | + | + | – | + |
| 13 | – | – | + | + |
| 14 | + | – | + | + |
| 15 | – | + | + | + |
| 16 | + | + | + | + |

(c) $2^4$ design

**Figure 5.2** Design matrix display.

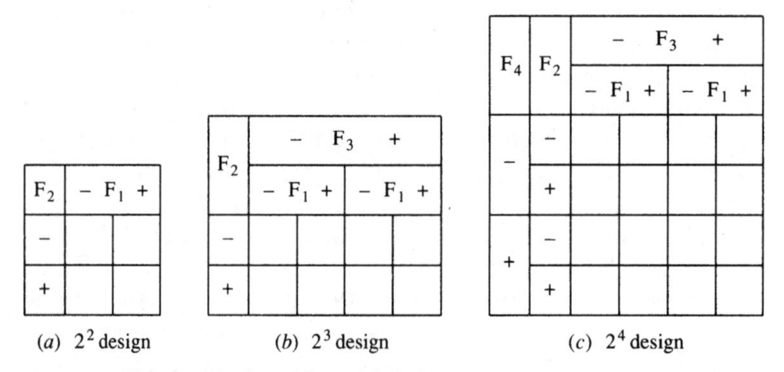


(*a*) $2^2$ design (*b*) $2^3$ design (*c*) $2^4$ design

**Figure 5.3** Tabular display of factorial designs.

sponds to a different set of combinations of the factors. The corners correspond to the cells in the tabular display. The value of the response variable obtained from each test is written at the appropriate corner. The geometric display, like the tabular display, is helpful for analysis. For the $2^3$ design, factor 1 can be studied by comparing the data on the left side of the cube to the data on the right side. Factor 2 is studied by comparing the bottom of the cube to the top, and factor 3 is studied by comparing front to back.

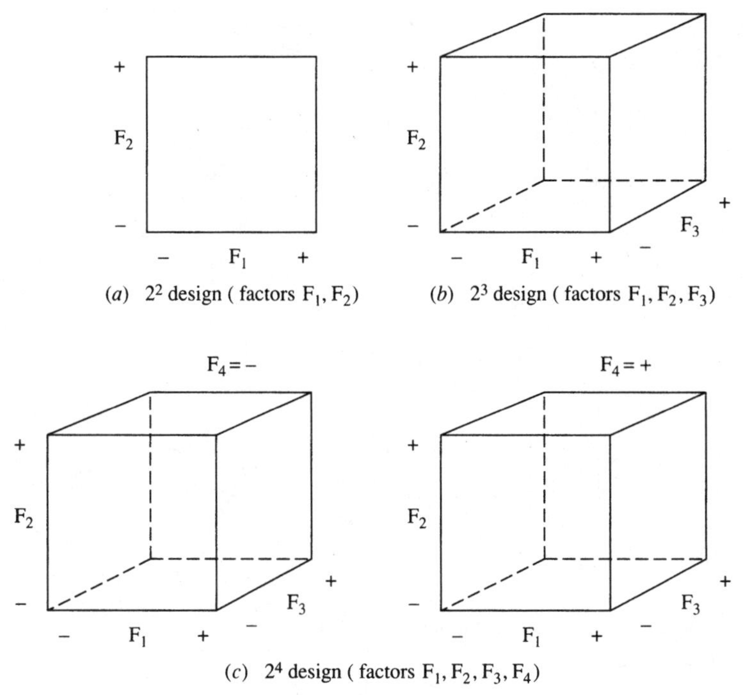


**Figure 5.4** Geometric display of a factorial design.

Figure 5.5 illustrates the sides of the cube that are used for each comparison. Clearly, the structure of factorial designs allows all the data from the experiment to be used to study each factor. The design and analysis of multifactor studies are developed in the next three examples.

**Example 5.1: A $2^2$ Design—Manufacture of Plasticizer** As was the case for the one-factor design, careful planning of the experiment is important in order to maximize the amount of information obtained for the expended resources. To aid the planning of experiments with multiple factors, the planning form introduced in Chap. 3 will be used. Figure 5.6 contains the form used to summarize an experiment performed to improve the process of manufacturing a certain type of plasticizer.

Normally the product is made in batches, and the reaction is allowed to continue until a certain viscosity is obtained. The unit manager desired that this reaction take from 7 to 9 h. The plant was experiencing other problems, which led to the reaction's proceeding too fast. The fast reaction time resulted in batches that were difficult to control and product of unacceptable viscosity. The purpose of the experiment was to find a combination of the percentage of a key ingredient X and the reaction temperature that would result in a reaction that proceeded at the desired rate.

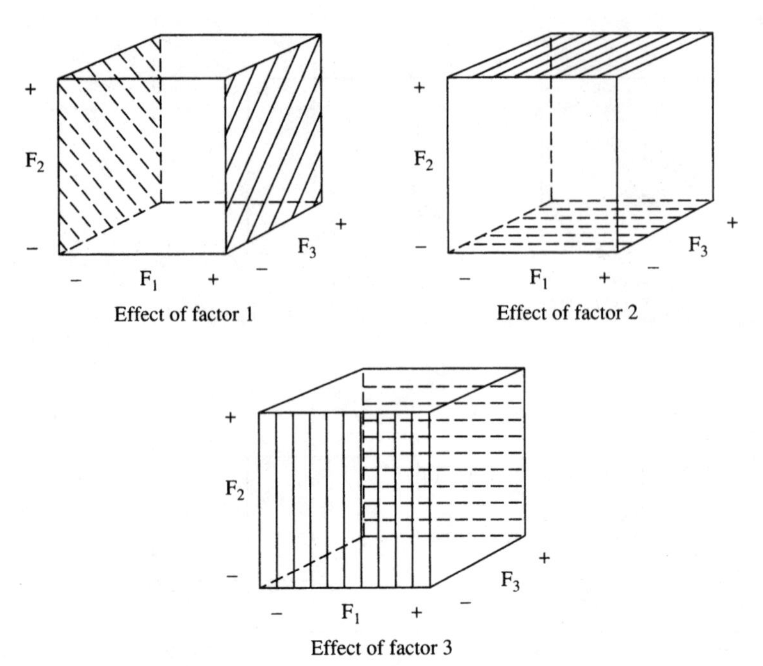


**Figure 5.5** Comparison of data on a cube.

The response variable was the reaction time necessary to reach the desired viscosity. There were two factors under study: the percentage of ingredient X and the reaction temperature. The levels for the ingredient were chosen at 42 and 48 percent. The temperature levels were chosen at 175 and 195°C.

Several background variables were considered. The experiment was initially conducted in the laboratory, so confirmation of the results in the plant would be necessary. The rate of heat-up was controlled by a temperature programmer. One laboratory operator was used. The difference in blends of X was handled by completing the experimental pattern using only one blend of X.

In analytic studies, it is important to run studies sequentially over a variety of conditions to build up the degree of belief. For this experiment, two replications of the experimental pattern were run. An adequate amount of ingredient X was obtained and mixed so that the four combinations of the two factors needed for one replication of a $2^2$ pattern could be performed using a homogeneous blend of ingredient X. The second replication of the pattern was performed using a blend of ingredient X from a different shipment. The order of the four runs in each replication was randomized separately, using a table of random permutations.

The first step in the analysis of data from a factorial design is a run chart of the data. This chart appears in Fig. 5.7. No obvious time

1. **Objective:**
   Find a combination of the amount of ingredient X and the reaction temperature to increase the reaction time to 7–9 hours in a batch process for the manufacture of plasticizer.

2. **Background information:**
   The reaction in plant production has often been proceeding too fast, resulting in batches that were difficult to control and of unacceptably high viscosity.

3. **Experimental variables:**

| A. Response variables | Measurement technique |
|---|---|
| Reaction time to reach desired viscosity | Viscometer and clock |

| B. Factors under study | Levels | |
|---|---|---|
| 1. Percentage of ingredient X | 42% (−) | 48% (+) |
| 2. Temperature (°C) | 175 (−) | 190 (+) |

| C. Background variables | Method of control |
|---|---|
| 1. Lab experiment | Confirmation on plant batches |
| 2. Rate of heat-up | Temperature programmer |
| 3. Operator | One lab technician |
| 4. Blend variation in X | Blocking |

4. **Replication:**
   Two replications of the experimental pattern, resulting in eight batches. Each replication used a different blend of ingredient X.

5. **Methods of randomization:**
   Randomize the order of the four runs within each blend of X using a table of random permutations.

6. **Design matrix:**
   (See Table 5.1.)

7. **Data collection forms:**
   (See Table 5.2.)

8. **Planned methods of statistical analysis:**
   - Run chart
   - Analysis of the square
   - Summary of effects if appropriate (response plot)

**Figure 5.6** Documentation of planned experiment, Example 5.1.

trends or outlying values are seen, so analysis of the effects of the factors can be performed. The data from a factorial design can be analyzed as a series of paired comparisons, one factor at a time. The comparisons are carried out under various conditions of the other factors, which increases the degree of belief in the results. The data are displayed on the square in Fig. 5.8.

Each corner of the square contains two reaction times. The first reaction time is the result from the first replication (the first blend of in-

**TABLE 5.1 Design Matrix for the Plasticizer Experiment**

| X | $T$ |
|---|---|
| − | − |
| + | − |
| − | + |
| + | + |

| Percentage of X | | Temperature, °C | |
|---|---|---|---|
| − | + | − | + |
| 42 | 48 | 175 | 195 |

**TABLE 5.2 Data Collection Form for the Plasticizer Experiment**

| Batch ID | Percentage of X | Temperature | Reaction time |
|---|---|---|---|
| | Replication 1 (Blend 1 of X) | | |
| 1 | 42 | 195 | 5.5 |
| 2 | 42 | 175 | 9.0 |
| 3 | 48 | 175 | 9.0 |
| 4 | 48 | 195 | 1.5 |
| | Replication 2 (Blend 2 of X) | | |
| 5 | 42 | 195 | 6.5 |
| 6 | 48 | 195 | 1.0 |
| 7 | 42 | 175 | 9.5 |
| 8 | 48 | 175 | 8.0 |

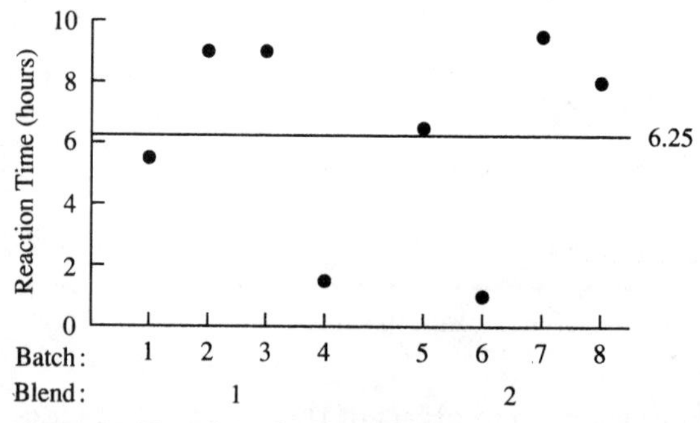


**Figure 5.7** Run chart for Example 5.1.

gredient X), and the second is the result from the second replication. For each replication, there are two paired comparisons of the effect of the temperature. The first comparison is performed by comparing the results at the top corners of the square for the first replication. This is

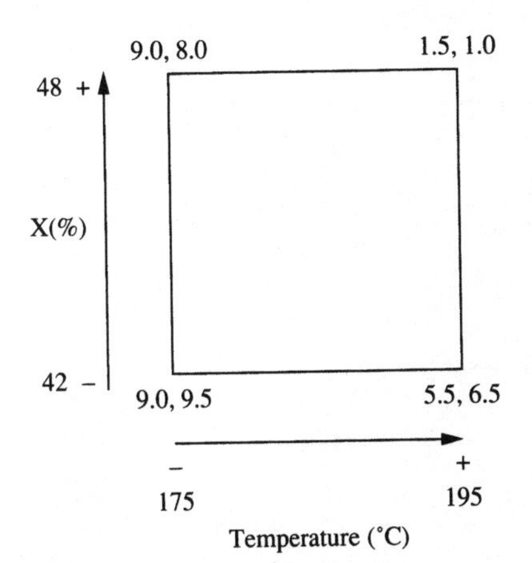


**Figure 5.8** Square for the plasticizer experiment.

done quantitatively by subtracting the reaction time for the low temperature from the reaction time for the high temperature, with X held constant at 48 percent, i.e.,

$$1.5 - 9.0 = -7.5$$

When the concentration of X is 42 percent, the effect of temperature is

$$5.5 - 9.0 = -3.5$$

Similarly, the effect of the percentage of X when temperature is held constant at 175°C comes from the left corners of the square:

$$9.0 - 9.0 = 0.0$$

The effect of the percentage of X when temperature is held constant at 195°C comes from the right corners:

$$1.5 - 5.5 = -4.0$$

The same comparisons can be made for the second replication. The results for both replications are given in Table 5.3. Based on these comparisons, some important observations can be made about how temperature and the percentage of X affect reaction time:

- The replications produced very similar results, indicating that there were no important effects of blends. All four cases resulted in lower temperature associated with longer reaction time. The magnitude of the difference was large enough to be of importance in the process.
- The temperature comparisons when X was at 48 percent produced larger differences than when X was at 42 percent. This observation

**TABLE 5.3 Results of Paired Comparisons for Plasticizer Experiment**

| Difference in reaction time for change from 175 to 195°C | Percentage of X at which comparison is made |
|---|---|
| Replication 1 (Blend 1) | |
| 1.5−9.0 = −7.5 | 48 |
| 5.5−9.0 = −3.5 | 42 |
| Replication 2 (Blend 2) | |
| 1.0−8.0 = −7.0 | 48 |
| 6.5−9.5 = −3.0 | 42 |

| Difference in reaction time for change from 42 to 48% | Temperature (°C) at which comparison is made |
|---|---|
| Replication 1 (Blend 1) | |
| 9.0−9.0 = 0.0 | 175 |
| 1.5−5.5 = −4.0 | 195 |
| Replication 2 (Blend 2) | |
| 8.0−9.5 = −1.5 | 175 |
| 1.0−6.5 = −5.5 | 195 |

along with the consistency of the replications suggests an interaction between temperature and percentage of X.

- Three of the four comparisons of the effect of high versus low percentage of X on reaction time resulted in longer reaction time associated with 42 percent of X. In one comparison, no effect was seen.

At this point, it is up to those familiar with the process to decide what action should be taken next based on their degree of belief. The next cycle could be further laboratory tests under different conditions or possibly confirmation of the results from this cycle in the plant.

Once the important effects have been determined, the relationship between the response and the important factors should be graphically displayed. Plots of this type were introduced earlier and are called *response plots*. The response plots will provide the analyst with further insight into the cause-and-effect relationships (especially when interaction is present) and will simplify the presentation of the important findings in the study. Since the factors temperature and percentage of X interact, the relationship between reaction time and temperature

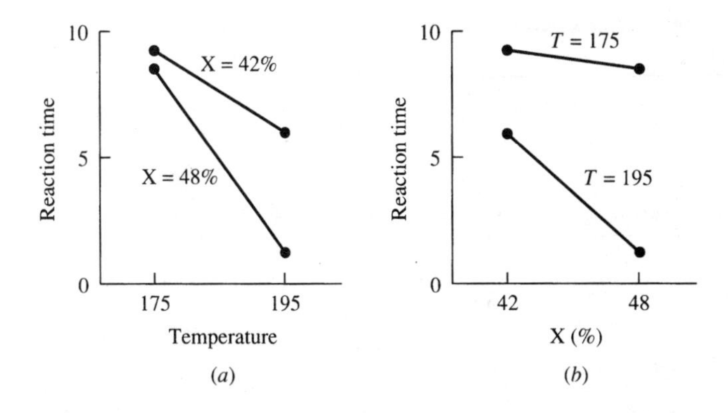


Average of replication tests for the four conditions

| X (%) | Temperature (°C) | |
|---|---|---|
| | 175 | 195 |
| 42 | 9.25 | 6.00 |
| 48 | 8.50 | 1.25 |

(c)

**Figure 5.9** Response plots for the plasticizer experiment.

should be plotted separately for each percentage of X. These plots are shown in Fig. 5.9*a*.

The points on the plot are obtained by averaging the two replications for each of the four conditions in the experimental pattern (i.e., averaging the two results at each corner of the square in Fig. 5.8). The relationship between reaction time and percentage of X for the two levels of temperature is plotted in Fig. 5.9*b*. The plots in Fig. 5.9*a* and *b* contain the same information displayed in two different ways, so only one is actually needed. (It is sometimes helpful, however, to view both plots.) The plots in Fig. 5.9 are especially useful in presenting the results of the study to those not directly involved in the factorial experiments.

The next example extends these concepts to an experiment with three factors.

**Example 5.2: A $2^3$ Design for a Dye Process** The data in Table 5.4 resulted from a $2^3$ factorial design on a dye process. The aim of the experiment was to quantify the effects of three factors on the shade of dyed material and to use the information to determine settings of the factors. The three factors studied were

Material quality $M$

Oxidation temperature $T$

Oven pressure $P$

**TABLE 5.4 Tabular Design Matrix (with Data) for Dye Process**

| | Material quality | | | |
|---|---|---|---|---|
| | A | | B | |
| | Oxidation temperature | | Oxidation temperature | |
| Oven pressure | Low | High | Low | High |
| Low | 189<br>(4) | 195<br>(8) | 228<br>(7) | 200<br>(6) |
| High | 218<br>(3) | 238<br>(2) | 259<br>(5) | 241<br>(1) |

Run order indicated in parentheses.

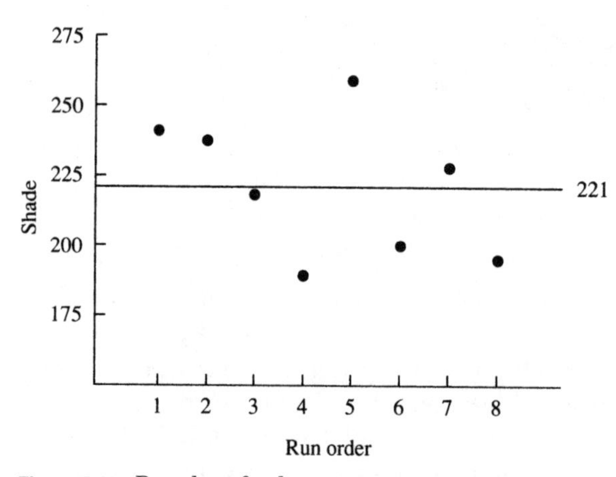


**Figure 5.10** Run chart for dye process experiment.

Important background variables were identified and held constant. The response variable was a measure of shade using an optical instrument. It was desired to choose settings of the factors to obtain a shade reading of 200. (*Note:* No units are given because the unit of measurement is somewhat arbitrary and depends on calibration of the instrument to reference standards.) In addition, it was desired to choose conditions to make the process as insensitive to variations in material as possible.

To begin the analysis, a run chart was made; it appears in Fig. 5.10. For an unreplicated $2^3$ design, only eight points are available to plot, so that only gross trends or outlying values can be identified from the run chart. No such values appear to be present in the chart in Fig. 5.10.

Figure 5.11 contains the cube labeled with the three factors studied in this experiment. At each corner, the value of the response variable at that set of combinations of the factors is given.

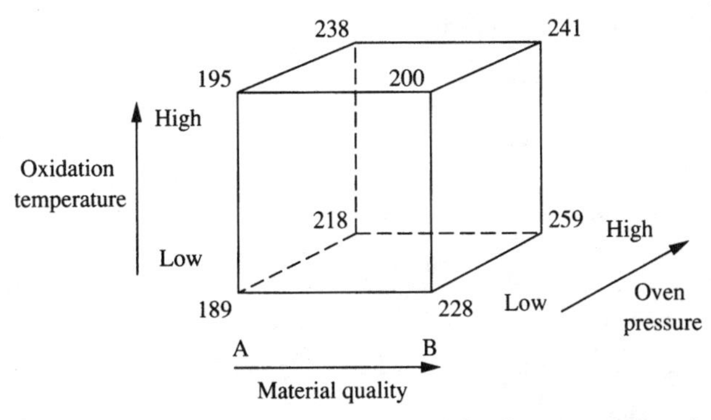


**Figure 5.11** Geometric display (cube) for the dye process experiment.

By appropriate analysis of the cube, the effects of the factors can be estimated. To study the effect of material quality, the four values on the left side of the cube are compared to the four values on the right side. The effect of oxidation temperature is obtained by comparing the top and bottom of the cube. Comparison of the back and front of the cube provides information on the effect of oven pressure.

A more detailed analysis of the data must be performed if interactions between factors or special causes of variation in the data are to be found. The $2^3$ factorial design can also be thought of as a series of paired comparisons. Each edge of the cube represents one of these comparisons. For example, the bottom front edge connects two corners between which the only difference in the test conditions is that material of quality A is used at the left corner and material of quality B is used at the right corner. At both corners, oxidation temperature is low and oven pressure is low. There are three other comparisons of the effect of material quality during which the other factors were held constant. These comparisons are performed by comparing the corners connected by the top front edge, the corners connected by the top back edge, and the corners connected by the bottom back edge.

Because of the symmetry of the $2^3$ design, there are also four pairs of tests that can be used to study the effect of oxidation temperature. These pairs are found at the corners connected by edges running from the top to the bottom of the cube. The four pairs used to study the effect of oven pressure are found along the edges connecting the front of the cube to the back. The computation of these effects from the paired comparisons is contained in Table 5.5. The paired comparisons can also be displayed graphically. These graphs are shown in Fig. 5.12. These paired-comparison graphs shown here do not usually need to be developed unless it is difficult to visualize the comparisons on the cube.

**TABLE 5.5 Computation of Effects from Paired Comparisons for Dye Process**

| | Combination for which comparison is made | |
|---|---|---|
| Effect of oven pressure | Oxidation temperature | Material |
| 218 − 189 = 29 | Low | A |
| 238 − 195 = 43 | High | A |
| 259 − 228 = 31 | Low | B |
| 241 − 200 = 41 | High | B |
| Effect of material quality | Oxidation temperature | Oven pressure |
| 228 − 189 = 39 | Low | Low |
| 200 − 195 = 5 | High | Low |
| 259 − 218 = 41 | Low | High |
| 241 − 238 = 3 | High | High |
| Effect of oxidation temperature | Oven pressure | Material |
| 195 − 189 = 6 | Low | A |
| 238 − 218 = 20 | High | A |
| 200 − 228 = −28 | Low | B |
| 241 − 259 = −18 | High | B |

From Table 5.5 and Fig. 5.12, it is apparent that as oven pressure increases, the measure of shade increases and that this increase is reasonably consistent (from 29 to 43) over the four sets of conditions. Since the results are consistent, the average effect of oven pressure is a useful summary of the data. The average effect of oven pressure is 36, computed by subtracting the average shade when oven pressure is low from the average shade when oven pressure is high:

$$239 - 203 = 36$$

This average effect can also be obtained by averaging the four paired comparisons.

On average, the shade value when material B is used is 22 units higher than that when material A is used. That is, the average effect of material is 22. As shown in Table 5.5 and Fig. 5.12, the results are not consistent over all the conditions. The effect of material is large when oxidation temperature is low. When oxidation temperature is high, material has little effect. That is, material quality and oxidation temperature interact. The effect of material cannot be given without first specifying the oxidation temperature. The average effect of mate-

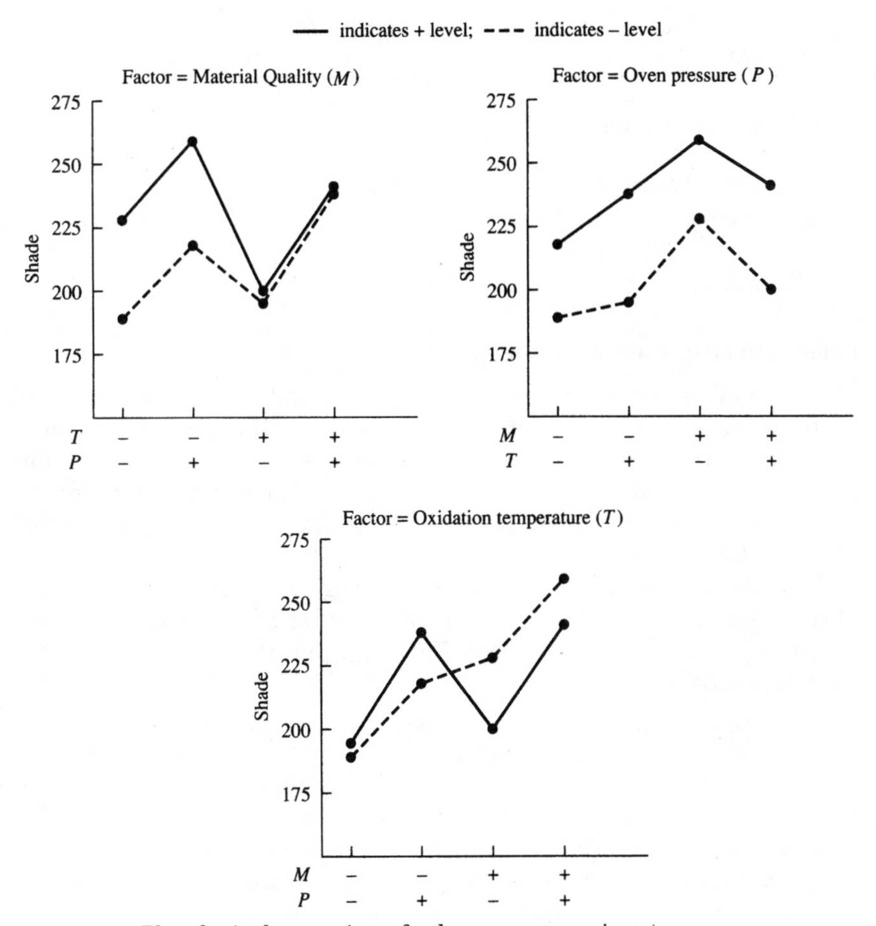


**Figure 5.12** Plot of paired comparisons for dye process experiment.

rial when oxidation temperature is low is 40. When oxidation temperature is high, the average effect of material is 4. The magnitude of the interaction is $-18$ and is computed according to the convention outlined in Table 5.6.

Next the effect of oxidation temperature is studied. In Table 5.5 and Fig. 5.12, it can be seen that shade increases as temperature increases in two cases and shade decreases as temperature increases in two cases. The positive effects of temperature were observed when material quality A was used, and the negative effects of temperature were seen when material of quality B was used. This is another way of viewing the material-temperature interaction.

**TABLE 5.6 Computation of Interaction Effect from Paired Comparisons**

| | | |
|---|---|---|
| Average effect of material when oxidation temperature is high | = 4 | Interaction = 4−22 = −18 |
| Average effect of material | = 22 | |
| Average effect of material when oxidation temperature is low | = 40 | |

## Estimating effects using the design matrix

The computation of the effects of the factors and the interactions can be performed by using an algorithm based on the extended design matrix. Table 5.7 illustrates the standard form of a design matrix for a $2^3$ design and the estimated effects. The columns corresponding to the various interactions are obtained by multiplying the signs for the factors contained in the interactions.

Each of the effects is estimated by adding or subtracting the value of the response variable, depending on whether the sign of the appropriate column is plus or minus. For example, the average effect of oven pressure is

$$\frac{-189 - 228 + 218 + 259 - 195 - 200 + 238 + 241}{4} = 36$$

It is easy to see that this computation is the same as subtracting the average shade value when oven pressure is low from the average shade value when oven pressure is high. The estimate of the interaction between material and oxidation temperature is

$$\frac{+189 - 228 + 218 - 259 - 195 + 200 - 238 + 241}{4} = -18$$

**TABLE 5.7 Design Matrix for Computation of Effects from Dye Process**

| Test | Run order | *M* | *P* | *T* | *MP* | *MT* | *PT* | *MPT* | Response |
|---|---|---|---|---|---|---|---|---|---|
| 1 | 4 | − | − | − | + | + | + | − | 189 |
| 2 | 7 | − | − | − | − | − | + | + | 228 |
| 3 | 3 | − | − | − | − | + | − | + | 218 |
| 4 | 5 | − | − | − | + | − | − | − | 259 |
| 5 | 8 | − | − | − | + | − | − | + | 195 |
| 6 | 6 | − | − | + | − | + | − | − | 200 |
| 7 | 2 | − | − | + | − | − | + | − | 238 |
| 8 | 1 | − | − | + | + | + | + | + | 241 |
| Divisor = 4 | | | | | | | | | |
| Effect | | 22 | 36 | −5 | 0 | −18 | 6 | −1 | |

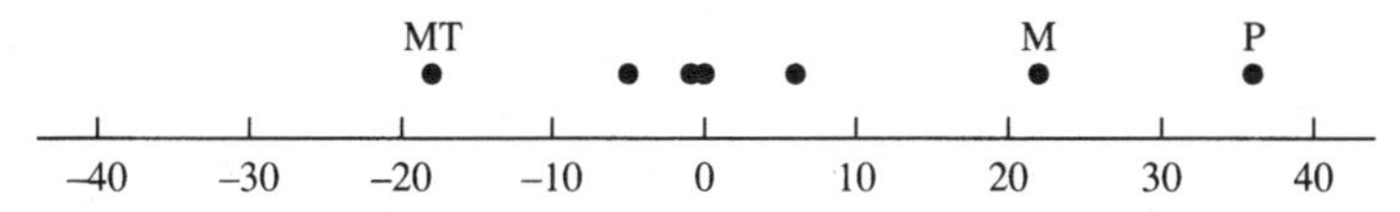


**Figure 5.13** Dot diagram of effects in dye process experiment.

This is the same as the estimate obtained in Table 5.6.

Once the estimates are obtained, they can be plotted to help determine which are the most important effects. The effects are plotted in Fig. 5.13. This plot is called a *dot diagram* by Box et al. (1978, p. 25). Effects clustered near zero on the diagram cannot be distinguished from variation due to nuisance variables. It is clear from the plot and the analysis of the cube that the most important effects are the average effect of oven pressure, the average effect of material, and the interaction between material quality and oxidation temperature.

During the analysis of data from any experiment, a close watch must be kept for evidence of special causes of variation due to background or nuisance variables. Since the effects computed from the design matrix are averages, they can be distorted by special causes. These estimates should be verified by a more detailed analysis of the data. The estimates of average effects from the design matrix should initially be used to provide some preliminary sense of which factors or interactions might be important. Usually in studying two or three factors, this method of estimating effects is not necessary since the analysis can be done using a square or cube. However, the approach is very useful for four or more factors.

Once the preliminary estimates are obtained, the data in the cube should be studied for consistency between the individual comparisons contained in the average effects. Plots of the individual comparisons such as those in Fig. 5.12 are also important for verifying the validity of the estimates of average effects.

#### Response plots

Once the important effects have been identified and estimated, the relationships can be graphically summarized using simple response plots such as those in Fig. 5.14. This figure shows three interaction plots, one for each of the three combinations of two factors.

Construction of the plots involves only some simple arithmetic. For example, Fig. 5.14*a* shows the effect of material quality and oxidation temperature on shade. To construct this plot, the $2 \times 2$ table shown under the plot is needed. Each entry in the table is the average of the two values of shade corresponding to the same level of material quality and oxidation temperature. For example, 203.5 is the average of the two values (189 and 218) obtained when temperature was low and

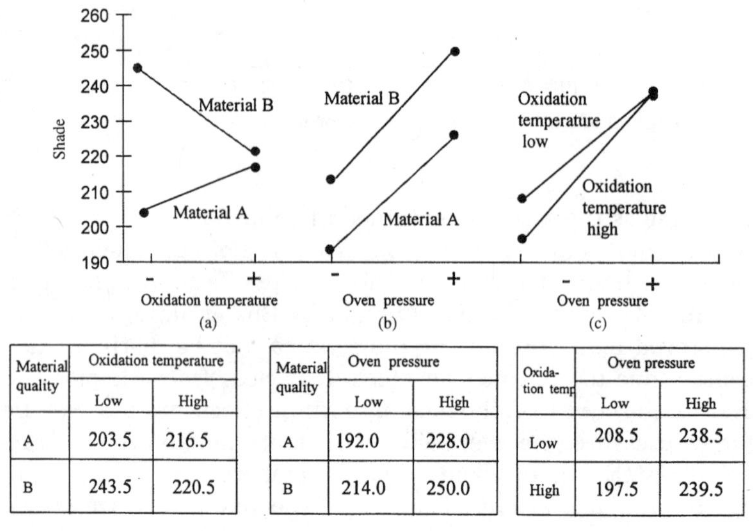


| Material quality | Oxidation temperature | |
|---|---|---|
| | Low | High |
| A | 203.5 | 216.5 |
| B | 243.5 | 220.5 |

| Material quality | Oven pressure | |
|---|---|---|
| | Low | High |
| A | 192.0 | 228.0 |
| B | 214.0 | 250.0 |

| Oxida-tion temp | Oven pressure | |
|---|---|---|
| | Low | High |
| Low | 208.5 | 238.5 |
| High | 197.5 | 239.5 |

**Figure 5.14** Interaction plots for factors in dye process experiment.

material quality was A. Referring to the cube for the experiment should assist in determining the values that correspond to the same levels of the factors.

The response plot is then constructed by plotting shade versus oxidation temperature separately for each material quality. The numbers 203.5 and 216.5 are connected by a straight line to display the relationship between shade and oxidation temperature when material quality A is used. A straight line between 243.5 and 220.5 displays the same relationship when material B is used. The usefulness of the plot depends on the assumption that shade increases approximately linearly between the two extremes of oxidation temperature for both levels of material quality.

The remaining two plots shown in Fig. 5.14 are constructed in the same manner. Each of the three plots could be shown with the factors reversed, as in Fig. 5.9.

The lines in the response plot shown in Fig. 5.14*a* are not parallel. This indicates that the factors of material quality and oxidation temperature interact. Since the lines in the response plots shown in Figs. 5.14*b* and *c* are approximately parallel, this indicates that oven pressure does not interact with either material quality or oxidation temperature. Oven pressure, however, does have an important effect by itself on shade. These results are consistent with those determined from the design matrix in Table 5.7.

Since oven pressure does not interact with either material quality or oxidation temperature, its effect can be shown alone on a response

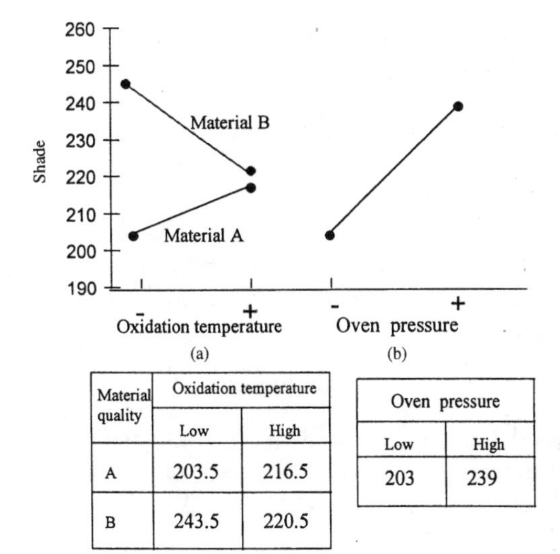


| Material quality | Oxidation temperature | |
|---|---|---|
| | Low | High |
| A | 203.5 | 216.5 |
| B | 243.5 | 220.5 |

| Oven pressure | |
|---|---|
| Low | High |
| 203 | 239 |

**Figure 5.15** Response plots for the important effects in dye process experiment.

plot. This response plot is obtained by plotting the average shade versus oven pressure when oven pressure is low and high and connecting the points with a straight line. The average shade when oven pressure is low, 203, is computed by averaging the four shade figures on the front face of the cube. The average shade when oven pressure is high is 239, obtained by averaging the four shade figures on the back side of the cube. This plot is shown in Fig. 5.15*a*.

As the number of factors included in the experiment increases, the number of response plots needed to plot all combinations of two factors becomes rather large. To reduce the number of plots necessary, only those effects determined to be important from the design matrix are plotted. Therefore, in this example, one plot of average shade versus oven pressure is needed since oven pressure is an important effect but does not interact with either of the other two factors. Another plot is needed to display the interaction between material quality and oxidation temperature. These plots are both shown in Fig. 5.15.

### Conclusions for Example 5.2

1. Increases in oven pressure increase shade at a rate independent of material quality and oxidation temperature (within the bounds of the experiment).

2. Oxidation temperature and material quality interact. Increases in oxidation temperature increase shade with material quality A but decrease shade with material B.

3. Running the process at high oxidation temperature would make the process less sensitive to variation in material quality and would result in a more uniform shade.

4. After the oxidation temperature is set high, the oven pressure could be set near the low level used in the experiment to obtain the desired dye shade level of 200.

5. Follow-up to this study should be a verification of the average shade level and of the reduced variation in shade as a result of using a high oxidation temperature to dampen the effect of variation in material.

The next example illustrates a study with four factors.

**Example 5.3: A $2^4$ Experiment for the Design of a Solenoid Valve** A manufacturer of solenoid valves designed a solenoid to be used with a pollution control device on an automotive engine. The solenoid was used to turn the pollution control device on and off. Engineers responsible for the design of the solenoid ran a $2^4$ factorial design to determine the effect of some of the important components in the solenoid valve on the flow (measured in cubic feet per minute, or cfm) of air from the valve. Flow is an important quality characteristic of the valve. The results of the study would be used to set specifications for components of the solenoid. The four factors that were studied and their levels are shown in Table 5.8. Figure 5.16 contains the planning form for the study.

The run charts for the averages and standard deviations of flow from the four tests of each of the 16 combinations appear in Fig. 5.17. Both charts have some patterns worth noting. The chart for averages shows eight points all near 0.72. All eight points are associated with a bobbin depth of 1.105, indicating the important effect of the factor of bobbin depth on flow. The relatively small variation of the eight points led the engineers to believe that some other component in the solenoid was preventing the flow from exceeding 0.73. Further tests were planned to investigate this possibility.

Three large standard deviations (0.1 cfm or greater) appear in the run chart. A check of data on the background variables did not identify any special causes of variation. The three were all associated with long

**TABLE 5.8 Factors for the Solenoid Experiment, Example 5.3**

| Factor | Level – | Level + |
|---|---|---|
| $A$ = Length of the armature, in | 0.595 | 0.605 |
| $S$ = Spring load, g | 70 | 100 |
| $B$ = Bobbin depth, in | 1.095 | 1.105 |
| $T$ = Length of the tube, in | 0.500 | 0.510 |

1. **Objective:**
Study the effects that four important components of the solenoid valve have on flow. This information will be used to determine manufacturing specifications for the components.

2. **Background information:**
Previous experiments using fractional factorial designs (see Chapter 6) indicated that the four components chosen for this experiment had the largest effect on flow. Experience in manufacturing has shown that bobbin depth is one of the most difficult dimensions to control.

3. **Experimental variables:**

A.

| Response variables | Measurement technique |
|---|---|
| 1. Average and standard deviation of flow (cfm) | Flow meter 65c |

B.

| Factors under study | Levels | |
|---|---|---|
| 1. Armature length (in) | 0.595 | 0.605 |
| 2. Spring load (g) | 70 | 100 |
| 3. Bobbin depth (in) | 1.095 | 1.105 |
| 4. Tube length (in) | 0.500 | 0.510 |

C.

| Background variables | Method of control |
|---|---|
| 1. Environment | Pressure, humidity, and temperature recorded at time of test |
| 2. Resistance of wire | Control chart used at the winders to monitor resistance |
| 3. Flow tester | Calibration checked at the beginning, middle, and end of the test |

4. **Replication:**
Four solenoids were assembled for each of the 16 conditions in the study.

5. **Methods of randomization:**
The order in which the 16 combinations were assembled was randomized using a random permutation table. The four solenoids for each combination were all assembled at the same time. The order of test was the same as the order of assembly to save time.

6. **Design matrix:** (attach copy)
(See Table 5.9.)

7. **Data collection forms:**
(Not shown here)

8. **Planned methods of statistical analysis:**
Compute the average and standard deviation of the four flow readings for each of the 16 combinations. Analyze the effects of the factors on both these statistics.

9. **Estimated cost, schedule, and other resources:**
Study can be run in one day along with normal production.

**Figure 5.16** Planning form for the solenoid study.

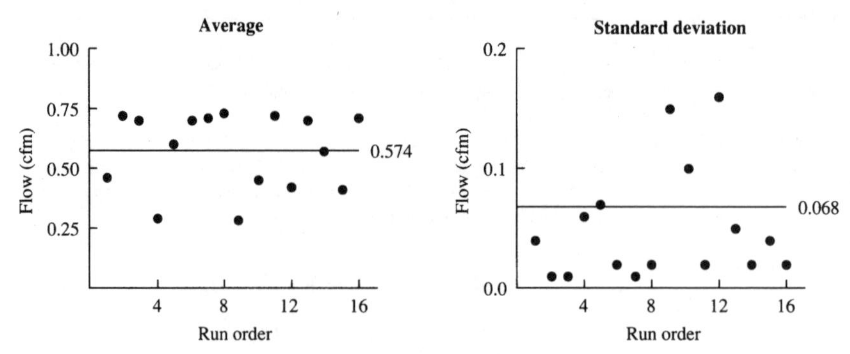


**Figure 5.17** Run charts for the solenoid study.

armature length and short bobbin depth, so the engineers attributed them to an interaction between the factors. This preliminary conclusion would be substantiated during subsequent analysis.

#### Design matrix and dot diagrams

Table 5.9 contains the design matrix for the solenoid experiment. Figure 5.18 shows the effects of the factors and the dot diagrams. Effects of the factors on both the average flow and the standard deviation of flow were estimated from the design matrix. From Fig. 5.18, it is seen that bobbin depth had a large effect on average flow. Bobbin depth, armature length, and their interaction were the most important factors affecting the standard deviation of flow. This interaction was observed in the run chart.

For a $2^4$ design, the dot diagrams can be modified to provide a display of the variation due to nuisance variables. The magnitudes of the four three-factor interactions and the single four-factor interaction can usually be assumed to be primarily the result of nuisance variables. The smaller the effect of nuisance variables, the closer these higher-order interactions will be to zero. Figure 5.19 contains a modified dot diagram of the effects on the standard deviation of flow. In the figure, the effects are spread out horizontally, beginning with the effects of the individual factors and ending with the four-factor interaction. The vertical line alerts the analyst to the range of variation in the effects that could be expected due to nuisance variables.

#### Analysis of the cubes

Figure 5.20 shows the two cubes representing the 16 combinations in the experiment. The paired comparisons do not indicate the presence of any special causes of variation and therefore confirm the effects that were found using the design matrix. The eight paired compar-

**TABLE 5.9 Design Matrix and Results for the Solenoid Study**

| Test | Run order | $A$ | $S$ | $B$ | $T$ | $AS$ | $AB$ | $AT$ | $SB$ | $ST$ | $BT$ | $ASB$ | $AST$ | $ABT$ | $SBT$ | $ASBT$ | Flow, cfm $\overline{X}$ | Flow, cfm $s$ |
|---|---|---|---|---|---|---|---|---|---|---|---|---|---|---|---|---|---|---|
| 1 | 1 | − | − | − | − | + | + | + | + | + | + | − | − | − | − | + | 0.46 | 0.04 |
| 2 | 12 | + | − | − | − | − | − | − | + | + | + | + | + | + | − | − | 0.42 | 0.16 |
| 3 | 14 | − | + | − | − | − | + | + | − | − | + | + | + | − | + | − | 0.57 | 0.02 |
| 4 | 10 | + | + | − | − | + | − | − | − | − | + | − | − | + | + | + | 0.45 | 0.10 |
| 5 | 8 | − | − | + | − | + | − | + | − | + | − | + | − | + | + | − | 0.73 | 0.02 |
| 6 | 7 | + | − | + | − | − | + | − | − | + | − | − | + | − | + | + | 0.71 | 0.01 |
| 7 | 13 | − | + | + | − | − | − | + | + | − | − | − | + | + | − | + | 0.70 | 0.05 |
| 8 | 3 | + | + | + | − | + | + | − | + | − | − | + | − | − | − | − | 0.70 | 0.01 |
| 9 | 15 | − | − | − | + | + | + | − | + | − | − | − | + | + | + | − | 0.42 | 0.04 |
| 10 | 9 | + | − | − | + | − | − | + | + | − | − | + | − | − | + | + | 0.28 | 0.15 |
| 11 | 5 | − | + | − | + | − | + | − | − | + | − | + | − | + | − | + | 0.60 | 0.07 |
| 12 | 4 | + | + | − | + | + | − | + | − | + | − | − | + | − | − | − | 0.29 | 0.06 |
| 13 | 6 | − | − | + | + | + | − | − | − | − | + | + | + | − | − | + | 0.70 | 0.02 |
| 14 | 16 | + | − | + | + | − | + | + | − | − | + | − | − | + | − | − | 0.71 | 0.02 |
| 15 | 11 | − | + | + | + | − | − | − | + | + | + | − | − | − | + | − | 0.72 | 0.02 |
| 16 | 2 | + | + | + | + | + | + | + | + | + | + | + | + | + | + | + | 0.72 | 0.01 |
| Divisor = 8 | | | | | | | | | | | | | | | | | $\overline{\overline{X}} = 0.574$ | $\overline{S} = 0.068$ |

$\overline{\overline{X}}$ = Overall average
$\overline{S}$ = Pooled standard deviation

| Factor or interaction | Estimate (cfm) $\bar{X}$ | $s$ |
|---|---|---|
| *A* | −.08 | .03 |
| *S* | .04 | −.02 |
| *B* | .28 | −.06 |
| *T* | −.04 | .00 |
| *AS* | −.03 | −.02 |
| *AB* | .07 | −.05 |
| *AT* | −.03 | −.01 |
| *SB* | −.04 | .02 |
| *ST* | .01 | .00 |
| *BT* | .04 | .00 |
| *ASB* | .03 | .02 |
| *AST* | −.01 | −.01 |
| *ABT* | .04 | .02 |
| *SBT* | .00 | −.01 |
| *ASBT* | .01 | .01 |

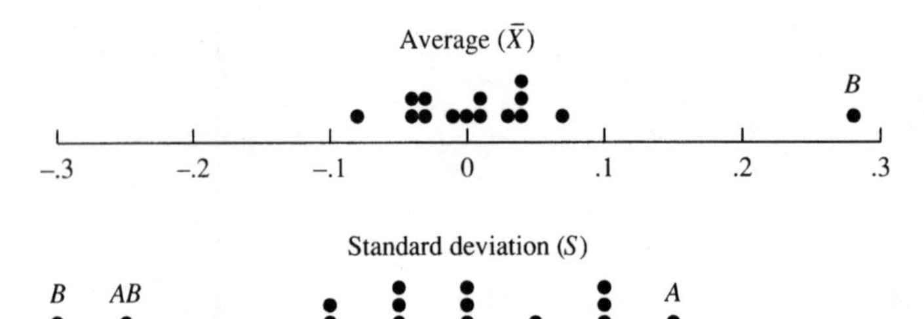


**Figure 5.18** Effects of factors on flow.

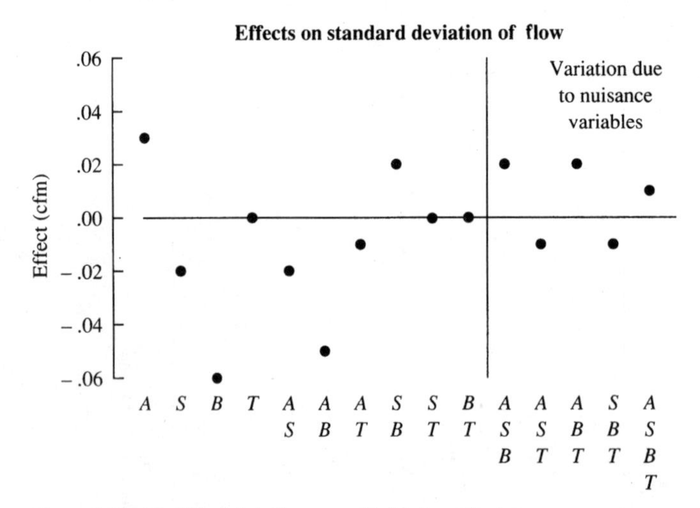


**Figure 5.19** Modified dot diagram of effects of factors.

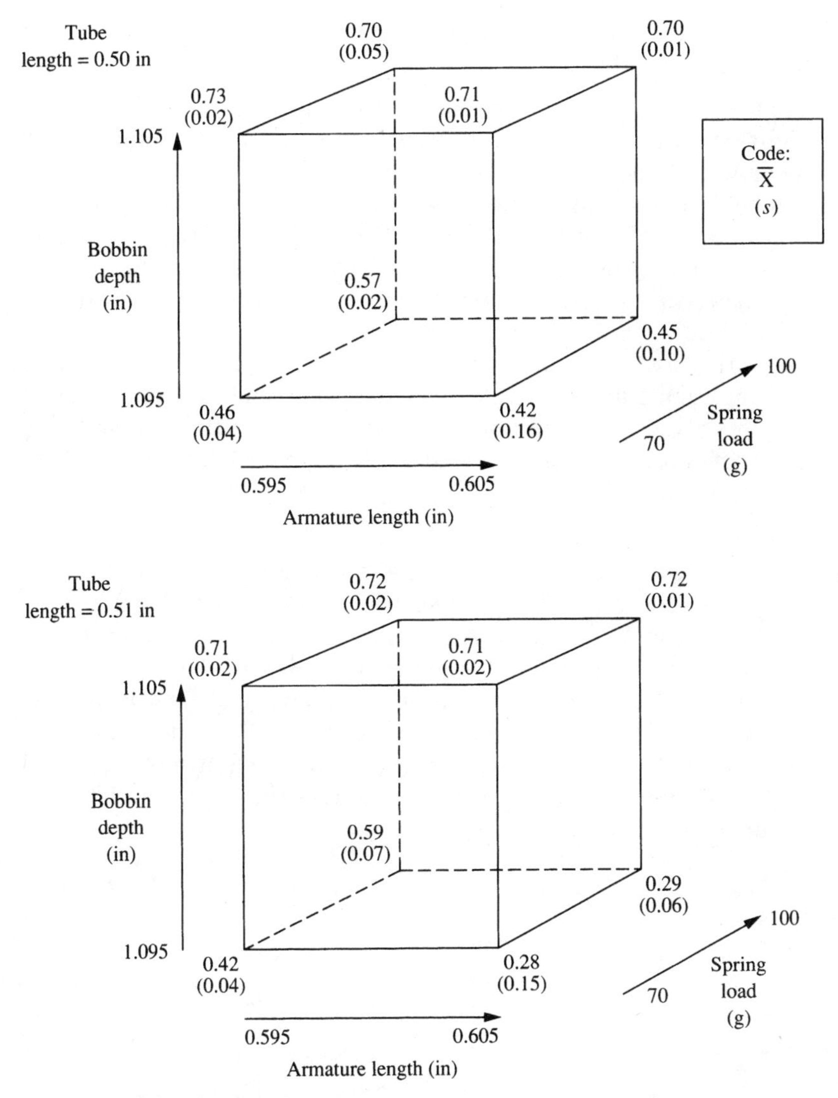


**Figure 5.20** Cubes for the solenoid experiment.

isons for each factor (four for each cube) are obtained in the same manner as for the $2^3$ design except for the factor of tube length. The paired comparisons for tube length are found by comparing the corresponding corners of the two cubes; e.g., compare the result at the bottom front left corner of the top cube, 0.46, to the result at the corresponding corner of the bottom cube, 0.42. The consistency of the effects can be evaluated by using these plots.

The following are examples of observations based on evaluating the consistency of the effects.

1. As bobbin depth increases, the average flow increases for each of the eight combinations of the factors. This indicates an important positive effect of bobbin depth on the average flow.
2. As bobbin depth increases, there is no effect on the standard deviation of flow for low levels of armature length, and a consistent positive effect for high levels of armature length. This indicates an important armature length–bobbin depth interaction on the standard deviation of flow.
3. As spring load increases, the standard deviation of flow increases for two conditions, decreases for five conditions, and is approximately zero for the remaining condition. This indicates that there is probably not an important effect of spring load on the standard deviation of flow.

### Response plots

Figure 5.21 shows the response plots summarizing the important results of the experiment. The response plots are constructed just as they were for the $2^3$ design. Since bobbin depth was the only factor with a substantial effect on average flow, there is a single response plot for this factor. When the standard deviation of flow is the response variable, there is an interaction between bobbin depth and armature length. Therefore, two plots of the standard deviation of flow versus bobbin depth are needed to summarize the data (one for each level of armature length).

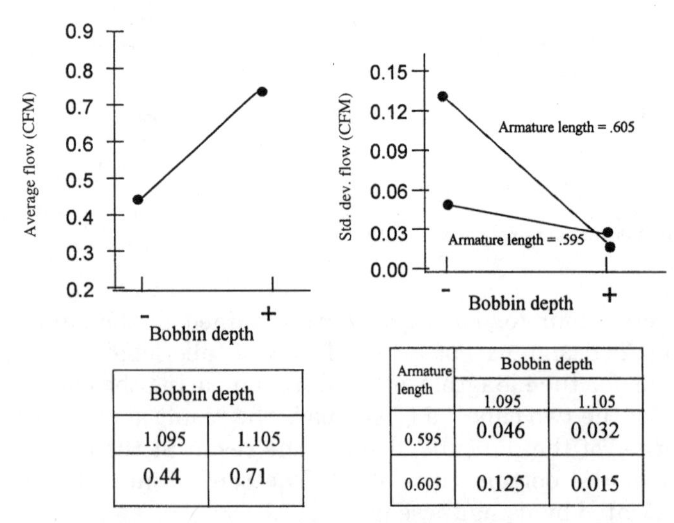


| Bobbin depth | |
|---|---|
| 1.095 | 1.105 |
| 0.44 | 0.71 |

| Armature length | Bobbin depth | |
|---|---|---|
| | 1.095 | 1.105 |
| 0.595 | 0.046 | 0.032 |
| 0.605 | 0.125 | 0.015 |

**Figure 5.21** Response plots for the solenoid experiment.

The four standard deviations corresponding to a given level of bobbin depth and armature length are pooled together rather than averaged. (For more details on pooling standard deviations, see the appendix to Chap. 7.)

Conclusions for Example 5.3, design of a solenoid valve:

1. There were no special causes of variation detected in the run chart or the analysis of the paired comparisons in the cube.
2. Bobbin depth is the only factor in the experiment that had a substantial effect on the average flow of the solenoid. The longer the bobbin depth, the greater the flow of the valve.
3. The standard deviation of flow was affected by bobbin depth and armature length. These two factors interacted with each other. The combination of a long armature and a short bobbin produced large variations in flow.
4. The control of bobbin depth will be critical in the manufacture of the solenoid. The armature length should be specified at 0.595 in to minimize variation.

## 5.2 Design of Factorial Experiments

In Chap. 3, the principles of experimental design were discussed. In this section, the discussion centers on how the following relate to factorial designs:

- Choosing the number of factors
- Choosing the levels for each factor
- Considering background variables in the design
- Selecting the amount of replication
- Randomizing the order of the tests

### Choosing the number of factors

The number of factors to be included in the experiment will obviously depend on the objective and the available resources. In the early stages of experimentation, when little is known about the process or the product, the objective of the experiment may be to screen out the unimportant variables. If it is reasonable to assume that only a few of the variables will be important, a fractional factorial design should be considered. These designs will be discussed in Chap. 6. There are some particularly useful fractional factorial designs for screening up to 16 factors.

If it is desired to study in depth the relationship between the factors and the response variables, including interactions between the factors, two to five factors should be chosen for the study. A full factorial design is most useful for two to four factors. A fractional factorial design should be considered for five factors.

If the objective of the experiment includes the study of the important factors under a range of conditions to increase degree of belief, a chunk variable could be included as one of the factors. A chunk factor was defined in Chap. 3 as a combination of background variables.

The difficulty of changing levels is another consideration for how many factors should be in an experiment. To change the levels of some factors may require little more than a turn of a dial, and their effect on the response is seen very quickly. Other factors are hard to change. A physical change to the equipment may take hours or days to accomplish. For some processes, it may take hours to reach equilibrium after changes in operating conditions have been made. It is usually unwise to run a study in which more than half of the factors are hard to change. In that case, it would be advisable to run a series of smaller experiments.

Table 5.10 contains a summary of these ideas.

### Choosing the levels for each factor

Choice of the specific levels used for each factor will be based on knowledge of the process or product and the conditions of the study. It is desirable to set the levels of the factors far enough apart that the effects of the factors will be large relative to the variation caused by the nuisance variables. However, the levels should not be so far apart that there is a good chance for trouble to develop. Examples of such trouble are the following:

- Conditions that make the experiment run unsafely
- Conditions that cause substantial disruption of a manufacturing facility

**TABLE 5.10 Choosing the Number of Factors**

| Objective | Number of factors | Design |
|---|---|---|
| Screen out unimportant factors | 5 or more | Fractional factorial |
| Study important factors in depth | 2–4<br>5 | Factorial<br>Fractional factorial |
| Increase degree of belief | 2–4 | Factorial with a chunk variable |
| | 5 | Fractional factorial with a chunk variable |

- Important nonlinearities or discontinuities hidden between the levels
- Substantially different cause-and-effect mechanisms for the different conditions in the experiment

In some experiments, it will not be possible to hold a factor constant at the desired level because the factor cannot be controlled that precisely. Although it is desirable, it is not necessary that a factor be held constant at the planned level. However, the variation in the factor should be small relative to the distance between the planned levels. If this cannot be achieved, the factor should be made a background variable and measured during the experiment.

When continuous variables such as temperature, pressure, and line speed are used as factors, then a run at the center of the design can be used to check for discontinuities or nonlinear effects. For example, suppose the following factors and levels were used in a factorial design:

| Factor | Level | |
|---|---|---|
| | − | + |
| Temperature (°C) | 220 | 240 |
| Pressure (lb/in$^2$) | 50 | 80 |
| Concentration (%) | 10 | 12 |

The center point of this design is

| | Center point |
|---|---|
| Temperature | 230 |
| Pressure | 65 |
| Concentration | 11 |

This point would be in the middle of the cube. (See Chap. 8 for more on the use of center points.)

The experimenter will also need to take into account the amount of unusable material generated during the study and the safety of those conducting the experiment. Box and Draper (1969) discuss ways of designing experiments for processes already in production. In their approach, levels are set close together to minimize unacceptable product. The small effect of a factor when the levels are close together is overcome by running many experimental units per factor combination. This usually can be done when the process is in full production. This approach is especially useful when most of the special causes have been removed from the process. Chap. 8 contains further discussion of this situation.

### Considering background variables in the design

As indicated in Chap. 3, there are two important decisions to be made concerning background variables:

1. How to control the background variables so that the effects of the factors are not distorted by them
2. How to use the background variables to establish a wide range of conditions for the study to increase degree of belief

When a sequential approach to a study is taken, it is often useful to hold background variables constant and study the effects of the factors. To design the next study, the background variables should be used to establish conditions for the study that differ substantially from those of the previous study. This will allow the experimenters to determine if the effects of the factors are consistent over a wide range of conditions. The following example illustrates the use of background variables to widen the range of conditions.

**Example 5.4: A Welding Process Experiment** Consider an experiment in a welding operation for assembling an automotive part. Previous improvement cycles using fractional factorial designs indicated that the two most important factors were pressure and vacuum. Improvement cycles using control charts indicated that day-to-day variation in the process, the environmental conditions, and operator technique were important background factors. The response variable of interest was a particular dimension of the assembled part.

The primary aim of the experiment was to study the effects of pressure and vacuum over a wide range of conditions to increase degree of belief in their effects. Table 5.11 identifies the design used to accomplish this aim.

Day-to-day variation and operator technique were incorporated into a chunk factor. This chunk factor is used along with pressure and vacuum to form a $2^3$ factorial design. Because of the nature of the chunk factor, randomization is restricted to each chunk. The four tests in chunk 1 are done in random order, followed by the four tests in chunk 2 in random order.

If neither pressure nor vacuum is found to interact with the chunk factor, then an increased degree of belief will result that the estimated effects of pressure and vacuum can be used in the future. That is, the effects can be used to set specifications for pressure and vacuum and also be used as a guide to adjust the process (e.g., to counteract a special cause).

If either pressure or vacuum is found to interact with the chunk factor, then the appropriate setting of either pressure or vacuum can be used to mitigate the effect of the chunk factor, i.e., day-to-day variation and operator technique. This will lead to improved consistency of the welding process.

The analysis of this experiment is discussed in Sec. 5.3.

**TABLE 5.11 Design for Welding Experiment**

*Response variable:* Dimension

| Factors | Levels | |
|---|---|---|
| Pressure, lb/in² | 30 | 45 |
| Vacuum, inHg | 8 | 10 |

*Background variables* (combined into one chunk factor):
Day-to-day variation
Operator technique

*Chunk factor:*
Level 1: Monday, operator 1
Level 2: Wednesday, operator 2

*Design:* $2^3$ factorial

| Factors | Levels | |
|---|---|---|
| Pressure | 30 | 45 |
| Vacuum | 8 | 10 |
| Chunk | 1 | 2 |

*Note:* Students of planned experimentation will recognize this as a split-plot design. The primary reason for considering the design as a split plot is to facilitate selection of the proper error term to use in the analysis of variance. The use of graphical methods of analysis reduces the importance of distinguishing the design as a split-plot arrangement.

### Selecting the amount of replication

Replication is the primary means of studying the stability of the effects of the factors and increasing the degree of belief in the effects. Different types of replication that can be used in an experiment were discussed in Chap. 3. In factorial experiments, replication is built into the experimental pattern. For example, in a $2^4$ design there are eight comparisons of the high and low levels of each factor. These comparisons are made under the eight different combinations of the other three factors.

The amount of replication that is feasible depends partly on how difficult it is to change the levels of the factors. When multiple factors are included in an experiment, between 8 and 16 runs are desirable. If the aim of the study is to improve a process, and the effects of the factors cannot be distinguished from the nuisance variables in 8 to 16 runs, then the wrong factors or levels have probably been chosen.

More important than the number of runs is the range of conditions included in the study. Replication over similar conditions provides little increase in the degree of belief.

### Randomizing the order of the tests

Randomization is the primary means of controlling the effect of nuisance variables. If blocks are constructed to control background variables, randomization should be performed within each block. Randomization should be considered for choosing the order of the tests, assigning the combinations of the factors to experimental units, and choosing the order in which the measurements are made. (Because of practical considerations, the order of measurements is sometimes constrained to be the same as the order of the tests.)

In factorial experiments, sometimes randomization is not practical because of the difficulty of changing the levels of some of the factors. In such a case, it is desirable to repeat one or more of the earlier runs at the end of the experiment. These replications can be used to check for special causes of variation that may have occurred during the experiment.

There is less need to randomize when experiments are conducted on stable processes. For processes dominated by special causes, randomization is an important tool. See Daniel (1976, pp. 22–26) for an excellent discussion of the use of randomization in factorial experiments.

## 5.3 Analysis of Factorial Experiments

In Sec. 5.1 analyses of data from factorial designs with two, three, and four factors were illustrated. This section provides more information on the analysis of factorial designs. The emphasis is on the analysis of factorial designs when things do not go as planned. Also, some additional details about the various graphical displays used in the earlier examples will be given.

The approach to the analysis of factorial designs in the previous examples followed four steps:

1. Plot the data in run order to look for trends and obvious special causes. If an uncontrolled background variable has been measured, then the data should also be plotted in increasing order of measurement of the background variable.

The purpose of the remaining three steps is to partition the variation seen in the run chart between the factors, background factors, and nuisance variables.

2. Estimate the effects of the factors using the design matrix, and plot the estimates on a dot diagram. This step provides a preliminary assessment of the factors that contribute most to the variation in the run chart.

3. Analyze the data using a square, a cube, or sets of cubes to study the variation in the paired comparisons that make up the effects estimated in step 2. A study of the paired comparisons will help identify the impact of nuisance variables on the variation and will indicate the presence of special causes.
4. Summarize the results of the analysis by preparing response plots for the important factors and interactions.

The aim of these four steps is to partition the variation in the measurements of the response variable among the factors, the background variables, the nuisance variables, and any interactions between them. This partitioning, along with other knowledge of the product or process, allows the experimenter to determine the most advantageous approach to improvement.

This analysis can be carried out on the individual measurements of the response variable or on a statistic chosen based on knowledge of the process or product. If multiple experimental units are measured for each combination of the factors, statistics such as the average or range may be computed. The effects of the factors on these statistics can then be estimated.

For example, in a machining operation it may be simple to obtain the desired average dimension but difficult to find ways to reduce the variation. In this case, several experimental units would be measured for each condition, the range of the measurements computed, and the four-step analysis carried out using the range as the response of interest.

Taguchi (1987) suggests combining various statistics into what he calls a *signal-to-noise ratio* and using the signal-to-noise ratio as the response of interest. Kackar (1985) provides an excellent summary of these ideas. Box (1988) discusses the use of signal-to-noise ratios and transformation.

### Graphical displays for the analysis of factorial designs

The following graphical displays have proven useful in analyzing data from factorial designs:

- Run charts
- Dot diagrams
- Geometric figures or other displays of paired comparisons
- Response plots

Each of these displays has been illustrated in the examples. In this section, some additional insight into the use of these displays will be given.

**Run chart.** The run chart is usually a display of the measurements of the response variable in the order that the tests were made. Other orderings, such as order of measurement or increasing order of a measured background variable, are also useful. The run chart displays the total variation in the response variable that is to be partitioned among the factors, background variables, and nuisance variables.

Analysis of the run chart begins this partitioning. Special causes of variation due to nuisance variables can be spotted. If some type of replication has been included in the design, the individual measurements can be plotted in the run chart. Then the run chart can be used to assess the magnitude of either common or special causes of variation resulting from nuisance variables. If the experiment has been run in blocks, the blocks should be designated on the run chart. The variation in the measurements of the response variable that is due to the background variables comprising the blocks can be assessed. (See, e.g., Fig. 5.7.) Sometimes the run chart will point out that the variation is primarily due to one dominant factor because the data on the run chart stratify into two groups. (See run chart for averages, Fig. 5.17.)

**Dot diagrams.** The second step in the analysis of data from a factorial experiment is to estimate the effects of the factors from the design matrix and to plot the estimates on a dot diagram. The dot diagram is used to obtain a finer partitioning of the variation than is possible in the run chart. The dot diagram identifies factors whose effects are clearly separated from the variation due to nuisance variables.

Data from a factorial design will vary because of the factors, background variables, and nuisance variables. Effects clustered near zero cannot be distinguished from variation due to nuisance variables. Nuisance variables and background variables can impact a study as either common or special causes of variation.

If a three- or four-factor interaction is one of the largest (in absolute value) effects in the dot diagram, then

- It may not be possible to separate any of the effects of the factors from variation due to nuisance variables.
- A special cause may be dominating the estimates of the effects (the run chart and the cubes should be checked).
- The interaction may be important (the least likely alternative).

For a $2^4$ design, a modified dot diagram can be used to better distinguish the effects of nuisance variables. (See Fig. 5.19.)

Box et al. (1978, p. 328) use reference distributions in combination with dot diagrams to separate variation due to nuisance variables from variation due to factors. Daniel (1976, chap. 5) uses normal probability plots to graphically accomplish this separation.

Regardless of the method of analysis, caution should be exercised in placing a high degree of belief in small, although discernible effects. If an effect is small compared to the variation caused by other factors, background variables, or nuisance variables, then it is liable to change substantially in the future because of interactions with the conditions of the study. A high degree of belief in small effects can be established only through a synthesis of knowledge of the subject matter and replication of the effect over a wide variety of conditions.

By following these simple guidelines one can make the dot diagram more useful for separating the important effects from variation due to nuisance variables:

1. Construct the scale on the diagram such that zero is at the center point and the endpoints of the scale are symmetric (for example, −50, 50). The endpoints should be chosen such that all effects can be plotted on a linear scale.
2. Effects too close together to be plotted side by side should be plotted one above the other, as in a histogram.
3. All effects that are separated from the cluster around zero should be labeled.
4. If a complete factorial design is replicated in two or more blocks, the effects should be computed separately for each block. The effects from each of the blocks are all plotted on the same dot diagram.

**Geometric figures and plots of paired comparisons.** Factorial designs can be depicted using a square or one or more cubes. From these geometric figures, the paired comparisons that make up the factorial designs can be identified and studied. Analysis of the paired comparisons determines the factors with the greatest effect on the response variable. This is the same information obtained from the dot diagram. However, study of the paired comparisons provides some important additional information.

By studying the individual paired comparisons for consistency, the experimenter can ascertain the presence of special causes in the data. The existence of special causes in the data may not be evident in the run chart because of the variation attributable to changes in the factors. If special causes of variation exist, they may also not be evident in the dot diagram, but they will influence the estimates of the effects. Example 5.7 (to follow) illustrates the existence of a special cause that was identified by analysis of the paired comparisons.

In Example 5.2, the paired comparisons were analyzed by identifying them on the cube, listing them (Table 5.5), and plotting them (Fig. 5.12). Whether the effects are listed or plotted, the experimenter can analyze them for consistency; it is not necessary to do both. The aim of the analysis is the same in each case: to confirm that special causes of variation are not distorting the effects estimated from the design matrix.

**Response plots.** Response plots are used to help experts in the subject matter visualize the impact of the effects of the factors found to be important in steps 2 and 3. The number of response plots constructed depends on the factors and interactions found to be important in steps 1, 2, and 3. The following items should be considered when response plots are constructed:

- If a factor is found to be important but does not interact with any other factor, the response plot consists of one line approximating the relationship between the factor and the response. (See, e.g., Fig. 5.15*a*.)
- If two factors A and B interact, their relationship to the response variable is displayed on one response plot by using two lines. The relationship between the response variable and factor A when B is at the minus level is displayed by one line. The relationship between the response variable and factor A when B is at the plus level is displayed by another line. (See, e.g., Fig. 5.15*b*.) Once this response plot has been constructed, there is no need to construct individual response plots of the type in Fig. 5.15*a* for each of factors A and B.
- If two factors A and B interact and, in addition, one or both of them interact with a third factor C, the response plot displaying the interaction of A and B must be done separately for the two levels of C (see, e.g., Fig. 6.11).

The response plots constructed for a particular experiment provide a visual model of the relationships between the important factors and the response variable. The response plots are interpreted as follows:

1. The response plots are only a linear approximation of the relationship; therefore, their usefulness is limited based on the linear relationship of the data. The use of center points to check for lack of linearity has been mentioned in this chapter and will be discussed in greater detail in Chap. 8.

2. The slope of the response plot is proportional to the effect estimated from the design matrix.

3. If two factors A and B interact and the interaction is displayed using two lines on the response plot (as in Fig. 5.15*b*), then the differ-

ence in the slopes is proportional to the interaction between A and B. If A and B do not interact, then the two lines will be parallel.

4. The slopes of the lines in a response plot are their most important aspect. The intercepts of the lines depend on the levels of other factors in the study that are not in the response plot. As constructed in this chapter, the intercepts correspond to the other factors being set midway between the minus and plus levels.

In some cases, the analysis of data from a factorial design can proceed in a straightforward manner through the four steps listed at the beginning of this section. Often, the analysis of the data is not so straightforward because of problems such as the presence of special causes of variation, a large amount of variation caused by nuisance variables, or one or more data points missing. The primary purpose of the remainder of this section is to provide some guidance for the analysis of data from factorial experiments when things "go wrong."

### Analysis of factorial designs when things go wrong

**Special causes in the run chart.** There are times when trends or other obvious special causes are seen in the run charts. It is not advisable to proceed further with the analysis in such cases until the special cause is determined. The potential interactions of the special cause with the factors should also be assessed. If the cause is identified, the data can possibly be adjusted to remove its effect. Regardless of the method of analysis, extreme care should be taken in extrapolating the results of the experiment. Verification of a conclusion by future experiments is almost always necessary.

**Example 5.5: Study of Automotive Emissions** A $2^4$ factorial design was run to study the effects of four factors on emissions from an automotive engine. The run charts for hydrocarbon (HC) and carbon dioxide (CO) concentrations are shown in Fig. 5.22. The plot of HC concentration shows no trends or special causes. The plot of CO concentration shows an obvious

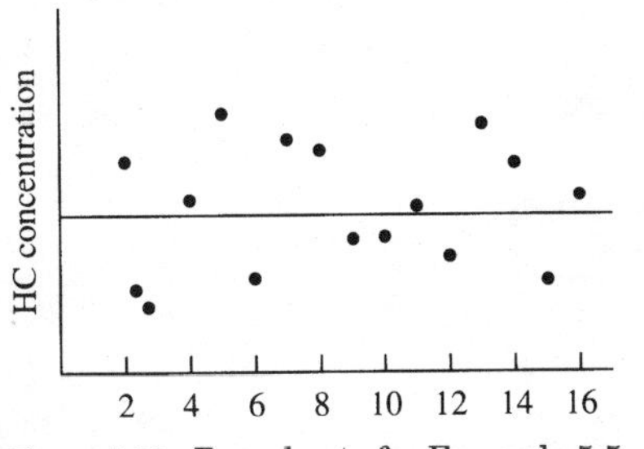


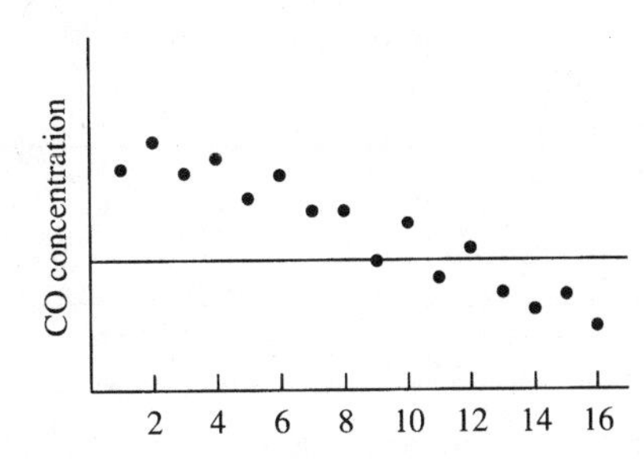


**Figure 5.22** Run charts for Example 5.5.

downward trend. This trend accounts for a large portion of the variation in the run chart.

Analysis of the data for HC concentration should proceed only on the advice of experts in the process. It is their responsibility to determine if the nuisance variable affecting the CO concentration could also be affecting the measurements of HC concentration.

If the nuisance variable affecting the CO readings can be identified, its effect could be quantified and the CO data adjusted to remove the trend.

**Example 5.6: Design of a Throttle Return Mechanism** In a study to determine the durability of a throttle return mechanism, the following three factors, which related to the pin in the mechanism, were included:

| Factor | Level | |
|---|---|---|
| Pin coated | No | Yes |
| Pin length | – | + |
| Pin diameter | – | + |

A $2^3$ design was used, in which the response variable was the number of cycles to failure on a stress test. Figure 5.23 contains the run chart. Figure 5.24 shows the data displayed on a cube.

From the run chart it is seen that the number of cycles obtained on the first run (corresponding to the conditions of pin coated, plus length, and plus diameter) differ substantially from the rest of the data. These conditions represent an extreme case because each of the factors is set at the level that the engineer believed would provide increased durability. It is not unusual that extreme points produce results of greater magnitude than the sum of the individual effects.

In this experiment a special fixture had to be set up to run the tests. Since the extreme result was the first run, questions were

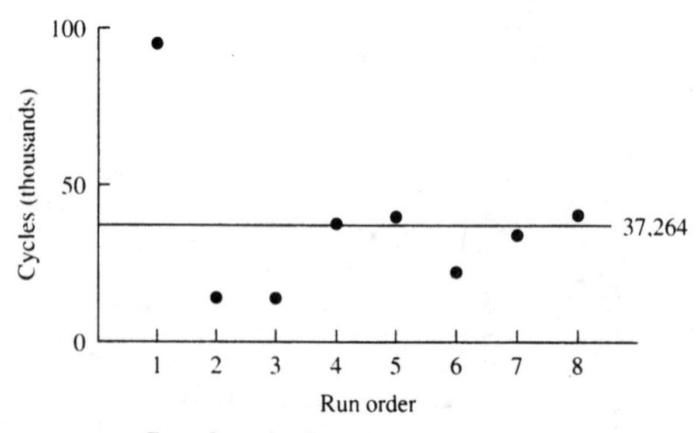


**Figure 5.23** Run chart for Example 5.6.

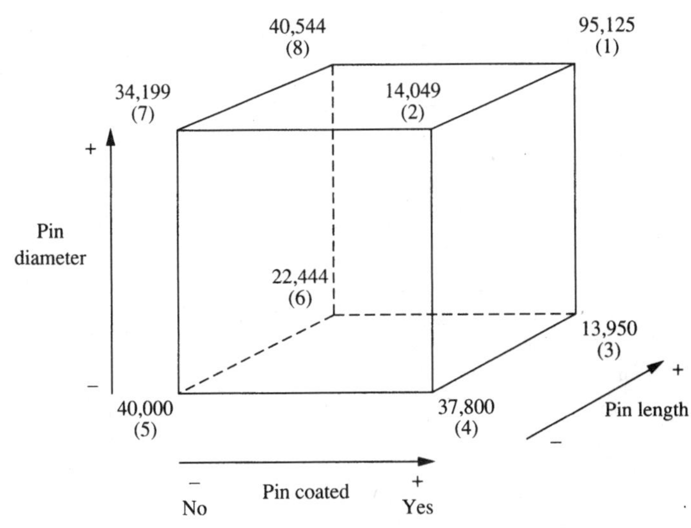


**Figure 5.24** Cube for the $2^3$ factorial design in Example 5.6.

raised about the setup and break-in of the fixture. It was decided to verify the results by running a paired comparison of the two extreme conditions, three pluses and one minus.

**Special causes buried in the data.** A special cause of variation in the data from a factorial experiment may not be evident in a run chart. Since the experimental conditions differ for each run, a point that appears in agreement with the rest of the data from the experiment may be quite abnormal. Analysis of the data using a square or a cube is the primary means of determining this type of special cause. An example will help clarify this idea.

**Example 5.7: Pilot Line for a Tile Process** On a pilot line built to study a new process of making flooring tile, a $2^3$ design was used to determine the effect of three factors on an important quality characteristic of the tile. The factors are related to the setup of the production line.

The pilot line was run for 1 h at each of the eight conditions. During each hour, 10 tiles were selected, and the quality characteristic was measured. Based on an analysis of the run chart of the readings for each condition, it was deemed appropriate to summarize the data for each condition by an average and a standard deviation.

Figure 5.25 is the run chart of the averages. No obvious special causes are seen in the run chart. The data are displayed on a cube in Fig. 5.26. The paired comparisons, plotted in Fig. 5.27, show that of the four comparisons for each factor, three are consistent and one is

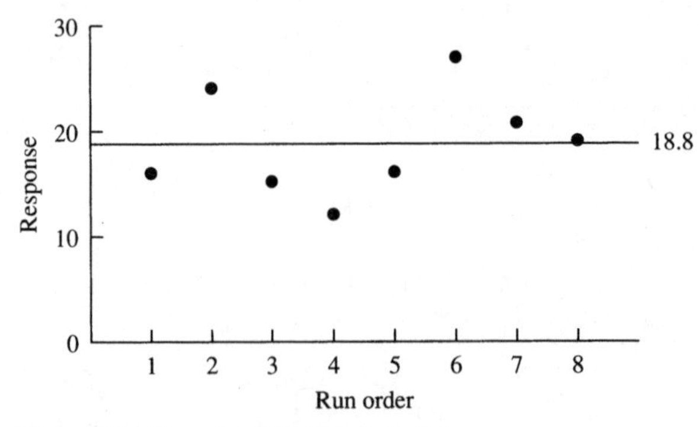


Figure 5.25 Run chart for Example 5.7.

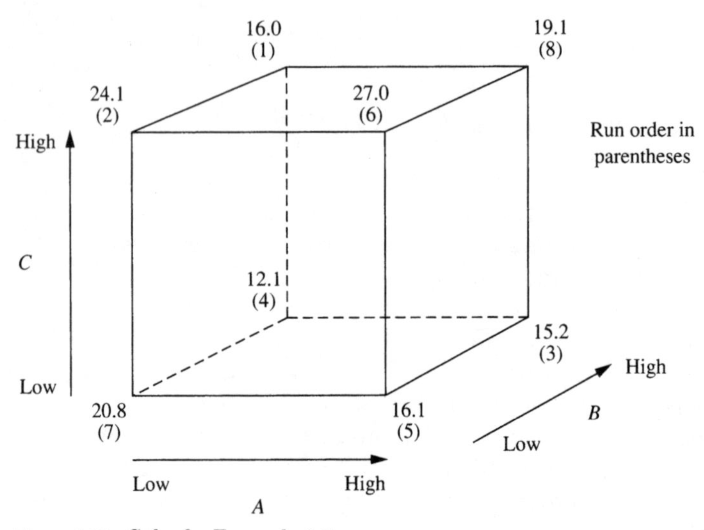


Figure 5.26 Cube for Example 5.7.

quite different. As they are plotted in Fig. 5.27, the first, second, and third comparisons for factors A, B, and C, respectively, are quite different from the other three in each plot. These comparisons have one point in common; it is the fifth run. This run was performed at the conditions A = +, B = −, C = − and resulted in an average response of 16.1. Based on an examination of the other comparisons, an average response of 23 to 24 would have been consistent with the rest of the data. After a check of the notes kept during the experiment, it was found that some trouble was encountered running the line at these conditions.

The analysis of the cube uncovered the possible existence of a special

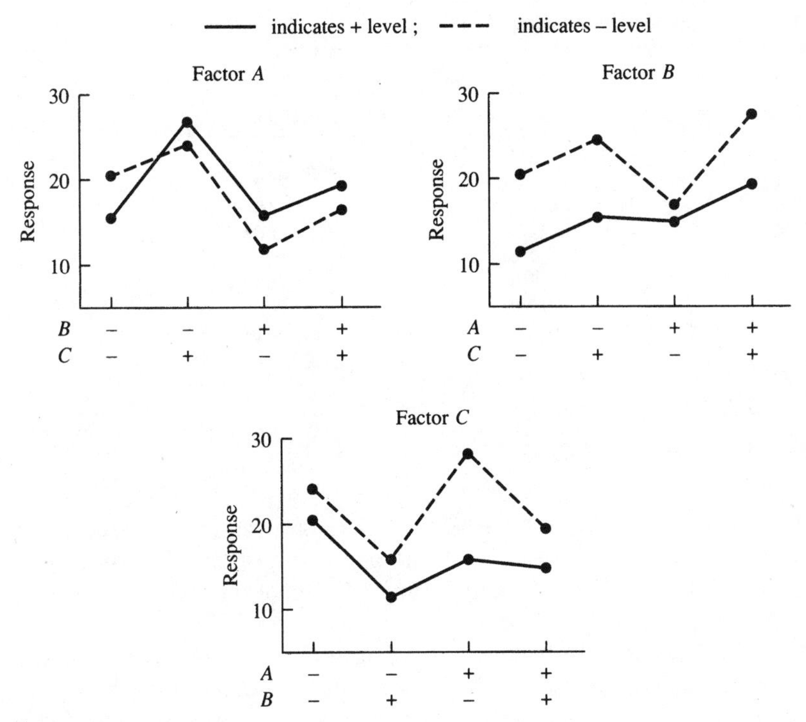


**Figure 5.27** Plot of paired comparisons in Example 5.7.

cause, although none was evident from the run chart. This provides an increased opportunity to learn about the process. This example illustrates the importance of detailed analysis of the cube prior to estimating the factor effects. If only the average effects estimated from the design matrix are used, some important information concerning special causes may be lost.

**Missing data.** It occasionally happens that the data from one of the experimental conditions are missing. When this happens, it is important to determine whether the fact that the data are missing is related to the experimental conditions. For example, the data may be missing because the product could not be made under those conditions.

If the data are missing because of the experimental conditions, the analysis should proceed, using the cube to determine better conditions under which to run the process.

Sometimes data are missing for reasons that have nothing to do with the experimental conditions (e.g., lost or damaged samples). In this case, the missing data can be estimated by using the highest-

**TABLE 5.12 Estimation of Missing Value**

| Test | (a) 123 | (a) Response | (b) 123 | (b) Response |
|---|---|---|---|---|
| 1 | − | 189 | − | 189 |
| 2 | + | 228 | + | 228 |
| 3 | + | 218 | + | 218 |
| 4 | − | 259 | − | 259 |
| 5 | + | 195 | + | $x$ |
| 6 | − | 200 | − | 200 |
| 7 | − | 238 | − | 238 |
| 8 | + | 241 | + | 241 |
| Divisor = 4 | −1 | | Set = 0 and solve for $x$ ($x$ = 199) | |

level interaction in the design matrix (e.g., a three-factor interaction for a $2^3$ design). Table 5.12*a* contains a portion of the design matrix from Example 5.2, the $2^3$ design for a dye process, which is taken from Table 5.7.

The three-factor interaction from Table 5.12*a* is −1. Table 5.12*b* shows a procedure for estimating a missing value. Assume, e.g., that the data for test 5 are missing, for reasons that have nothing to do with the experimental conditions of that test. Since the data are missing, the design matrix cannot be used to estimate effects. Since the value of the interaction is expected to be small, the missing data value $x$ is estimated by setting $x$ equal to a value that will make the interaction equal to 0. In this case, that value is 199. This is close to 195, the actual value measured. After the missing value is estimated, the analysis can proceed as usual.

**Nuisance variables dominate.** There are occasions when the variation in the data from a factorial experiment is caused primarily by nuisance variables rather than by changes in the factors under study. The nuisance variables may be either common or special causes of variation.

The experimenter can be alerted to the presence of nuisance variables dominating the variation in the data by the presence of one or more of the following conditions:

- Variation in the run chart that is of similar magnitude to the variation seen in an existing control chart for the process
- A measurement system with variation that is large relative to the variation in the data

- Three- or four-factor interactions that are large relative to average effects of the factors
- Unexplained inconsistencies in the paired comparisons embedded in the cube
- Estimates of effects that run counter to the theory that is held with a high degree of belief

When it is suspected that nuisance variables dominate the variation, the following steps can be taken:

- Examine the measurement system. If it is stable but too imprecise, average multiple measurements per experimental unit. If it is unstable, remove the special causes and repeat the experiment.
- If the variation caused by the nuisance variables is from common causes (use the run chart and the cube to determine this), average multiple experimental units per combination of the factors.
- If there are special causes present, identify them and control them in subsequent experiments.
- Widen the range between the levels of the factors.
- Use statistical process control to reduce variation in the process.

The following example illustrates a situation where the nuisance variables in an experiment are suspected of dominating the variation.

**Example 5.8: Study of a Process for Batch Mixing** An experiment was conducted in a process of batch mixing to determine the effects of three factors on the increase in viscosity of the material during mixing. The three factors were mixing speed *S*, mixing time *T*, and the viscosity of the material before mixing *V*. Figure 5.28 contains the run chart. No obvious time trends

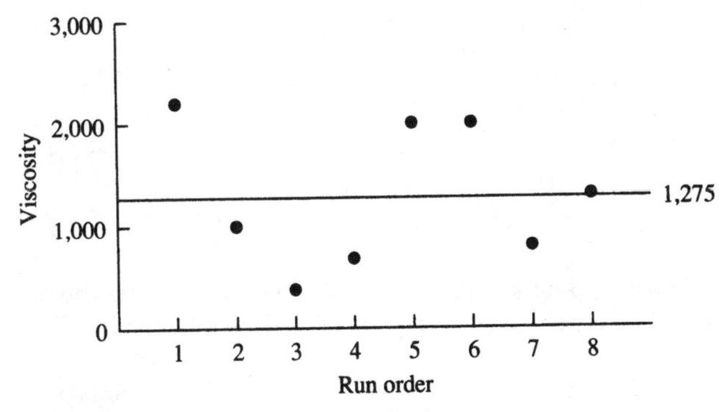


**Figure 5.28** Run chart for Example 5.8.

**TABLE 5.13 Design Matrix for the Experiment on Batch Mixing, Example 5.8**

| Test | Run order | *S* | *T* | *V* | *ST* | *SV* | *TV* | *STV* | Viscosity |
|---|---|---|---|---|---|---|---|---|---|
| 1 | 3 | − | − | − | + | + | + | − | 380 |
| 2 | 1 | + | − | − | − | − | + | + | 2200 |
| 3 | 5 | − | + | − | − | + | − | + | 2000 |
| 4 | 2 | + | + | − | + | − | − | − | 1000 |
| 5 | 8 | − | − | + | + | − | − | + | 1300 |
| 6 | 6 | + | − | + | − | + | − | − | 2000 |
| 7 | 4 | − | + | + | − | − | + | − | 680 |
| 8 | 7 | + | + | + | + | + | + | + | 800 |
| Divisor | = 4 | | | | | | | | |
| Effect | | 410 | −350 | −200 | −850 | 0 | −560 | 560 | |

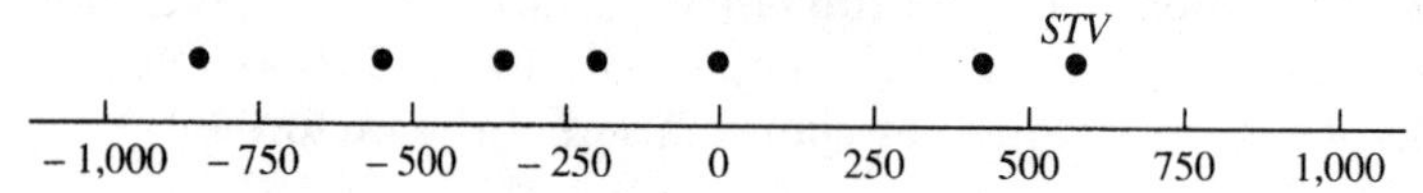


**Figure 5.29** Dot diagram for Example 5.8.

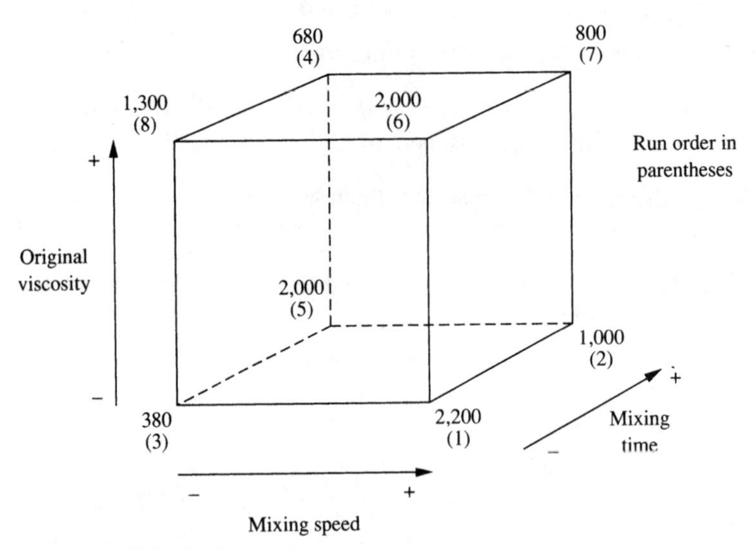


**Figure 5.30** Cube for Example 5.8.

or outlying values are observed on the run chart. Table 5.13 contains the design matrix, and Fig. 5.29 shows the dot diagram. The data are displayed on a cube in Fig. 5.30.

Based on the magnitude of the three-factor interaction, the inconsistencies of the paired comparisons in Fig. 5.33, and the fact that there is no strong evidence of one run's being abnormal, nuisance variables are suspected of dominating the variation. To confirm the domination of nuisance variables, the experiment could be repeated, or a control chart for batches mixed at constant settings of the factors could be developed.

**Factors interact with conditions.** The conditions under which an experiment is run can never be exactly duplicated. Certainly, conditions under which the results are used will not match precisely the conditions of the experiments. The need to run experiments in analytic studies over a wide range of conditions was stressed in Chap. 3. In Sec. 5.2, the combining of background variables into a chunk variable to include a wide range of conditions was discussed. An illustration of this method was given in Example 5.4, which is continued now.

**Example 5.4 (Continued): A Welding Process** This example examines two factors, vacuum and pressure, to determine their effects on a dimension (in thousandths of an inch) of a welded part. In addition, a chunk variable was formed from the background variables, operator technique and day-to-day variation. The chunk variable was used as the third factor in a $2^3$ design. For each of the eight combinations of the factors, parts were welded. Table 5.14 shows the design matrix, results, and estimated effects for the study.

Figure 5.31 contains the initial run chart of the dimension for each part. No evidence of trends or special causes unrelated to the combinations of the factors is seen. However, significant variation between the eight parts within each combination of factors is present.

Since the variation within each combination of factors is rather large, the analysis was carried out on both the average and the standard deviation of the eight parts. The run charts for the averages and standard deviations appear in Fig. 5.32. No special causes seem to be present.

**TABLE 5.14 Design Matrix for Welding Experiment**

| Test | Run order | $P$ | $V$ | $C$ | $PV$ | $PC$ | $VC$ | $PCV$ | $\overline{X}$ | $s$ |
|---|---|---|---|---|---|---|---|---|---|---|
| 1 | 1 | − | − | − | + | + | + | − | 505 | 8.8 |
| 2 | 3 | + | − | − | − | − | + | + | 514 | 5.5 |
| 3 | 4 | − | + | − | − | + | − | + | 504 | 9.8 |
| 4 | 2 | + | + | − | + | − | − | − | 511 | 6.8 |
| 5 | 6 | − | − | + | + | − | − | + | 505 | 9.4 |
| 6 | 8 | + | − | + | − | + | − | − | 508 | 4.5 |
| 7 | 7 | − | + | + | − | − | + | − | 503 | 6.4 |
| 8 | 5 | + | + | + | + | + | + | + | 505 | 4.7 |
| Divisor | = 4 | | | | | | | | | |
| Effect: | $\overline{X}$ | 5.3 | −2.3 | −3.3 | −0.8 | −2.8 | −0.3 | 0.3 | | |
| | $s$ | −3.2 | −0.1 | −1.5 | 0.9 | −0.1 | −1.3 | 0.7 | | |

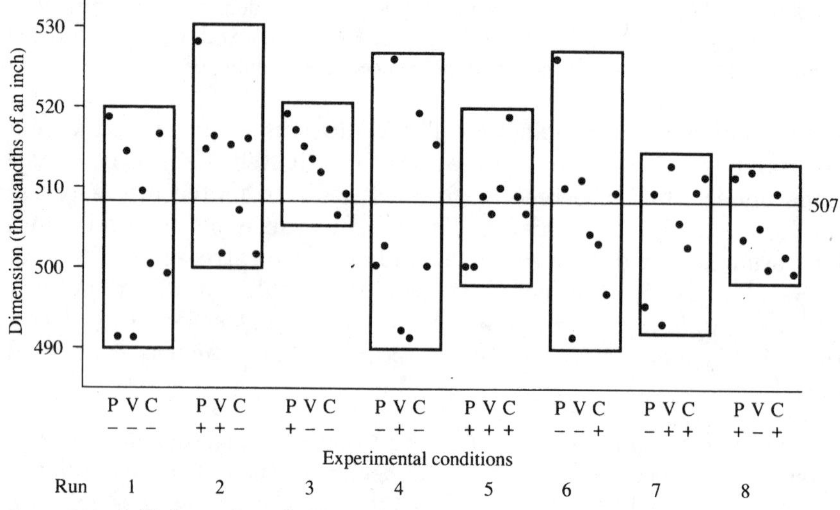


**Figure 5.31** Initial run chart for Example 5.4.

Figure 5.33 contains the dot diagrams for the estimated effects for both the average and standard deviation of the dimension. The averages and standard deviations for each run are displayed on a cube in Fig. 5.34.

For the averages, it is seen in Table 5.14 and Fig. 5.33 that pressure, the chunk factor, and the interaction between chunk and pressure are the largest effects. In all cases, increased pressure resulted in a larger average dimension. However, the magnitude of the effect depends on the setting of the chunk factor. The chunk factor has less of an effect at the minus level of pressure (30 lb/in$^2$) than at the plus level (45 lb/in$^2$). Therefore, the effect of the chunk factor could be mitigated by setting the pressure at 30 lb/in$^2$.

In Table 5.14 and Fig. 5.33, it can also be seen that pressure is the factor that has the largest effect on the standard deviation of the dimension. Since pressure does not interact with the chunk, a high degree of belief is established that increasing pressure will reduce short-term variation. Unfortunately, this is the opposite of the pressure setting needed to mitigate the effect of the chunk factor.

The most desirable result would have been that the same setting of pressure reduced both the effect of the chunk factor and the short-term variation. A better consequence than the present result would have been if pressure were an important factor in only one of these situations; then, at least, a setting for pressure could be determined to take advantage of that fact.

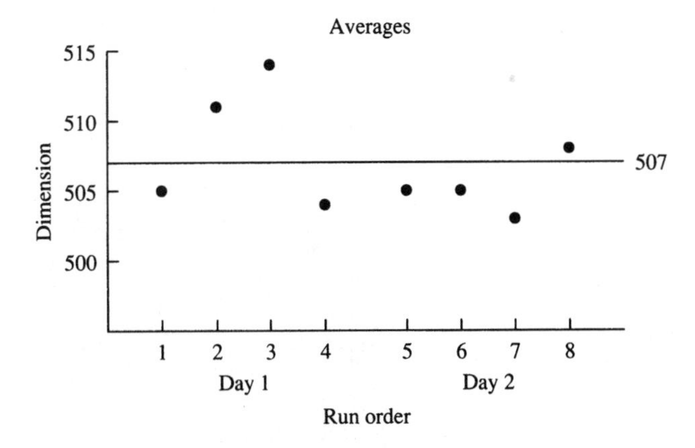


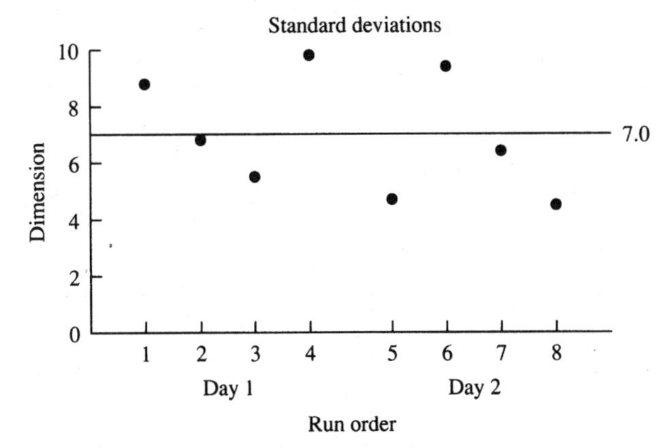


**Figure 5.32** Run charts of averages and standard deviation for Example 5.4.

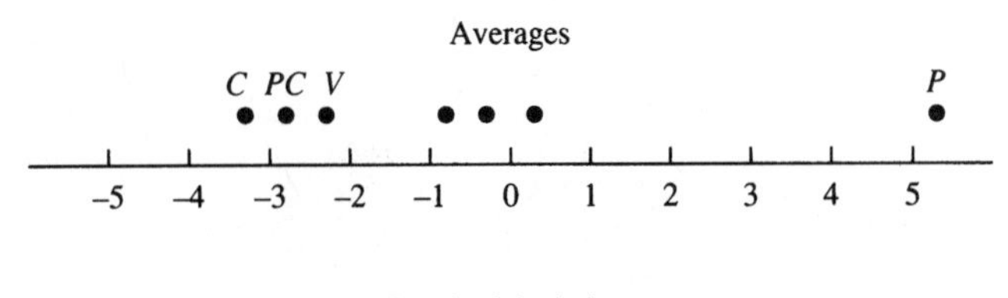


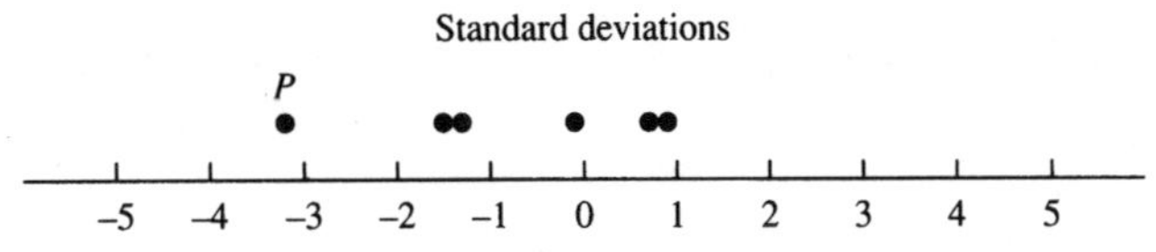


**Figure 5.33** Dot diagrams for Example 5.4.

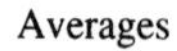


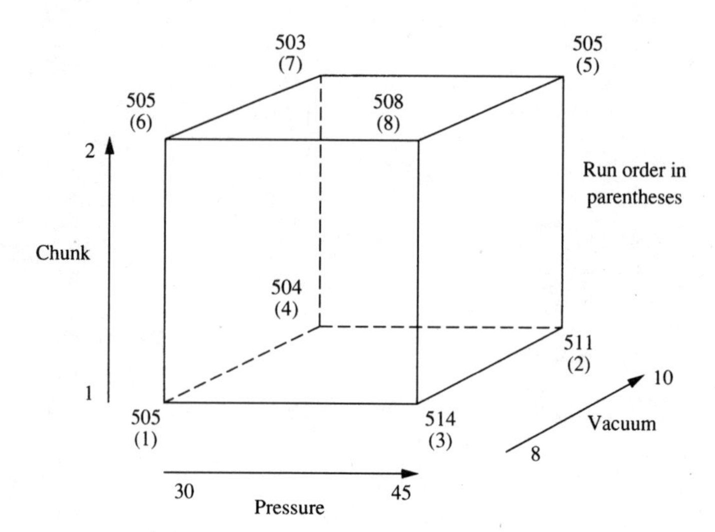


Standard deviations

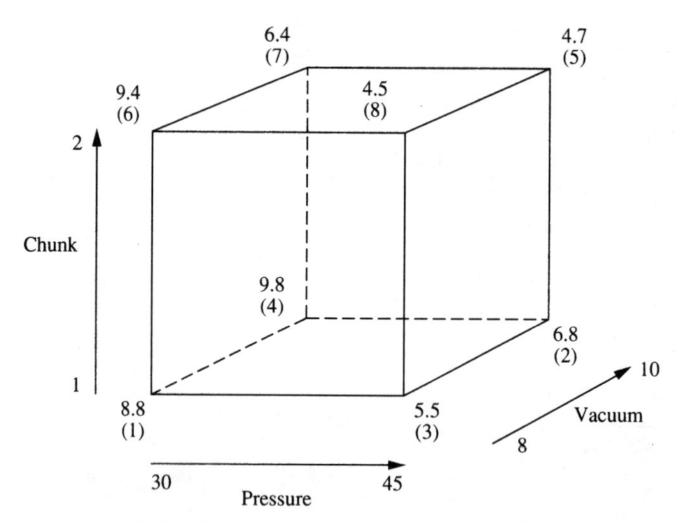


**Figure 5.34** Cubes for welding experiment.

However, based on the results of this experiment, an easy decision on the setting for pressure is not possible. It might be advisable to begin by setting the pressure at the plus level (30 lb/in$^2$) to reduce the short-term variation and then to monitor the variation caused by the chunk factor (day-to-day variation and operator technique) using a control chart. The variation due to the chunk factor would then be reduced if special causes of variation could be identified and removed.

Figure 5.35 contains the response plots for the averages and standard deviations, which summarize the results of the experiment.

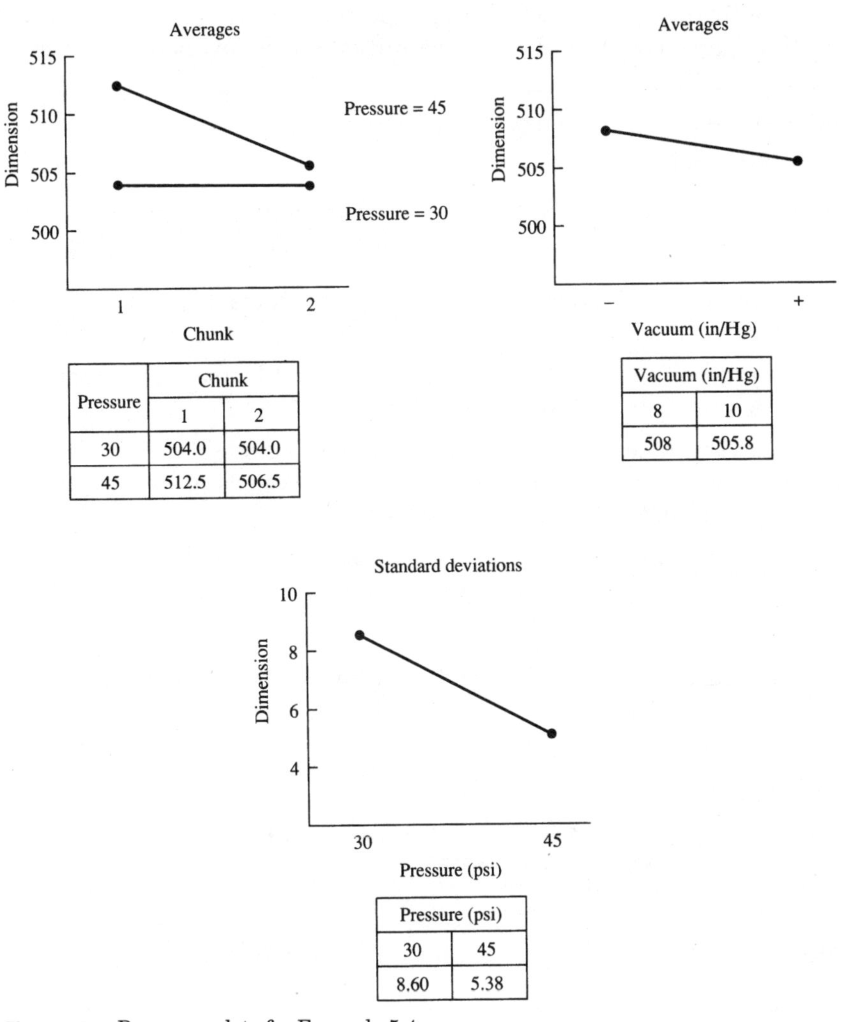


| Pressure | Chunk | |
|---|---|---|
| | 1 | 2 |
| 30 | 504.0 | 504.0 |
| 45 | 512.5 | 506.5 |

| Vacuum (in/Hg) | |
|---|---|
| 8 | 10 |
| 508 | 505.8 |

| Pressure (psi) | |
|---|---|
| 30 | 45 |
| 8.60 | 5.38 |

**Figure 5.35** Response plots for Example 5.4.

When a factor interacts with a chunk variable, as was the case in the preceding example, there are several possible courses of action. It may be that the effects of background and nuisance variables on the process must be reduced before a reasonable degree of belief in the estimates of the effects of the factors can be established. This usually can be done through the use of control charts.

It may also be the case that certain settings of the factors can mitigate the effect of background and nuisance variables. This idea was pioneered by Taguchi (1987). In Fig. 5.35, the response plots show the result that the chunk factor has less of an effect when pressure is low. If it is feasible in this case to set the pressure at the low level and confirm this result by future experiments, the effect of variation in the chunk factor could then be mitigated by low settings of pressure.

## 5.4 Summary

In this chapter, factorial designs of experiments were presented. Some of the important features of factorial designs are as follows:

1. They are most useful to determine the effects of multiple factors (two, three, or four) on a response variable.
2. In addition to the effects of each individual factor, interactions between the factors can be studied.
3. Studying factors at two levels requires relatively few runs, leads to a simple analysis, and meets most of the needs of experiments to make improvements in processes and systems.
4. Graphical displays (run charts, cubes, paired comparisons, and response plots) can be used with factorial designs to make the analysis easy to understand.

The references that follow contain additional discussion of factorial designs.

## References

Box, G. (1988): "Signal-to-Noise Ratios, Performance Criteria, and Transformations," *Technometrics,* vol. 30, no. 1, February.

Box, G., and N. Draper (1969): *Evolutionary Operation,* Wiley, New York.

Box, G., W. Hunter, and J. S. Hunter (1978): *Statistics for Experimenters,* Wiley, New York.

Daniel, C. (1976): *Applications of Statistics to Industrial Experimentation,* Wiley, New York.

Kackar, R. N. (1985): "Offline Quality Control, Parameter Design, and the Taguchi Method," *Journal of Quality Technology,* vol. 17, pp. 176–188.

Taguchi, G. (1987): *System of Experimental Design,* Unipub/Krause International Publications, White Plains, N.Y.

## Exercises

**5.1** Choose a product or process with which you are familiar. For an important quality characteristic, list potential causes that affect either the average level or the variation of the quality characteristic. (Consider use of a cause-and-effect diagram to do this.) Plan a factorial experiment to determine the effects of some of the potential causes. Use the planning form.

**5.2** The data in the following table were the result of an experiment to determine the effect that three factors—holding pressure, booster pressure, and screw speed—had on part shrinkage in an injection molding process. The data in the table are coded values of part shrinkage.

| | Holding pressure | | | |
|---|---|---|---|---|
| | − | | + | |
| | Booster pressure | | Booster pressure | |
| Screw speed | − | + | − | + |
| − | 21.9<br>20.3<br>(2) | 15.9<br>16.7<br>(8) | 22.3<br>21.5<br>(1) | 17.1<br>17.5<br>(6) |
| + | 16.8<br>15.4<br>(7) | 14.0<br>15.0<br>(3) | 27.6<br>27.4<br>(5) | 24.0<br>22.6<br>(4) |

(*a*) Analyze the data to determine the important effects.

(*b*) Within the bounds of the study, which settings of the three factors minimize shrinkage?

(*c*) Based on the analysis in (*b*), suggest another design to see if better settings can be found.

(*d*) Assume that uniformity of the parts is the goal, with minimum shrinkage a secondary objective. Assume also that the inability to keep holding pressure constant during operation is a major cause of variation in the dimension of the parts. Choose settings to increase uniformity of the parts by desensitizing the process to variations in holding pressure.

**5.3** In the solenoid experiment (Example 5.3) an additional response variable was measured. The response variable was pull-in voltage, which is the voltage needed to activate the solenoid. The average and the standard deviation of the pull-in voltage for the four solenoids at each combination of the factors are indicated in the design matrix below. Analyze the data to determine the effects of the four factors.

| Test | Run order | *A* | *S* | *B* | *T* | Pull-in voltage Average | Standard deviation |
|---|---|---|---|---|---|---|---|
| 1 | 1 | − | − | − | − | 5.66 | 0.26 |
| 2 | 12 | + | − | − | − | 5.77 | 0.15 |
| 3 | 14 | − | + | − | − | 7.96 | 0.21 |
| 4 | 10 | + | + | − | − | 8.18 | 0.52 |
| 5 | 8 | − | − | + | − | 7.33 | 0.18 |
| 6 | 7 | + | − | + | − | 6.92 | 0.15 |
| 7 | 13 | − | + | + | − | 10.79 | 0.39 |
| 8 | 3 | + | + | + | − | 10.30 | 0.54 |
| 9 | 15 | − | − | − | + | 5.50 | 0.22 |
| 10 | 9 | + | − | − | + | 6.19 | 0.56 |
| 11 | 5 | − | + | − | + | 8.27 | 0.44 |
| 12 | 4 | + | + | − | + | 8.42 | 0.11 |
| 13 | 6 | − | − | + | + | 8.83 | 0.22 |
| 14 | 16 | + | − | + | + | 6.93 | 0.11 |
| 15 | 11 | − | + | + | + | 10.57 | 0.42 |
| 16 | 2 | + | + | + | + | 10.42 | 0.20 |

**5.4** List some factors related to your customers (internal or external) that could potentially affect their needs. These factors could be demographic, such as age, sex, work experience, and place of residence. The factors might pertain to how the customer uses your product or service. Choose two to four of these factors, and define two levels for each. Construct a factorial pattern using the factors and add a center point. Use the pattern as a basis for selecting customers to interview to determine their present and future needs.

**5.5** The following artificial data are taken from Daniel (1976, p. 65) to illustrate the effect of one "bad value" in a factorial design. Find the test that seems to be influenced by a special cause. First make a run chart and identify the most likely special cause. Then put the data on a cube and identify the point. Why is it easier to find the special cause on the cube? What value would be more consistent with the rest of the data?

| Test | Run order | Factor A | B | C | Response |
|---|---|---|---|---|---|
| 1 | 4 | − | − | − | 158 |
| 2 | 5 | + | − | − | 132 |
| 3 | 2 | − | + | − | 212 |
| 4 | 8 | + | + | − | 136 |
| 5 | 3 | − | − | + | 264 |
| 6 | 7 | + | − | + | 188 |
| 7 | 1 | − | + | + | 318 |
| 8 | 6 | + | + | + | 242 |

Chapter

# 6

# Reducing the Size of Experiments

The number of runs required by a full factorial design increases geometrically as the number of factors increases. Factorial designs with two or three factors make efficient use of resources by using all the data to estimate the average effects and interactions. This is one of the strengths of factorial designs. However, as the number of factors increases, an increasing proportion of the data is used to estimate higher-order interactions. These interactions are usually negligible and therefore are of little interest to the experimenter.

For example, for a $2^4$ factorial design, estimates of four main effects, six two-factor interactions, four three-factor interactions, and one four-factor interaction can be made from the design. Only 10 of the 15 estimates available are likely to be of interest. Fractional factorial designs are an important class of experimental designs that allow the size of factorial experiments to be kept practical while still enabling the estimation of important effects.

The notation used for the experimental patterns in fractional factorial designs differs slightly from that used for full factorial designs. Whereas $2^7$ indicates a full factorial pattern using seven factors in 128 runs, $2^{7-4}$ indicates a 1/16 fraction of a $2^7$ pattern, using seven factors in eight runs. The notation for describing fractional factorial patterns comes from the following:

$$1/16 \times 2^7 = \frac{2^7}{2^4} = 2^{7-4}$$

Figure 6.1 shows two tabular displays for fractional factorial patterns involving seven factors. These patterns are (1) a 1/8 fraction of a $2^7$ pattern ($2^{7-3}$) in 16 runs and (2) a 1/16 fraction of a $2^7$ pattern ($2^{7-4}$) in 8 runs. The letters $A$ through $G$ indicate the seven factors,

A 1/8 fraction of a $2^7$ factorial design ($2^{7-3}$)

| | | | A − | | | | | | | | A + | | | | | | | |
|---|---|---|---|---|---|---|---|---|---|---|---|---|---|---|---|---|---|---|
| | | | B − | | | | B + | | | | B − | | | | B + | | | |
| | | | C − | | C + | | C − | | C + | | C − | | C + | | C − | | C + | |
| E | F | G | D − | D + | D − | D + | D − | D + | D − | D + | D − | D + | D − | D + | D − | D + | D − | D + |
| − | − | − | ■ | | | | | | | | | | | | | | | ■ |
| | | + | | | | ■ | | | | | | | | | ■ | | | |
| | + | − | | | | | | ■ | | | | | ■ | | | | | |
| | | + | | | | | | | ■ | | | ■ | | | | | | |
| + | − | − | | | | | | | ■ | | | ■ | | | | | | |
| | | + | | | | | | ■ | | | | | ■ | | | | | |
| | + | − | | | | ■ | | | | | | | | | ■ | | | |
| | | + | ■ | | | | | | | | | | | | | | | ■ |

A 1/16 fraction of a $2^7$ factorial design ($2^{7-4}$)

| | | | A − | | | | | | | | A + | | | | | | | |
|---|---|---|---|---|---|---|---|---|---|---|---|---|---|---|---|---|---|---|
| | | | B − | | | | B + | | | | B − | | | | B + | | | |
| | | | C − | | C + | | C − | | C + | | C − | | C + | | C − | | C + | |
| E | F | G | D − | D + | D − | D + | D − | D + | D − | D + | D − | D + | D − | D + | D − | D + | D − | D + |
| − | − | − | | | | | | | | | | | | | | ■ | | |
| | | + | | | | ■ | | | | | | | | | | | | |
| | + | − | | | | | | | ■ | | | | | | | | | |
| | | + | | | | | | | | | ■ | | | | | | | |
| + | − | − | | | | | | | | | | | ■ | | | | | |
| | | + | | | | | ■ | | | | | | | | | | | |
| | + | − | | ■ | | | | | | | | | | | | | | |
| | | + | | | | | | | | | | | | | | | | ■ |

**Figure 6.1** Tabular displays for fractional factorial patterns.

and the symbols − and + indicate the two levels of each factor. The darkened squares indicate the appropriate tests that would be included in the design matrix for the fractional factorial design.

The choice to use a full factorial design or a specific fractional factorial design depends on the level of current knowledge of the process or product. A *high level of knowledge* is indicated when all the factors in the experiment are known to have a substantial effect on the response. When there is a high level of knowledge, the aim of the study is usually to obtain detailed information on the effects of the factors and their interactions. This is best done with a full factorial design.

A *moderate level of knowledge* is indicated by a strong belief that most of, but probably not all, the factors have a substantial effect on the response. Fractional factorial designs for these cases will be discussed in this chapter. A *low level of knowledge* is indicated by a belief that only a small proportion of the factors has a substantial effect on the response, and it is not known which ones they are. Fractional factorial designs that are useful for screening out the unimportant factors will also be discussed in this chapter.

The analysis of data from a fractional factorial design follows the methods used for a full factorial design, given in Chap. 5:

- The run chart is used to screen for special causes of variation from nuisance variables.
- Effects are next estimated from the design matrix.
- Geometric displays (e.g., cubes) are used to identify special causes.
- Response plots are prepared for the important factors or interactions.

Section 6.1 presents the conceptual development of the experimental pattern for fractional factorial designs. An example of a study discussed in Chap. 5 is used to illustrate the analysis of fractional factorial designs. Section 6.2 presents fractional factorial designs to be used when current knowledge of the significance and interrelationship of the factors under study is at a moderate level. Section 6.3 presents screening designs for use when the level of current knowledge is low. Section 6.4 discusses the use of incomplete blocks with factorial and fractional factorial designs.

## 6.1 Introduction to Fractional Factorial Designs

Suppose the experimenters would like to reduce the size of the $2^4$ pattern in Table 6.1 to eight tests. It is important to choose the eight tests wisely so that as much information as possible about the effects of the factors is preserved. A simplistic approach is to choose the first eight tests in the design matrix. Inspection of the design matrix reveals that this results in a $2^3$ factorial pattern for factors 1, 2, and 3, with factor 4 held constant at its minus level. Another choice, choosing the even-numbered tests, results in a $2^3$ factorial pattern in factors 2, 3, and 4 with factor 1 set at the plus level.

Both of these approaches lead to unsatisfactory loss of information about one of the four factors. How should the eight tests be chosen? Intuitively, it seems desirable that the eight tests be chosen so that each of the columns in the design matrix has four minuses and four pluses. This selection would provide a balance to the design. It turns out that this balance can be achieved for 14 of the 15 columns in the design matrix in Table 6.1. This balance is obtained by giving up the ability to estimate one of the 15 effects represented by the column headings of the design matrix.

**TABLE 6.1 Design Matrix for a $2^4$ Factorial Pattern**

| Test | 1 | 2 | 3 | 4 | 12 | 13 | 14 | 23 | 24 | 34 | 123 | 124 | 134 | 234 | 1234 |
|---|---|---|---|---|---|---|---|---|---|---|---|---|---|---|---|
| 1 | − | − | − | − | + | + | + | + | + | + | − | − | − | − | + |
| 2 | + | − | − | − | − | − | − | + | + | + | + | + | + | − | − |
| 3 | − | + | − | − | − | + | + | − | − | + | + | + | − | + | − |
| 4 | + | + | − | − | + | − | − | − | − | + | − | − | + | + | + |
| 5 | − | − | + | − | + | − | + | − | + | − | + | − | + | + | − |
| 6 | + | − | + | − | − | + | − | − | + | − | − | + | − | + | + |
| 7 | − | + | + | − | − | − | + | + | − | − | − | + | + | − | + |
| 8 | + | + | + | − | + | + | − | + | − | − | + | − | − | − | − |
| 9 | − | − | − | + | + | + | − | + | − | − | − | + | + | + | − |
| 10 | + | − | − | + | − | − | + | + | − | − | + | − | − | + | + |
| 11 | − | + | − | + | − | + | − | − | + | − | + | − | + | − | + |
| 12 | + | + | − | + | + | − | + | − | + | − | − | + | − | − | − |
| 13 | − | − | + | + | + | − | − | − | − | + | + | + | − | − | + |
| 14 | + | − | + | + | − | + | + | − | − | + | − | − | + | − | − |
| 15 | − | + | + | + | − | − | − | + | + | + | − | − | − | + | − |
| 16 | + | + | + | + | + | + | + | + | + | + | + | + | + | + | + |

The least important estimate obtained from a full $2^4$ design is the four-factor interaction. This estimate would usually be readily given up to reduce the size of the experiment. Fortunately, there is a simple way to choose the eight tests to provide the desired balance. The eight tests are chosen by selecting only the rows in the design matrix in which the sign in the column headed 1234 is plus (or, alternatively, minus). This results in the design matrix in Table 6.2. From this design matrix, it is seen that the desired balance, four minuses and four pluses in each column, has been obtained.

It would be fortunate if the only information lost by reducing the pattern in this manner were the estimate of the four-factor interaction and some precision of the remaining estimates. This is not the case. From Table 6.2 it is seen that columns 14 and 23 are identical, which means that the estimates of the two effects will be identical. The estimate provided by either column is actually an estimate of the

**TABLE 6.2 Design Matrix for a 1/2 Fraction of a $2^4$ Pattern**

| Test | 1 | 2 | 3 | 4 | 12 | 13 | 14 | 23 | 24 | 34 | 123 | 124 | 134 | 234 | 1234 |
|---|---|---|---|---|---|---|---|---|---|---|---|---|---|---|---|
| 1 | − | − | − | − | + | + | + | + | + | + | − | − | − | − | + |
| 4 | + | + | − | − | + | − | − | − | − | + | − | − | + | + | + |
| 6 | + | − | + | − | − | + | − | − | + | − | − | + | − | + | + |
| 7 | − | + | + | − | − | − | + | + | − | − | − | + | + | − | + |
| 10 | + | − | − | + | − | − | + | + | − | − | + | − | − | + | + |
| 11 | − | + | − | + | − | + | − | − | + | − | + | − | + | − | + |
| 13 | − | − | + | + | + | − | − | − | − | + | + | + | − | − | + |
| 16 | + | + | + | + | + | + | + | + | + | + | + | + | + | + | + |

sum of the 14 and the 23 interactions. These two interactions are said to be confounded, or confused, with each other.

Further inspection of Table 6.2 reveals other pairs of columns that are identical and therefore produce estimates that are actually sums of two effects. The confounding pattern for a 1/2 fraction of a $2^4$ design is

$$1 + 234 \qquad 12 + 34$$

$$2 + 134 \qquad 13 + 24$$

$$3 + 124 \qquad 14 + 23$$

$$4 + 123$$

Estimates of the single-factor effects can be obtained that are confounded only by three-factor interactions. A more serious confounding pattern exists for the two-factor interactions.

Another justification for using this pattern is that any three of the four factors form a full factorial design. If any one of the factors has a negligible effect on the response, then the other three can be analyzed as a full $2^3$ factorial design in the manner laid out in Chap. 5. This analysis is performed under the assumption that the factor having a negligible average effect also does not interact with any of the other factors. The lack of interaction cannot be determined by analysis of the data but is an assumption that must be made by experts in the subject matter.

Example 6.1 illustrates the analysis of a fractional factorial design.

**Example 6.1 The Solenoid Experiment as a $2^{4-1}$ Design** Suppose that the solenoid experiment in Chap. 5 were run as a $2^{4-1}$ design rather than as a full $2^4$. The experiment would then have included 8 tests rather than 16 tests. Table 6.3 contains the design matrix assuming a $2^{4-1}$ design was run. The headings of the design matrix give the confounding pattern for the design. The property that any three factors form a full factorial pattern is given in parentheses below the title. The standard deviations of flow of the four solenoids at each of the eight factor combinations that would have been run if the pattern had only been a 1/2 fraction are listed. The run order from the full factorial experiment is also provided.

Figure 6.2 contains a list of effects estimated from the design matrix in Table 6.3 and a plot of these effects on a dot diagram. The substantial effect of bobbin depth on the standard deviation of flow is seen, as is the interaction of bobbin depth and armature length. This interaction is now confounded with the *ST* interaction.

Figure 6.2 also contains the dot diagram from the full factorial design of Chap. 5. Comparison of the two diagrams illustrates what is lost when the experimental design uses a $2^{4-1}$ pattern rather than a full $2^4$ pattern.

**TABLE 6.3 Design Matrix for the $2^{4-1}$ Solenoid Experiment**
(Any three factors form a full factorial pattern)

| Test | Run order | *A* | *S* | *B* | *T* | *AT* *SB* | *ST* *AB* | *BT* *AS* | Standard deviation of flow (cfm) |
|---|---|---|---|---|---|---|---|---|---|
| 1 | 1 | − | − | − | − | + | + | + | .04 |
| 2 | 9 | + | − | − | + | + | − | − | .15 |
| 3 | 5 | − | + | − | + | − | + | − | .07 |
| 4 | 10 | + | + | − | − | − | − | + | .10 |
| 5 | 6 | − | − | + | + | − | − | + | .02 |
| 6 | 7 | + | − | + | − | − | + | − | .01 |
| 7 | 13 | − | + | + | − | + | − | − | .05 |
| 8 | 2 | + | + | + | + | + | + | + | .01 |
| Divisor = 4 | | | | | | | | | |

| Effect | Estimate (cfm) | Effect | Estimate (cfm) |
|---|---|---|---|
| *A* | .02 | *AT* + *SB* | .01 |
| *S* | .00 | *ST* + *AB* | −.05 |
| *B* | −.07 | *BT* + *AS* | −.03 |
| *T* | .01 | | |

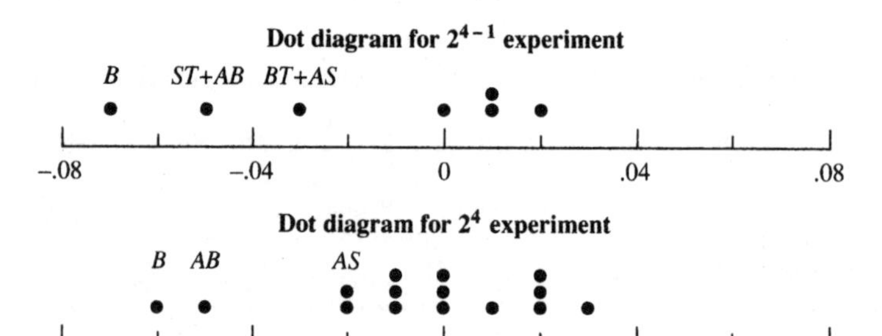


**Figure 6.2** Effects of factors on standard deviation of flow.

1. The dots corresponding to the third- and fourth-order interactions (clustered around zero in the $2^4$ dot diagram) are not available. These dots provided information on the effects of nuisance variables and are an aid to determine which are the important factors.

2. Two-factor interactions are confounded with each other. To allow further analysis of interactions, it will take an assumption by experts in the subject matter that one of the two interactions added together

is small. Methods to separate the confounded interactions by using additional tests are given in Box et al. (1978, p. 413).

3. Since there are only 8 tests in the fractional factorial pattern rather than the 16 tests in the full factorial pattern, there is less of an opportunity to average out the common causes of variation. Thus, it is more difficult to separate the effects of the nuisance variables from the factors. (*Note:* Comparison of the two dot diagrams in Fig. 6.2 does not give a complete picture of the impact of this loss because the data for the fractional factorial pattern are a subset of the full factorial pattern, and therefore the two diagrams tend to look more alike than if the tests using the two patterns were independent.)

The experimenter can proceed in several ways from this point. These alternatives will be discussed later in this chapter. Knowledge of the product or the process and analysis of the data are needed to choose among the alternatives. At this point it seems reasonable that the experimenter should be pleased to have detected that the combination of a small bobbin depth along with a long armature is to be avoided. Figure 6.3 contains the square obtained by considering only bobbin depth and armature length. (Either tube length or spring load could also have been included and the data displayed on a cube.) Figure 6.4 contains the response plot for the $AB$ interaction.

Future experiments could be used to confirm that the large interaction is in fact $AB$ rather than $ST$. Also the experimenter may use future experiments to determine more about the moderate interaction ($-0.03$) associated with $BT + AS$.

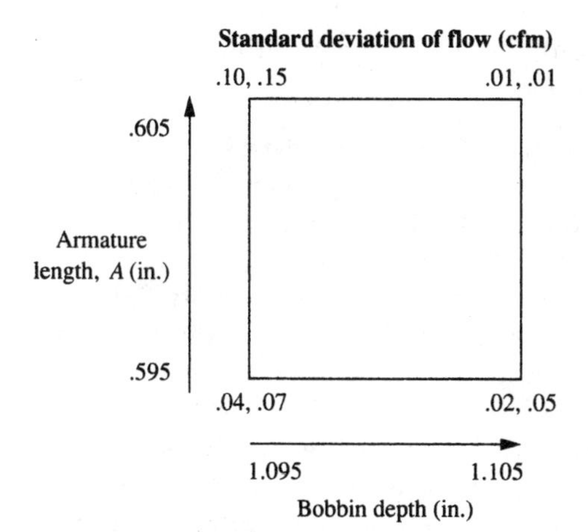


**Figure 6.3** Square for the solenoid experiment.

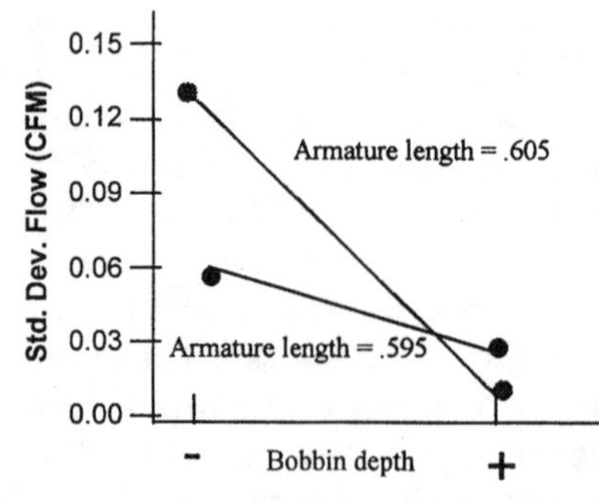


| Armature length | Bobbin depth | |
|---|---|---|
| | 1.095 | 1.105 |
| .595 | 0.055 | 0.035 |
| .605 | 0.125 | 0.010 |

**Figure 6.4** Response plots for the solenoid experiment.

## 6.2 Fractional Factorial Designs—Moderate Current Knowledge

A team of experimenters is said to have a *moderate level of knowledge* if they possess a strong belief that most of, but probably not all, the factors have a substantial effect on the response. This belief may turn out to be wrong, resulting in an experiment that is less than optimal. Nevertheless, experimenters should try to assess the current level of knowledge and should base the choice of a particular fractional factorial design in part on this assessment.

Given a moderate level of knowledge about the process or product, four fractional factorial patterns are particularly useful. These patterns are recommended for three reasons:

1. They can accommodate 4 to 16 factors.
2. They can be performed in 16 tests or fewer for up to 8 factors and 32 tests for 9 to 16 factors.
3. The effects of the individual factors are confounded with only third-order or higher interactions.

The four patterns along with some of their properties are listed in Table 6.4. The $2^{4-1}$ pattern has been discussed previously, and its design matrix is shown in Table 6.2. Some of the alternatives for subsequent experimentation are contained in Table 6.5.

The design matrix for the $2^{5-1}$ pattern is given in Table 6.6. This design is a particularly useful one. Any four factors form a full factorial design. Also, individual factors and two-factor interactions are only confounded with three-factor or higher interactions. Because this design is so powerful, it is rare that a full $2^5$ design would be a better choice. The $2^{5-1}$ design can be used for situations in which the experimenters have either moderate or high current knowledge of the pro-

**TABLE 6.4 Fractional Factorial Designs—Moderate Current Knowledge**

| Experimental pattern | Number of factors | Number of runs | Confounding* | Number of factors forming a full factorial design |
|---|---|---|---|---|
| $2^{4-1}$ | 4 | 8 | 2fi with 2fi | Any 3 |
| $2^{5-1}$ | 5 | 16 | None | Any 4 |
| $2^{8-4}$ | 6, 7, 8 | 16 | 2fi with 2 fi | Any 3 and some groups of 4 |
| $2^{16-11}$ | 9–16 | 32 | 2fi with 2fi | Any 3 and some groups of 4 or 5 |

*Only confounding of individual factors or their two-factor interactions (2fi) has been considered.

**TABLE 6.5 Follow-up to a $2^{4-1}$ Experiment**

| Number of important factors* | Alternatives |
|---|---|
| Less than four | 1. Analyze as a full factorial and run other full factorial designs to study interactions and find better operating conditions. |
| | 2. Run another $2^{4-1}$ at different levels of the factors to increase degree of belief in the results and find better operating conditions. |
| All four | 1. Run a $2^4$ design at different levels of the factors to find better operating conditions and estimate interactions. |
| | 2. Run the other half of the original $2^{4-1}$ to estimate interactions. This results in a full $2^4$ design in two blocks of eight. (See Sec. 6.4 for details on blocking.) |

*The number of factors found to have a substantial effect on the response variable based on estimating these effects from the design matrix.

cess. Subsequent experiments are usually needed only to establish optimal settings of the factors or to increase the degree of belief by including different background variables.

The design matrix for a $2^{8-4}$ pattern is contained in Table 6.7. Any three factors form a full factorial pattern. Although not all subsets of four factors form a full factorial pattern, some subsets of four do. One of these subsets contains the four factors associated with the first four columns of the design matrix. Therefore, the four factors thought to be most likely to have a substantial effect on the response should be listed in the first four columns. The individual factors are confounded

**TABLE 6.6 Design Matrix for a $2^{5-1}$ Pattern**
(Any four factors form a full factorial pattern)

| Test | 1 | 2 | 3 | 4 | 5 | 12 | 13 | 14 | 15 | 23 | 24 | 25 | 34 | 35 | 45 |
|---|---|---|---|---|---|---|---|---|---|---|---|---|---|---|---|
| 1 | − | − | − | − | + | + | + | + | − | + | + | − | + | − | − |
| 2 | + | − | − | − | − | − | − | − | − | + | + | + | + | + | + |
| 3 | − | + | − | − | − | − | + | + | + | − | − | − | + | + | + |
| 4 | + | + | − | − | + | + | − | − | + | − | − | + | + | − | − |
| 5 | − | − | + | − | − | + | − | + | + | − | + | + | − | − | + |
| 6 | + | − | + | − | + | − | + | − | + | − | + | − | − | + | − |
| 7 | − | + | + | − | + | − | − | + | − | + | − | + | − | + | − |
| 8 | + | + | + | − | − | + | + | − | − | + | − | − | − | − | + |
| 9 | − | − | − | + | − | + | + | − | + | + | − | + | − | + | − |
| 10 | + | − | − | + | + | − | − | + | + | + | − | − | − | − | + |
| 11 | − | + | − | + | + | − | + | − | − | − | + | + | − | − | + |
| 12 | + | + | − | + | − | + | − | + | − | − | + | − | − | + | − |
| 13 | − | − | + | + | + | + | − | − | − | − | − | − | + | + | + |
| 14 | + | − | + | + | − | − | + | + | − | − | − | + | + | − | − |
| 15 | − | + | + | + | − | − | − | − | + | + | + | − | + | − | − |
| 16 | + | + | + | + | + | + | + | + | + | + | + | + | + | + | + |

Divisor = 8

**TABLE 6.7 Design Matrix for a $2^{8-4}$ Pattern**

(Any three factors form a full factorial pattern)

| Test | 1 | 2 | 3 | 4 | 5 | 6 | 7 | 8 | 12<br>37<br>56<br>48 | 13<br>27<br>46<br>58 | 14<br>36<br>57<br>28 | 15<br>26<br>47<br>38 | 16<br>25<br>34<br>78 | 17<br>23<br>45<br>68 | 24<br>35<br>67<br>18 |
|---|---|---|---|---|---|---|---|---|---|---|---|---|---|---|---|
| 1 | − | − | − | + | + | + | − | + | + | + | − | − | − | + | − |
| 2 | + | − | − | − | − | + | + | + | − | − | − | − | + | + | + |
| 3 | − | + | − | − | + | − | + | + | − | + | + | − | + | − | − |
| 4 | + | + | − | + | − | − | − | + | + | − | + | − | − | − | + |
| 5 | − | − | + | + | − | − | + | + | + | − | − | + | + | − | − |
| 6 | + | − | + | − | + | − | − | + | − | + | − | + | − | − | + |
| 7 | − | + | + | − | − | + | − | + | − | − | + | + | − | + | − |
| 8 | + | + | + | + | + | + | + | + | + | + | + | + | + | + | + |
| 9 | + | + | + | − | − | − | + | − | + | + | − | − | − | + | − |
| 10 | − | + | + | + | + | − | − | − | − | − | − | − | + | + | + |
| 11 | + | − | + | + | − | + | − | − | − | + | + | − | + | − | − |
| 12 | − | − | + | − | + | + | + | − | + | − | + | − | − | − | + |
| 13 | + | + | − | − | + | + | − | − | + | − | − | + | + | − | − |
| 14 | − | + | − | + | − | + | + | − | − | + | − | + | − | − | + |
| 15 | + | − | − | + | + | − | + | − | − | − | + | + | − | + | − |
| 16 | − | − | − | − | − | − | − | − | + | + | + | + | + | + | + |

Divisor = 8

with only three-factor or higher interactions. The two-factor interactions are confounded with other two-factor interactions.

Although the design matrix stipulates eight factors, this experimental pattern can be used for six, seven, or eight factors. For example, if the pattern is used to study seven factors, the 16 tests are given by the first seven columns. The pattern is then a $2^{7-3}$. Column 8 becomes a measure of the effect of nuisance variables since the effect estimated from this column consists of three-factor or higher interactions. The confounding pattern is studied by ignoring any two-factor interactions that contain an 8.

If the pattern is used for six factors, the first six columns provide the 16 tests and the pattern is a $2^{6-2}$. Columns 7 and 8 provide a measure of the effect of nuisance variables. The confounding pattern is studied by ignoring any interactions containing a 7 or an 8.

There are numerous alternatives after the data from this design have been analyzed. Some of these alternatives are shown in Table 6.8.

The design matrix for a $2^{16-11}$ pattern is contained in Table 6.9. Any three factors form a full factorial pattern. Also, some subsets of four or five factors form a full factorial pattern. In particular, the five factors corresponding to the first five factors in the design matrix form a full factorial pattern. Therefore, the five factors thought to be the most likely to have a substantial effect on the response should be listed in the first five columns. The individual factors are confounded with only three-factor or higher interactions.

The two-factor interactions are confounded with other two-factor interactions. Each of the last 15 columns of the design matrix corresponds to the sum of eight two-factor interactions. The eight interactions for each column have not been enumerated. For other patterns

**TABLE 6.8 Follow-up to a $2^{8-4}$ Experiment**

| Number of important factors* | Alternatives |
|---|---|
| Any 3 or any subsets of 4 that form a full factorial pattern | Analyze as a full factorial and run other full factorial designs to improve levels or increase degree of belief. |
| Any 4 not forming a full factorial | Run a $2^{4-1}$ to study better levels or a $2^4$ to study better levels and interactions. |
| 5 | Run a $2^{5-1}$ or a full factorial to study better levels and interactions. |
| 6, 7, or 8 | Run a $2^{8-4}$ to study better levels or focus on a subset of 5 or fewer to study better levels or interactions. |

*The number of factors found to have a substantial effect on the response variable based on estimating these effects from the design matrix.

**TABLE 6.9 Design Matrix for a $2^{16-11}$ Pattern**

(Any three factors form a full factorial design)

| Test | 1 | 2 | 3 | 4 | 5 | 6 | 7 | 8 | 9 | 10 | 11 | 12 | 13 | 14 | 15 | 16 | ..........................Two-factor interactions.......................... | | | | | | | | | | | | | | |
|---|---|---|---|---|---|---|---|---|---|---|---|---|---|---|---|---|---|---|---|---|---|---|---|---|---|---|---|---|---|---|---|
| 1 | − | − | − | − | + | + | + | + | + | + | − | − | − | − | + | + | − | − | − | − | + | + | + | + | + | + | − | − | − | − | + |
| 2 | + | − | − | − | − | − | − | + | + | + | + | + | + | − | − | + | + | − | − | − | − | − | − | + | + | + | + | + | + | − | − |
| 3 | − | + | − | − | − | + | + | − | − | + | + | + | − | + | − | + | − | + | − | − | − | + | + | − | − | + | + | + | − | + | − |
| 4 | + | + | − | − | + | − | − | − | − | + | − | − | + | + | + | + | + | + | − | − | + | − | − | − | − | + | − | − | + | + | + |
| 5 | − | − | + | − | + | − | + | − | + | − | + | − | + | + | − | + | − | − | + | − | + | − | + | − | + | − | + | − | + | + | − |
| 6 | + | − | + | − | − | + | − | − | + | − | − | + | − | + | + | + | + | − | + | − | − | + | − | − | + | − | − | + | − | + | + |
| 7 | − | + | + | − | − | − | + | + | − | − | − | + | + | − | + | + | − | + | + | − | − | − | + | + | − | − | − | + | + | − | + |
| 8 | + | + | + | − | + | + | − | + | − | − | + | − | − | − | − | + | + | + | + | − | + | + | − | + | − | − | + | − | − | − | − |
| 9 | − | − | − | + | + | + | − | + | − | − | − | + | + | + | − | + | − | − | − | + | + | + | − | + | − | − | − | + | + | + | − |
| 10 | + | − | − | + | − | − | + | + | − | − | + | − | − | + | + | + | + | − | − | + | − | − | + | + | − | − | + | − | − | + | + |
| 11 | − | + | − | + | − | + | − | − | + | − | + | − | + | − | + | + | − | + | − | + | − | + | − | − | + | − | + | − | + | − | + |
| 12 | + | + | − | + | + | − | + | − | + | − | − | + | − | − | − | + | + | + | − | + | + | − | + | − | + | − | − | + | − | − | − |
| 13 | − | − | + | + | + | − | − | − | − | + | + | + | − | − | + | + | − | − | + | + | + | − | − | − | − | + | + | + | − | − | + |
| 14 | + | − | + | + | − | + | + | − | − | + | − | − | + | − | − | + | + | − | + | + | − | + | + | − | − | + | − | − | + | − | − |
| 15 | − | + | + | + | − | − | − | + | + | + | − | − | − | + | − | + | − | + | + | + | − | − | − | + | + | + | − | − | − | + | − |
| 16 | + | + | + | + | + | + | + | + | + | + | + | + | + | + | + | + | + | + | + | + | + | + | + | + | + | + | + | + | + | + | + |
| 17 | + | + | + | + | − | − | − | − | − | − | + | + | + | + | − | − | − | − | − | − | + | + | + | + | + | + | − | − | − | − | + |
| 18 | − | + | + | + | + | + | + | − | − | − | − | − | − | + | + | − | + | − | − | − | − | − | − | + | + | + | + | + | + | − | − |
| 19 | + | − | + | + | + | − | − | + | + | − | − | − | + | − | + | − | − | + | − | − | − | + | + | − | − | + | + | + | − | + | − |
| 20 | − | − | + | + | − | + | + | + | + | − | + | + | − | − | − | − | + | + | − | − | + | − | − | − | − | + | − | − | + | + | + |
| 21 | + | + | − | + | − | + | − | + | − | + | − | + | − | − | + | − | − | − | + | − | + | − | + | − | + | − | + | − | + | + | − |
| 22 | − | + | − | + | + | − | + | + | − | + | + | − | + | − | − | − | + | − | + | − | − | + | − | − | + | − | − | + | − | + | + |
| 23 | + | − | − | + | + | + | − | − | + | + | + | − | − | + | − | − | − | + | + | − | − | − | + | + | − | − | − | + | + | − | + |
| 24 | − | − | − | + | − | − | + | − | + | + | − | + | + | + | + | − | + | + | + | − | + | + | − | + | − | − | + | − | − | − | − |
| 25 | + | + | + | − | − | − | + | − | + | + | + | − | − | − | + | − | − | − | − | + | + | + | − | + | − | − | − | + | + | + | − |
| 26 | − | + | + | − | + | + | − | − | + | + | − | + | + | − | − | − | + | − | − | + | − | − | + | + | − | − | + | − | − | + | + |
| 27 | + | − | + | − | + | − | + | + | − | + | − | + | − | + | − | − | − | + | − | + | − | + | − | − | + | − | + | − | + | − | + |
| 28 | − | − | + | − | − | + | − | + | − | + | + | − | + | + | + | − | + | + | − | + | + | − | + | − | + | − | − | + | − | − | − |
| 29 | + | + | − | − | − | + | + | + | + | − | − | − | + | + | − | − | − | − | + | + | + | − | − | − | − | + | + | + | − | − | + |
| 30 | − | + | − | − | + | − | − | + | + | − | + | + | − | − | + | − | + | − | + | + | − | + | + | − | − | + | − | − | + | − | − |
| 31 | + | − | − | − | + | + | + | − | − | − | + | + | + | − | + | − | − | + | + | + | − | − | − | + | + | + | − | − | − | + | − |
| 32 | − | − | − | − | − | − | − | − | − | − | − | − | − | − | − | − | + | + | + | + | + | + | + | + | + | + | + | + | + | + | + |

Divisor = 16

**TABLE 6.10 Follow-up to a $2^{16-11}$ Experiment**

| Number of important factors | Alternatives |
|---|---|
| Any 3 or any subset of 4 or 5 that form a full factorial pattern | Analyze as a full factorial design and run other full or fractional factorial designs to improve levels or increase degree of belief. |
| Any 4 not forming a full factorial pattern | Run a $2^{4-1}$ to study better levels or a $2^4$ to study better levels or interactions. |
| Any 5 not forming a full factorial pattern | Run a $2^{5-1}$ to study better levels or interactions. |
| 6 through 16 | Run a fractional factorial design to increase degree of belief or study better levels. Run a full factorial design on a subset to study better levels or interactions. |

where two-factor interactions were confounded, the confounding patterns were listed. This was done to allow the study of interactions to proceed if the experts in the subject matter were willing to assume that factors having negligible effect on the response do not interact with the important factors. Because there are eight interactions confounded with one another, it is usually impractical to make that assumption. The effects for each of the columns should still be computed and plotted on the dot diagram.

The design matrix stipulates 16 factors, but it can also be used for 9 to 15 factors. The design matrix is used for fewer than 16 factors in the same way that the design matrix for a $2^{8-4}$ pattern was used for fewer than 8 factors. The alternatives for follow-up to this design are given in Table 6.10.

Example 6.2 illustrates the analysis of a $2^{5-1}$ fractional factorial design.

**Example 6.2 A $2^{5-1}$ Design for a Welding Process** To assemble a part for an automotive engine, two sets of components are welded together. An important quality characteristic of the part is the force it took to pull the two sets of components apart after they had been welded together (pull force). Through the use of a control chart, the process had been stabilized with respect to the pull force. Steps had been taken to remove special causes of variation associated with the measurement and the cleanliness of the components as they came to the welding station.

Based on the control chart, the process average and standard deviation were determined to be 1100 and 61 inch-pounds (in·lb), respectively. If it remained stable, the process was capable of meeting the lower specification of 900 in·lb. However the plant personnel desired to raise the average pull force to decrease the chances that a special cause would result in unacceptable product. They attempted to do this by studying the effect that variables in the welding process had on the pull force.

The five factors studied were as follows:

| Factor | Level | |
|---|---|---|
| | − | + |
| $H$ = heat (°C) | 80 | 95 |
| $P$ = pressure (psi) | 50 | 75 |
| $W$ = weld time (s) | 2.5 | 3.0 |
| $T$ = hold time (s) | 1.0 | 2.0 |
| $S$ = squeeze time (s) | 4.5 | 5.5 |

The levels of the factors were set approximately equidistant from the current operating conditions.

Figure 6.5 contains the planning form for the experiment.

### Run charts

Since the parts were assembled first and then tested for the pull force later, the orders of assembly and of test were determined by using two different sets of random numbers. Therefore, two different run charts were plotted. The run chart in order of assembly is shown in Fig. 6.6. The run chart in order of test is contained in Fig. 6.7.

### Design matrix and dot diagram

Table 6.11 contains the design matrix for the experiment. Figure 6.8 contains the effects of the factors computed from the design matrix as well as the dot diagram. Effects of the factors on the pull force were estimated.

### Analysis of the cube

Since a characteristic of a $2^{5-1}$ design is that any four or fewer factors form a full factorial design, the results using the important factors—heat, pressure, and squeeze time—are analyzed by using a cube. Figure 6.9 shows the cube. The 16 responses of pull force from the experiment are placed on the cube. There are two responses corresponding to each combination of the factors. Since the variation between the two responses at each factor combination is consistent with the variation previously seen on the control chart, the average of the two is used in the analysis of the paired comparisons.

Analysis of the paired comparisons on the cube indicates that the pair of measurements 871, 920 corresponding to $P = 50-$, $H = 95+$, and $S = 4.5-$ are primarily responsible for the relatively large magnitude of the effect of squeeze time and its interactions with heat and pressure. If these measurements were around 1300 rather than 900,

## Documentation of the welding experiment

1. **Objective:**

Study the effects that five factors have on pull force. Use the results to set operating conditions for the process of resistance welding of cans to brackets.

2. **Background information:**

The process is stable with average 1100 in-lbs and standard deviation 61 in-lbs. The measurement process and the lack of cleanliness have been found to be sources of special causes of variation in pull force in the past.

3. **Experimental variables:**

| A. Response variables | Measurement technique | |
|---|---|---|
| 1. Pull force | Dillon force gage | |

| B. Factors under study | Levels | |
|---|---|---|
| 1. Heat (°C) | 80 | 95 |
| 2. Pressure (psi) | 50 | 75 |
| 3. Weld Time (sec.) | 2.5 | 3.0 |
| 4. Hold Time (sec.) | 1.0 | 2.0 |
| 5. Squeeze Time (sec.) | 4.5 | 5.5 |

| C. Background variables | Method of control |
|---|---|
| 1. Conditions of can and bracket | Clean before welding. |
| 2. Condition of electrodes | Start with new ones. |
| 3. Operator pace | Record temperature of electrodes before each part. |
| 4. Ambient temperature | Record. |

4. **Replication:**

One part per factor combination

5. **Methods of randomization:**

Randomly select components, the order of assembling the parts, and the order of testing the parts.

6. **Design matrix:** (attach copy)

$2^{5-1}$ (see Table 7.11.)

7. **Data collection forms:** (not shown)

8. **Planned methods of statistical analysis:**

Analyze the effects of the five factors using the design matrix, and collapse to a full factorial design if one or more factors are unimportant.

9. **Estimated cost, schedule, and other resources:**

Three person-hours administrative time to prepare the components, instructions for the operator and data collection sheets. Less than one half hour of assembly time.

**Figure 6.5** Documentation of the welding experiment.

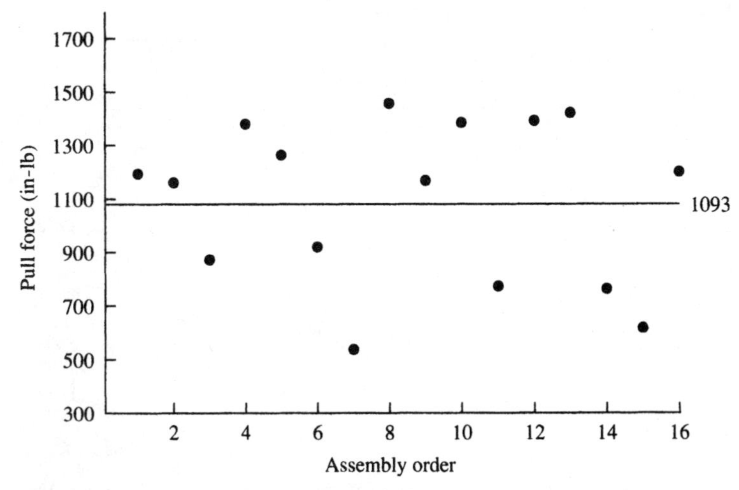


**Figure 6.6** Run chart in order of assembly.

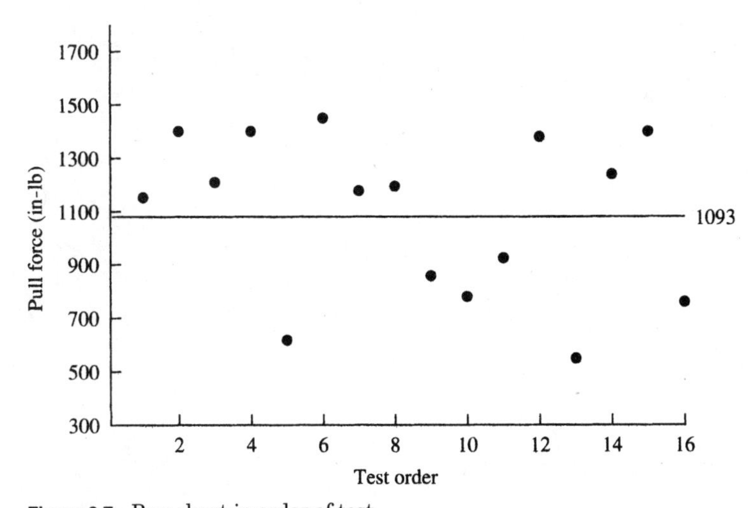


**Figure 6.7** Run chart in order of test.

the variation in the response could be explained simply by an interaction between pressure and heat.

The process may be as complex (three important interactions) as the original analysis suggests, or the apparent complexity could be the result of a special cause. To further investigate this potential special cause, checks of the measurement process and the assembly process should be made. The two measurements were made at a simi-

**TABLE 6.11 Design Matrix for the Welding Study**

| Test | Build order | Test order | $H$ | $P$ | $W$ | $T$ | $S$ | $HP$ | $HW$ | $HT$ | $HS$ | $PW$ | $PT$ | $PS$ | $WT$ | $WS$ | $TS$ | Pull force |
|---|---|---|---|---|---|---|---|---|---|---|---|---|---|---|---|---|---|---|
| 1 | 16 | 8 | − | − | − | − | + | + | + | + | − | + | + | − | + | − | − | 1194 |
| 2 | 3 | 9 | + | − | − | − | − | − | − | − | − | + | + | + | + | + | + | 871 |
| 3 | 14 | 16 | − | + | − | − | − | − | + | + | + | − | − | − | + | + | + | 764 |
| 4 | 8 | 6 | + | + | − | − | + | + | − | − | + | − | − | + | + | − | − | 1463 |
| 5 | 1 | 3 | − | − | + | − | − | + | − | + | + | − | + | + | − | − | + | 1205 |
| 6 | 5 | 14 | + | − | + | − | + | − | + | − | + | − | + | − | − | + | − | 1256 |
| 7 | 15 | 5 | − | + | + | − | + | − | − | + | − | + | − | + | − | + | − | 616 |
| 8 | 4 | 12 | + | + | + | − | − | + | + | − | − | + | − | − | − | − | + | 1384 |
| 9 | 2 | 1 | − | − | − | + | − | + | + | − | + | + | − | + | − | + | − | 1152 |
| 10 | 12 | 4 | + | − | − | + | + | − | − | + | + | + | − | − | − | − | + | 1398 |
| 11 | 7 | 13 | − | + | − | + | + | − | + | − | − | − | + | + | − | − | + | 533 |
| 12 | 10 | 2 | + | + | − | + | − | + | − | + | − | − | + | − | − | + | − | 1382 |
| 13 | 9 | 7 | − | − | + | + | + | + | − | − | − | − | − | − | + | + | + | 1170 |
| 14 | 6 | 11 | + | − | + | + | − | − | + | + | − | − | − | + | + | − | − | 920 |
| 15 | 11 | 10 | − | + | + | + | − | − | − | − | + | + | + | − | + | − | − | 776 |
| 16 | 13 | 15 | + | + | + | + | + | + | + | + | + | + | + | + | + | + | + | 1410 |

Divisor = 8

| Factor | Effect |
|---|---|
| $H$ | 334 |
| $P$ | −105 |
| $W$ | −3 |
| $T$ | −2 |
| $S$ | 73 |

| Interaction | Effect |
|---|---|
| $HP$ | 403 |
| $HW$ | −34 |
| $HT$ | 35 |
| $HS$ | 169 |
| $PW$ | 14 |
| $PT$ | −30 |
| $PS$ | −144 |
| $WT$ | −45 |
| $WS$ | −31 |
| $TS$ | −3 |

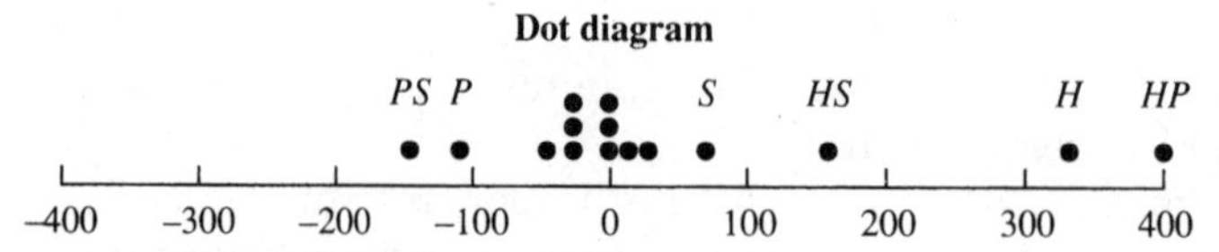


**Figure 6.8** Effect of factors on the pull force.

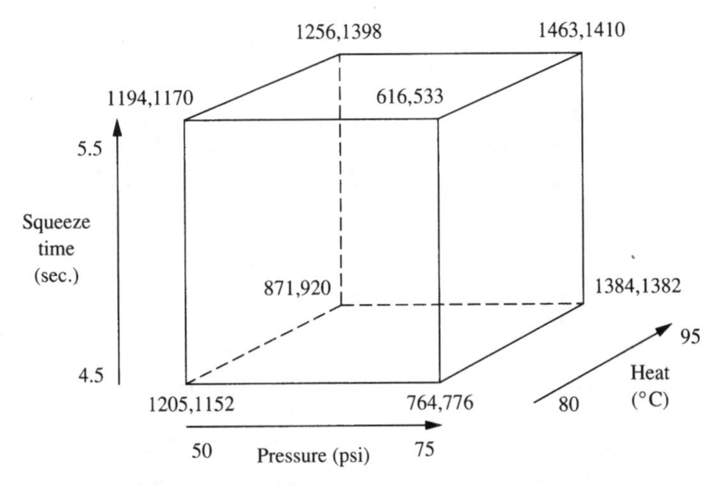


**Figure 6.9** Cube for the welding experiment.

lar time (9th and 11th), and they were built at a similar time (3d and 6th). Was there anything specific that occurred during that time period? How do the results of this experiment compare with the predictions that were made before it was run?

### Response plots

Figure 6.10 contains the response plots that summarize the results of the experiment. The validity of these plots depends on the assumption that there was no special cause of variation present during the experiment. Since the three important factors all interact in pairs, the three factors are used on the response plots to better display the relationship between them. The plots show the relationship between heat and pressure separately when squeeze time is 4.5 s and when squeeze time is 5.5 s.

### Conclusions for Example 6.2

1. Three factors and three interactions between the factors were found to be important. The important factors were heat, pressure, and squeeze time.

2. The response plots displayed the interactions between the important factors. From these plots it is seen that the prediction of the average pull force that would be obtained at the current operating conditions (halfway between the levels chosen for the experiment) is consistent with the 1100 in·lb determined from the control chart. The

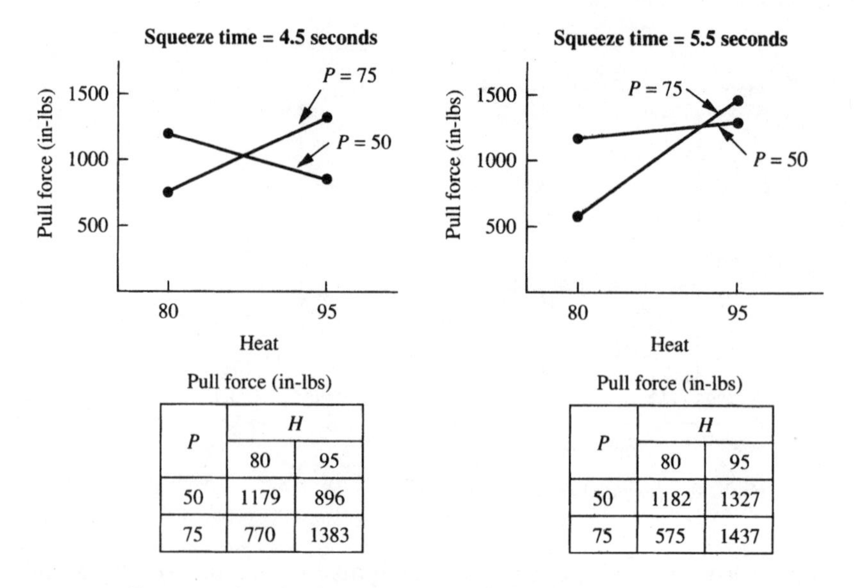


Pull force (in-lbs)

| P | H 80 | H 95 |
|---|---|---|
| 50 | 1179 | 896 |
| 75 | 770 | 1383 |

Pull force (in-lbs)

| P | H 80 | H 95 |
|---|---|---|
| 50 | 1182 | 1327 |
| 75 | 575 | 1437 |

**Figure 6.10** Response plots for the welding experiment—three-factor interaction.

plots also show that a substantial increase in pull force can be obtained by using high squeeze time and high heat. The combination of high heat and high pressure along with low squeeze time also results in considerable increase in pull force.

3. Another plausible explanation of the variation in the data is that pressure and heat interact, and a special cause was present during the experiment. Past and future tests and knowledge of the process should be used to determine which of the two explanations provides the better model of the cause-and-effect system in the process.

## 6.3 Fractional Factorial Designs—Low Current Knowledge

A low level of knowledge is indicated by a belief that only a small proportion of the possible factors has a substantial effect on the response, and it is not known which ones they are. When a team of experimenters has a low level of knowledge, the aim of the first experiment is usually to screen out unimportant factors. Subsequent experiments can then be planned to study the important factors in greater depth. In this section, fractional factorial designs to screen out unimportant factors will be given.

Three fractional factorial patterns are particularly useful to screen out unimportant factors. These patterns are chosen primarily for three reasons:

1. They require relatively few runs (16 or fewer).
2. They can accommodate up to 15 factors.
3. If the assumption is incorrect that only a small number of factors are important, then another fractional factorial design is available to combine with the original design. This combination of designs will then have a confounding pattern similar to that of the designs for moderate knowledge given in Sec. 6.2.

These designs and their properties are given in Table 6.12.

The design matrix for the $2^{3-1}$ design is contained in Table 6.13. Any two factors form a full factorial design. The individual factors are confounded with the two-factor interactions. Because of the small number of runs in this design, the design is useful for applications to the design of a product. Since it is often expensive to build prototypes for a product, testing of alternative, possibly improved designs is sometimes not done. The $2^{3-1}$ design provides a relatively inexpensive way to test different product designs.

If one of the factors turns out to be unimportant, then the remaining two factors can be analyzed as a full factorial design. If all three factors turn out to be important, then four more tests can be made. These four tests can be chosen so that when they are analyzed in combination with the original four tests, the individual factors are not confounded with the two-factor interactions. With the addition of these four runs, the pattern becomes a $2^3$ in two blocks. The design

**TABLE 6.12 Fractional Factorial Designs—Low Current Knowledge**

| Experimental pattern | Number* of factors | Number of runs | Confounding† | Number of factors forming a full factorial design |
|---|---|---|---|---|
| $2^{3-1}$ | 3 | 4 | Individual factors with 2fi | 2 |
| $2^{7-4}$ | 5, 6, 7 | 8 | Individual factors with 2fi | Any 2 and some groups of 3 |
| $2^{15-11}$ | 8–15 | 16 | Individual factors with 2fi | Any 2 and some groups of 3 or 4 |

*With four factors, use a $2^{4-1}$ design (given in Sec. 6.2).
†Only confounding of individual factors or their two-factor interactions (2fi) have been given.

**TABLE 6.13 Design Matrix for a $2^{3-1}$ Pattern**
(Any two factors form a full factorial pattern)

| Test | 1<br>23 | 2<br>13 | 3<br>12 |
|---|---|---|---|
| 1 | − | − | + |
| 2 | + | − | − |
| 3 | − | + | − |
| 4 | + | + | + |
| Divisor = 2 | | | |

matrix and the properties of the $2^3$ design in two blocks are given in Sec. 6.4.

The design matrix for a $2^{7-4}$ design is given in Table 6.14. Any two factors form a full factorial pattern. Although any arbitrary group of three factors does not form a full factorial pattern, some groups of three factors do. In particular, the first three factors form a full factorial pattern.

The three factors most likely to have a substantial effect on the response should be listed in the first three columns. The effects of individual factors are confounded with two-factor interactions. Using this design, the experimenters are looking for two or three factors with large effects relative to the others in the study.

If more factors than expected turn out to be important or if there is other uncertainty that needs to be resolved, eight more tests can be made. These eight tests can be chosen so that when they are analyzed

**TABLE 6.14 Design Matrix for a $2^{7-4}$ Pattern**
(Any two factors form a full factorial design)

| Test | 1<br>24<br>35<br>67 | 2<br>14<br>36<br>57 | 3<br>15<br>26<br>47 | 4<br>12<br>56<br>37 | 5<br>13<br>46<br>27 | 6<br>23<br>45<br>17 | 7<br>34<br>25<br>16 |
|---|---|---|---|---|---|---|---|
| 1 | − | − | − | + | + | + | − |
| 2 | + | − | − | − | − | + | + |
| 3 | − | + | − | − | + | − | + |
| 4 | + | + | − | + | − | − | − |
| 5 | − | − | + | + | − | − | + |
| 6 | + | − | + | − | + | − | − |
| 7 | − | + | + | − | − | + | − |
| 8 | + | + | + | + | + | + | + |
| Divisor = 4 | | | | | | | |

in combination with the original eight tests, the individual factors are not confounded with the two-factor interactions. With the addition of these eight tests, the pattern becomes a $2^{7-3}$ in two blocks. The design matrix for this pattern and its properties are given in Sec. 6.4.

There are many alternatives for subsequent experiments after the data from a $2^{7-4}$ is analyzed. Table 6.15 contains some of these alternatives.

Table 6.16 contains the design matrix for a $2^{15-11}$ fractional factorial pattern. Any two factors and some groups of three or four factors form a full factorial pattern. In particular, the first four factors form a full factorial pattern. The four factors that the experimenters think are most likely to have a substantial effect on the response variable should be listed in the first four columns of the design matrix.

Each of the individual factors is confounded with seven two-factor interactions. The confounding between the individual factors and the two-factor interactions can be resolved by adding another block of eight tests. The pattern then becomes a $2^{15-10}$ in two blocks. The design matrix and the properties of this pattern are given in Sec. 6.4.

**TABLE 6.15 Follow-up to a $2^{7-4}$ Experiment**

| Number of important factors* | Alternatives |
|---|---|
| Any 2 or any subset of 3 that forms a full factorial pattern | Analyze as a full factorial design and run other full factorial designs to improve levels or increase degree of belief. |
| | Run another block of eight tests to resolve uncertainty about which are the important factors (see Sec. 6.4). |
| Any 3 that do not form a full factorial pattern, or 4 | Run another block of eight tests to resolve uncertainty about which are the important factors. |
| | Run a $2^{4-1}$ to study better levels. |
| | Run a full factorial design to study better levels and interactions. |
| 5 | Run a $2^{5-1}$ to study better levels or interactions. |
| | Run another block of eight tests to resolve uncertainty about which are the important factors. |
| 6 or 7 | Run another block of eight tests to resolve uncertainty about which are the important factors. |

*The number of factors found to have a substantial effect on the response variable based on estimating these effects from the design matrix.

**TABLE 6.16 Design Matrix for a $2^{15-11}$ Pattern**
(Any two factors form a full factorial pattern)

| Test | 1 | 2 | 3 | 4 | 5 | 6 | 7 | 8 | 9 | 10 | 11 | 12 | 13 | 14 | 15 |
|---|---|---|---|---|---|---|---|---|---|---|---|---|---|---|---|
| | ............ | | | | | Two-factor interactions............ | | | | | | | | | |
| 1 | − | − | − | − | + | + | + | + | + | + | − | − | − | − | + |
| 2 | + | − | − | − | − | − | − | + | + | + | + | + | + | − | − |
| 3 | − | + | − | − | − | + | + | − | − | + | + | + | − | + | − |
| 4 | + | + | − | − | + | − | − | − | − | + | − | − | + | + | + |
| 5 | − | − | + | − | + | − | + | − | + | − | + | − | + | + | − |
| 6 | + | − | + | − | − | + | − | − | + | − | − | + | − | + | + |
| 7 | − | + | + | − | − | − | + | + | − | − | − | + | + | − | + |
| 8 | + | + | + | − | + | + | − | + | − | − | + | − | − | − | − |
| 9 | − | − | − | + | + | + | − | + | − | − | − | + | + | + | − |
| 10 | + | − | − | + | − | − | + | + | − | − | + | − | − | + | + |
| 11 | − | + | − | + | − | + | − | − | + | − | + | − | + | − | + |
| 12 | + | + | − | + | + | − | + | − | + | − | − | + | − | − | − |
| 13 | − | − | + | + | + | − | − | − | − | + | + | + | − | − | + |
| 14 | + | − | + | + | − | + | + | − | − | + | − | − | + | − | − |
| 15 | − | + | + | + | − | − | − | + | + | + | − | − | − | + | − |
| 15 | + | + | + | + | + | + | + | + | + | + | + | + | + | + | + |
| Divisor = 8 | | | | | | | | | | | | | | | |

Alternatives for subsequent experiments after a $2^{15-11}$ design has been run are shown in Table 6.17.

Example 6.3 illustrates the use of a fractional factorial design to screen out unimportant factors.

**Example 6.3 A $2^{7-4}$ Design for Tensile Strength of Rivets** An automotive part contained several components that were riveted together. The tensile strength of the part was an important quality characteristic. The product engineers responsible for the design of the part planned an experiment to determine how various configurations of the components affected tensile strength. Figure 6.11 contains the documentation of the experiment. Table 6.18 contains the design matrix for the experiment.

### Run charts

No special causes seem to be present in the run chart in Fig. 6.12. Since the variation within each replication was small compared to the overall variation in the run chart, the four values of tensile strength at each factor combination were averaged and the averages were used in the design matrix to determine the effects.

### Design matrix and dot diagrams

Table 6.18 shows the effects of the factors calculated by using the design matrix for the experiment, and Fig. 6.13 contains the dot diagram.

**TABLE 6.17 Follow-up to a $2^{15-11}$ Experiment**

| Number of important factors* | Alternatives |
|---|---|
| Any 2 or any group of 3 or 4 that form a full factorial pattern | Analyze as full factorial design and run other full factorial designs to improve levels or increase degree of belief. |
| | Run another block of 16 tests to resolve uncertainty about which are the important factors (see Sec. 6.4). |
| Any 3 or 4 that form a full factorial pattern | Run a full factorial design to study better levels and interactions. |
| | Run a $2^{4-1}$ to study better levels. |
| | Run another block of 16 tests to resolve uncertainty about which are the important factors. |
| 5 | Run a $2^{5-1}$ to study better levels or interactions. |
| | Run another block of 16 tests to resolve uncertainty about which are the important factors. |
| 6 through 15 | Run another block of 16 tests to resolve uncertainty about which are the important factors. |

*The number of factors found to have a substantial effect on the response variable based on estimating these effects from the design matrix.

### Analysis of the cube

A characteristic of a $2^{7-4}$ pattern is that there are some groups of three factors that form a full factorial pattern. In this case, the important factors—plate thickness, stem rivet length, and bushing thickness—do form a full factorial pattern. Therefore, they can be analyzed on a cube. The cube is illustrated in Fig. 6.14.

The analysis of the paired comparisons on the cube does not indicate the presence of any special causes of variation and therefore confirms the effects that were found by using the design matrix.

Table 6.15 contained the alternatives for subsequent experiments after a $2^{7-4}$ when three important factors formed a full factorial pattern. It was decided to follow up the $2^{7-4}$ by running another block of eight tests to separate the confounding between factors and interactions and, therefore, to resolve uncertainty about which are the important factors. With the addition of these eight tests, the pattern becomes a $2^{7-3}$ in two blocks. This pattern is discussed in Sec. 6.4 with further elaboration of Example 6.3.

1. **Objective:**

Study the effects that seven factors have on the tensile strength of stem diaphragm plate rivets. The results will be used to determine the nominal values and tolerances for the important components.

2. **Background information:**

The customer's specification for tensile strength was 480 in-lbs minimum. Prototype parts built to nominal dimensions were able to meet this specification. Little information is available on the effect on tensile strength of variation in the components.

3. **Experimental variables:**

| A. Response variables | Measurement technique | |
|---|---|---|
| 1. Tensile strength (in-lbs) | Pull tester | |

| B. Factors under study | Levels | |
|---|---|---|
| 1. *RH* = Rivet height (in.) | .015 | .025 |
| 2. *PD* = plate i.d. (in.) | .128 | .132 |
| 3. *PT* = Plate thickness (in.) | .030 | .036 |
| 4. *SD* = Stem rivet diameter (in.) | .123 | .125 |
| 5. *SL* = Stem rivet length (in.) | .200 | .210 |
| 6. *BD* = Bushing i.d. (in.) | .129 | .133 |
| 7. *BT* = Bushing thickness (in.) | .095 | .105 |

| C. Background variables | Method of control |
|---|---|
| 1. Operators (riveters and testing) | Hold constant |
| 2. Time | Randomization |

4. **Replication:**

Four for each factor combination.

5. **Methods of randomization:**

The order of the eight factor combinations was randomized using a random number table.

6. **Design matrix:** (attach copy)

$2^{7-4}$. see Table 6.18.

7. **Data collection forms:** (not shown here)

8. **Planned methods of statistical analysis:**

The variation within each factor combination is expected to be small relative to the effects of the factors. If so the analysis will be performed on the averages of the four tensile strength readings for each of eight factor combinations.

9. **Estimated cost, schedule, and other resource considerations:**

One day of administrative time is needed to organize the running of the experiment. Two hours of an operator's time will be needed to assemble the parts.

**Figure 6.11** Documentation of experiment for Example 6.3.

**TABLE 6.18 Design Matrix for the Rivet Study**

| | | 1 = $RH$<br>2 = $PD$ | | 3 = $PT$<br>4 = $SD$ | | 5 = $SL$<br>6 = $BD$ | | 7 = $BT$ | |
|---|---|---|---|---|---|---|---|---|---|
| Test | Run order | 1<br>24<br>35<br>67 | 2<br>14<br>36<br>57 | 3<br>15<br>26<br>47 | 4<br>12<br>56<br>37 | 5<br>13<br>46<br>27 | 6<br>23<br>45<br>17 | 7<br>34<br>25<br>16 | Tensile strength $\overline{X}$ |
| 1 | 8 | − | − | − | + | + | + | − | 513 |
| 2 | 4 | + | − | − | − | − | + | + | 461 |
| 3 | 6 | − | + | − | − | + | − | + | 488 |
| 4 | 2 | + | + | − | + | − | − | − | 481 |
| 5 | 7 | − | − | + | + | − | − | + | 523 |
| 6 | 3 | + | − | + | − | + | − | − | 558 |
| 7 | 5 | − | + | + | − | − | + | − | 532 |
| 8 | 1 | + | + | + | + | + | + | + | 546 |
| Divisor = 4 | | | | | | | | | |
| Effect | | −2.5 | −2.0 | 54.0 | 6.0 | 27.0 | 0.5 | −16.5 | |

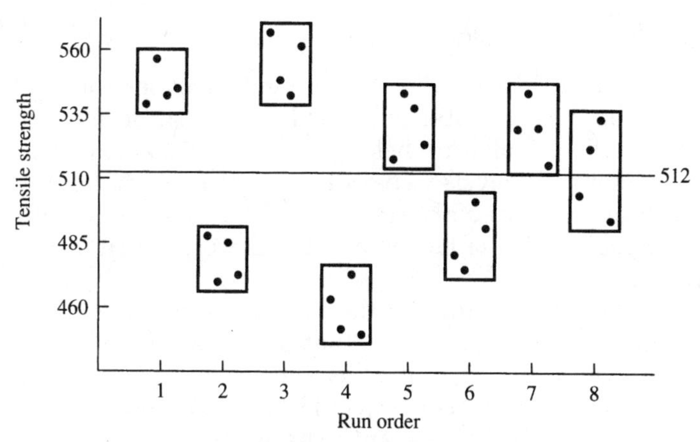


**Figure 6.12** Run chart for the rivet experiment.

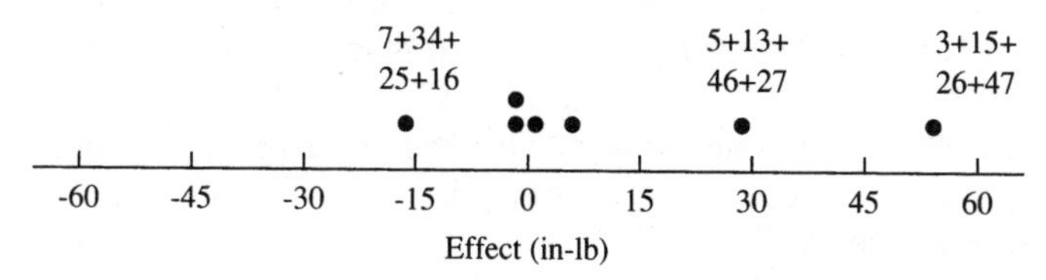


**Figure 6.13** Dot diagram for the rivet experiment.

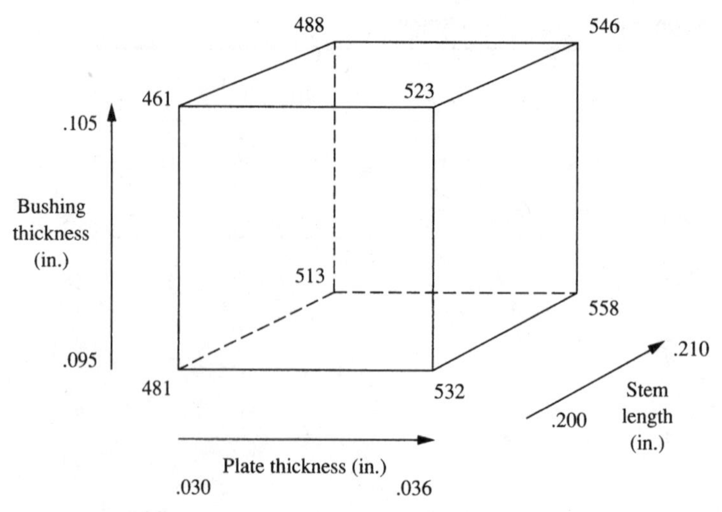


**Figure 6.14** Cube for the rivet experiment ($2^{7-4}$ design).

## 6.4 Blocking in Factorial Designs

Blocks are usually formed by holding background variables constant for all or part of the experimental pattern. In Chap. 4, designs for which the background variables are held constant for the entire set of combinations of factors and levels were called *complete* block designs. Designs for which the background variables are held constant for only a portion of the combinations of factors and levels were called *incomplete* block designs.

Factorial designs can be run in either complete or incomplete blocks. For example, consider a $2^3$ design that was to be run in a manufacturing facility. "Shift" was identified as a major background variable. Time permitting, the experiment could be run on one shift to keep the conditions relatively constant. The background variable "shift" defines a block. The experimenters may choose to repeat the experiment on another shift to increase the degree of belief in the results. If they do, the total study will consist of two replications of the full $2^3$ design. Each replication will be carried out in a complete block defined by the shift. The analysis of the effects of the factors should be performed separately for each block. The results from each block can then be compared to determine if there are any interactions between the blocks and the factors.

Although it may be desirable to carry out the full $2^3$ all on one shift, it may not be possible. Perhaps each test takes 2 hours to complete. Then only four of the eight tests could be completed on one shift. Running the remaining four tests on the next shift or waiting until the next day creates the potential to have background variables and nuisance variables confuse the results of the experiment.

Even if it is possible to run all eight tests on one shift, we may not want to do so. By running the tests on more than one shift, a wider range of conditions is included in the study. This increases the degree of belief that the results provide a basis for action. The use of incomplete blocks provides a method to reduce the effect of background variables when the factors are not studied under uniform conditions.

In this section the use of incomplete blocks with factorial or fractional factorial designs is discussed. Four particularly useful experimental patterns that incorporate incomplete blocks are given in some detail. These patterns are

- A $2^3$ pattern in two blocks of four tests each
- A $2^4$ pattern in two blocks of eight tests each
- A $2^{7-3}$ pattern in two blocks of eight tests each
- A $2^{15-10}$ pattern in two blocks of 16 tests each

Also, a method to develop other experimental patterns by using incomplete blocks is given.

### A $2^3$ design in two blocks of size 4

If it is not possible to run all a $2^3$ factorial pattern in one block, two blocks with four tests in each block could be used. How should the eight tests be split between the two blocks? Intuitively, it seems desirable to have two tests at the low level and two tests at the high level of each factor in each block. For example, we would not want all tests at the low level of a particular factor run on shift 1 and all tests at the high level of the factor run on shift 2. Then it would be impossible to separate the effect of the factor from the effect of the shift. This discussion is similar to the discussion concerning how to chose the eight tests in a $2^{4-1}$ pattern. There is a direct link between fractional factorial patterns and incomplete blocks.

To divide the eight tests in a $2^3$ pattern into two blocks of four tests each, a blocking variable is used. This blocking variable is usually a background variable or several background variables that are combined into a chunk. A $2^{4-1}$ pattern is used to accommodate the three factors and the blocking variable in eight tests. The blocking variable is used to set up the two blocks by treating it as if it were a factor. All tests that have a plus level of the blocking variable are included in block 1, and all tests that have a minus level of the blocking variable are included in block 2. The difference between the $2^{4-1}$ pattern and a $2^3$ pattern in two blocks lies in the randomization of the eight tests. The randomization is done separately in each of the two blocks. The

**TABLE 6.19 A $2^3$ Design in Two Blocks—Dye Process**
(Any three factors form a full factorial pattern)

| Test | Run order | $M$ | $P$ | $T$ | $b$* | $Mb$ $PT$ | $Pb$ $MT$ | $Tb$ $PM$ | Response |
|---|---|---|---|---|---|---|---|---|---|
| Block 1 | | | | | | | | | |
| 1 | 3 | − | − | + | + | − | − | + | 195 |
| 2 | 1 | + | − | − | + | + | − | − | 228 |
| 3 | 4 | − | + | − | + | − | + | − | 218 |
| 4 | 2 | + | + | + | + | + | + | + | 241 |
| Block 2 | | | | | | | | | |
| 5 | 6 | + | + | − | − | − | − | + | 259 |
| 6 | 7 | − | + | + | − | + | − | − | 238 |
| 7 | 8 | + | − | + | − | − | + | − | 200 |
| 8 | 5 | − | − | − | − | + | + | + | 189 |
| Divisor = 4 | | | | | | | | | |

*$b$ = Blocking variable.

analysis of data from a $2^3$ design in two blocks is carried out as if it were a $2^{4-1}$ pattern.

**Example 6.4 A $2^3$ Design in Two Blocks for the Dye Process** Suppose that it was either necessary for practical reasons or desirable for increased degree of belief to run the $2^3$ pattern in the dyeing process described in Example 5.2 in two blocks of four tests each. Background variables would first have to be combined into a chunk variable. Table 6.19 contains the design matrix for the $2^3$ design in two blocks. The randomization is done separately within the two blocks.

## A $2^4$ design in two blocks of size 8

The $2^4$ factorial pattern can be separated into two blocks of eight tests each by setting up a blocking variable just as was done for the $2^3$ pattern. This blocking variable becomes the fifth factor to be included in the 16 tests and is used to set up the blocks. A $2^{5-1}$ pattern is used to accommodate the four factors plus the blocking variable.

**Example 6.5 The Solenoid Experiment in Two Blocks** Table 6.20 contains the design matrix for the $2^4$ experiment on the solenoid described in Example 5.3 as if it were run in two blocks. Figure 6.15 contains the run charts for the averages and standard deviations. The randomization for this design is done separately in each block.

The analysis of the data from a $2^4$ design in two blocks is carried out as if it were a $2^{5-1}$ pattern. Figure 6.16 contains the effects of the factors computed from the design matrix contained in Table 6.20, and it shows the dot diagram as well. A review of Fig. 6.16 indicates that block $b$ has an insignificant effect on both the average and standard deviation of flow. Since the remaining four factors form a full factorial design, the analysis can proceed in the same fashion as in Example 5.3.

**TABLE 6.20 A $2^4$ Design in Two Blocks of 8—Solenoid Experiment**
(Any four factors form a full factorial pattern)

| Test | Run order | $A$ | $S$ | $B$ | $T$ | $b$* | $A$ $S$ | $A$ $B$ | $A$ $T$ | $A$ $b$ | $S$ $B$ | $S$ $T$ | $S$ $b$ | $B$ $T$ | $B$ $b$ | $T$ $b$ | $\overline{X}$, | $s$ |
|---|---|---|---|---|---|---|---|---|---|---|---|---|---|---|---|---|---|---|
| Block 1 | | | | | | | | | | | | | | | | | | |
| 1 | 5 | − | − | − | − | + | + | + | + | − | + | + | − | + | − | − | .46, | .04 |
| 2 | 3 | + | + | − | − | + | + | − | − | + | − | − | + | + | − | − | .45, | .10 |
| 3 | 7 | + | − | + | − | + | − | + | − | + | − | + | − | − | + | − | .71, | .01 |
| 4 | 1 | − | + | + | − | + | − | − | + | − | + | − | + | − | + | − | .70, | .05 |
| 5 | 4 | + | − | − | + | + | − | − | + | + | + | − | − | − | − | + | .28, | .15 |
| 6 | 6 | − | + | − | + | + | − | + | − | − | − | + | + | − | − | + | .60, | .07 |
| 7 | 8 | − | − | + | + | + | + | − | − | − | − | − | − | + | + | + | .70, | .02 |
| 8 | 2 | + | + | + | + | + | + | + | + | + | + | + | + | + | + | + | .72, | .01 |
| Block 2 | | | | | | | | | | | | | | | | | | |
| 9 | 15 | − | + | − | − | − | − | + | + | + | − | − | − | + | + | + | .57, | .02 |
| 10 | 10 | − | − | + | − | − | + | − | + | + | − | + | + | − | − | + | .73, | .02 |
| 11 | 9 | + | + | + | − | − | + | + | − | − | + | − | − | − | − | + | .70, | .01 |
| 12 | 13 | − | − | − | + | − | + | + | − | + | + | − | + | − | + | − | .42, | .04 |
| 13 | 16 | + | + | − | + | − | + | − | + | − | − | + | − | − | + | − | .29, | .06 |
| 14 | 11 | + | − | + | + | − | − | + | + | − | − | − | + | + | − | − | .71, | .02 |
| 15 | 14 | − | + | + | + | − | − | − | − | + | + | + | − | + | − | − | .72, | .02 |
| 16 | 12 | + | − | − | − | − | − | − | − | − | + | + | + | + | + | + | .42, | .16 |
| Divisor = 8 | | | | | | | | | | | | | | | | | | |

*$b$ = Blocking variable.

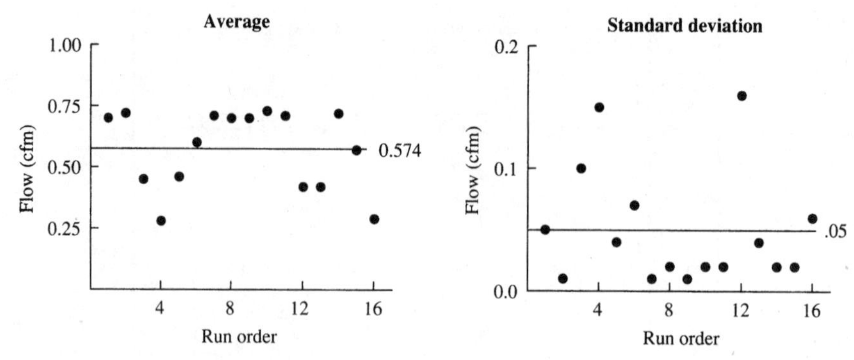


**Figure 6.15** Run charts for the solenoid study.

| Effect | Estimate (cfm) $\bar{x}$ | $s$ |
|---|---|---|
| $A$ | −.08 | .03 |
| $S$ | .04 | −.02 |
| $B$ | .28 | −.06 |
| $T$ | −.04 | .00 |
| $b$ | .01 | .01 |
| $AS$ | −.03 | −.02 |
| $AB$ | .07 | −.05 |
| $AT$ | -.03 | −.01 |

| Effect | Estimate (cfm) $\bar{x}$ | $s$ |
|---|---|---|
| $Ab$ | .00 | −.01 |
| $SB$ | −.04 | .02 |
| $ST$ | .01 | .00 |
| $Sb$ | .04 | .02 |
| $BT$ | .04 | .00 |
| $Bb$ | −.01 | −.01 |
| $Tb$ | .03 | .02 |

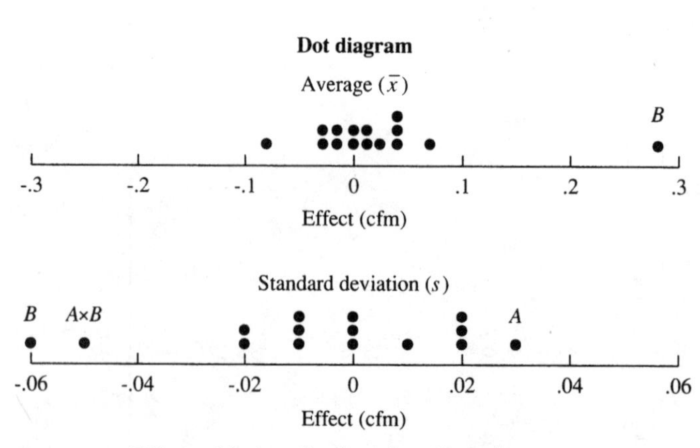


**Figure 6.16** Effects of factors for the solenoid study.

### A $2^{7-3}$ design in two blocks of size 8

The $2^{7-3}$ factorial pattern can be separated into two blocks of eight tests each by setting up a blocking variable just as was done for the $2^3$ and $2^4$ patterns. This blocking variable becomes the eighth factor to be included in the 16 tests and is used to set up the blocks. A $2^{8-4}$ pattern is used to accommodate the seven factors plus the blocking variable. Table 6.21 contains the design matrix for the $2^{7-3}$ pattern in two blocks.

### A $2^{15-10}$ design in two blocks of size 16

The $2^{15-10}$ factorial pattern can be separated into two blocks of 16 tests each by setting up a blocking variable just as was done for the $2^3$, $2^4$, and $2^{7-3}$ patterns. This blocking variable becomes the 16th factor to be included in the 32 tests and is used to set up the blocks. A $2^{16-11}$ pattern is used to accommodate the 15 factors plus the blocking variable. Table 6.22 contains the design matrix for the $2^{15-10}$ pattern in two blocks.

**Example 6.3 (*Continued*)** ***Tensile Strength of Rivets*** To determine the factors that had an effect on the tensile strength of rivets, a $2^{7-4}$ experimental pattern was used initially. The design matrix for a $2^{7-4}$ pattern is contained in Table 6.18. After the results of this design were analyzed, eight more tests were run to separate the confounding between the factors and interactions. The additional tests are those in block 2 of the design matrix for the $2^{7-3}$ pattern in two blocks, shown in Table 6.21.

The design matrix with the data for the rivet study is contained in Table 6.23. The run chart for the 16 factor combinations is shown in Fig. 6.17. No special causes seem to be present. Figure 6.18 shows the effect of the factors on the tensile strength and the dot diagram of the effects.

From Fig. 6.18, it can be seen that the three factors—plate thickness, stem rivet length, and bushing thickness—have a significant effect on tensile strength. These were the same three factors found to be important by using the $2^{7-4}$ design. This indicates that the confounded two-factor interactions were not the source of the large effects.

### Analysis of the cube

A characteristic of a $2^{7-3}$ design in two blocks is that any three factors form a full factorial design. The three important factors were analyzed on a cube in Fig. 6.19. The two values at each corner of the cube are the tensile strengths for the appropriate factor combination in each block. The analysis of the paired comparisons from the cube does not indicate the presence of any special causes of variation and therefore confirms the effects that were found by using the design matrix.

### Response plots

Figure 6.20 contains the response plots that summarize the results of the experiment. Since the factors of plate thickness, stem rivet length,

**TABLE 6.21 Design Matrix for a $2^{7-3}$ Pattern in Two Blocks**
(Any three factors form a full factorial pattern)

| Test | 1 | 2 | 3 | 4 | 5 | 6 | 7 | b | 12<br>37<br>56<br>4b | 13<br>27<br>46<br>5b | 14<br>36<br>57<br>2b | 15<br>26<br>47<br>3b | 16<br>25<br>34<br>7b | 17<br>23<br>45<br>6b | 24<br>35<br>67<br>1b |
|---|---|---|---|---|---|---|---|---|---|---|---|---|---|---|---|
| Block 1 | | | | | | | | | | | | | | | |
| 1 | − | − | − | + | + | + | − | + | + | + | − | − | − | + | − |
| 2 | + | − | − | − | − | + | + | + | − | − | − | − | + | + | + |
| 3 | − | + | − | − | + | − | + | + | − | + | + | − | + | − | − |
| 4 | + | + | − | + | − | − | − | + | + | − | + | − | − | − | + |
| 5 | − | − | + | + | − | − | + | + | + | − | − | + | + | − | − |
| 6 | + | − | + | − | + | − | − | + | − | + | − | + | − | − | + |
| 7 | − | + | + | − | − | + | − | + | − | − | + | + | − | + | − |
| 8 | + | + | + | + | + | + | + | + | + | + | + | + | + | + | + |
| Block 2 | | | | | | | | | | | | | | | |
| 9 | + | + | + | − | − | − | + | − | + | + | − | − | − | + | − |
| 10 | − | + | + | + | + | − | − | − | − | − | − | − | + | + | + |
| 11 | + | − | + | + | − | + | − | − | − | + | + | − | + | − | − |
| 12 | − | − | + | − | + | + | + | − | + | − | + | − | − | − | + |
| 13 | + | + | − | − | + | + | − | − | + | − | − | + | + | − | − |
| 14 | − | + | − | + | − | + | + | − | − | + | − | + | − | − | + |
| 15 | + | − | − | + | + | − | + | − | − | − | + | + | − | + | − |
| 16 | − | − | − | − | − | − | − | − | + | + | + | + | + | + | + |
| Divisor = 8 | | | | | | | | | | | | | | | |

**TABLE 6.22 Design Matrix for a $2^{15-10}$ Pattern in Two Blocks**

(Any three factors form a full factorial pattern)

| Test | 1 | 2 | 3 | 4 | 5 | 6 | 7 | 8 | 9 | 10 | 11 | 12 | 13 | 14 | 15 | b | ........................Two-factor interactions........................ | | | | | | | | | | | | | | |
|---|---|---|---|---|---|---|---|---|---|---|---|---|---|---|---|---|---|---|---|---|---|---|---|---|---|---|---|---|---|---|---|
| Block 1 | | | | | | | | | | | | | | | | | | | | | | | | | | | | | | | |
| 1 | − | − | − | − | + | + | + | + | + | + | − | − | − | − | + | + | − | − | − | − | + | + | + | + | + | + | − | − | − | − | + |
| 2 | + | − | − | − | − | − | − | + | + | + | + | + | + | − | − | + | + | − | − | − | − | − | − | + | + | + | + | + | + | − | − |
| 3 | − | + | − | − | − | + | + | − | − | + | + | + | − | + | − | + | − | + | − | − | − | + | + | − | − | + | + | + | − | + | − |
| 4 | + | + | − | − | + | − | − | − | − | + | − | − | + | + | + | + | + | + | − | − | + | − | − | − | − | + | − | − | + | + | + |
| 5 | − | − | + | − | + | − | + | − | + | − | + | − | + | + | − | + | − | − | + | − | + | − | + | − | + | − | + | − | + | + | − |
| 6 | + | − | + | − | − | + | − | − | + | − | − | + | − | + | + | + | + | − | + | − | − | + | − | − | + | − | − | + | − | + | + |
| 7 | − | + | + | − | − | − | + | + | − | − | − | + | + | − | + | + | − | + | + | − | − | − | + | + | − | − | − | + | + | − | + |
| 8 | + | + | + | − | + | + | − | + | − | − | + | − | − | − | − | + | + | + | + | − | + | + | − | + | − | − | + | − | − | − | − |
| 9 | − | − | − | + | + | + | − | + | − | − | − | + | + | + | − | + | − | − | − | + | + | + | − | + | − | − | − | + | + | + | − |
| 10 | + | − | − | + | − | − | + | + | − | − | + | − | − | + | + | + | + | − | − | + | − | − | + | + | − | − | + | − | − | + | + |
| 11 | − | + | − | + | − | + | − | − | + | − | + | − | + | − | + | + | − | + | − | + | − | + | − | − | + | − | + | − | + | − | + |
| 12 | + | + | − | + | + | − | + | − | + | − | − | + | − | − | − | + | + | + | − | + | + | − | + | − | + | − | − | + | − | − | − |
| 13 | − | − | + | + | + | − | − | − | − | + | + | + | − | − | + | + | − | − | + | + | + | − | − | − | − | + | + | + | − | − | + |
| 14 | + | − | + | + | − | + | + | − | − | + | − | − | + | − | − | + | + | − | + | + | − | + | + | − | − | + | − | − | + | − | − |
| 15 | − | + | + | + | − | − | − | + | + | + | − | − | − | + | − | + | − | + | + | + | − | − | − | + | + | + | − | − | − | + | − |
| 16 | + | + | + | + | + | + | + | + | + | + | + | + | + | + | + | + | + | + | + | + | + | + | + | + | + | + | + | + | + | + | + |

**TABLE 6.22 Design Matrix for a $2^{15-10}$ Pattern in Two Blocks (*Continued*)**
(Any three factors form a full factorial pattern)

| Test | 1 | 2 | 3 | 4 | 5 | 6 | 7 | 8 | 9 | 10 | 11 | 12 | 13 | 14 | 15 | b | ..........Two-factor interactions.......... | | | | | | | | | | | | | | |
|---|---|---|---|---|---|---|---|---|---|---|---|---|---|---|---|---|---|---|---|---|---|---|---|---|---|---|---|---|---|---|---|
| Block 2 | | | | | | | | | | | | | | | | | | | | | | | | | | | | | | | |
| 17 | + | + | + | + | − | − | − | − | − | − | + | + | + | + | − | − | − | − | − | − | + | + | + | + | + | + | − | − | − | − | + |
| 18 | − | + | + | + | + | + | + | − | − | − | − | − | − | + | + | − | + | − | − | − | − | − | − | + | + | + | + | + | + | − | − |
| 19 | + | − | + | + | + | − | − | + | + | − | − | − | + | − | + | − | − | + | − | − | − | + | + | − | − | + | + | + | − | + | − |
| 20 | − | − | + | + | − | + | + | + | + | − | + | + | − | − | − | − | + | + | − | − | + | − | − | − | − | + | − | − | + | + | + |
| 21 | + | + | − | + | − | + | − | + | − | + | − | + | − | − | + | − | − | − | + | − | + | − | + | − | + | − | + | − | + | + | − |
| 22 | − | + | − | + | + | − | + | + | − | + | + | − | + | − | − | − | + | − | + | − | − | + | − | − | + | − | − | + | − | + | + |
| 23 | + | − | − | + | + | + | − | − | + | + | + | − | − | + | − | − | − | + | + | − | − | − | + | + | − | − | − | + | + | − | + |
| 24 | − | − | − | + | − | − | + | − | + | + | − | + | + | + | + | − | + | + | + | − | + | + | − | + | − | − | + | − | − | − | − |
| 25 | + | + | + | − | − | − | + | − | + | + | + | − | − | − | + | − | − | − | − | + | + | + | − | + | − | − | − | + | + | + | − |
| 26 | − | + | + | − | + | + | − | − | + | + | − | + | + | − | − | − | + | − | − | + | − | − | + | + | − | − | + | − | − | + | + |
| 27 | + | − | + | − | + | − | + | + | − | + | − | + | − | + | − | − | − | + | − | + | − | + | − | − | + | − | + | − | + | − | + |
| 28 | − | − | + | − | − | + | − | + | − | + | + | − | + | + | + | − | + | + | − | + | + | − | + | − | + | − | − | + | − | − | − |
| 29 | + | + | − | − | − | + | + | + | + | − | − | − | + | + | − | − | − | − | + | + | + | − | − | − | − | + | + | + | − | − | + |
| 30 | − | + | − | − | + | − | − | + | + | − | + | + | − | + | + | − | + | − | + | + | − | + | + | − | − | + | − | − | + | − | − |
| 31 | + | − | − | − | + | + | + | − | − | − | + | + | + | − | + | − | − | + | + | + | − | − | − | + | + | + | − | − | − | + | − |
| 32 | − | − | − | − | − | − | − | − | − | − | − | − | − | − | − | − | + | + | + | + | + | + | + | + | + | + | + | + | + | + | + |
| Divisor = 16 | | | | | | | | | | | | | | | | | | | | | | | | | | | | | | | |

**TABLE 6.23 Design Matrix for the Rivet Study**

| | | 1 = *RH* 2 = *PD* | | 3 = *PT* 4 = *SD* | | 5 = *SL* 6 = *BD* | | 7 = *BT* 8 = block | | | | | | | | | |
|---|---|---|---|---|---|---|---|---|---|---|---|---|---|---|---|---|---|
| Test | Run order | 1 | 2 | 3 | 4 | 5 | 6 | 7 | 8 | 12 37 56 48 | 13 27 46 58 | 14 36 57 28 | 15 26 47 38 | 16 25 34 78 | 17 23 45 68 | 24 35 67 18 | Tensile strength $\overline{X}$ |
| Block 1 | | | | | | | | | | | | | | | | | |
| 1 | 8 | − | − | − | + | + | + | − | + | + | + | − | − | − | + | − | 513 |
| 2 | 4 | + | − | − | − | − | + | + | + | − | − | − | − | + | + | + | 461 |
| 3 | 6 | − | + | − | − | + | − | + | + | − | + | + | − | + | − | − | 488 |
| 4 | 2 | + | + | − | + | − | − | − | + | + | − | + | − | − | − | + | 481 |
| 5 | 7 | − | − | + | + | − | − | + | + | + | − | − | + | + | − | − | 523 |
| 6 | 3 | + | − | + | − | + | − | − | + | − | + | − | + | − | − | + | 558 |
| 7 | 5 | − | + | + | − | − | + | − | + | − | − | + | + | − | + | − | 532 |
| 8 | 1 | + | + | + | + | + | + | + | + | + | + | + | + | + | + | + | 546 |
| Block 2 | | | | | | | | | | | | | | | | | |
| 9 | 14 | + | + | + | − | − | − | + | − | + | + | − | − | − | + | − | 508 |
| 10 | 12 | − | + | + | + | + | − | − | − | − | − | − | − | + | + | + | 558 |
| 11 | 9 | + | − | + | + | − | + | − | − | − | + | + | − | + | − | − | 545 |
| 12 | 15 | − | − | + | − | + | + | + | − | + | − | + | − | − | − | + | 530 |
| 13 | 10 | + | + | − | − | + | + | − | − | + | − | − | + | + | − | − | 499 |
| 14 | 13 | − | + | − | + | − | + | + | − | − | + | − | + | − | − | + | 463 |
| 15 | 16 | + | − | − | + | + | − | + | − | − | − | + | + | − | + | − | 492 |
| 16 | 11 | − | − | − | − | − | − | − | − | + | + | + | + | + | + | + | 481 |

Divisor = 8

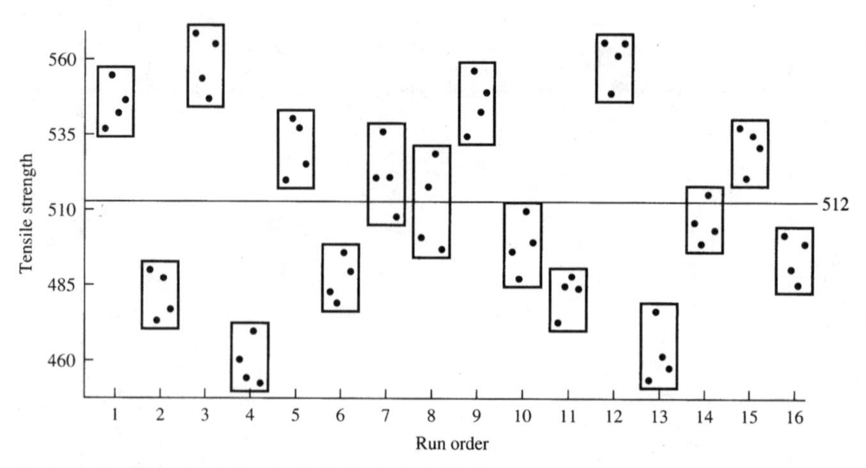


**Figure 6.17** Run chart for rivet study in two blocks.

1 = *RH* 3 = *PT* 5 = *SL* 7 = *BT*
2 = *PD* 4 = *SD* 6 = *BD* 8 = block

| Factor | Effect | Interactions | Effect |
|---|---|---|---|
| 1 (*RH*) | 0.3 | 12+37+56+48 | –5.0 |
| 2 (*PD*) | –3.5 | 13+27+46+58 | 3.3 |
| 3 (*PT*) | 52.8 | 14+36+57+28 | 1.5 |
| 4 (*SD*) | 5.0 | 15+26+47+38 | 1.3 |
| 5 (*SL*) | 23.8 | 16+25+34+78 | 3.0 |
| 6 (*BD*) | 0.0 | 17+23+45+18 | .5 |
| 7 (*BT*) | –19.8 | 24+35+67+18 | –2.8 |
| 8 (*block*) | 3.3 | | |

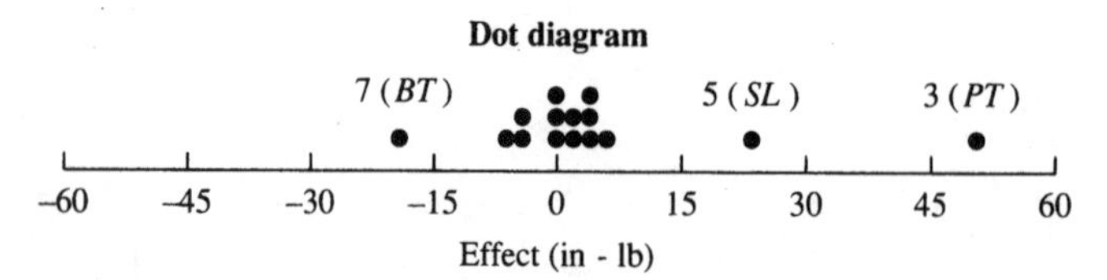


**Figure 6.18** Effects of factors—rivet study in two blocks.

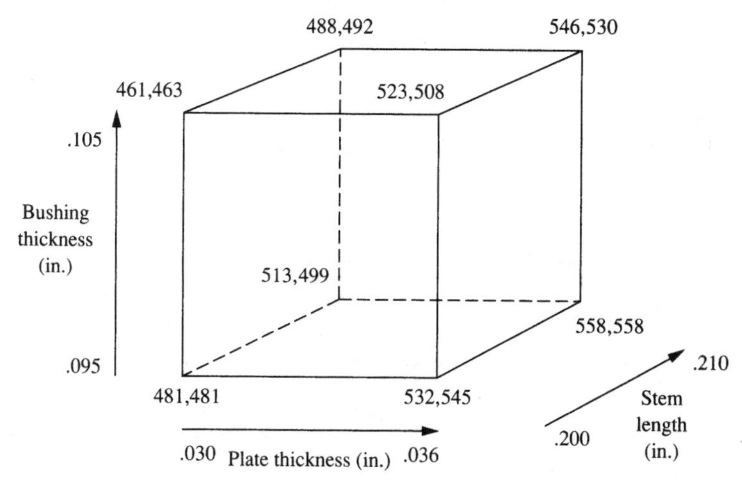


**Figure 6.19** Cube for the rivet experiment.

and bushing thickness had a substantial effect on tensile strength but showed no large interactions, three separate response plots were constructed. The responses from each block were averaged since the block-to-block effect was small.

### Conclusions for the continuation of Example 6.3

1. There were no special causes of variation detected in the run chart or the analysis of the paired comparisons in the cube.
2. Plate thickness, stem rivet length, and bushing thickness had a substantial effect on the average tensile strength of the stem diaphragm plate rivets. As the plate thickness or stem length increased, the tensile strength increased; and as the bushing thickness increased, the tensile strength decreased.
3. The important factors did not interact.
4. Very similar results were seen in each block.

### Development of other blocking patterns

There are many useful experimental patterns besides the three given above that arrange factorial or fractional factorial patterns in incomplete blocks. There is a strong connection between fractional factorial patterns and patterns using incomplete blocks. In this section, a method is described to develop incomplete blocking arrangements using the fractional factorial patterns discussed previously in this chapter.

Table 6.24 contains some of the useful incomplete blocking arrange-

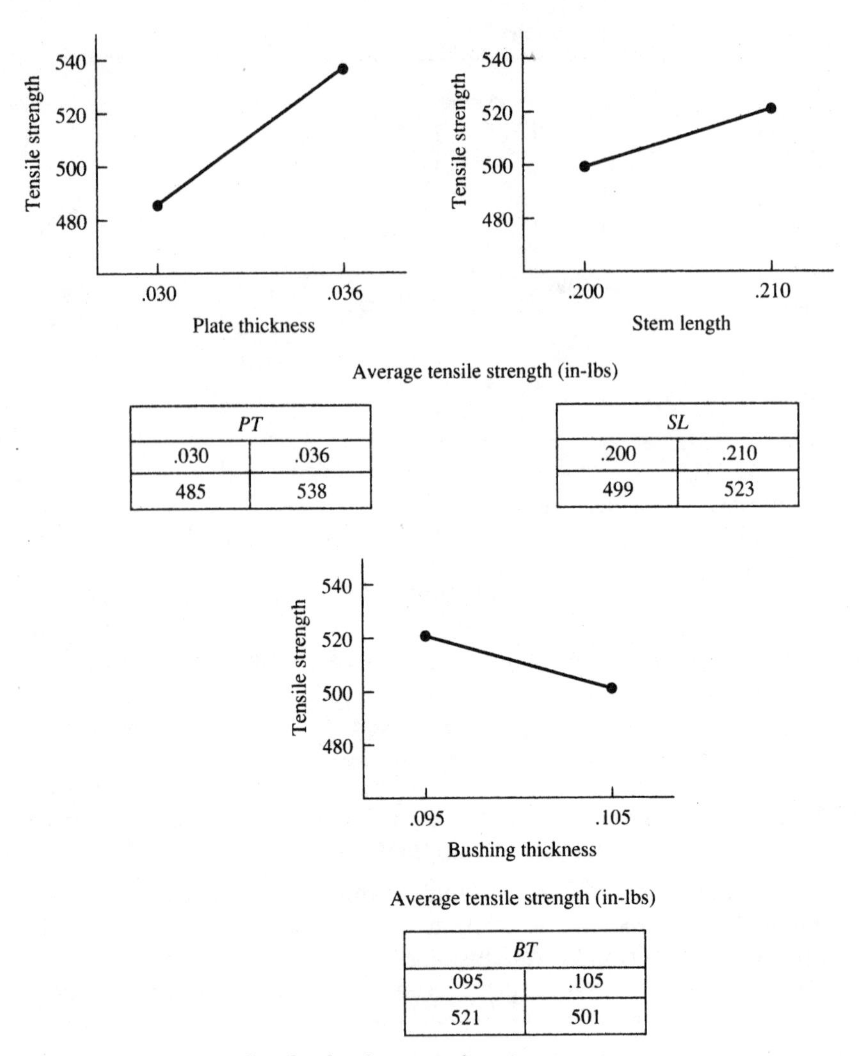


| *PT* | |
|---|---|
| .030 | .036 |
| 485 | 538 |

| *SL* | |
|---|---|
| .200 | .210 |
| 499 | 523 |

| *BT* | |
|---|---|
| .095 | .105 |
| 521 | 501 |

**Figure 6.20** Response plots for the rivet experiment.

ments that can be developed from the fractional factorial patterns in this chapter. Columns 1, 2, and 3 describe the number of factors, the fraction of a full factorial pattern that is to be run, and the total number of tests. Columns 4 and 5 contain the blocking arrangement. Column 6 contains the method to obtain the desired experimental pattern.

The first pattern listed in Table 6.24 is a $2^3$ factorial pattern arranged in two blocks of four tests. This pattern is obtained by using a $2^{4-1}$ pattern. The first three factors in the $2^{4-1}$ pattern correspond to the three factors in the $2^3$ pattern. A blocking variable is then substituted for the fourth factor and is used to set up the blocks. All tests for

**TABLE 6.24 Incomplete Blocking Arrangements**

| Number of factors | Fraction | Number of tests | Number of blocks | Size of blocks | Development of blocks* |
|---|---|---|---|---|---|
| 3 | Full | 8 | 2 | 4 | 1 b.v., $2^{4-1}$ |
| | | | 4 | 2 | 2 b.v., $2^{7-4}$ |
| 4 | Full | 16 | 2 | 8 | 1 b.v., $2^{5-1}$ |
| | | | 4 | 4 | 2 b.v., $2^{8-4}$ |
| | | | 8 | 2 | 3 b.v., $2^{8-4}$ |
| 5 | 1/2 | 16 | 2 | 8 | 1 b.v., $2^{8-4}$ |
| | | | 4 | 4 | 2 b.v., $2^{8-4}$ |
| | 1/4 | 8 | 2 | 4 | 1 b.v., $2^{7-4}$ |
| 6 | 1/4 | 16 | 2 | 8 | 1 b.v., $2^{8-4}$ |
| | | | 4 | 4 | 2 b.v., $2^{8-4}$ |
| | 1/8 | 8 | 2 | 4 | 1 b.v., $2^{7-4}$ |
| 7 | 1/8 | 16 | 2 | 8 | 1 b.v., $2^{8-4}$ |
| 8–13 | 1/16–1/512 | 16 | 2 | 8 | 1 b.v., $2^{15-11}$ |
| | | | 4 | 4 | 2 b.v., $2^{15-11}$ |
| 14 | 1/1024 | 16 | 2 | 8 | 1 b.v., $2^{15-11}$ |

*The information in this column is the number of blocking variables (b.v.) needed to set up the desired number of blocks and the design matrix that is used to arrange the original factorial or fractional pattern into incomplete blocks.

which the blocking variable is negative are put in one block, and all tests for which the blocking variable is positive are put in the other block.

Sometimes it is desirable to arrange the factorial or fractional factorial pattern in smaller blocks, and thus more than two blocks are needed. For example, consider the $2^4$ pattern arranged in four blocks of four tests each. To develop this pattern, start with a $2^{8-4}$ pattern. This pattern is contained in Table 6.7. Label the first four factors as the four factors in the original $2^4$ pattern. Label factors 5 and 6 as blocking variables 1 and 2. Arrange the blocks so that the tests in each block have the same levels as the two blocking variables. Remembering that column 5 and the design matrix now refer to the two blocking variables, the four blocks correspond to column 5 and 6 being $++$, $+-$, $-+$, and $--$. For example, the first block contains tests 1, 8, 12, and 13 in Table 6.7. These four tests are in the same block because the two blocking variables (columns 5 and 6) are at the $++$ level for each of the tests.

The randomization is done separately in each of the four blocks. The response data are analyzed as if they resulted from a fractional factorial pattern with six factors in 16 tests. The confounding pattern is found as before, by crossing out any interactions containing a 7 or an 8. The effects computed from columns 7 and 8 measure the impact of nuisance variables.

Table 6.25 contains the design matrix for a $2^4$ pattern in four blocks of size 4.

**TABLE 6.25 Design Matrix for a $2^4$ Pattern in Four Blocks**
(Any three factors form a full factorial pattern)

| Test* | 1 | 2 | 3 | 4 | 5† | 6† | N‡ | N‡ | 12<br>56 | 13<br>46 | 14<br>36 | 15<br>26 | 16<br>25<br>34 | 23<br>45 | 24<br>35 |
|---|---|---|---|---|---|---|---|---|---|---|---|---|---|---|---|
| Block 1 | | | | | | | | | | | | | | | |
| 1 (1) | − | − | − | + | + | + | − | + | + | + | − | − | − | + | − |
| 2 (8) | + | + | + | + | + | + | + | + | + | + | + | + | + | + | + |
| 3 (12) | − | − | + | − | + | + | + | − | + | − | + | − | − | − | + |
| 4 (13) | + | + | − | − | + | + | − | − | + | − | − | + | + | − | − |
| Block 2 | | | | | | | | | | | | | | | |
| 5 (3) | − | + | − | − | + | − | + | + | − | + | + | − | + | − | − |
| 6 (6) | + | − | + | − | + | − | − | + | − | + | − | + | − | − | + |
| 7 (10) | − | + | + | + | + | − | − | − | − | − | − | − | + | + | + |
| 8 (15) | + | − | − | + | + | − | + | − | − | − | + | + | − | + | − |
| Block 3 | | | | | | | | | | | | | | | |
| 9 (2) | + | − | − | − | − | + | + | + | − | − | − | − | + | + | + |
| 10 (7) | − | + | + | − | − | + | − | + | − | − | + | + | − | + | − |
| 11 (11) | + | − | + | + | − | + | − | − | − | + | + | − | + | − | − |
| 12 (14) | − | + | − | + | − | + | + | − | − | + | − | + | − | − | + |
| Block 4 | | | | | | | | | | | | | | | |
| 13 (4) | + | + | − | + | − | − | − | + | + | − | + | − | − | − | + |
| 14 (5) | − | − | + | + | − | − | + | + | + | − | − | + | + | − | − |
| 15 (9) | + | + | + | − | − | − | + | − | + | + | − | − | − | + | − |
| 16 (16) | − | − | − | − | − | − | − | − | + | + | + | + | + | + | + |
| Divisor = 8 | | | | | | | | | | | | | | | |

*The number in parentheses refers to the test number from the $2^{8-4}$ pattern contained in Table 6.7.
†5 represents the first blocking variable and 6 represents the second blocking variable.
‡These columns measure the effect of nuisance variables.

**TABLE 6.26 Summary of Designs of Fractional Factorial Patterns**

| Experimental pattern | Number of factors | Number of runs | Confounding | Number of factors forming a full factorial design |
|---|---|---|---|---|
| | | Moderate Level of Knowledge | | |
| $2^{4-1}$ | 4 | 8 | 2fi with 2fi | 3 |
| $2^{5-1}$ | 5 | 16 | None | 4 |
| $2^{8-4}$ | 6, 7, 8 | 16 | 2fi with 2fi | Any 3 and some groups of 4 |
| $2^{16-11}$ | 9–16 | 32 | 2fi with 2fi | Any 3 and some groups of 4 or 5 |
| | | Low Level of Knowledge | | |
| $2^{3-1}$ | 3 | 4 | Individual factors with 2fi | 2 |
| $2^{7-4}$ | 5, 6, 7 | 8 | Individual factors with 2fi | Any 2 and some groups of 3 |
| $2^{15-11}$ | 8–15 | 16 | Individual factors with 2fi | Any 2 and some groups of 3 or 4 |

## 6.5 Summary

In this chapter, a small set of fractional factorial designs has been presented. These designs, summarized in Table 6.26, have been separated into designs that are appropriate when experimenters have a moderate level of knowledge about the process and the factors involved and designs that are appropriate when there is a low level of knowledge.

Emphasis has been placed on the sequential use of fractional factorial designs. Some guidance has been given as to what to do after a particular fractional factorial design has been run. Finally, the arrangement of factorial or fractional factorial patterns in incomplete blocks has been discussed.

Fractional factorial designs provide experimenters with a powerful means of learning about their products and processes. The major advantages of these designs are that

- They allow a large number of factors to be studied in relatively few tests.
- They allow each factor to be studied over a wide range of conditions.
- They can be easily used sequentially to build knowledge.

- They take advantage of the Pareto principle that relatively few factors have most of the influence on the response variable.

Fractional factorial designs are not without some disadvantages. The major disadvantages are that

- They are vulnerable to special causes of variation and missing values.
- It is difficult to detect the influence of special causes.
- The theory underlying them is not as easily understood by experimenters as the theory for full factorial designs.
- There is a loss of information relative to full factorial designs due to the confounding of effects.

The small set of designs included in this chapter will meet most of the needs of experimenters whose aim it is to learn more about their products or processes. Other fractional factorial designs that have not been presented may be better in a particular application. For more on other fractional factorial designs, how to construct fractional factorial designs, and how to add other tests to eliminate the confounding between individual factors and interactions, see the References.

## References

Box, G., W. Hunter, and J. S. Hunter (1978): *Statistics for Experimenters,* John Wiley & Sons, New York.

Daniel, C. (1976): *Applications of Statistics to Industrial Experimentation,* John Wiley & Sons, New York.

## Exercises

**6.1** Choose a product or process with which you are familiar. For an important quality characteristic, list potential causes that affect either the average level or the variation of the quality characteristic. (Consider use of a cause-and-effect diagram to do this.) Plan a fractional factorial experiment to determine the effects of some of the potential causes. Use the planning form.

**6.2** In Example 6.2, tests 2 and 14 in Table 6.11 showed evidence of being affected by a special cause of variation. Two additional tests were made at these conditions to see if these results were repeatable. The pull force for test 2 was 1275 and the pull force for test 14 was 1325. Plot these results on the run chart for the original experiment. Substitute these results for the original results of 871 and 920. Analyze the data from the experiment, using these two values, and compare the results to the original analysis. Since the two additional tests were run after the original experiment was performed, these tests could have been influenced by special causes occurring after the original test was completed. Suggest another design to build on the knowledge gained in the first 18 tests.

**6.3** Use the three entries in Table 6.24 for six factors to set up the three experimental patterns. Randomize the tests for each pattern.

**6.4** The manager of an administrative group supporting the sales department of a manufacturing company was concerned about the large number of notes of credit that were sent to customers to correct errors in invoices. A preliminary investigation was done to determine the major sources of the errors. A source of data was a database that was historically kept by the administrative group on every shipment of product. The database contained information on customers, type of product, price, size of shipment, and the like. Also, an entry was made if a letter of credit was needed to correct the invoice.

The group identified four factors relating to a shipment of product and defined two levels for each factor:

| Factor | Level | |
|---|---|---|
| Customer (C) | Minor (−) | Major (+) |
| Customer location (L) | Foreign (−) | Domestic (+) |
| Type of product (T) | Commodity (−) | Specialty (+) |
| Size of shipment (S) | Small (−) | Large (+) |

The group then set up a $2^{4-1}$ pattern to use to sample the database. For each of the eight cells in the pattern, 100 entries in the database were randomly selected, and the percentage of notes of credit for the invoices was recorded. The results of the sampling are contained in the following table:

| $C$ | $L$ | $T$ | $S$ | Percentage needing notes of credit |
|---|---|---|---|---|
| − | − | − | − | 15 |
| + | − | − | + | 18 |
| − | + | − | + | 6 |
| + | + | − | − | 2 |
| − | − | + | + | 19 |
| + | − | + | − | 23 |
| − | + | + | − | 16 |
| + | + | + | + | 21 |

Analyze the data to determine major factors causing the notes of credit.

**6.5** A chemical company had just started production of a new product in its new batch process. After some initial tests, it was found that increasing the yield and decreasing the variation in the viscosity were necessary for a successful product. The batches were run in 9-day campaigns with two batches made each day. Holding tanks were available for raw materials with capacities for enough materials for 18 batches. There were four reactors in the unit that could be used for the new product.

Viscosity was measured in the laboratory using a composite sample from each batch. Yield was calculated through a material balance for each batch. The operators in the unit identified a number of factors that could affect either yield or viscosity:

| Factor | Levels (maximum range for operation) |
|---|---|
| Reactor | A, B, C, or D |
| Reactor temperature | 150 to 270 (°C) |
| Reactor pressure | 130 to 180 (psi) |
| Reaction time | 6 to 8 (h) |
| Raw material supplier | Quality, A-1, Discount |
| Agitation rate | 100 to 150 (rpm) |
| Distillation time | 2 to 4 (h) |
| Catalyst concentration | 6 to 12 (percent) |

Plan an experiment for this process with the following constraints:

1. Learn something about each potential factor.
2. Complete the experiment during the next campaign of the product of interest.
3. Develop the experimental pattern, blocking, replication, and randomization for the study. Complete an experimental design planning form.

Chapter

# 7

# Evaluating Sources of Variation

The experimental designs studied in the last two chapters assumed that the factor combinations were interchangeable, which meant that any combination of factors and levels could be tested. Sometimes, however, the factors of interest are not interchangeable. It may not be meaningful to compare each factor at each level of the other factors. A nested or hierarchical pattern is used to accommodate these types of factors. In a nested design, levels of different factors are studied within a given level of another factor.

Figure 7.1 shows examples of nested patterns for three different studies with two, three, and four factors, respectively. In each case, a factorial arrangement of the factors would not make sense. In the first example, patients are associated with a particular hospital. It would not be reasonable or desirable to interchange the patients with the hospitals.

In the second example, cavity *a* in mold 1 is a physical location, not related to cavity *a* in the other injection mold. If, instead of cavity, the factor were position in the mold, then a factorial arrangement for position in the mold would be appropriate. Cavity *a* might be the position farthest from the center in each of the molds. In the nested experimental pattern, no commonality between the cavities in different molds is assumed.

In the third example, both plants have a day and night shift, but there is no relationship between the two day shifts or the two night shifts. Three operators are selected from each shift, but again there is nothing in common between the four "operator 2s" in the study. Two assemblies are selected from each operator, but "assembly 1" has no meaning beyond being the first assembly selected from a particular operator.

A. Two factors: Hospital (3 levels)
Patients within a hospital (five levels)

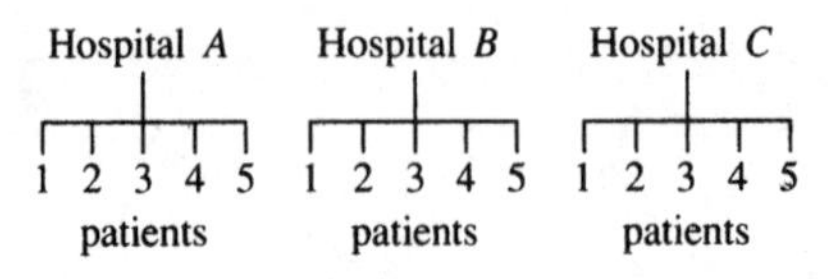


B. Three factors: Injection mold (2 levels)
Cavity within a mold (4 levels)
Parts within a cavity (3 levels)

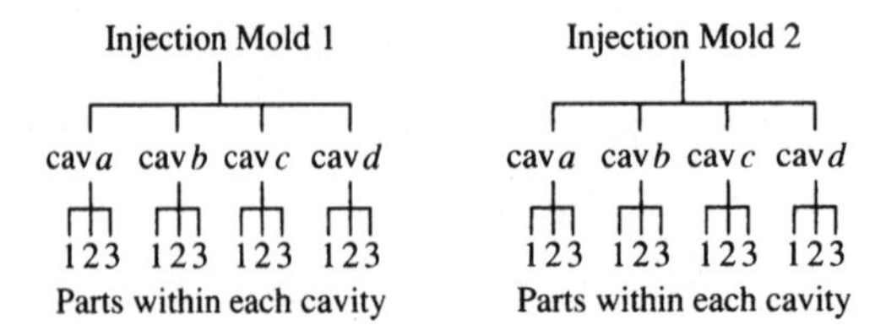


C. Four factors: Plants (2 levels)
Shifts within a plant (2 levels)
Operators ($O_i$) within a shift (3 levels)
Assemblies within an operator (2 levels)

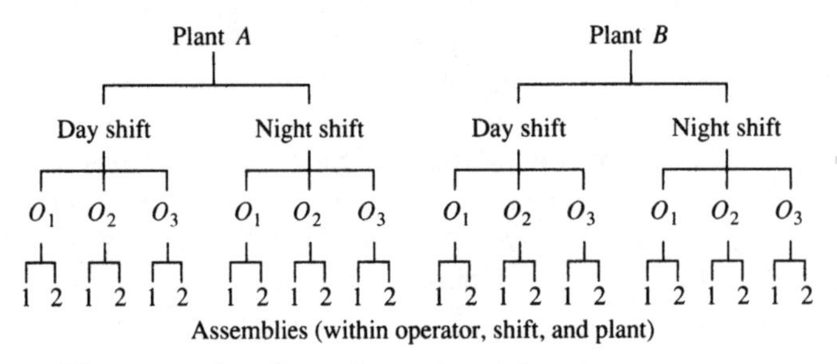


**Figure 7.1** Three examples of nested experimental patterns.

Nested experiments are commonly used to identify the important sources of variation in a system. Many nested studies involve the evaluation of sampling and testing strategies. The analysis of data from a nested design can usually be done with a dot-frequency diagram (Snee, 1983), which is discussed in the next section. As with factorial designs, a run chart is used to evaluate the impact of nuisance variables. A method for quantifying the magnitude of effects of the factors can be used, if needed.

This chapter begins by showing that a control chart can be considered as a two-factor nested design. Designs for more than two factors are then discussed. Section 7.4 summarizes the planning and analysis of a study with nested factors. Section 7.6 discusses a study with both nested and crossed factors.

## 7.1 The Control Chart as a Nested Design

The Shewhart $\overline{X}$ and $R$ control chart is an example of a nested experimental design. The two factors in the design of the control chart are (1) between subgroups and (2) within subgroups. These are chunk-type factors that potentially represent a number of process variables. Shewhart's concept of rational subgrouping was to organize data from the process in a way that is likely to give the greatest chance for the data in each subgroup to be alike and for the data in other subgroups to be different.

Figure 7.2 shows a schematic of an $\overline{X}$ and $R$ control chart as a nested pattern. To analyze data from the control chart, the average and range of the data from each subgroup are calculated. The range is used to evaluate the variation within a subgroup. The variation of the averages is used to evaluate the variation between subgroups. If the averages are within statistical control, then there is no important variation between subgroups in the process.

A dot-frequency diagram could be used to analyze data collected for the $\overline{X}$ and $R$ control chart. The bottom half of Fig. 7.2 shows a schematic of such an analysis. The different subgroups are shown on the horizontal axis, and a scale for the measurements is on the verti-

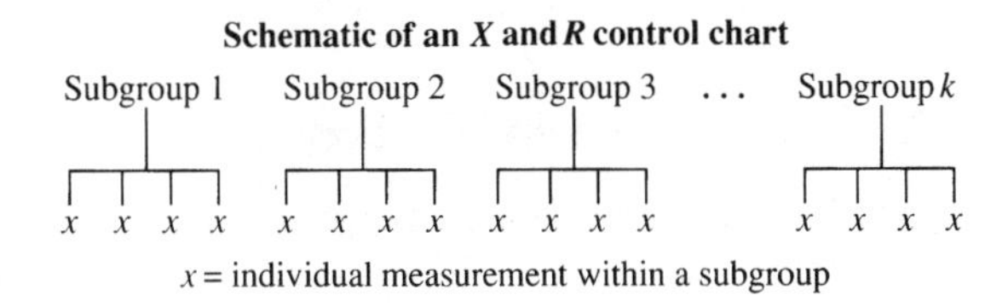


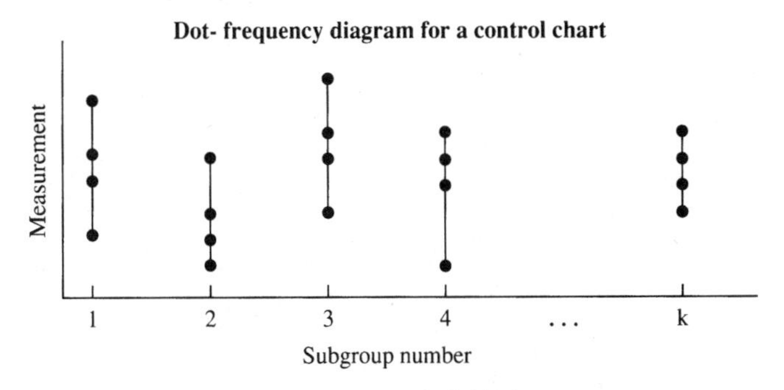


**Figure 7.2** A control chart as a nested design.

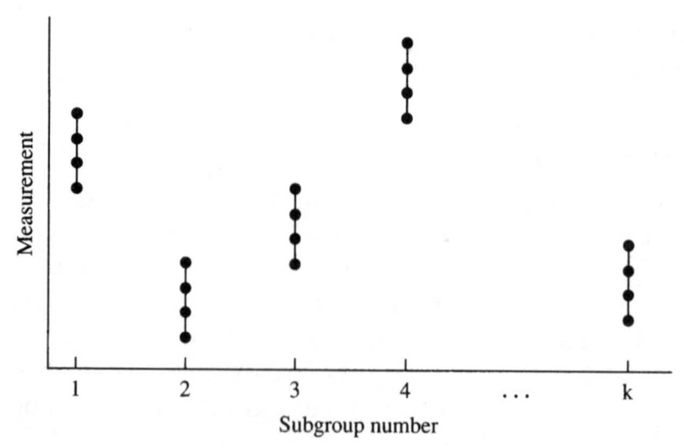


**Figure 7.3** Dot-frequency diagram for a control chart that is not stable.

cal axis. A point is plotted for each measurement above the appropriate subgroup. In this example, the variation within a subgroup is the key source of variation. Each of the lines representing a subgroup overlaps all the other lines. A control chart for these data would be in statistical control.

Figure 7.3 shows a dot-frequency diagram for a control chart that is not stable. In this diagram, the variation within a subgroup is small relative to the variation between subgroups. Some of the lines representing individual subgroups do not overlap. Individual subgroup averages would be outside of the control limits calculated for these data.

## 7.2 Nested Design to Study Measurement Variation

The $\overline{X}$ and $R$ control charts for a process can be modified to include an evaluation of the measurement process used to generate the data. Figure 7.4 shows a schematic of the design for this modified chart.

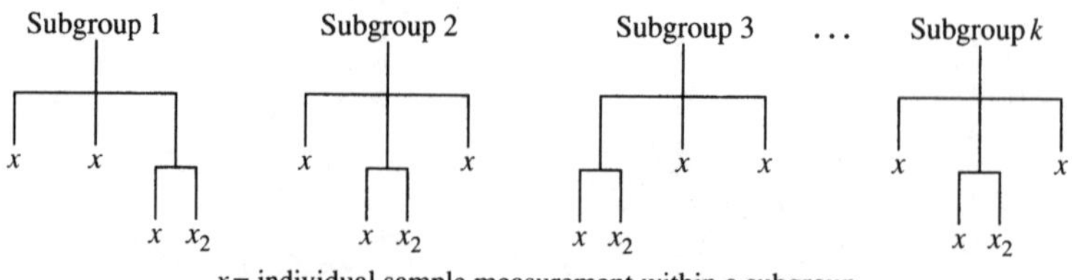


**Figure 7.4** Modified $\overline{X}$ and $R$ chart to evaluate measurement variation.

The three factors in this study are (1) variation between subgroups, (2) variation within subgroups, and (3) variation from the measurement process. Three statistics can then be calculated from the data for each subgroup:

1. The average of the $x$'s for each subgroup
2. The range of the $x$'s for each subgroup
3. The range for the two measurements of the same sample for each subgroup

The average and range of the sample measurements (excluding the additional measurements) are plotted on the usual $\overline{X}$ and $R$ charts. The range of the two measurements on the same sample is plotted on a third chart. Figure 7.5 is an example of such a set of control charts.

A dot-frequency diagram for the data in Fig. 7.5 is shown in Fig. 7.6. To prepare this chart, the vertical axis is scaled to include the range of all the data. The subgroup and sample (part) number are labeled on the horizontal axis. The points plotted on this chart represent a single measurement. The repeated measurements for the first sample are connected by a line. A box is drawn around the points in each subgroup. This is an incomplete or unbalanced nested design—incomplete in the sense that measurements are repeated for only one of the samples in each subgroup. The dot-frequency diagram shows that the subgroups (boxes) overlap the centerline and one another. This indicates that the most important variation is within subgroups. The length of the lines indicates the magnitude of measurement variation. The measurement variation appears small relative to the sample-to-sample variation.

Since all three control charts are stable (in statistical control), the variation in the data can be summarized by estimating standard deviations for the process and for the measurement system. Then the percentage of variation in the process attributable to the measurement system and the percentage of variation attributable to the samples (parts) can be determined. Table 7.1 summarizes these calculations, which are called a *variance component analysis.* For the two factors in this study (variation within a subgroup and measurement variation), the variance component analysis can be viewed geometrically as a right triangle. The sum of the squares of the two sides is equal to the square of the hypotenuse (see figure at bottom of Table 7.1).

From Table 7.1, the measurement variation represents about 19 percent of the variation in the process while the variation of the samples represents about 81 percent. The calculations of the percentage variation due to each source of variation are based on the squares of the standard deviations, called *variance components.*

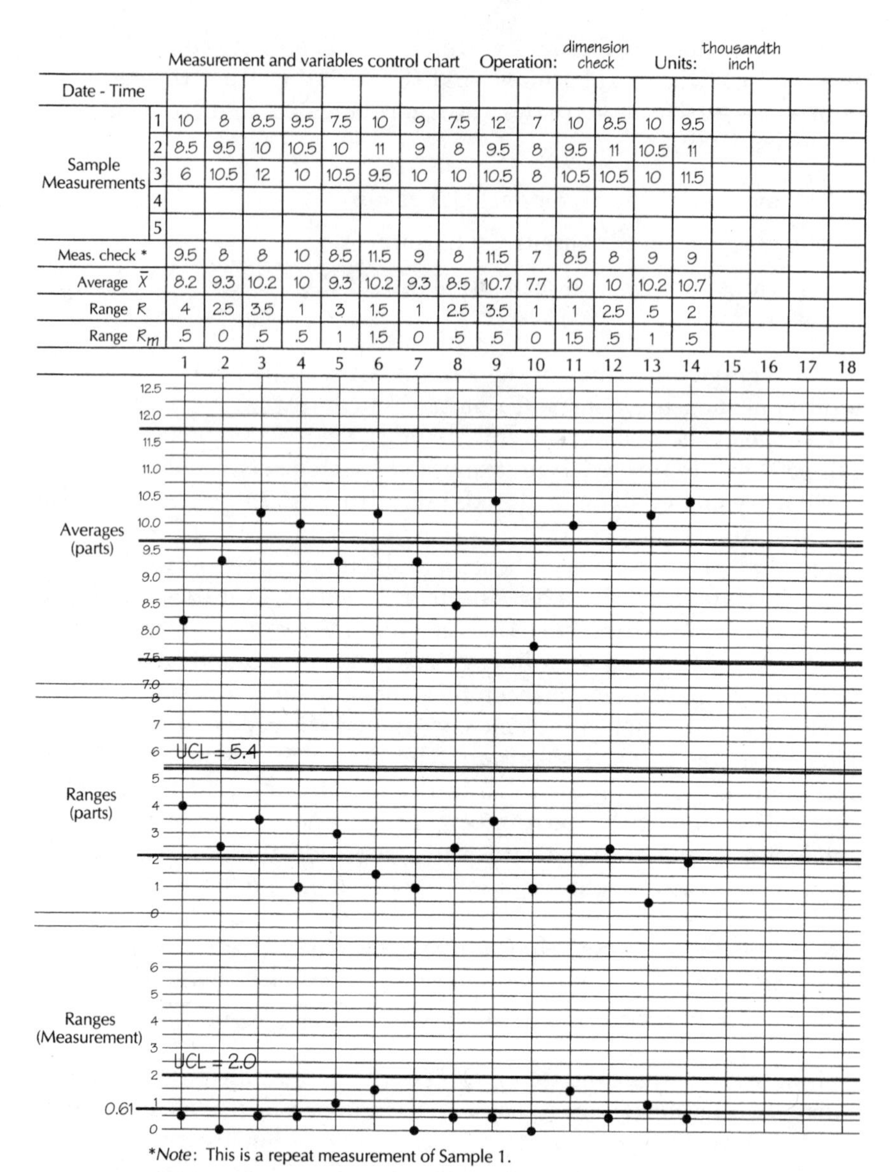
Measurement and variables control chart Operation: dimension check Units: thousandth inch

| | | 1 | 2 | 3 | 4 | 5 | 6 | 7 | 8 | 9 | 10 | 11 | 12 | 13 | 14 | 15 | 16 | 17 | 18 |
|---|---|---|---|---|---|---|---|---|---|---|---|---|---|---|---|---|---|---|---|
| Date - Time | | | | | | | | | | | | | | | | | | | |
| Sample Measurements | 1 | 10 | 8 | 8.5 | 9.5 | 7.5 | 10 | 9 | 7.5 | 12 | 7 | 10 | 8.5 | 10 | 9.5 | | | | |
| | 2 | 8.5 | 9.5 | 10 | 10.5 | 10 | 11 | 9 | 8 | 9.5 | 8 | 9.5 | 11 | 10.5 | 11 | | | | |
| | 3 | 6 | 10.5 | 12 | 10 | 10.5 | 9.5 | 10 | 10 | 10.5 | 8 | 10.5 | 10.5 | 10 | 11.5 | | | | |
| | 4 | | | | | | | | | | | | | | | | | | |
| | 5 | | | | | | | | | | | | | | | | | | |
| Meas. check * | | 9.5 | 8 | 8 | 10 | 8.5 | 11.5 | 9 | 8 | 11.5 | 7 | 8.5 | 8 | 9 | 9 | | | | |
| Average $\overline{X}$ | | 8.2 | 9.3 | 10.2 | 10 | 9.3 | 10.2 | 9.3 | 8.5 | 10.7 | 7.7 | 10 | 10 | 10.2 | 10.7 | | | | |
| Range $R$ | | 4 | 2.5 | 3.5 | 1 | 3 | 1.5 | 1 | 2.5 | 3.5 | 1 | 1 | 2.5 | .5 | 2 | | | | |
| Range $R_m$ | | .5 | 0 | .5 | .5 | 1 | 1.5 | 0 | .5 | .5 | 0 | 1.5 | .5 | 1 | .5 | | | | |



*Note: This is a repeat measurement of Sample 1.

**Figure 7.5** Example of $\overline{X}$, $R$, and measurement control charts.

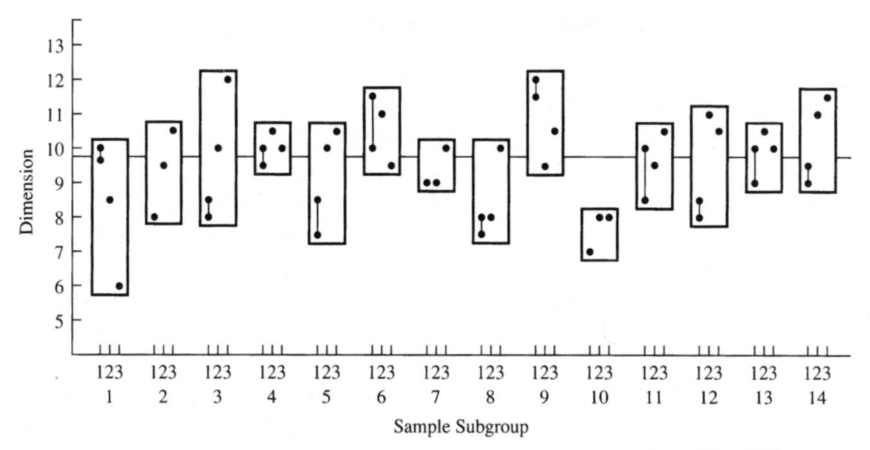


**Figure 7.6** Dot-frequency diagram for measurement control chart data (Fig. 7.5).

**TABLE 7.1 Summary of Range Control Charts**

Variation for process:

$$\bar{R} = 2.10 \qquad \hat{\sigma}_p = \frac{\bar{R}}{d_2} = \frac{2.10}{1.693} = 1.24$$

Variation of measurement process:

$$\bar{R}_m = 0.61 \qquad \hat{\sigma}_m = \frac{\bar{R}_m}{d_2} = \frac{0.61}{1.128} = 0.54$$

Variation of product:

$$\hat{\sigma}_{\text{product}} = \sqrt{(\hat{\sigma}_{\text{process}})^2 - (\hat{\sigma}_{\text{measurement}})^2}$$

$$\hat{\sigma}_{\text{product}} = \sqrt{(1.24)^2 - (0.54)^2} = 1.12$$

| | Summary of variation (units = 0.001 inch) | | |
|---|---|---|---|
| Source of variation | Standard deviation $\hat{\sigma}$ | Variance component $\hat{\sigma}^2$ | Percentage of variation |
| Product | 1.12 | 1.25 | 81.1 |
| Measurement | 0.54 | 0.29 | 18.9 |
| Total (process) | 1.24 | 1.54 | 100.0 |

Geometric relationship of variance components

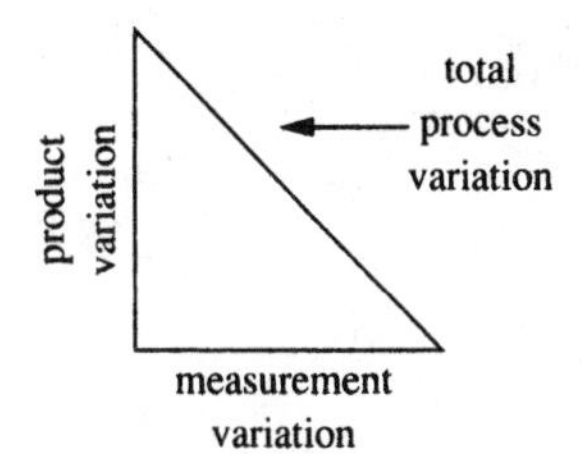


$$\hat{\sigma}^2 \text{ process} = \hat{\sigma}^2 \text{ product} + \hat{\sigma}^2 \text{ measurement}$$

## 7.3 A Three-Factor Nested Experiment

Figure 7.7 shows a completed planning form for a three-factor nested experiment. The objective was to evaluate the measurement system used in the process. The measurement process involved two key steps—setup of the part to be measured and the gaging of the part by using calipers. A total of 54 measurements were made for the study (6 parts × 3 setups × 3 measurements). The data for the completed study are shown on the data collection form in Fig. 7.8.

Figure 7.9 shows a run chart of the data. No obvious trends or other special causes are seen on the run chart. The low measurements on the sixth test were the first setup on part 4. The other two setups for part 4 (runs 13 and 15) also showed low results. A dot-frequency diagram that partitions the variation in the gap width between the three factors in the study is shown in Fig. 7.10.

The dot-frequency diagram indicates that part-to-part differences are the largest source of variation. Each box in the diagram represents a part. Part 4 has a much smaller gap than the other five parts. The variation within each part can be evaluated by studying the lines drawn in each box. The length of the lines represents the measurement variation, while the difference between the lines represents the setup variation. For parts 1, 2, 4, and 6, the setup differences are large relative to the measurement variation.

The factors in the study can be ordered by their contribution to the variation:

1. Part—greatest source of variation
2. Setup within a part
3. Measurement—smallest source of variation

Reductions in variation in the process could come from improving the setup procedure in the measurement process and then concentrating on factors that cause the part-to-part variation.

The percentage of variation attributable to each of the factors can be quantified by extending the variance component analysis procedure presented in the previous example. This procedure is presented in App. 7A. The variance component analysis verifies the visual analysis of the dot-frequency diagram:

- 72 percent of the variation is attributable to part differences.
- 18 percent is attributable to the setup.
- 10 percent is due to the measurement procedure.

## Form for documentation of a planned experiment

1. **Objective:**
   Evaluate the variability of the measurement system and determine which part of the measurement process contributes most of the variation.
2. **Background information:**
   This measurement system had not been evaluated since new calipers were purchased. Previous studies of similar systems had shown the setup procedure to be important.
3. **Experimental variables:**

| | | |
|---|---|---|
| A. | Response variables | Measurement technique |
| 1. | Width of gap (millimeters) | QC standard procedure using calipers |
| B. | Factors under study | Levels |
| 1. | Parts | Six parts selected. |
| 2. | Setup within parts | Each part setup three times. |
| 3. | Measurement within setup | Measurement made by three operators within each setup |
| C. | Background variables | Method of Control |
| 1. | Time | Parts selected from an hour period when free of special causes. |
| 2. | Operators | Three volunteers used in study. |
| 3. | Calibration | Calipers checked at the beginning and end of study. |

4. **Replication:**
   The study has to be completed in one afternoon. Each setup takes 15 minutes, so less than 20 setups can be done. The replication is determined by the levels of each factor in the study. Only one replication of the experimental pattern will be done.
5. **Methods of randomization:**
   The 18 setups (3 setups per part) were randomly ordered by putting the numbers 1 to 6 in a hat and drawing with replacement until 3 setups for each part were selected. The measurements by the three operators were done in the same order each time.
6. **Design matrix:** (see Figure 7.8)
7. **Data collection forms:** (see Figure 7.8)
8. **Planned methods of statisical analysis:**
   Run chart, dot-frequency diagram, estimate of components of variation.
9. **Estimated cost, schedule, and other resource considerations:**
   Study can be completed in 5-hour period using 3 operators.

**Figure 7.7** Form for documentation of three-factor nested experiment.

**Design matrix (experimental pattern)**

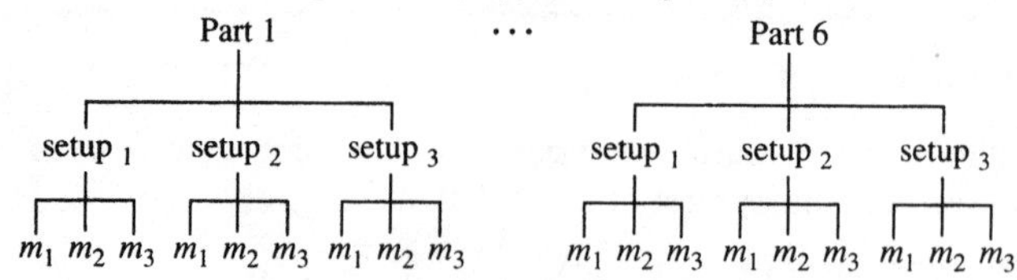


$m$ = Measurements by three operators after each setup.
Randomize the order of the 18 setups in the study.

**Data collection form (with completed data)**
Measurement by three operators (gap width in millimeters)

| Test order | Part | Setup | Meas 1 | Meas 2 | Meas 3 | Notes |
|---|---|---|---|---|---|---|
| 1 | 3 | 1 | 4.30 | 4.70 | 4.70 | |
| 2 | 2 | 1 | 4.55 | 4.70 | 4.70 | |
| 3 | 5 | 1 | 3.90 | 4.85 | 4.60 | |
| 4 | 2 | 2 | 4.30 | 4.20 | 4.50 | |
| 5 | 5 | 2 | 4.35 | 4.90 | 4.35 | |
| 6 | 4 | 1 | 2.60 | 2.65 | 2.50 | verified that readings were okay after initial low result |
| 7 | 5 | 3 | 4.10 | 4.00 | 3.95 | |
| 8 | 6 | 1 | 3.95 | 4.10 | 4.00 | |
| 9 | 1 | 1 | 3.40 | 3.45 | 3.70 | |
| 10 | 1 | 2 | 3.95 | 3.80 | 4.00 | |
| 11 | 3 | 2 | 4.50 | 4.55 | 4.65 | took 15 minute break after 12th setup |
| 12 | 6 | 2 | 3.90 | 3.90 | 4.00 | |
| 13 | 4 | 2 | 3.00 | 3.15 | 3.15 | |
| 14 | 2 | 3 | 4.75 | 5.15 | 5.20 | |
| 15 | 4 | 3 | 3.40 | 3.55 | 3.40 | |
| 16 | 6 | 3 | 4.40 | 4.30 | 4.40 | |
| 17 | 3 | 3 | 4.15 | 4.45 | 4.50 | |
| 18 | 1 | 3 | 3.95 | 4.00 | 4.15 | |

**Figure 7.8** Design matrix and data collection form for a three-factor nested study.

## 7.4 Planning and Analyzing an Experiment with Nested Factors

The previous three sections have given examples of experimental situations in which the factors were nested. The control chart and the study of measurement variation are common examples of nested studies. This section summarizes the design and analysis of experiments with nested factors.

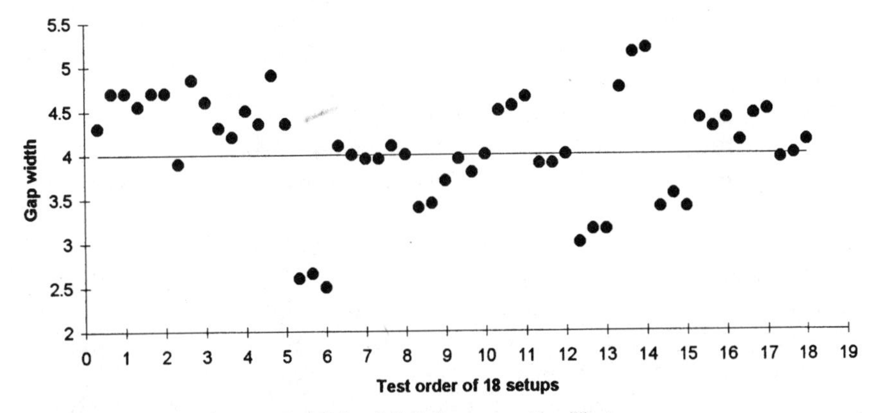


**Figure 7.9** Run chart for three-factor nested study.

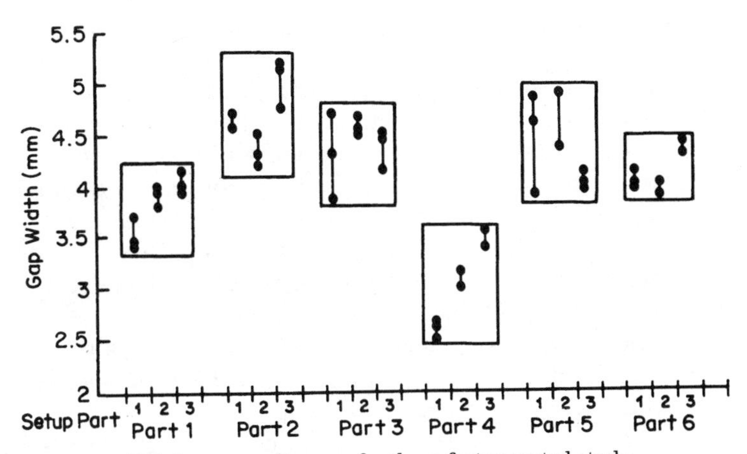


**Figure 7.10** Dot-frequency diagram for three-factor nested study.

## Planning a nested experiment

The basic steps to plan an experiment with nested factors are no different from those for a factorial experiment. Section 5.2 discussed these steps. The form entitled "Documentation of a Planned Experiment" is useful in the planning. The objective of a nested experiment often focuses on understanding sources of variability in a process. A nested experimental design is often used as a screening study to direct focus for activities to improve a process. One of the most common applications of a nested design is to study measurement procedures and sampling strategies.

The selection of factors and factor levels is often straightforward in a nested study. In some studies, the physical layout of a process defines the factors and levels. For a study of a process with three molds and six cavities in each mold, the factors molds and cavities within a mold are obvious, with the levels fixed for both factors.

The amount of information available to evaluate the importance of each factor is another consideration in the assignment of levels. The hierarchical structure provides more information for factors at the lower levels of the hierarchy. In the three-factor example in Sec. 7.3, there were only six different parts in the study (the factor at the highest level in the hierarchy), but 54 different measurements were used to evaluate the measurement effect (the lowest factor in the hierarchy). When possible, more levels should be assigned to the highest factors in the hierarchy and fewer levels (usually two) should be used for the lowest factors in the hierarchy. This will balance the amount of information available about the contribution to variation from each factor.

Background variables such as operator or raw material supplier can be incorporated into a nested design in the same way as they are incorporated into a factorial experiment. Blocks can be established as a chunk variable by using a number of background variables and then can be treated as a factor in the nested design. For example, the nested factor machine within plant might include operator, setup, configuration, and the like. Also, blocks can be set up, and the nested experimental pattern can be replicated in each block.

The experimental pattern for a nested experiment displays the hierarchical nature of the factors. Figures 7.1, 7.4, and 7.8 show examples of experimental patterns for nested studies.

The amount of replication is important in a nested design. The levels of the factors in the study often dictate the replication that is done. Balanced replication (an equal number of levels for each evaluation of a nested factor) makes the analysis much easier but is not always an efficient use of test resources. The $\overline{X}$ and $R$ control charts are common examples. It is desirable to select an equal number of samples or measurements within each subgroup. A variable subgroup size can be used, but this complicates the analysis. Complete replication is not necessary in a nested experiment. For example, in the evaluation of a measurement system (Sec. 7.2), the measurement was repeated on only one of the three samples in the subgroup. Other forms of incomplete replication can be easily incorporated into a nested design (Bainbridge, 1965).

Randomization in a nested experiment is applicable to sampling, the order of running the tests, and the order of making measurements on the experimental units. In contrast to factorial designs, the hierarchical structure of the factors in a nested study often leads to some restrictions on the randomization. For example, once a machine is set up, it may be desirable to complete all the tests within that machine before moving to the next (see the measurement example in Sec. 7.3). In a medical study, hospital and patients within a hospital might be two nested factors. Randomization could be used in selecting the hospitals from a list and the patients from those currently in the chosen hospitals. The order of visiting each hospital and of surveying the selected patients within each hospital could also be randomized.

### Analyzing a nested experiment

The examples in the previous sections of this chapter contained some of the analysis procedures for nested studies. The recommended approach to the analysis comprises three steps:

1. Plot the data in run order to evaluate stability during the study. Look for obvious trends and other types of special causes in this run chart.
2. Prepare a dot-frequency diagram for the basic data. The dot-frequency diagram shows all the observations organized by the associated nested factors.
3. Study the dot-frequency diagram and summarize the information for each factor. Additional dot-frequency diagrams with a different ordering of factors may be desirable to highlight the most important factors. It also may be useful to prepare $\overline{X}$ and $R$ control charts, with appropriate subgrouping, to further evaluate the importance of selected factors.

**Step 1: Run chart.** An example of a run chart for a nested experiment was given in Fig. 7.9. This chart is prepared to evaluate the impact of nuisance variables in the study. Alternatively, $\overline{X}$ and $R$ charts can be prepared by using the variation of the lowest factor in the hierarchy to calculate the range. Note that the $\overline{X}$ chart may be out of control if any of the factors higher in the hierarchy are important.

It is important to consider any trends, runs, or cycles in the run order plot before a dot-frequency diagram is prepared. Certain patterns due to special causes from nuisance variables may distort the dot-frequency diagram. The effect of the nuisance variables that caused the nonrandom pattern will often be attributed to one or more of the factors. Individual points that are affected by special causes should also be removed or adjusted prior to preparing the dot-frequency diagram, since this diagram is designed to focus on the factors in the study. If these individual points are not removed, however, they can usually be detected in the dot-frequency diagram.

**Step 2: Dot-frequency diagram.** The dot-frequency diagram is designed to partition the variation in the original data among the factors in the study. This allows a visual analysis of the data. The construction of a dot-frequency diagram is as follows:

1. Set up a vertical scale to include the highest and lowest numerical values obtained in the study.
2. Develop a horizontal scale using the factors in the study. There will be an identifier row for all the factors except the lowest factor in the hierarchy. The lowest factor will be represented by different dots plotted directly above the other identifiers. The other factors should be ordered with the highest factor in the hierarchy at the bottom of the scale.
3. Plot the original data in the location on the plot identified by the factor levels associated with each value.
4. For the lowest factor in the hierarchy, draw a vertical line to connect the dots.
5. Draw a box around the lines for the second-lowest factor in the hierarchy. Additional boxes can be drawn to identify other factors in the hierarchy.

**Step 3: Analysis of the dot-frequency diagram.** Figure 7.11 contains illustrations of dot-frequency diagrams for a three-factor nested experiment (factors $A$, $B$ within $A$, and $C$ within $B$ and $A$). Four different diagrams are shown, illustrating the following situations:

1. Factor $A$ most important
2. Factor $B$ most important
3. Factor $C$ most important
4. All factors equally important

Special causes can impact the interpretation of a dot-frequency diagram. Figure 7.12 illustrates dot-frequency diagrams with two types of special causes:

1. The special cause affects a single value.
2. The special cause is a trend across all values.

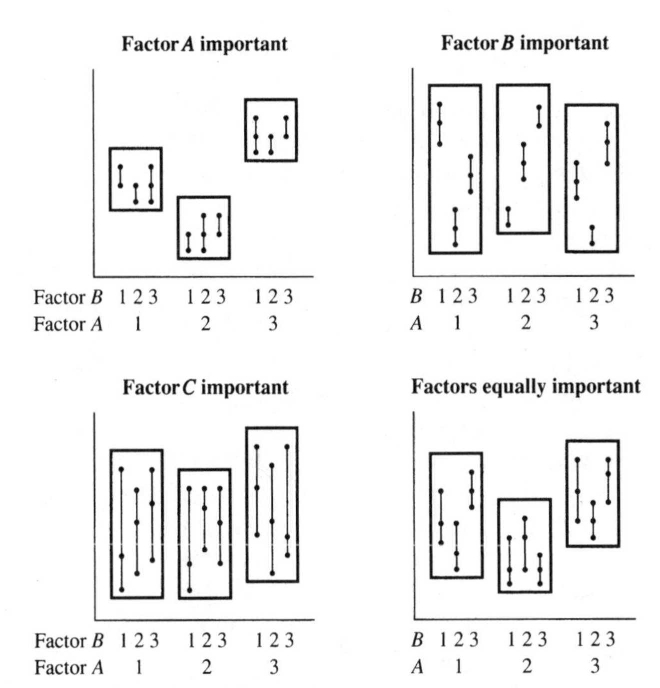


**Figure 7.11** Dot-frequency diagrams.

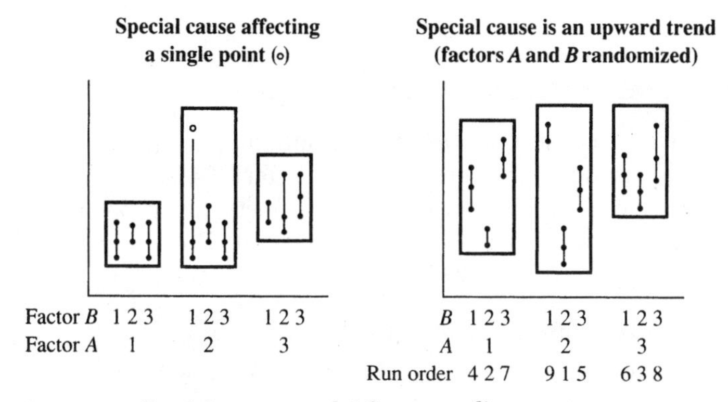


**Figure 7.12** Special cause on a dot-frequency diagram.

The impact of the trend will depend on the amount of randomization done. The trend will be associated with the lowest factor in the hierarchy for which the order was randomized. In the example in Fig. 7.12, the response increases linearly with run order. Since factor $B$ is the lowest factor in the hierarchy that was randomized, the trend makes factor $B$ appear to have a large effect.

Sometimes it is desirable to prepare a dot-diagram to summarize the key factors in the study. The factors can be reordered to highlight the most important factors, and factors not important can be eliminated from the diagram. The example in Sec. 7.5 can be used to illustrate this point. In foundry 2, both the day effect and the part effect are insignificant. A dot-frequency diagram based on the six heats (ignoring days) and the nine measurements of hardness within each heat (ignoring parts) would highlight the important factors.

The $\overline{X}$ and $R$ control charts can be prepared to verify observations on the dot-frequency diagram. The data can be organized in rational subgroups with the least important factors included within subgroups and the important factors included between subgroups. In the three-factor example in Sec. 7.3, a control chart could be prepared with subgroups defined by the parts, and the averages of the three measurements on each setup within a part could be used to calculate the average and range. The subgroup size (three) would be used to calculate control limits. The control chart would verify the importance of the variation between parts by showing part averages outside the control limits.

For some nested studies, it may be desirable to quantify the magnitude of each of the factors. The method of variance component analysis is available to estimate the component of variation associated with each factor. The measurement study examples in Secs. 7.2 and 7.3 illustrated this analysis (see Table 7.1). This analysis is further described in App. 7A.

## 7.5 More Than Three Factors in a Nested Design

The procedures used to evaluate the measurement process in Secs. 7.2 and 7.3 can be extended to more than three factors. A study was done to determine the most important factors affecting the hardness of parts. The specification for hardness was 30 to 40 Rockwell units. The following factors were considered in the study:

1. Foundries (2 different suppliers of parts)
2. Days within foundries (3 days selected for each foundry)
3. Heat within a day (2 heats selected for each day)
4. Parts within a heat (3 parts selected from each heat)
5. Measurements within a part (hardness measured 3 times)

A complete nested design was conducted with a resultant 108 ($2 \times 3 \times 2 \times 3 \times 3$) data points. The design matrix is shown in Fig. 7.13. Each part was collected and labeled, and then the hardness tests were conducted in random order. A run chart indicating the test order is also shown in Fig. 7.13. The run chart does not indicate any special causes.

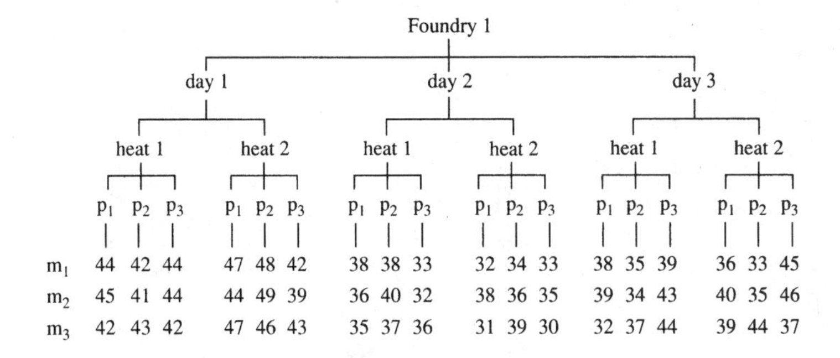


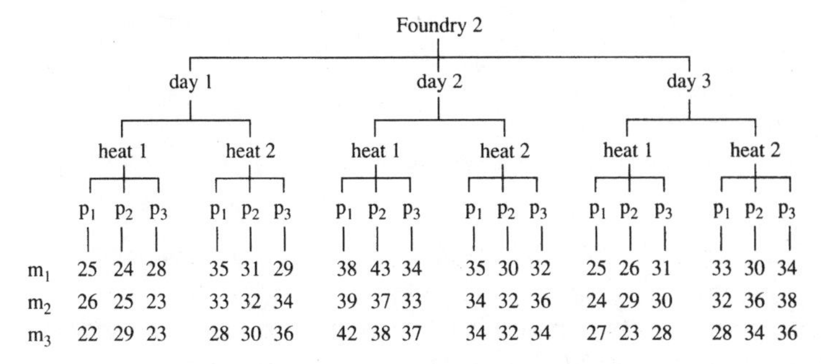


$p_1, p_2, p_3$ = three parts within each heat
$m_1, m_2, m_3$ = three measurements of each part
*Note*: Measurements are in Rockwell hardness scale.

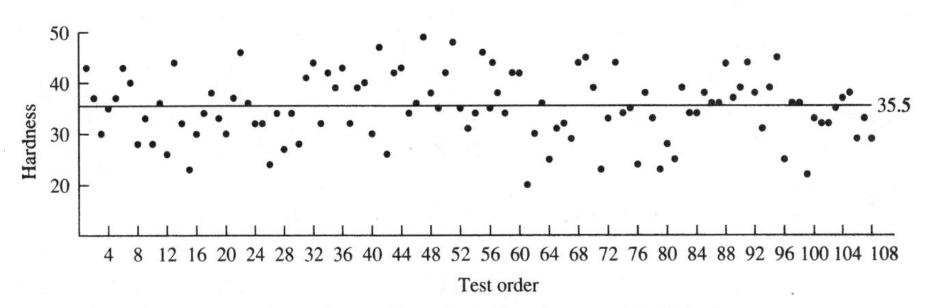


**Figure 7.13** Design matrix and run chart for a five-factor nested design.

A dot-frequency diagram for the study is shown in Fig. 7.14. Each line on the dot-frequency diagram represents an individual part. Each box shown on the graph represents a particular heat of parts. Additional boxes can be drawn to represent each day and each foundry.

The dot-frequency diagram allows one to visually assign the variation in the response variable to each of the factors. The length of the lines drawn represents the measurement variation. The length of the lines ranges from 2 to 12 units and "averages" about 4 or 5 units. The height of the boxes (and the differences in the lines within a box) rep-

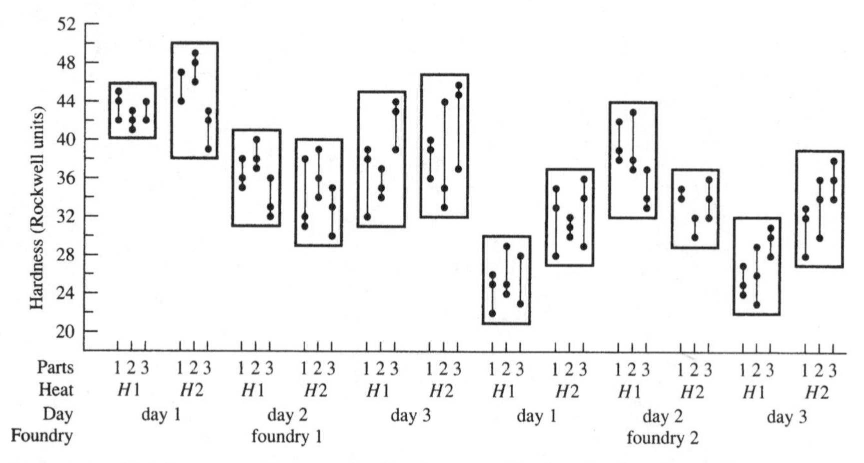


**Figure 7.14** Dot-frequency diagram for five factors affecting the hardness of parts.

resents the part-to-part variation. Note that the "average" height of the boxes is greater than the specification width of 10 units.

The difference between the consecutive pairs of boxes indicates the heat-to-heat variation within a day. For foundry 1 there is almost no difference in these pairs of boxes, while for foundry 2 the pairs are very different. The variation due to heats is insignificant in foundry 1, but it is a very important source of variation in foundry 2.

The day-to-day variation within a foundry can best be seen by drawing a box around the two heats in each day (Fig. 7.15*a*). Note that there is some day-to-day variation in foundry 1, but the day-to-day variation in foundry 2 is no greater than the heat-to-heat variation.

The variation due to foundry differences can be seen by comparing all the measurements on the left half of the dot-frequency diagram with those on the right half (Fig. 7.15*b*). The measurements average near the upper specification of 40 for foundry 1 and near the lower specification of 30 for foundry 2. Significant improvements would result if foundry 1 could target hardness 5 units less and foundry 2 could target hardness 5 units greater.

To summarize the dot-frequency diagrams, reduction in the variation in hardness will require different actions in the two foundries. For foundry 1, the process should first be changed to produce parts that average about 5 hardness units less. Next, the focus should be on the day-to-day variation. The procedure used to measure hardness should also be reviewed for improvement. The heat-to-heat and the part-to-part variations are not important sources of hardness variation.

For foundry 2, the process should first be changed to produce parts

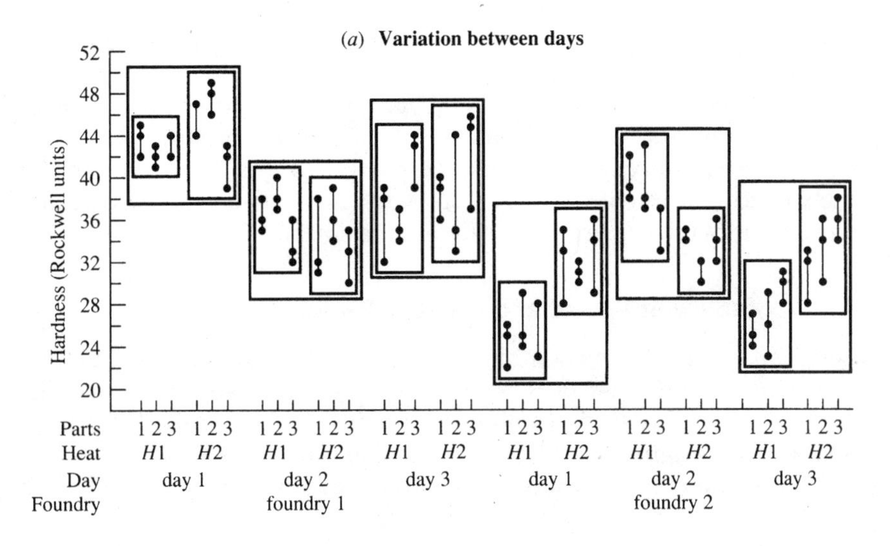


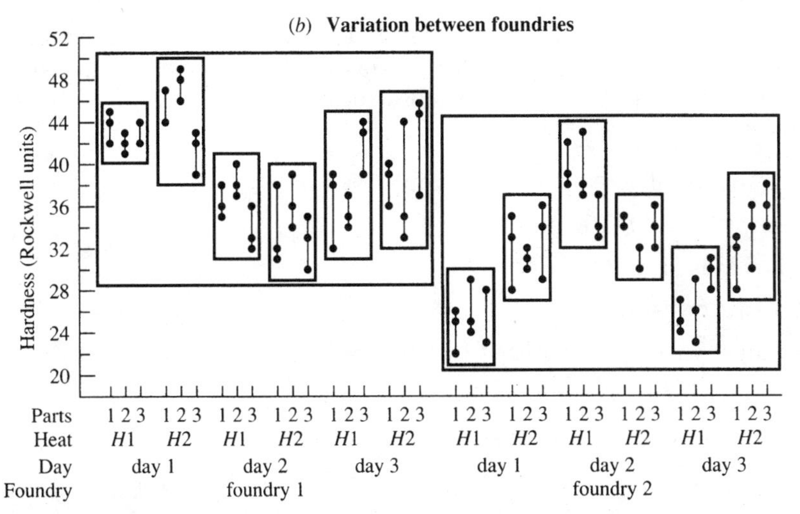


**Figure 7.15** Dot-frequency diagram (hardness of parts) for five-factor nested design with different factors emphasized.

that average about 5 hardness units greater. Then the focus should be on the variation from heat to heat. The measurement process is also a significant source of variation for the parts from foundry 2. The part-to-part variation and the day-to-day variation are not important sources of hardness variation.

The amount of variation due to each of the factors in the study can be quantified by a variance component analysis. This analysis is given in App. 7A.

The summary of the variance components supports the conclusions from the dot-frequency diagrams. The action required for improvement is different in the two foundries. In both foundries, the measurement process represents about 21 percent of the total variation. In foundry 1 the variation between days accounts for 67 percent of the variation, while in foundry 2 the variation between heats within a day accounts for 69 percent of the total variation.

## 7.6 A Study with Nested and Crossed Factors

Sometimes a study will contain some factors that are crossed (a factorial pattern) and other factors that are nested (a nested pattern). The analysis of such a study combines graphical tools from both factorial and nested designs. The following example illustrates this.

An interlaboratory study was done to evaluate a new analytical method to determine particle size, a method that would be used in three laboratories within one company. The response variables were the weights of materials that passed through various size screens. The key response variable was the amount of material not passing through a 200-mesh screen. The following seven factors were included in the study:

1. Alignment of screens (stack *A* and stack *B*)
2. Sample weights (50 and 100 g)
3. Flow aids (1 and 2)
4. Laboratories (*A, B, C*)
5. Analyst within laboratory (2 within each lab)
6. Days with analyst (2 days for each analyst)
7. Measurement within days (2 measurements each day)

The first three factors are across the entire study and thus form a factorial design with each factor at two levels. The last four factors form a nested design. The design matrix for the entire experiment is shown in Fig. 7.16. The factors are organized on the design matrix to highlight the focus of the study. The three factorial factors represent different ways to run the test, while the three nested factors represent different conditions under which a given test method will be run in the future. One of the objectives of the study was to choose a test condition that would minimize the measurement variation across different laboratories and analysts.

Table 7.2 contains the data obtained from the study. Run charts were prepared for the data from each laboratory. Excessive variation

**Factorial design matrix for three crossed factors**

| | Alignment - stack *A* | | Alignment - stack *B* | |
|---|---|---|---|---|
| | Flow aid 1 | Flow aid 2 | Flow aid 1 | Flow aid 2 |
| Sample size = 50 g | 22 tests (see design below) | 22 tests | 22 tests | 22 tests |
| Sample size = 100 g | 22 tests | 22 tests | 22 tests | 22 tests |

**Experimental pattern in each of the eight cells in the factorial design matrix**

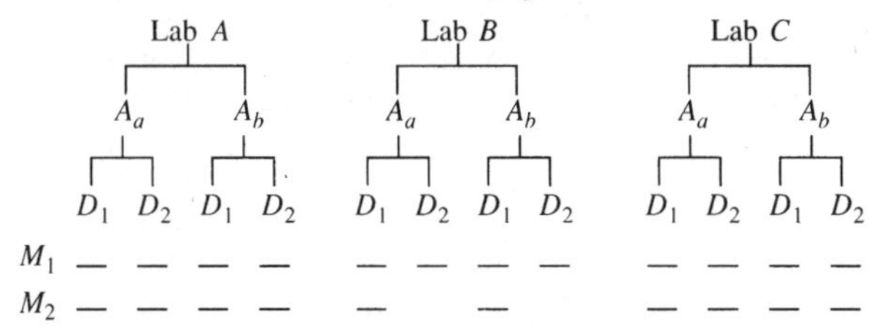


Lab: laboratory in which tests will be run.
$A_i$: Analyst *a* or analyst *b* in each laboratory.
$D_i$: First day of tests or second day of tests by each analyst in each lab.
$M_i$: First or second measurement each day by each analyst.

**Figure 7.16** Design matrix for nested/factorial study to evaluate new method of determining particle size.

was noted in laboratory *C* for analyst *B*. Figure 7.17 shows eight dot-frequency diagrams for the four nested factors. There is one diagram for each of the eight factorial combinations.

The dot-frequency diagrams indicate that laboratory *C* has much more variable results than the other laboratories. The repeat tests by the same analyst on the same day are usually consistent, except for 5 of the 56 cases. The variation between days (repeated tests on different days) can be studied by comparing the lines within each box. These lines overlap in all but 7 of the 32 cases in laboratories *A* and *B*. Variation between analysts in the same laboratory (comparison between consecutive pairs of boxes) is usually small. The second-day results for analyst *A* in laboratory *A* for the conditions stack *A*, flow aid 2, and the 50-g sample are inconsistent with other data and were not used in further analysis. In summary, for the four nested factors:

**TABLE 7.2 Data from Interlaboratory Study**

| | | | | Analyst $A$ | | | | Analyst $B$ | | | |
|---|---|---|---|---|---|---|---|---|---|---|---|
| | | | | Day 1 | | Day 2 | | Day 1 | | Day 2 | |
| Stk | Fa | Sample size | Lab | $M_1$ | $M_2$ | $M_1$ | $M_2$ | $M_1$ | $M_2$ | $M_1$ | $M_2$ |
| $A$ | 1 | 50 | $A$ | 29.4 | 28.2 | 27.3 | 31.6 | 30.2 | 26.0 | 28.0 | 27.5 |
| | | | $B$ | 26.6 | 26.7 | 26.4 | — | 27.3 | 29.4 | 27.2 | — |
| | | | $C$ | 29. | 34. | 50. | 52. | 25. | 30. | 42. | 50. |
| $A$ | 1 | 100 | $A$ | 31.3 | 32.0 | 29.0 | 27.5 | 29.7 | 32.3 | 36.5 | 35.2 |
| | | | $B$ | 35.4 | 43.6 | 44.0 | — | 27.7 | 28.7 | 28.0 | — |
| | | | $C$ | 31. | 31. | 34. | 32. | 57. | 43. | 46. | 48. |
| $A$ | 2 | 50 | $A$ | 33.7 | 32.3 | 44.8 | 41.7 | 27.4 | 29.0 | 30.4 | 31.4 |
| | | | $B$ | 32.8 | 30.5 | 33.0 | — | 31.4 | 27.3 | 29.2 | — |
| | | | $C$ | 31. | 31. | 30. | 30. | 43. | 49. | 30. | 30. |
| $A$ | 2 | 100 | $A$ | 27.8 | 27.8 | 34.4 | 36.3 | 36.8 | 37.1 | 37.7 | 37.4 |
| | | | $B$ | 40.5 | 40.3 | 37.8 | — | 39.8 | 42.1 | 40.3 | — |
| | | | $C$ | 30. | 27. | 36. | 35. | 50. | 51. | 35. | 34. |
| $B$ | 1 | 50 | $A$ | 31.7 | 27.5 | 27.3 | 31.6 | 31.2 | 27.2 | 27.4 | 27.4 |
| | | | $B$ | 26.3 | 26.6 | 27.1 | — | 26.3 | 26.6 | 26.8 | — |
| | | | $C$ | 35. | 34. | 33. | 34. | 27. | 27. | 27. | 28. |
| $B$ | 1 | 100 | $A$ | 28.0 | 27.9 | 29.0 | 28.2 | 32.4 | 30.7 | 28.8 | 29.6 |
| | | | $B$ | 34.0 | 34.2 | 32.5 | — | 35.0 | 28.5 | 29.6 | — |
| | | | $C$ | 39. | 40. | 31. | 32. | 43. | 33. | 31. | 29. |
| $B$ | 2 | 50 | $A$ | 29.7 | 32.1 | 32.4 | 31.9 | 27.8 | 29.8 | 28.8 | 28.9 |
| | | | $B$ | 29.6 | 28.5 | 29.3 | — | 27.1 | 27.7 | 27.1 | — |
| | | | $C$ | 36. | 34. | 30. | 51. | 30. | 29. | 28. | 29. |
| $B$ | 2 | 100 | $A$ | 45.1 | 42.2 | 34.5 | 35.5 | 32.6 | 36.1 | 35.6 | 34.7 |
| | | | $B$ | 35.2 | 38.4 | 29.2 | — | 32.0 | 33.1 | 33.1 | — |
| | | | $C$ | 56. | 56. | 32. | 36. | 44. | 50. | 33. | 38. |

*Notes:*
Stk: Stack $A$ or stack $B$ (alignment of screens).
Fa: Flow aid 1 or 2.
Size: Sample size of 50 or 100 g.
Lab: Laboratory in which tests were done ($A$, $B$, or $C$).
$M_i$: Individual measurements of percent over 200-mesh screen. (No repeat analysis in laboratory $B$ on the second day; data from laboratory $C$ reported in whole numbers.)

1. Laboratories $A$ and $B$ show good agreement on most tests. The test results for laboratory $C$ show much more variation and are not used in further analysis.
2. Analysts within a laboratory show generally good agreement. The differences within laboratories are not consistent with any one analyst.
3. The variation from day to day is not much greater than the variation from repeated tests. Since the variation between laboratories $A$ and $B$ is small, all the results can be pooled together in further analysis.

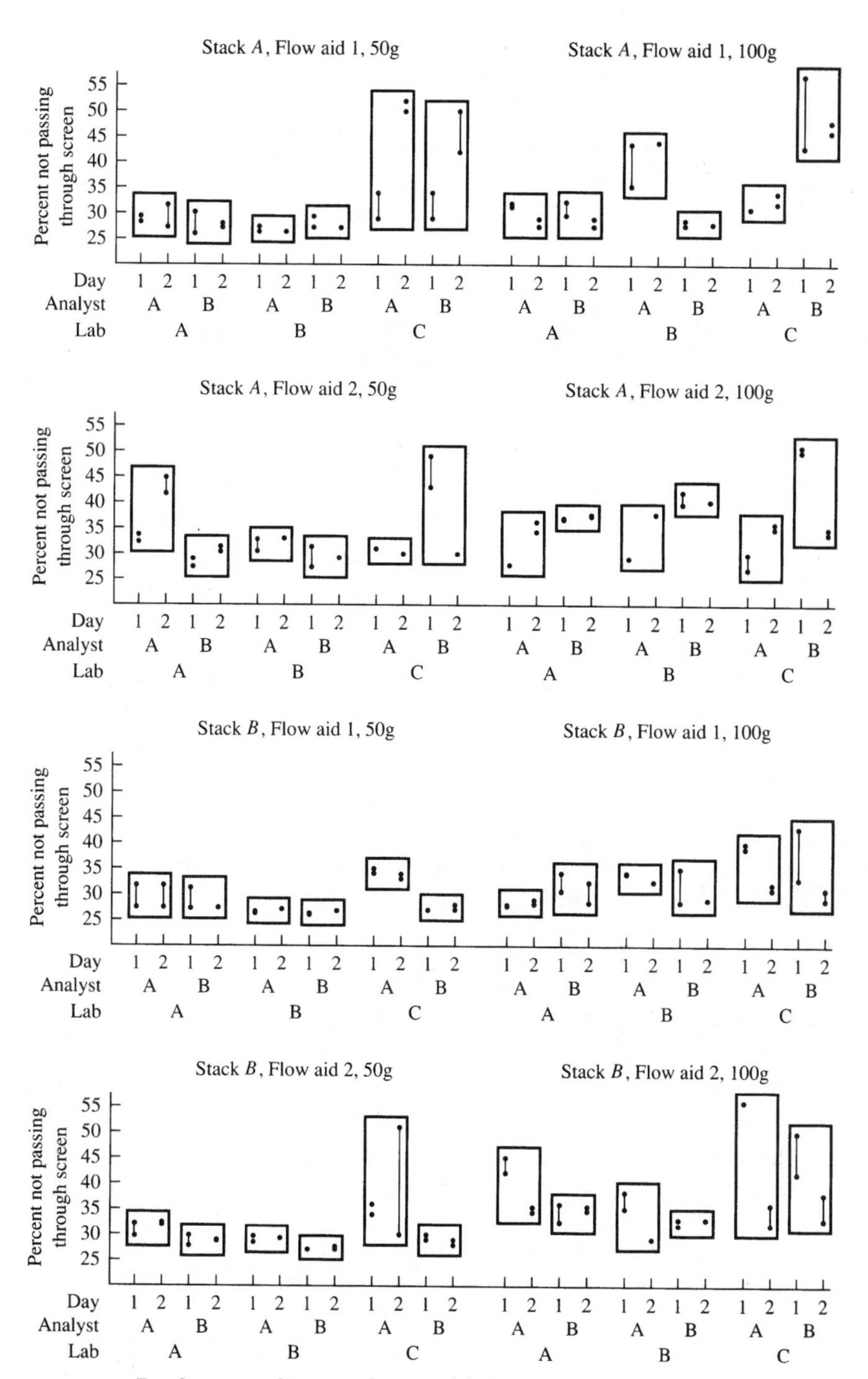


Figure 7.17 Dot-frequency diagrams for nested factors in interlaboratory study.

| | Alignment - stack $A$ | | Alignment - stack $B$ | |
|---|---|---|---|---|
| | Flow aid 1 | Flow aid 2 | Flow aid 1 | Flow aid 2 |
| Sample size = 50 g | $\bar{x}$ = 27.99<br>$s$ = 1.61 | $\bar{x}$ = 30.70*<br>$s$ = 2.13* | $\bar{x}$ = 27.93<br>$s$ = 1.98 | $\bar{x}$ = 29.34<br>$s$ = 1.76 |
| Sample size = 100 g | $\bar{x}$ = 32.92<br>$s$ = 5.47 | $\bar{x}$ = 36.86<br>$s$ = 4.35 | $\bar{x}$ = 30.60<br>$s$ = 2.52 | $\bar{x}$ = 35.52<br>$s$ = 4.11 |

$x$ = percent of material not through 200-mesh screen.
$\bar{x}$ = average of 14 test results from laboratories $A$ and $B$.
$s$ = standard deviation of 14 test results from labs $A$ and $B$.
*Data for lab $A$, analyst $A$, day 2 not used.

**Figure 7.18** Statistical summary of data for laboratories $A$ and $B$.

To analyze the crossed factors, first the data are summarized. Only data from laboratories $A$ and $B$ were used in calculating the summary statistics shown in Fig. 7.18.

The effects of the factors on both the average and the standard deviation are important in this study. The standard deviation is a measure of the consistency of the test method on different days, by different analysts, and in different laboratories. Figure 7.19 shows these statistics displayed on cubes (see Chap. 5).

Figure 7.20 contains estimates of the effects for the factors and interactions as well as response plots for the important factors.

The following conclusions can be drawn from this study:

1. Laboratory $C$ reported inconsistent results in the study. The data were only reported to whole numbers. Large differences between analysts and between days under the same test conditions were observed. Training in the test method should be done in laboratory $C$.

2. The agreement between laboratories $A$ and $B$ was good. Variations due to analyst and day were also small in these laboratories.

3. The sample size (50 or 100 g) was the most important factor in the study. Increasing the sample size from 50 to 100 g increased the percentage of material not passing through a 200-mesh screen by about 5 percent. The standard deviation of repeated tests (different laboratories, days, and analysts) more than doubles from 1.88 to 4.25 percent.

4. Using flow aid 2 results in about a 4.2 percent increase over flow aid 1 for percentage of material not passing through a 200-mesh screen.

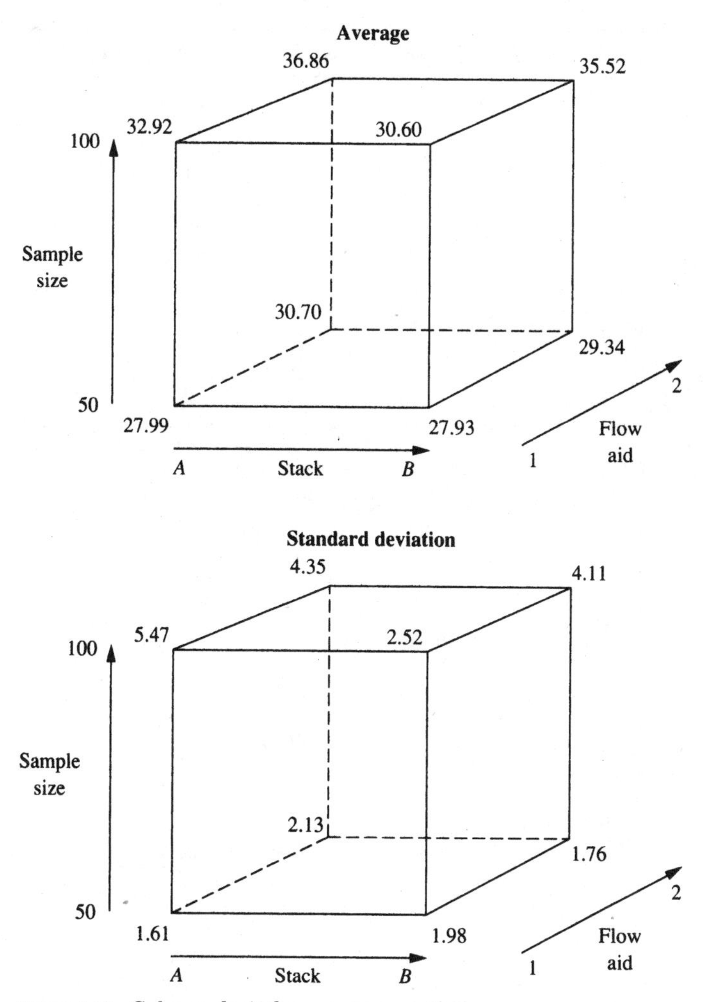


**Figure 7.19** Cube analysis for summary statistics.

In other nested/factorial experiments, it may be appropriate to analyze first the crossed factors and then the nested factors.

## 7.7 Summary

This chapter has discussed planned experimentation for nested factors used to evaluate the important sources of variation. Physical and location constraints and sampling and testing considerations often lead to nested factors. The Shewhart control chart can be viewed as a two-factor nested design. Many of the examples in this chapter dealt with measurement processes.

| Factors (interactions) | Average effects (percent) | Standard deviation effects (percent) |
|---|---|---|
| Sample size ($S$) | 4.98 | 2.24 |
| Flow-aid ($F$) | 3.25 | 0.19 |
| Screen alignment ($A$) | −1.27 | −0.08 |
| $S \times F$ | 1.19 | 0.04 |
| $S \times A$ | −0.56 | −0.80 |
| $F \times A$ | −0.08 | 0.49 |
| $S \times F \times A$ | −0.57 | 0.86 |

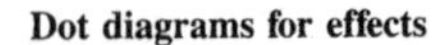


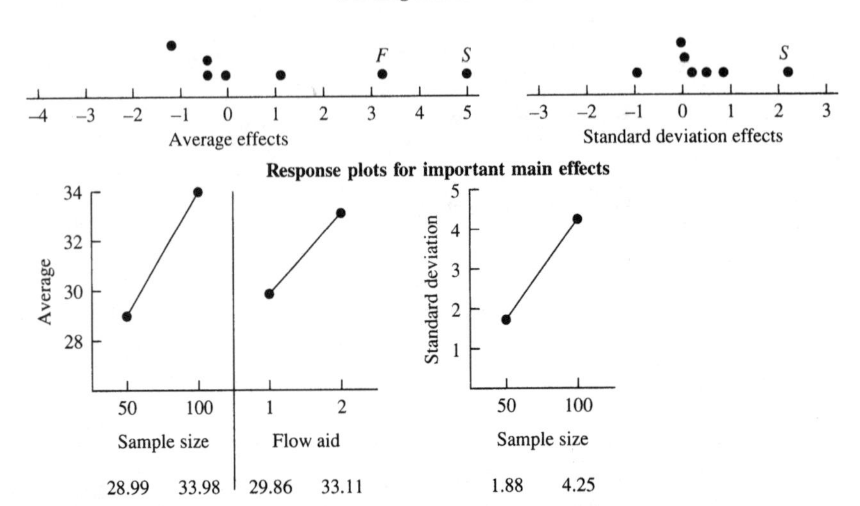


**Figure 7.20** Estimate of factor effects and interactions (interlaboratory study).

Planning a nested study is similar to planning a factorial design, but one difference lies in how the randomization is performed. The dot-frequency diagram is the primary analysis tool to partition the variation in the response variables among the factors. Components of variance to quantify the amount of variation associated with each factor can be estimated. Appendix 7A contains details of these calculations.

Nested factors can be mixed with crossed factors to form more complex experimental patterns.

## Appendix 7A: Calculation of Variance Components

Appendix 7A presents methods for quantifying the importance of the factors in a nested study. These procedures are known as a variance component analysis. This analysis is only meaningful for a stable process. With an unstable process, special causes are present, so the analysis can give misleading results.

The basic steps in the procedure are given, and the analysis is illustrated for two of the examples in this chapter. The variance component analysis is an extension of the procedure used to separate measurement and product variation from total process variation (Table 7.1), presented in Sec. 7.2.

Estimates of the variation due to each factor are called variance components. Ranges can be calculated from the data and from averages of the data to obtain estimates of these variance components. But the ranges calculated from the averages will also include variation from the sources that were averaged. It is therefore necessary to understand exactly which components of variance are estimated from each quantity and then to solve for the appropriate variance components. The procedure given here to estimate the components of variance for each factor has been described by Box et al. (1978) and Snee (1983).

The following steps summarize the calculation of the variance components:

1. Calculate ranges and averages for the factor in the lowest hierarchy. (*Note:* The standard deviation can be used instead of the range at each step in this analysis. For factors with greater than five levels, the standard deviation is preferred.)
2. Using the averages from step 1, calculate ranges and averages for the second-lowest factor in the nested design.
3. Continue calculating ranges and averages for each factor in the design, using the averages from the previous step.
4. For the highest factor in the hierarchy, the range will be based on only one subgroup of averages.
5. Calculate the average range and an upper control limit for each set of ranges calculated in the first four steps. Any ranges greater than the upper control limit should be investigated for special causes. Corrections should be made, or the range eliminated and the average range recalculated.
6. Estimate the standard deviation for each average range using the tabled values of $d_2^*$ from the table in App. 7B. (*Note:* If the standard deviation is used instead of the range, the pooled standard deviation is calculated in place of the average range. No factor is needed.)
7. Square the standard deviation to obtain a variance associated with each factor. Write out the quantity estimated from each calculated variance. In general, each variance will estimate the variance component associated with the averages from which the range was calculated and some fraction of the variance component for all factors

lower in the hierarchy. The fraction will be 1 over the total number of measurements of the factor in the averages used to calculate the range.

8. Use the form of the calculated variance to calculate each variance component estimate. Begin with the lowest factor in the hierarchy, and substitute the calculated components in determining the component for the next factor in the hierarchy. It is possible to obtain a negative value for a variance component by using this procedure. If a negative value is calculated, an estimate of zero should be used for that component.

9. Summarize the variance components by calculating the percentage of the total variation associated with each nested factor.

The following two examples illustrate these steps. In the first example, the standard deviation is used instead of the range. The second example uses the range.

#### Measurement study example (Sec. 7.3)

In Sec. 7.3, a three-factor nested study of a measurement process was described. The three factors studied were parts, setups within parts, and measurement within a setup. Table 7.3 shows the summary calculations required to do a variance component analysis for the data from this study. First, averages and standard deviations of the three measurements are calculated. Then the average and standard deviation are calculated for each set of three setup averages. Control limits for the standard deviations are calculated to evaluate stability of the individual estimates, and pooled averages of the individual standard deviations are calculated. Finally, the standard deviation of the six part averages is calculated.

Table 7.4 shows the calculation to partition the variation among the factors. The three standard deviations obtained from Table 7.3 are used to obtain the estimates of the variance components. An expression is written for each standard deviation and then is solved to compute the components. This variance component analysis verifies the visual analysis of the dot-frequency diagram given in Sec. 7.3:

- 72 percent of the variation is attributable to part differences.
- 18 percent is attributable to the setup.
- 10 percent is due to the measurement procedure.

#### Foundry example (Sec. 7.5)

In Sec. 7.5, a five-factor nested study for hardness (in Rockwell units) of parts from two foundries was described. The amount of variation

**TABLE 7.3 Analysis of Three-Factor Nested Study**

| Part | Setup | Measurements 1 | 2 | 3 | Summary statistics $\overline{X}_1$ | $S_1$ | $\overline{X}_2$ | $S_2$ |
|---|---|---|---|---|---|---|---|---|
| 1 | 1 | 3.40 | 3.45 | 3.70 | 3.52 | .16 | | |
| | 2 | 3.95 | 3.80 | 4.00 | 3.92 | .10 | | |
| | 3 | 3.95 | 4.00 | 4.15 | 4.03 | .10 | 3.82 | .268 |
| 2 | 1 | 4.55 | 4.70 | 4.70 | 4.65 | .09 | | |
| | 2 | 4.30 | 4.20 | 4.50 | 4.33 | .15 | | |
| | 3 | 4.75 | 5.15 | 5.20 | 5.03 | .25 | 4.67 | .350 |
| 3 | 1 | 4.30 | 4.70 | 4.70 | 4.57 | .23 | | |
| | 2 | 4.50 | 4.55 | 4.65 | 4.57 | .08 | | |
| | 3 | 4.15 | 4.45 | 4.50 | 4.37 | .19 | 4.50 | .116 |
| 4 | 1 | 2.60 | 2.65 | 2.50 | 2.58 | .08 | | |
| | 2 | 3.00 | 3.15 | 3.15 | 3.10 | .09 | | |
| | 3 | 3.40 | 3.55 | 3.40 | 3.45 | .09 | 3.04 | .438 |
| 5 | 1 | 3.90 | 4.85 | 4.60 | 4.45 | .49 | | |
| | 2 | 4.35 | 4.90 | 4.35 | 4.53 | .32 | | |
| | 3 | 4.10 | 4.00 | 3.95 | 4.02 | .08 | 4.33 | .274 |
| 6 | 1 | 3.95 | 4.10 | 4.00 | 4.02 | .08 | | |
| | 2 | 3.90 | 3.90 | 4.00 | 3.93 | .06 | | |
| | 3 | 4.40 | 4.30 | 4.40 | 4.37 | .06 | 4.11 | .232 |
| | | | $\overline{\overline{X}} = 4.08$ | | $\overline{S}_1 = .186$ UCL = .478 | | $S_3 = .589$ | $\overline{S}_2 = .297$ UCL = .76 |

*Notes:*

$\overline{X}_1$ = average of the three measurements for each setup.
$S_1$ = standard deviation of the three measurements for each setup.
$\overline{X}_2$ = average of the three setup averages for each part.
$S_2$ = standard deviation of the three setup averages for each part.
$\overline{\overline{X}}$ = overall average of data.
$\overline{S}_1$ = pooled standard deviation of measurements.
$\overline{S}_2$ = pooled standard deviation of averages for each setup.
$S_3$ = standard deviation of the averages for each part.
UCL = upper control limits for the individual standard deviations used to compute the pooled values.

due to each of the factors in the study can be quantified by a variance component analysis. Tables 7.5 through 7.8 contain these calculations for the five-factor nested study.

The dot-frequency diagrams in Sec. 7.5 indicated that the important factors were different in the two foundries. Therefore, the variance component analysis is done separately for each foundry. The variance components for each foundry (Table 7.8) are compared to the visual analysis of the dot-frequency diagrams in Sec. 7.5.

**TABLE 7.4 Calculation of Components of Variation**

Summary statistics from Table 7.3 (units = mm)

$\overline{S}_1 = .186$ = pooled standard deviation of measurements.

$\overline{S}_2 = .297$ = pooled standard deviation of setup averages.

$S_3 = .589$ = standard deviation of the averages for each part.

Standard deviation of measurement ($\hat{\sigma}_m$)

$$\hat{\sigma}_m = \overline{S}_1 = .186$$

Standard deviation of setup ($\hat{\sigma}_s$)

$$\overline{s}_2^2 = \hat{\sigma}_s^2 + \sigma_m^2/3$$

$$\hat{\sigma}_s = \sqrt{\overline{s}_2^2 - \hat{\sigma}_m^2/3}$$

$$\hat{\sigma}_s = \sqrt{.297^2 - .186^2/3}$$

$$\hat{\sigma}_s = .277$$

Standard deviation of parts ($\hat{\sigma}_p$)

$$s_3^2 = \sigma_p^2 + \sigma_s^2/3 + \sigma_m^2/9$$

$$\hat{\sigma}_p = \sqrt{s_3^2 - \hat{\sigma}_s^2/3 - \hat{\sigma}_m^2/9}$$

$$\hat{\sigma}_p = \sqrt{.589^2 - .277^2/3 - .186^2/9}$$

$$\hat{\sigma}_p = .564$$

| | Summary of Sources of Variation (Units = mm) | | |
|---|---|---|---|
| Source of variation | Standard deviation | Variance component | Percentage of variation |
| Part to part | .564 | .318 | 74.0 |
| Setup within a part | .277 | .077 | 17.9 |
| Measurement | .186 | .035 | 8.1 |
| Total | .656 | .430 | 100.0 |

**TABLE 7.5 Analysis of Hardness Study Data—Foundry 1**

| | | | Measurements | | | | | | | | | |
|---|---|---|---|---|---|---|---|---|---|---|---|---|
| Day | Heat | Part | 1 | 2 | 3 | $\overline{X}_a$ | $R_a$ | $\overline{X}_b$ | $R_b$ | $\overline{X}_c$ | $R_c$ | |
| 1 | 1 | 1 | 44 | 45 | 42 | 43.6 | 3 | | | | | |
| | | 2 | 42 | 42 | 43 | 42.3 | 1 | | | | | |
| | | 3 | 44 | 44 | 42 | 43.3 | 2 | 43.1 | 1.3 | | | |
| | 2 | 1 | 47 | 44 | 47 | 46.0 | 3 | | | | | |
| | | 2 | 48 | 49 | 46 | 47.7 | 3 | | | | | |
| | | 3 | 42 | 39 | 43 | 41.3 | 4 | 45.0 | 6.4 | 44.1 | 1.9 | |
| 2 | 1 | 1 | 38 | 36 | 35 | 36.3 | 3 | | | | | |
| | | 2 | 38 | 40 | 37 | 38.3 | 3 | | | | | |
| | | 3 | 33 | 32 | 36 | 33.7 | 4 | 36.1 | 4.6 | | | |
| | 2 | 1 | 32 | 38 | 31 | 33.7 | 7 | | | | | |
| | | 2 | 34 | 36 | 39 | 36.3 | 5 | | | | | |
| | | 3 | 33 | 35 | 30 | 32.7 | 5 | 34.2 | 3.6 | 35.2 | 1.9 | |
| 3 | 1 | 1 | 38 | 39 | 32 | 36.3 | 7 | | | | | |
| | | 2 | 35 | 34 | 37 | 35.3 | 3 | | | | | |
| | | 3 | 39 | 43 | 44 | 42.0 | 5 | 37.9 | 6.7 | | | |
| | 2 | 1 | 36 | 40 | 39 | 38.3 | 4 | | | | | |
| | | 2 | 33 | 35 | 44 | 37.3 | 11 | | | | | |
| | | 3 | 45 | 46 | 37 | 42.7 | 9 | 39.4 | 5.4 | 38.7 | 1.5 | |
| | | | | | | $\overline{\overline{X}} = 39.3$ | $\overline{R}_a = 4.56$ | | $\overline{R}_b = 4.67$ | | $\overline{R}_c = 1.77$ | $R_d = 8.9$ |
| | | | | | | | UCL = 11.7 | | UCL = 12.0 | | UCL = 5.8 | |

| | | | | |
|---|---|---|---|---|
| Number of subgroups $k$ | 18 | 6 | 3 | 1 |
| Measurements/subgroup $n$ | 3 | 3 | 2 | 3 |
| Standard deviation factor $d_2^*$ | 1.69 | 1.73 | 1.23 | 1.91 |
| Standard deviation $S$ | 2.70 | 2.70 | 1.44 | 4.66 |
| Variance $S^2$ | 7.29 | 7.29 | 2.07 | 21.71 |
| | $S_a^2$ | $S_b^2$ | $S_c^2$ | $S_d^2$ |

$\overline{X}_a$ = Part averages, average of the three measurements.
$R_a$ = Range of the three measurements.
$\overline{X}_b$ = Heat averages, average of the three part averages.
$R_b$ = Range of the three part averages.
$\overline{X}_c$ = Day averages, average of the two heat averages.
$R_c$ = Range of the two heat averages.
$\overline{\overline{X}}$ = Foundry average, average of the three day averages.
$R_d$ = Range of the three day averages.
$S_a^2$ = Variance based on measurements.
$S_b^2$ = Variance based on part averages.
$S_c^2$ = Variance based on heat averages.
$S_d^2$ = Variance based on day averages.
UCL = Upper control limit for the ranges in this column.
$d_2^*$ = Standard deviation factor (from App. 7B).

**TABLE 7.6 Analysis of Hardness Study Data—Foundry 2**

| | | | Measurements | | | | | | | | | |
|---|---|---|---|---|---|---|---|---|---|---|---|---|
| Day | Heat | Part | 1 | 2 | 3 | $\bar{X}_a$ | $R_a$ | $\bar{X}_b$ | $R_b$ | $\bar{X}_c$ | $R_c$ | |
| 1 | 1 | 1 | 25 | 26 | 22 | 24.3 | 4 | | | | | |
| | | 2 | 24 | 25 | 29 | 26.0 | 5 | | | | | |
| | | 3 | 28 | 23 | 23 | 24.7 | 5 | 25.0 | 1.7 | | | |
| | 2 | 1 | 35 | 33 | 28 | 32.0 | 7 | | | | | |
| | | 2 | 31 | 32 | 30 | 31.0 | 2 | | | | | |
| | | 3 | 29 | 34 | 36 | 33.0 | 7 | 32.0 | 2.0 | 28.5 | 7.0 | |
| 2 | 1 | 1 | 38 | 39 | 42 | 39.7 | 4 | | | | | |
| | | 2 | 43 | 37 | 38 | 39.3 | 6 | | | | | |
| | | 3 | 34 | 33 | 37 | 34.7 | 4 | 37.9 | 5.0 | | | |
| | 2 | 1 | 35 | 34 | 34 | 34.3 | 1 | | | | | |
| | | 2 | 30 | 31 | 32 | 31.0 | 2 | | | | | |
| | | 3 | 32 | 36 | 34 | 34.0 | 4 | 33.1 | 3.3 | 35.5 | 4.8 | |
| 3 | 1 | 1 | 25 | 24 | 27 | 25.3 | 3 | | | | | |
| | | 2 | 26 | 29 | 23 | 26.0 | 6 | | | | | |
| | | 3 | 31 | 30 | 28 | 29.7 | 3 | 27.0 | 4.4 | | | |
| | 2 | 1 | 33 | 32 | 28 | 31.0 | 5 | | | | | |
| | | 2 | 30 | 36 | 34 | 33.3 | 6 | | | | | |
| | | 3 | 34 | 38 | 36 | 36.0 | 4 | 33.4 | 5.0 | 30.2 | 6.4 | |
| | | | | | | $\bar{\bar{X}} = 31.4$ | $\bar{R}_a = 4.33$ | | $\bar{R}_b = 3.57$ | | $\bar{R}_c = 6.07$ | $R_d = 7.00$ |
| | | | | | | | UCL = 11.3 | | UCL = 9.3 | | UCL = 19.8 | |
| Number of subgroups $k$ | | | | | | | 18 | | 6 | | 3 | 1 |
| Measurements/subgroup $n$ | | | | | | | 3 | | 3 | | 2 | 3 |
| Standard deviation factor $d_2^*$ | | | | | | | 1.69 | | 1.73 | | 1.23 | 1.91 |
| Standard deviation $S$ | | | | | | | 2.56 | | 2.06 | | 4.93 | 3.66 |
| Variance $S^2$ | | | | | | | 6.55 | | 4.24 | | 24.35 | 13.43 |
| | | | | | | | $S_a^2$ | | $S_b^2$ | | $S_c^2$ | $S_d^2$ |

$\bar{X}_a$ = Part averages, average of the three measurements.
$R_a$ = Range of the three measurements.
$\bar{X}_b$ = Heat averages, average of the three part averages.
$R_b$ = Range of the three part averages.
$\bar{X}_c$ = Day averages, average of the two heat averages.
$R_c$ = Range of the two heat averages.
$\bar{\bar{X}}$ = Foundry average, average of the three day averages.
$R_d$ = Range of the three day averages.
$S_a^2$ = Variance based on measurements.
$S_b^2$ = Variance based on part averages.
$S_c^2$ = Variance based on heat averages.
$S_d^2$ = Variance based on day averages.
UCL = Upper control limit for the ranges in this column.
$d_2^*$ = Standard deviation factor (from App. 7B).

**TABLE 7.7 Calculation of Variance Components—Five-Factor Hardness Data Study**

| Variance | Foundry 1 | Foundry 2 | Form of the calculated variance |
|---|---|---|---|
| $S_a^2$ | 7.29 | 6.55 | $\hat{\sigma}_m^2$ |
| $S_b^2$ | 7.29 | 4.24 | $\hat{\sigma}_m^2/3$ |
| $S_c^2$ | 2.07 | 24.35 | $\hat{\sigma}_m^2/9 + \hat{\sigma}_p^2/3$ |
| $S_d^2$ | 21.71 | 13.43 | $\hat{\sigma}_m^2/18 + \hat{\sigma}_p^2/6 + \hat{\sigma}_h^2/2 + \hat{\sigma}_d^2$ |

Calculation of variance components for foundry 1

$\sigma^2 = S_a^2$
$\hat{\sigma}^2 = 7.29$
$\hat{\sigma} = 2.70$

$\hat{\sigma}_p^2 = S_b^2 - \hat{\sigma}_m^2/3$
$\hat{\sigma}_p^2 = 7.29 - 7.29/3$
$\hat{\sigma}_p^2 = 4.86$
$\hat{\sigma}_p = 2.20$

$\hat{\sigma}_h^2 = S_c^2 - \hat{\sigma}_p^2/3 - \hat{\sigma}_m^2/9$
$\hat{\sigma}_d^2 = 2.07 - 4.86/3 - 7.29/9$
$\hat{\sigma}_h^2 = 0.36$
$\hat{\sigma}_h = 0.0$

$\hat{\sigma}_d^2 = S_d^2 - \hat{\sigma}_c^2/2 - \hat{\sigma}_p^2/6 - \hat{\sigma}_m^2/18$
$\hat{\sigma}_d^2 = 21.71 - 0.0/2 - 4.86/6 - 7.29/18$
$\hat{\sigma}_d^2 = 20.50$
$\hat{\sigma}_d = 4.53$

Calculation of variance components for foundry 2

$\hat{\sigma}^2 = S_a^2$
$\hat{\sigma}^2 = 6.55$
$\hat{\sigma} = 2.56$

$\hat{\sigma}_p^2 = S_b{}^2 - \hat{\sigma}_m^2/3$
$\hat{\sigma}_p^2 = 4.24 - 6.55/3$
$\hat{\sigma}_p^2 = 2.06$
$\hat{\sigma}_p = 1.44$

$\hat{\sigma}_h^2 = S_c^2 - \hat{\sigma}_p^2/3 - \hat{\sigma}_m^2/9$
$\hat{\sigma}_d^2 = 24.35 - 2.06/3 - 6.55/9$
$\hat{\sigma}_h^2 = 22.94$
$\hat{\sigma}_h = 4.79$

$\hat{\sigma}_d^2 = S_d^2 - \hat{\sigma}_c^2/2 - \hat{\sigma}_p^2/6 - \hat{\sigma}_m^2/18$
$\hat{\sigma}_d^2 = 13.43 - 22.94/2 - 2.06/6 - 6.55/18$
$\hat{\sigma}_d^2 = 1.25$
$\hat{\sigma}_d = 1.12$

**TABLE 7.8 Summary of Variance Components for Hardness Data**

| Source of variation | Standard deviation $\hat{\sigma}$ | Variance component $\hat{\sigma}^2$ | Percentage of variation |
|---|---|---|---|
| Variance Components for Foundry 1 | | | |
| Days within foundry | 4.53 | 20.50 | 62.8 |
| Heats within days | 0.0 | 0.0 | 0.0 |
| Parts within heats | 2.20 | 4.86 | 14.9 |
| Measurement within parts | 2.70 | 7.29 | 22.3 |
| Total | | 32.64 | 100.0 |
| Variance Components for Foundry 2 | | | |
| Days within foundry | 1.12 | 1.25 | 3.8 |
| Heats within days | 4.79 | 22.94 | 69.9 |
| Parts within heats | 1.44 | 2.06 | 6.3 |
| Measurement within parts | 2.56 | 6.55 | 20.0 |
| Total | | 32.80 | 100.0 |

## Appendix 7B: Calculating and Combining Statistics ($\overline{\overline{X}}$, $\overline{R}$, or $\overline{S}$)

In this chapter, statistics were frequently combined to obtain an overall statistic. Appendix 7B summarizes the formula for combining averages, ranges, and standard deviations. A table of factors for estimating the standard deviation from the range is also included.

When $n$ is equal for all $k$ values of the statistic,

$$\overline{R} = \Sigma R/k$$

$$\overline{\overline{x}} = \Sigma \overline{x}/k$$

$$\overline{S} = \sqrt{\Sigma S^2/k}$$

When $n$ is different for some of the statistics (the range should not be used with unequal groups of data),

$$\overline{\overline{x}} = \frac{n_1\overline{x}_1 + n_2\overline{x}_2 + \cdots + n_k\overline{x}_k}{n_1 + n_2 + \cdots + n_k}$$

$$\overline{s}^2 = \frac{(n_1 - 1)s_1^2 + (n_2 - 1)s_2^2 + \cdots + (n_k - 1)s_k^2}{(n_1 - 1) + (n_2 - 1) + \cdots + (n_k - 1)}$$

$$\sigma = \frac{\overline{R}}{d_2}$$

(use $d_2^*$ when less than 10 subgroups).

**Table of $d_2$ and $d_2^*$ values**

| $n$ | $d_2$ | $k = 10$ | $k = 8$ | $k = 6$ | $k = 5$ | $k = 4$ | $k = 3$ | $k = 2$ | $k = 1$ |
|---|---|---|---|---|---|---|---|---|---|
| 2 | 1.128 | 1.16 | 1.17 | 1.18 | 1.19 | 1.21 | 1.23 | 1.28 | 1.41 |
| 3 | 1.693 | 1.72 | 1.72 | 1.73 | 1.74 | 1.75 | 1.77 | 1.81 | 1.91 |
| 4 | 2.059 | 2.08 | 2.08 | 2.09 | 2.10 | 2.11 | 2.12 | 2.15 | 2.24 |
| 5 | 2.326 | 2.34 | 2.35 | 2.35 | 2.36 | 2.37 | 2.38 | 2.40 | 2.48 |
| 6 | 2.534 | 2.55 | 2.55 | 2.56 | 2.56 | 2.57 | 2.58 | 2.60 | 2.67 |
| 7 | 2.704 | 2.72 | 2.72 | 2.73 | 2.73 | 2.74 | 2.75 | 2.77 | 2.83 |
| 8 | 2.847 | | | | | | | | |
| 9 | 2.970 | | | | (Use $d_2$) | | | | |
| 10 | 3.078 | | | | | | | | |

## References

Bainbridge, T. R. (1965): "Staggered, Nested Designs for Estimating Variance Components," *Industrial Quality Control,* no. 22, pp. 12–20.

Box, G. E. P., W. G. Hunter, and J. S. Hunter (1978): *Statistics for Experimenters,* Wiley-Interscience, New York, chap. 17.

Snee, R. D. (1983): "Graphical Analysis of Process Variation Studies," *Journal of Quality Technology,* vol. 15, no. 2, April, pp. 76–88.

Trout, R. (1985): "Design and Analysis of Experiments to Estimate Components of Variation—Two Case Studies," *Experiments in Industry,* Chemical and Process Industries Division, American Society for Quality Control, Quality Press, Milwaukee, Wis.

## Exercises

**7.1** Develop a design matrix for a three-factor nested design with the following number of levels for each factor:

| Factor | Levels |
|---|---|
| $A$ | 4 |
| $B$ within $A$ | 3 |
| $C$ within $B$ | 2 |

**7.2** Select a control chart that you are familiar with. Describe the control chart as a nested design. What do each of the factors represent?

**7.3** How many runs are required for a four-factor factorial design with each factor at two levels? How many runs are required for a four-factor nested design with each factor at two levels? What are the key differences in designing these two studies? What are important differences in analyzing the results of the two types of studies?

**7.4** Describe a process in which the physical layout would require a nested design to study the process. What are the factors, and how many levels does each factor have? Develop a design matrix for a study of the process.

**7.5** Consider a four-factor nested design with each factor at three levels. How many runs are required for a complete replication of the design? Sketch a dot-frequency diagram from such a study where the second factor in the hierarchy (say, $B$ within $A$) is the dominant factor. The other three factors are of very little importance.

**7.6** Develop a design matrix for a three-factor nested design with each factor at two levels. Assume that the study was completed but the order of testing was not randomized. The chemical tests to measure the response variable were done in the order given on the design matrix. None of the three factors were important, but the test apparatus had a steady drift during the study that was undetected. Sketch a dot-frequency diagram for the results of this study.

**7.7** Consider the measurement study example in Sec. 7.3. If four measurements of each production part could be made, how many setups should be made to obtain the most precise average measurement of the part?

**7.8** *Delays in expense reports.* The accounting department quality improvement team is studying problems in handling expense reports. Many employees have complained about long delays in receiving reimbursement of expenses. A number of factors were identified that might cause varying processing times. The following three factors were selected as the most likely causes:

1. Reports originated in the field versus headquarters.
2. Processing group—two groups handled field reports, and two other groups handled reports from headquarters.
3. Variation in the clerks processing the expense reports.

Data were collected for the 18 expense reports processed during the next week to investigate the problem.

Analyze these data as a three-factor nested design. Prepare a run order plot and a dot-frequency diagram. (*Note:* This example was developed by Rob Stiratelli of Rohm & Haas Company.)

| Expense report | Originating location | Processing group | Clerk | Processing time (days) |
|---|---|---|---|---|
| 1 | Field | *B* | 2 | 6 |
| 2 | Headquarters | *C* | 1 | 4 |
| 3 | Headquarters | *C* | 2 | 2 |
| 4 | Headquarters | *D* | 2 | 9 |
| 5 | Field | *A* | 2 | 6 |
| 6 | Field | *A* | 1 | 2 |
| 7 | Field | *B* | 1 | 9 |
| 8 | Field | *B* | 2 | 8 |
| 9 | Headquarters | *D* | 1 | 1 |
| 10 | Headquarters | *D* | 2 | 7 |
| 11 | Field | *B* | 3 | 5 |
| 12 | Field | *B* | 3 | 8 |
| 13 | Headquarters | *C* | 1 | 2 |
| 14 | Headquarters | *C* | 2 | 5 |
| 15 | Field | *A* | 2 | 3 |
| 16 | Headquarters | *D* | 1 | 4 |
| 17 | Field | *B* | 1 | 5 |
| 18 | Field | *A* | 1 | 5 |

**7.9** *Measurement variation study.* A study was done to evaluate the measurement of the flushness of a fitted assembly. The measurement process simulated the assembly by clamping the key part into a fixture constructed for the purpose. Then a micrometer was used to measure the flushness (the gap at a critical location between the part and the fixture). Measurements were in fractions of a millimeter.

Ten parts were selected from the process during a period when the process was stable. Each part was placed in the fixture (a setup) two different times, and the micrometer measurement was performed twice after each setup. The 20 setups were done in a random order. The following data were obtained.

| | | | Measurement | |
|---|---|---|---|---|
| Run order | Part | Setup | First | Second |
| 1 | 6 | 1 | 0.75 | 0.80 |
| 2 | 2 | 1 | 0.50 | 0.55 |
| 3 | 10 | 1 | 0.65 | 0.70 |
| 4 | 3 | 1 | 0.40 | 0.50 |
| 5 | 6 | 2 | 0.60 | 0.60 |
| 6 | 9 | 1 | 1.95 | 1.80 |
| 7 | 3 | 2 | 1.80 | 1.90 |
| 8 | 9 | 2 | 1.90 | 1.75 |
| 9 | 1 | 1 | 1.70 | 1.70 |
| 10 | 4 | 1 | 1.65 | 1.75 |
| 11 | 5 | 1 | 1.50 | 1.40 |
| 12 | 7 | 1 | 1.30 | 1.40 |
| 13 | 2 | 2 | 1.15 | 0.95 |
| 14 | 4 | 2 | 0.95 | 1.00 |
| 15 | 5 | 2 | 1.00 | 0.90 |
| 16 | 10 | 2 | 0.85 | 1.80 |
| 17 | 7 | 2 | 0.70 | 0.85 |
| 18 | 8 | 1 | 0.70 | 0.65 |
| 19 | 1 | 2 | 0.65 | 0.65 |
| 20 | 8 | 2 | 0.50 | 0.55 |

Analyze these data as a three-factor nested design. Prepare a run chart and a dot-frequency diagram. What is the biggest contributor to variation of the measurement procedure?

**7.10** *Variation in a milling process.* A milling process was used to obtain the correct width of a metal part. Both the width and the microfinish (smoothness) of the part were important quality characteristics. The process was run by two operators, each responsible for two mills. Each mill had two spindles, and there were two part positions (left and right) for each spindle. Thus, each machine could mill four parts at a time (two spindles with two positions each).

To study the important sources of variation in this process, the process supervisor selected two parts from each position during a 1-hour period. The pairs of parts were labeled by location and then randomly ordered for measurement of width (inches) and microfinish (micro units). The following data were obtained:

| | | | | | Width | | Microfinish | |
|---|---|---|---|---|---|---|---|---|
| Order | Operator | Mill | Spindle | Position | Part 1 | Part 2 | Part 1 | Part 2 |
| 1 | *A* | 101 | 1 | *L* | 2.003 | 1.998 | 110 | 122 |
| 2 | *A* | 123 | 1 | *L* | 2.011 | 2.007 | 112 | 115 |
| 3 | *B* | 220 | 2 | *L* | 1.991 | 1.989 | 123 | 113 |
| 4 | *A* | 123 | 2 | *R* | 2.009 | 2.008 | 130 | 126 |
| 5 | *B* | 220 | 1 | *L* | 2.001 | 1.998 | 138 | 130 |
| 6 | *B* | 285 | 1 | *R* | 2.015 | 2.012 | 155 | 120 |
| 7 | *B* | 285 | 1 | *L* | 2.001 | 1.999 | 148 | 125 |
| 8 | *A* | 101 | 2 | *L* | 1.996 | 2.001 | 160 | 146 |
| 9 | *B* | 220 | 2 | *R* | 1.990 | 1.988 | 158 | 136 |
| 10 | *A* | 123 | 1 | *R* | 2.012 | 2.010 | 171 | 155 |
| 11 | *A* | 101 | 2 | *R* | 2.001 | 2.004 | 182 | 164 |
| 12 | *B* | 285 | 2 | *L* | 2.003 | 1.999 | 168 | 160 |
| 13 | *B* | 220 | 1 | *R* | 1.999 | 2.000 | 180 | 172 |
| 14 | *A* | 123 | 2 | *L* | 2.009 | 2.007 | 182 | 166 |
| 15 | *A* | 101 | 1 | *R* | 1.997 | 1.999 | 198 | 183 |
| 16 | *B* | 285 | 2 | *R* | 2.000 | 2.002 | 186 | 192 |

Develop a design matrix for this study. Show the results of the study on the design matrix. Analyze the data for each quality characteristic with run charts and dot-frequency diagrams. For each characteristic, list the factors in order of contribution to variation.

What action is needed to reduce variation in this process?

Chapter

# 8

# Experiments for Special Situations

This chapter presents experimental designs for some special situations that occur in activities to make improvements. The focus of the chapter is on factorial patterns with more than two levels and designs for assemblies with interchangeable parts. An overview of designs for mixtures is given in Sec. 8.5. A brief description of other special designs is included in Sec. 8.6 and complex systems in Sec. 8.7.

## 8.1 Factorial Designs with More Than Two Levels

Chapter 5 discussed factorial designs at two levels. Usually, these designs can be used in sequence to understand the effects of factors, but sometimes it is desirable to study a factor in an experiment at more than two levels. For example, three companies supply raw material to a manufacturer. An experiment to determine the effect of a raw material on process yield will contain all three companies as levels of the factor supplier.

If the factor is a quantitative variable, such as temperature or speed, then an infinite number of levels are possible. The motivation for using more than two levels of a quantitative factor is usually to study the nonlinear effects of the factor.

Sometimes the levels of a factor appear to be qualitative but have some underlying continuum. For example, three different types of chemicals are usually considered as three levels of a qualitative factor. But if the primary difference between the chemicals is in the concentration of an inhibitor, then the factor could be thought of as quantitative. Another example is three different operators with different amounts of experience in a study. Rather than consider the operators to be qualitative levels, they could represent quantitative levels of the factor *experience.* It is usually better to treat the factor as quantitative when it is possible to do so.

Designs for qualitative factors with more than two levels are discussed in this section. Designs for quantitative factors with more than two levels are discussed in Secs. 8.2 and 8.3.

The following example describes a study with qualitative variables at more than two levels.

**Example 8.1 Microfinish Tolerances** A study was conducted to help set tolerances for the microfinish of a precision part. Needs of the customer suggested that the microfinish should be specified less than 30 units. Three factors thought by the operators to affect microfinish were studied. One of the factors, tool type, was studied at four levels since four different types of tools could be used in the process. The second factor, supplier, was studied at three levels since castings could be purchased from three sources. The third factor, coolant level, was quantitative and was studied at two levels.

Figure 8.1 shows the design matrix (a $4 \times 3 \times 2$ factorial design) and the data collected for this study. A run chart is also shown. The experimental design was run once in the morning and then was repeated in the afternoon. A random order of the 24 tests was chosen each time. The run chart indicates no special causes or problems with the data. Microfinish results ranged from a low of 5 units to a high of 43 units.

To better study the importance of the common-cause variation in the study, the run chart was next rearranged to compare the morning and afternoon results for each combination of factor levels. Figure 8.2 contains this plot. No systematic differences between the morning and afternoon are apparent from the data. The differences between the morning and afternoon test results (the common-cause variation) are small relative to the effects of the factors. No special causes are obvious in the data. Because of the consistent results, the morning and afternoon test results can be averaged to study the factor effects. Figure 8.3 shows these averages.

If the variability between morning and afternoon had been greater, it might have been interesting to study the effect of the factors on this variability. The ranges between morning and afternoon could have been studied as a response variable.

It is possible to compute estimates of the factor effects from a design matrix when there are more than two levels for a factor. The calculation procedure is, however, beyond the scope of this text. We will primarily use response plots to study the effects of factors at more than two levels. Often, with qualitative variables the average response at a particular setting is more interesting than the factor effect.

Response plots for the averages were developed next to study factor effects and two-factor interactions. These response plots are shown in Fig. 8.4. The following observations were made from these plots:

Tool type

| Coolant | Supplier | T1 | | T2 | | T3 | | T4 | |
|---|---|---|---|---|---|---|---|---|---|
| Low | A | 6 (22) | 5 (5) | 12 (8) | 15 (4) | 12 (7) | 16 (22) | 37 (19) | 33 (11) |
| Low | B | 17 (14) | 21 (24) | 24 (5) | 20 (19) | 20 (21) | 20 (3) | 25 (4) | 26 (21) |
| Low | C | 22 (15) | 24 (1) | 30 (24) | 31 (8) | 28 (9) | 26 (17) | 42 (16) | 39 (12) |
| High | A | 9 (2) | 5 (18) | 18 (10) | 16 (23) | 18 (3) | 22 (2) | 27 (6) | 21 (10) |
| High | B | 22 (20) | 22 (15) | 25 (11) | 28 (13) | 26 (12) | 27 (14) | 20 (18) | 20 (9) |
| High | C | 31 (17) | 26 (16) | 34 (1) | 39 (7) | 37 (23) | 35 (20) | 35 (13) | 37 (6) |

Legend for design matrix

| $R_1$ | $R_2$ |
|---|---|
| $(O_1)$ | $(O_2)$ |

$R_1$ = microfinish for morning test
$R_2$ = microfinish for afternoon test
$O_1$ = order in morning
$O_2$ = order in afternoon

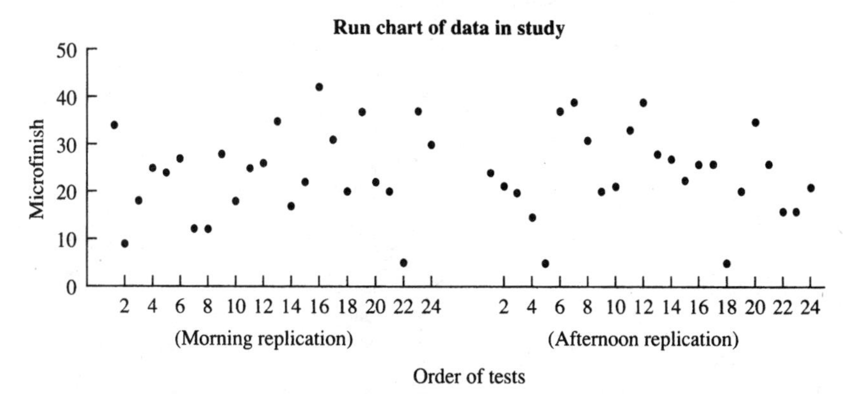


**Figure 8.1** Design matrix and data for $4 \times 3 \times 2$ factorial design.

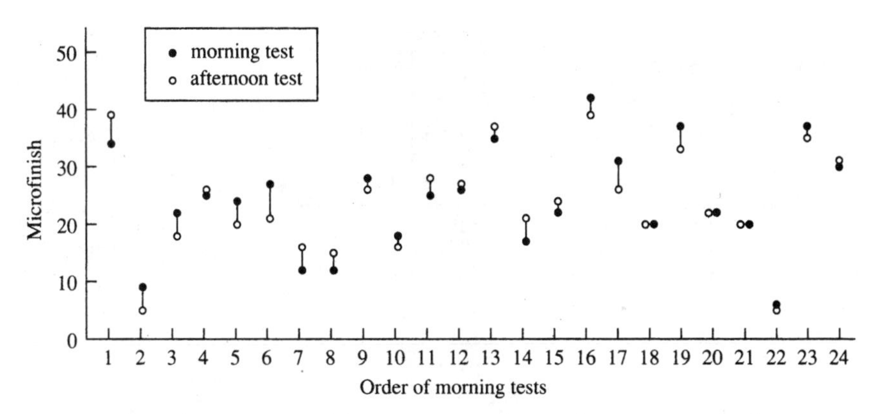


**Figure 8.2** Run chart rearranged to compare morning to afternoon test results.

| Coolant | | Supplier | Tool type: *T*1 | *T*2 | *T*3 | *T*4 |
|---|---|---|---|---|---|---|
| Low | Supplier | *A* | 5.5 | 13.5 | 14.0 | 35.0 |
| | | *B* | 19.0 | 22.0 | 20.0 | 25.5 |
| | | *C* | 23.0 | 30.5 | 27.0 | 40.5 |
| High | Supplier | *A* | 7.0 | 17.0 | 20.0 | 24.0 |
| | | *B* | 22.0 | 26.5 | 26.5 | 20.0 |
| | | *C* | 28.5 | 36.5 | 36.0 | 36.0 |

**Figure 8.3** Averages of microfinish data on design matrix.

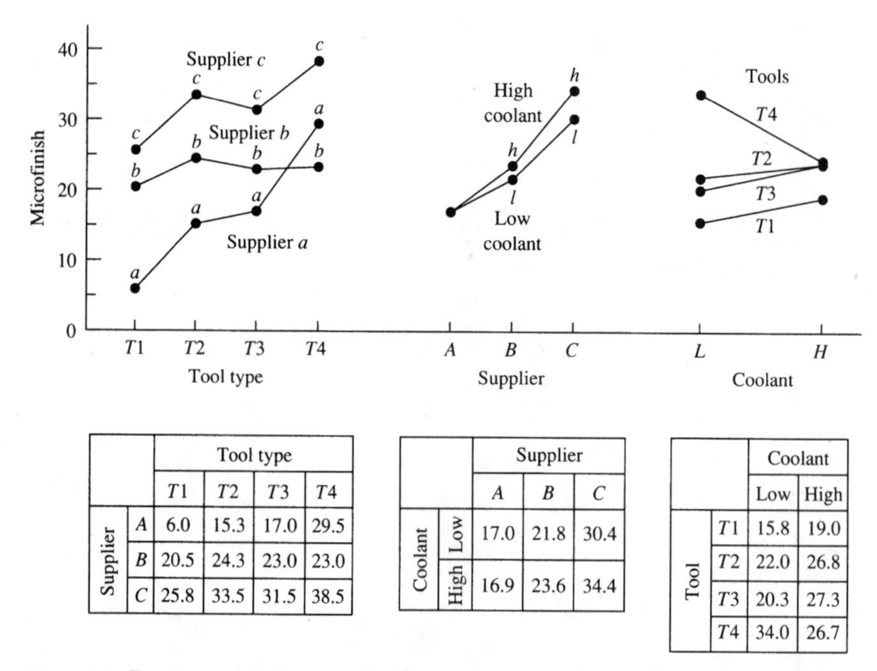


| Supplier | *T*1 | *T*2 | *T*3 | *T*4 |
|---|---|---|---|---|
| *A* | 6.0 | 15.3 | 17.0 | 29.5 |
| *B* | 20.5 | 24.3 | 23.0 | 23.0 |
| *C* | 25.8 | 33.5 | 31.5 | 38.5 |

| Coolant | Supplier *A* | *B* | *C* |
|---|---|---|---|
| Low | 17.0 | 21.8 | 30.4 |
| High | 16.9 | 23.6 | 34.4 |

| Tool | Coolant: Low | High |
|---|---|---|
| *T*1 | 15.8 | 19.0 |
| *T*2 | 22.0 | 26.8 |
| *T*3 | 20.3 | 27.3 |
| *T*4 | 34.0 | 26.7 |

**Figure 8.4** Response plots for microfinish averages.

1. The effects of both tool type and supplier are large. The coolant effect is smaller.
2. There are numerous interactions between the three factors that can be exploited to minimize microfinish.
3. The parts made from supplier *C* material have consistently higher microfinish. Supplier *A* had the best performance except for tool type *T*4 where the parts from supplier *A* averaged 29.5 units and the parts from supplier *B* averaged 23 units. If tool *T*4 could be eliminated, supplier *A* could be a single supplier.
4. Tool type *T*1 generally resulted in lower microfinish readings than the other tools.
5. If it is necessary to use all four tools, the following conditions will give the lowest expected microfinish:

| Tool type | Casting supplier | Coolant level |
|---|---|---|
| *T*1 | *A* or *B* | Low |
| *T*2 | *A* | Low |
| *T*3 | *A* or *B* | Low |
| *T*4 | *B* | High |

6. To obtain the lowest microfinish readings, use tool *T*1, supplier *A*, and either coolant level. These conditions should be run and control charts developed to confirm the improved process capability for microfinish.

The response plot will be the most important graphical representation of the data in studies with more than two levels. Replication is important in this type of study so that effects of nuisance variables can be evaluated prior to the study of the response plots. Replication also makes it possible to study the effect of the factors on the variation of the response variable due to nuisance variables.

If replication is not practical, it is more difficult to assess the impact of nuisance variables by using the run charts or a rearranged run chart. In this case, more detailed response plots can be prepared to evaluate the consistency of the factor effects. One way to do this is to prepare a response plot for each two-factor interaction at each level of the other factors. If the plots are similar for each level, then nuisance variables are relatively unimportant. If the response plots at each level of the other factors are not similar, then either nuisance variables are important or there is a complex system of interactions among the factors. In either case, replication or some other type of confirmation will be necessary.

Figure 8.5 shows response plots for the morning replication of the microfinish experiment (data from Fig. 8.1). Three plots are prepared, one for each factor in the study. The data from the experiment are ordered by the levels of the two other factors in the study. The following conclusions can be drawn from a study of these plots:

1. *Supplier effect:* The lowest microfinish is for parts from supplier $A$, except for tool type $T4$. Supplier $C$ is consistently the worst (eight out of eight cases). The supplier effect is not much different for the two coolant levels.
2. *Tool-type effect:* Tool $T1$ is consistently better, and tool $T4$ consistently the worst. The tool-type effect is different at different coolant levels.
3. *Coolant effect:* Low coolant is slightly better, except for tool type $T4$.

Based on these observations, summary response plots of the tool type/supplier interaction and the tool/coolant interaction would be prepared to summarize the findings. These response plots would be similar to those in Fig. 8.4.

## 8.2 Augmenting $2^k$ Factorial Designs with Center Points

Sometimes it is desirable to study quantitative factors at more than two levels. Some examples of situations that would lead to three or more levels are as follows:

1. The effect of the factor is expected to be nonlinear in the range of interest.
2. The purpose of the study is to find the optimum level at which to set the factor.
3. It is important to include the current levels of the factors in the study as well as the high and low levels.

A possible approach in these cases is the use of three-level factorial designs ($3^k$ factorials). With three or more factors, however, the $3^k$ factorials become very large (three factors, 27 tests; four factors, 81 tests; five factors, 243 tests). Designs based on fractions of $3^k$ have been developed to make these designs somewhat more manageable. Section 8.3 discusses $3^k$ factorial and fractional factorial designs.

An alternative to three-level factorial designs is a type of experimental pattern called *composite designs* (Box and Wilson, 1951). The simplest composite design is a $2^k$ factorial design with a center point. Example 8.2 shows how to analyze this design. Other types of com-

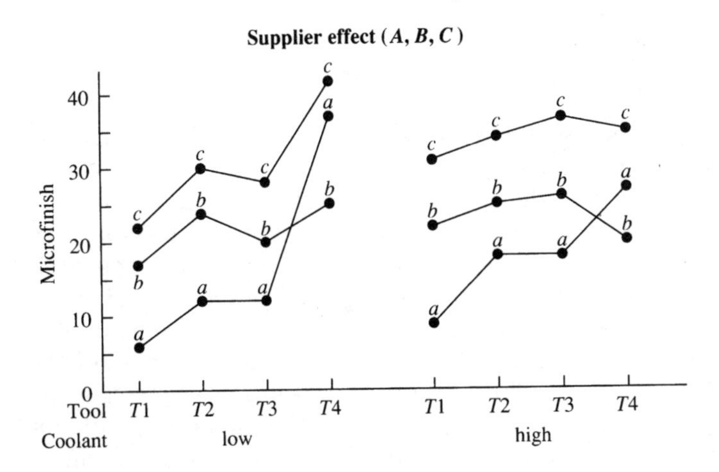


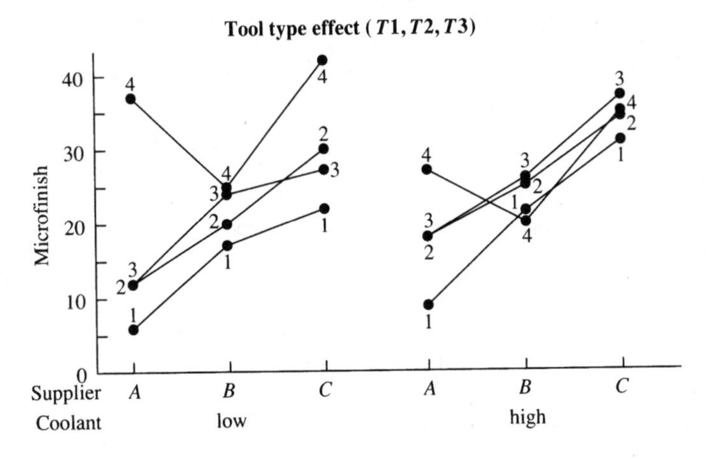


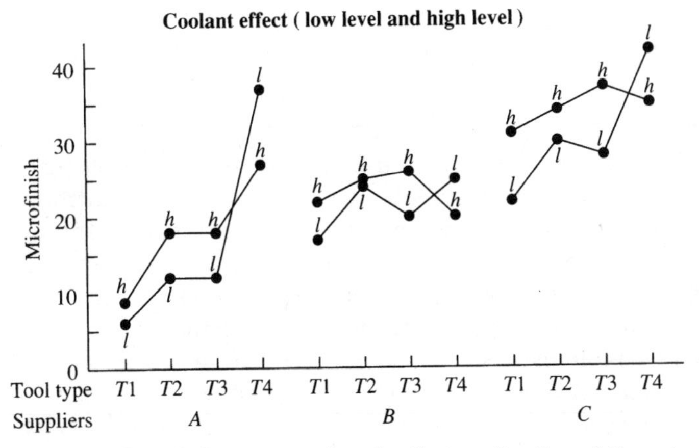


**Figure 8.5** Detailed response plots for first replication of the microfinish study.

posite designs can be developed by adding more runs to the $2^k$ or $2^{k-p}$ factorial designs. These designs are discussed briefly at the end of this section.

When center points are added to a factorial design, the importance of curvature or the presence of nonlinear effects can be evaluated. A center point is located at the midlevel for each of the factors in the factorial design. The curvature cannot be assigned to a particular factor, but the center point gives an indication of whether it is appropriate to interpolate between the factorial points. The magnitude of the curvature effect can be directly compared to the factor effects for relative importance if the number of center points is equal to one-half the number of factorial points. Thus, two tests at the center point of the design are recommended for a two-factor design, four center points for a three-factor design, and eight center points for a four-factor design.

If fewer runs than required for direct comparison are run, the nonlinear effect can still be subjectively evaluated. The responses at the center points can be compared to the other responses to evaluate gross departures from linearity. For example, if the center-point results are close to the lowest or highest response, the nonlinearity is probably important.

**Example 8.2** The operators in the filling process wanted to study variables in the process that affected the variation in the fill weights of their products. A number of factors that affected the average fill weight had been identified by using control charts, but identification of factors that affected the range had been inconsistent. The operators suspected that interactions between some of the factors might be the key to reducing variation of the fill weights. They proposed to run a $2^3$ factorial design on the following three factors:

| Factor | Current level | Low level | High level | Units |
|---|---|---|---|---|
| Product temperature | 200 | 180 | 220 | Degrees Fahrenheit |
| Line speed | 350 | 300 | 400 | Cans per minute |
| Product consistency | 15 | 10 | 20 | Viscosity (coded) |

The levels of the factor were selected by adding and subtracting an equal amount from the current levels of the process. In addition to the eight runs in the factorial pattern, four tests were run at the levels of the factors currently used in the process. Figure 8.6 shows the design matrix and the results of the study. The ranges for two five-can samples collected at each test condition were calculated.

The response variable used was the average of these two ranges.

A run chart of the average range is shown in Fig. 8.7. Note that the center points were spread throughout the study, while the order of the other runs was randomized. The repeated center points give a measure of the importance of nuisance variables in the study. From the

| Factor | low level (−) | high level (+) | current level (0) |
|---|---|---|---|
| Temperature (T) | 180 | 220 | 200 |
| Speed (S) | 300 | 400 | 350 |
| Consistency (C) | 10 | 20 | 15 |

| | | | | | | | | | Range of five cans (grams) | | |
|---|---|---|---|---|---|---|---|---|---|---|---|
| Test | Run order | *T* | *S* | *C* | *TS* | *TC* | *SC* | *TSC* | 1st | 2nd | $\bar{R}$ |
| 1 | 5 | − | − | − | + | + | + | − | 2.7 | 3.5 | 3.10 |
| 2 | 7 | + | − | − | − | − | + | + | 2.0 | 4.1 | 3.05 |
| 3 | 2 | − | + | − | − | + | − | + | 6.8 | 5.4 | 6.10 |
| 4 | 11 | + | + | − | + | − | − | − | 5.6 | 6.9 | 6.25 |
| 5 | 10 | − | − | + | + | − | − | + | 5.4 | 5.8 | 5.60 |
| 6 | 3 | + | − | + | − | + | − | − | 2.2 | 3.4 | 2.80 |
| 7 | 6 | − | + | + | − | + | − | − | 8.6 | 9.4 | 9.00 |
| 8 | 9 | + | + | + | + | + | + | + | 6.5 | 5.8 | 6.15 |
| 9 | 1 | 0 | 0 | 0 | 0 | 0 | 0 | 0 | 3.4 | 4.3 | 3.85 |
| 10 | 4 | 0 | 0 | 0 | 0 | 0 | 0 | 0 | 3.0 | 5.2 | 4.10 |
| 11 | 8 | 0 | 0 | 0 | 0 | 0 | 0 | 0 | 4.6 | 3.2 | 3.90 |
| 12 | 12 | 0 | 0 | 0 | 0 | 0 | 0 | 0 | 4.1 | 4.2 | 4.15 |

**Figure 8.6** Design matrix for fill weight variation study.

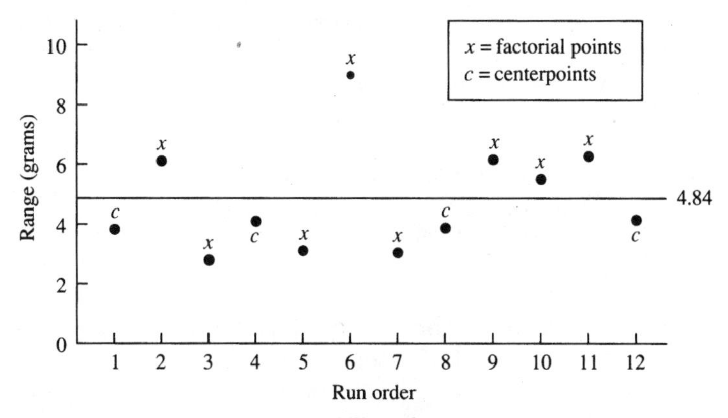


**Figure 8.7** Run chart of range of fill weights.

run chart, the effect of nuisance variables appears small relative to the factor effects.

Estimates of the effects were made and plotted on a dot-frequency diagram (Fig. 8.8). The curvature effect is the difference between the average of the four tests at the center point and the average of the eight factorial points. Since four center points were run (one-half the number of factorial points), the curvature effect *N* can be directly

| Factor | low level (−) | high level (+) | current level (0) |
|---|---|---|---|
| Temperature (*T*) | 180 | 220 | 200 |
| Speed (*S*) | 300 | 400 | 350 |
| Consistency (*C*) | 10 | 20 | 15 |

| Test | Run order | *T* | *S* | *C* | *TS* | *TC* | *SC* | *TSC* | Range of five cans (grams) 1st | 2nd | $\overline{R}$ |
|---|---|---|---|---|---|---|---|---|---|---|---|
| 1 | 5 | − | − | − | + | + | + | − | 2.7 | 3.5 | 3.10 |
| 2 | 7 | + | − | − | − | − | + | + | 2.0 | 4.1 | 3.05 |
| 3 | 2 | − | + | − | − | + | − | + | 6.8 | 5.4 | 6.10 |
| 4 | 11 | + | + | − | + | − | − | − | 5.6 | 6.9 | 6.25 |
| 5 | 10 | − | − | + | + | − | − | + | 5.4 | 5.8 | 5.60 |
| 6 | 3 | + | − | + | − | + | − | − | 2.2 | 3.4 | 2.80 |
| 7 | 6 | − | + | + | − | + | − | − | 8.6 | 9.4 | 9.00 |
| 8 | 9 | + | + | + | + | + | + | + | 6.5 | 5.8 | 6.15 |
| Effects | | −1.39 | 3.24 | 1.26 | .04 | −1.43 | .14 | −.06 | | Average $\overline{R}$ = | 5.26 |
| 9 | 1 | 0 | 0 | 0 | 0 | 0 | 0 | 0 | 3.4 | 4.3 | 3.85 |
| 10 | 4 | 0 | 0 | 0 | 0 | 0 | 0 | 0 | 3.0 | 5.2 | 4.10 |
| 11 | 8 | 0 | 0 | 0 | 0 | 0 | 0 | 0 | 4.6 | 3.2 | 3.90 |
| 12 | 12 | 0 | 0 | 0 | 0 | 0 | 0 | 0 | 4.1 | 4.2 | 4.15 |

Average of center points = 4.00
Nonlinear effect (N): 5.26 − 4.00 = 1.26

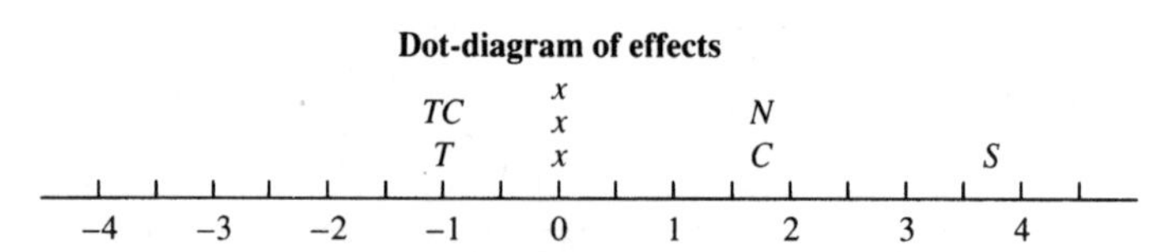


**Figure 8.8** Analysis of fill weight variation data.

compared to the other effects on the dot-frequency diagram. The important effects are the line speed, the temperature-consistency interaction, and the curvature.

The average ranges are shown on a cube in Fig. 8.9. The average of the center points is shown in the center of the cube. The important effects are also shown on response plots in Fig. 8.9. The response plot for the speed effect indicates less variation in the fill weights at the lower line speed. But the center point indicates little increase in variation from 300 to 350 cans per minute. The response plot for the temperature-consistency interaction indicates that there is no temperature effect for the low-consistency product but that variation is much greater at low temperature for the high-consistency product. Again, the results at the center point indicate that the temperature-consis-

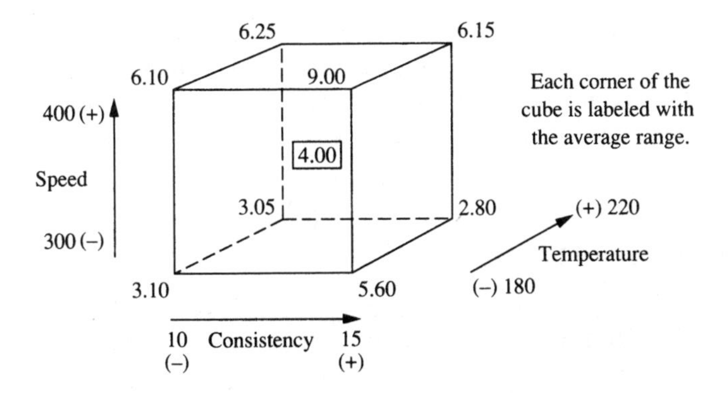


**Response plots for important effects**

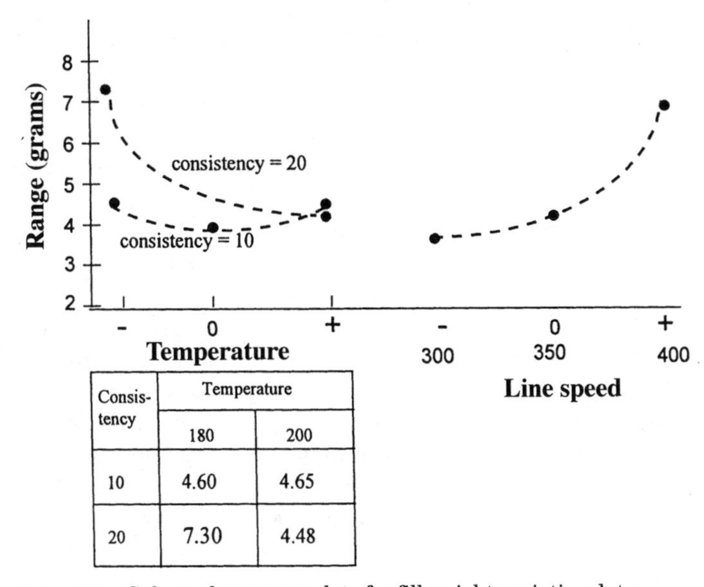


| Consis-tency | Temperature | |
|---|---|---|
| | 180 | 200 |
| 10 | 4.60 | 4.65 |
| 20 | 7.30 | 4.48 |

**Figure 8.9** Cube and response plots for fill weight variation data.

tency effects may not be linear. Additional tests would be required to describe the important nonlinear effects.

If nonlinear effects are important, there are a number of alternative ways to design future studies. The three-level factorial designs discussed in the next section are one alternative.

More complex, composite designs can be used to study factors with nonlinear effects (Box and Draper, 1987). In addition to the factorial points and center points, these composite designs include additional tests at other levels of the factors.

These composites are more efficient (fewer tests required) than the three-level factorial designs to be discussed in the next section. But the analysis of data based on these designs requires the use of a statistical model, and thus is beyond the scope of this book. The book by Box and Draper (1987) presents the design and analysis of these composite designs.

## 8.3 Three-Level Factorial Designs

As discussed in Sec. 8.2, factorial designs with factors studied at three levels are of limited use because of the large size of the experiments. It is also not possible to develop small fractions of most of these designs that enable main effects to be separated from two-factor interactions. It may be desirable to study three levels of a factor if previous two-level factorials have indicated a nonlinear effect on the response variable and if the factor interacts with other important factors. Three-level factorial designs could also be used to study a process when it is known that the selected factors are important and when their effects are expected to be nonlinear.

Table 8.1 summarizes recommended experimental designs when it is desirable to study factors at three levels. The designs can incorporate from two to five factors. If more than five factors are being considered, two-level fractional factorial designs (Chap. 6) should be used, augmented with center points to evaluate the importance of nonlinearity. Each of these designs can be used to evaluate the main effects and two-factor interactions of each of the factors (except where noted in Fig. 8.10). The design matrix for each of these designs is shown in Fig. 8.10. Cochran and Cox (1957) contains more information on these designs. Run charts (or control charts) and response plots of interactions or main effects can be used to analyze the results of these designs.

Example 8.3 illustrates fractional factorial design for four factors, each at three levels. The analysis of the other $3^k$ designs is similar.

**Example 8.3** A study in a chemical process was done to understand the effects of four factors in the region of the current operating conditions. Previous studies had shown that each of these factors had an effect on the response (yield), and the effects had been nonlinear in some of the studies. One of the

**TABLE 8.1 Recommended Three-Level Designs**

| Number of factors | Experimental pattern | Number of runs |
|---|---|---|
| 2 | $3^2$ | 9 |
| 3 | $3^3$ | 27 |
| 4 | $3^{4-1}$ | 27 |
| 5 | $3^{5-1}$ | 81 |
| >5 | Use $2^{k-p}$ with center points | |

| Code | Interpretation |
|---|---|
| – | Low level of factor |
| 0 | Middle level of factor |
| + | High level of factor |

**Two-factor design ($3^2$)**

| Test | A | B |
|---|---|---|
| 1 | – | – |
| 2 | 0 | – |
| 3 | + | – |
| 4 | – | 0 |
| 5 | 0 | 0 |
| 6 | + | 0 |
| 7 | – | + |
| 8 | 0 | + |
| 9 | + | + |

**Three-factor design ($3^3$)**

| Test | A | B | C |
|---|---|---|---|
| 1 | – | – | – |
| 2 | 0 | – | – |
| 3 | + | – | – |
| 4 | – | 0 | – |
| 5 | 0 | 0 | – |
| 6 | + | 0 | – |
| 7 | – | + | – |
| 8 | 0 | + | – |
| 9 | + | + | – |
| 10 | – | – | 0 |
| 11 | 0 | – | 0 |
| 12 | + | – | 0 |
| 13 | – | 0 | 0 |
| 14 | 0 | 0 | 0 |
| 15 | + | 0 | 0 |
| 16 | – | + | 0 |
| 17 | 0 | + | 0 |
| 18 | + | + | 0 |
| 19 | – | – | + |
| 20 | 0 | – | + |
| 21 | + | – | + |
| 22 | – | 0 | + |
| 23 | 0 | 0 | + |
| 24 | + | 0 | + |
| 25 | – | + | + |
| 26 | 0 | + | + |
| 27 | + | + | + |

**Four-factor design ($3^{4-1}$)**

| Test | A | B | C | D |
|---|---|---|---|---|
| 1 | – | – | – | – |
| 2 | + | + | 0 | – |
| 3 | 0 | 0 | + | – |
| 4 | 0 | + | – | – |
| 5 | – | 0 | 0 | – |
| 6 | + | – | + | – |
| 7 | + | 0 | – | – |
| 8 | 0 | – | 0 | – |
| 9 | – | + | + | – |
| 10 | + | + | – | 0 |
| 11 | 0 | 0 | 0 | 0 |
| 12 | – | – | + | 0 |
| 13 | – | 0 | – | 0 |
| 14 | + | – | 0 | 0 |
| 15 | 0 | + | + | 0 |
| 16 | 0 | – | – | 0 |
| 17 | – | + | 0 | 0 |
| 18 | + | 0 | + | 0 |
| 19 | 0 | 0 | – | + |
| 20 | – | – | 0 | + |
| 21 | + | + | + | + |
| 22 | + | – | – | + |
| 23 | 0 | + | 0 | + |
| 24 | – | 0 | + | + |
| 25 | – | + | – | + |
| 26 | + | 0 | 0 | + |
| 27 | 0 | – | + | + |

*Note:* For the $3^{4-1}$ design, most of the information on 2-factor interactions is clear. The following effects have some ambiguity:

***AB*** with ***CD***
***AC*** with ***BD***
***AD*** with ***BC***.

If the interactions' effects on factor ***A*** can be considered negligible, then the remaining interactions are clear.

**Figure 8.10** Design matrices for three-level factorial designs.

| Code | Interpretation |
|---|---|
| − | Low level of factor |
| 0 | Middle level of factor |
| + | High level of factor |

**Five-factor design ($3^{5-1}$)**

| Test | A | B | C | D | E | Test | A | B | C | D | E | Test | A | B | C | D | E |
|---|---|---|---|---|---|---|---|---|---|---|---|---|---|---|---|---|---|
| 1 | − | − | − | − | − | 28 | 0 | 0 | − | − | 0 | 55 | + | + | − | − | + |
| 2 | + | − | 0 | − | − | 29 | − | 0 | 0 | − | 0 | 56 | 0 | + | 0 | − | + |
| 3 | 0 | − | + | − | − | 30 | + | 0 | + | − | 0 | 57 | − | + | + | − | + |
| 4 | 0 | + | − | − | − | 31 | + | − | − | − | 0 | 58 | − | 0 | − | − | + |
| 5 | − | + | 0 | − | − | 32 | 0 | − | 0 | − | 0 | 59 | + | 0 | 0 | − | + |
| 6 | + | + | + | − | − | 33 | − | − | + | − | 0 | 60 | 0 | 0 | + | − | + |
| 7 | 0 | 0 | 0 | − | − | 34 | + | + | 0 | − | 0 | 61 | − | − | 0 | − | + |
| 8 | − | 0 | + | − | − | 35 | 0 | + | + | − | 0 | 62 | + | − | + | − | + |
| 9 | + | 0 | − | − | − | 36 | − | + | − | − | 0 | 63 | 0 | − | − | − | + |
| 10 | 0 | + | + | 0 | − | 37 | + | − | + | 0 | 0 | 64 | − | 0 | + | 0 | + |
| 11 | − | + | − | 0 | − | 38 | 0 | − | − | 0 | 0 | 65 | + | 0 | − | 0 | + |
| 12 | + | + | 0 | 0 | − | 39 | − | − | 0 | 0 | 0 | 66 | 0 | 0 | 0 | 0 | + |
| 13 | + | 0 | + | 0 | − | 40 | − | + | + | 0 | 0 | 67 | 0 | − | + | 0 | + |
| 14 | − | 0 | − | 0 | − | 41 | + | + | − | 0 | 0 | 68 | − | − | − | 0 | + |
| 15 | 0 | 0 | 0 | 0 | − | 42 | 0 | + | 0 | 0 | 0 | 69 | + | − | 0 | 0 | + |
| 16 | − | − | − | 0 | − | 43 | − | 0 | − | 0 | 0 | 70 | 0 | + | − | 0 | + |
| 17 | 0 | − | 0 | 0 | − | 44 | + | 0 | 0 | 0 | 0 | 71 | − | + | 0 | 0 | + |
| 18 | + | − | + | 0 | − | 45 | 0 | 0 | + | 0 | 0 | 72 | + | + | + | 0 | + |
| 19 | + | 0 | 0 | + | − | 46 | − | + | 0 | + | 0 | 73 | 0 | − | 0 | + | + |
| 20 | 0 | 0 | + | + | − | 47 | + | + | + | + | 0 | 74 | − | − | + | + | + |
| 21 | − | 0 | − | + | − | 48 | 0 | + | − | + | 0 | 75 | + | − | − | + | + |
| 22 | − | − | 0 | + | − | 49 | 0 | 0 | 0 | + | 0 | 76 | + | + | 0 | + | + |
| 23 | + | − | + | + | − | 50 | − | 0 | + | + | 0 | 77 | 0 | + | + | + | + |
| 24 | 0 | − | − | + | − | 51 | + | 0 | − | + | 0 | 78 | − | + | − | + | + |
| 25 | − | + | + | + | − | 52 | 0 | − | + | + | 0 | 79 | + | 0 | + | + | + |
| 26 | + | + | − | + | − | 53 | − | − | − | + | 0 | 80 | 0 | 0 | − | + | + |
| 27 | 0 | + | 0 | + | − | 54 | + | − | 0 | + | 0 | 81 | − | 0 | 0 | + | + |

*Note:* All two-factor interactions are clear in this design

**Figure 8.10** *(Continued)*

factors, catalyst concentration, did not interact with the other factors in the previous studies. The one-third replicate of a $3^4$ factorial was used for the study (see Fig. 8.10). Figure 8.11 shows the design matrix for this study.

The factor levels were chosen by making the current process level the middle level. The high and low levels were then determined by increasing or decreasing this level by an amount that would not unduly upset the process. Catalyst was designated the *A* factor since the previous studies had indicated it was unlikely to interact with the other factors.

The order of the 27 tests was randomized. The four factors were set to the required level, and then the process was allowed to stabilize (from 1 to 4 h). The yield of the process based on on-line measurements was then determined for a 20-h period. Figure 8.12 shows a run chart for the percentage yield for each test. The yields did not show any trends or special causes during the month-long study. The average yield of 95.6 percent was slightly lower than the process av-

| Factor | Code | Low (−) | Middle (0) | High () |
|---|---|---|---|---|
| Temperature (°F) | Temp | 190 | 200 | 210 |
| Pressure (psi) | Pres | 300 | 350 | 400 |
| Inhibitor level (ppm) | Inhb | 40 | 50 | 60 |
| Catalyst concentration (lbs.) | Catl | 8 | 10 | 12 |

Four-factor design ($3^{4-1}$)

| Test | Catl | Inhb | Pres | Temp | Run Order | Yield Percentage |
|---|---|---|---|---|---|---|
| 1 | − | − | − | − | 17 | 93.1 |
| 2 | + | + | 0 | − | 12 | 94.3 |
| 3 | 0 | 0 | + | − | 7 | 96.8 |
| 4 | 0 | + | − | − | 15 | 91.9 |
| 5 | − | 0 | 0 | − | 16 | 93.6 |
| 6 | + | − | + | − | 1 | 98.5 |
| 7 | + | 0 | − | − | 5 | 92.2 |
| 8 | 0 | − | 0 | − | 19 | 97.7 |
| 9 | − | + | + | − | 27 | 94.3 |
| 10 | + | + | − | 0 | 10 | 93.6 |
| 11 | 0 | 0 | 0 | 0 | 22 | 96.4 |
| 12 | − | − | + | 0 | 9 | 97.1 |
| 13 | − | 0 | − | 0 | 25 | 92.7 |
| 14 | + | − | 0 | 0 | 18 | 97.8 |
| 15 | 0 | + | + | 0 | 13 | 96.0 |
| 16 | 0 | − | − | 0 | 23 | 97.4 |
| 17 | − | + | 0 | 0 | 2 | 93.5 |
| 18 | + | 0 | + | 0 | 26 | 96.6 |
| 19 | 0 | 0 | − | + | 4 | 97.0 |
| 20 | − | − | 0 | + | 11 | 96.8 |
| 21 | + | + | + | + | 21 | 95.5 |
| 22 | + | − | − | + | 14 | 98.4 |
| 23 | 0 | + | 0 | + | 3 | 96.1 |
| 24 | − | 0 | + | + | 8 | 95.0 |
| 25 | − | + | − | + | 20 | 93.7 |
| 26 | + | 0 | 0 | + | 6 | 96.7 |
| 27 | 0 | − | + | + | 24 | 98.8 |

**Figure 8.11** Design matrix for $3^{4-1}$ factorial study of yields.

erage before the study (96 percent). A few of the tests had yields greater than 98 percent.

Response plots of the three factors that were expected to interact were made. Figure 8.13 shows these plots. The factors temperature and pressure did interact. The effect of the inhibitor level was independent of temperature and pressure.

The effects of each of the factors are nonlinear over the range studied. The pressure effect is about 4 percent at the low temperature, but it is insignificant at the high temperature. At low pressure the temperature is also important, but at high pressure a variation in tempera-

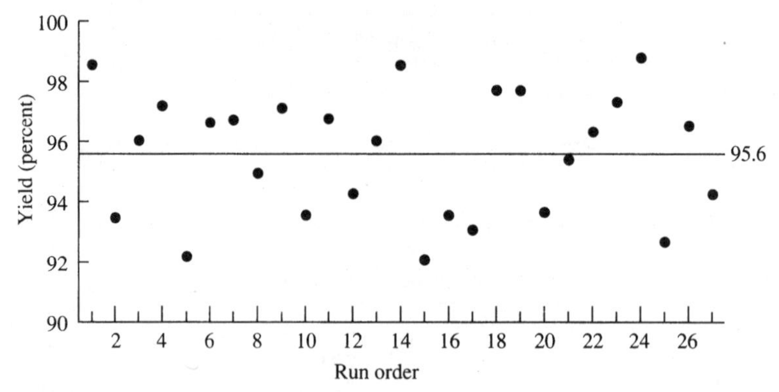


**Figure 8.12** Run chart for percentage yield.

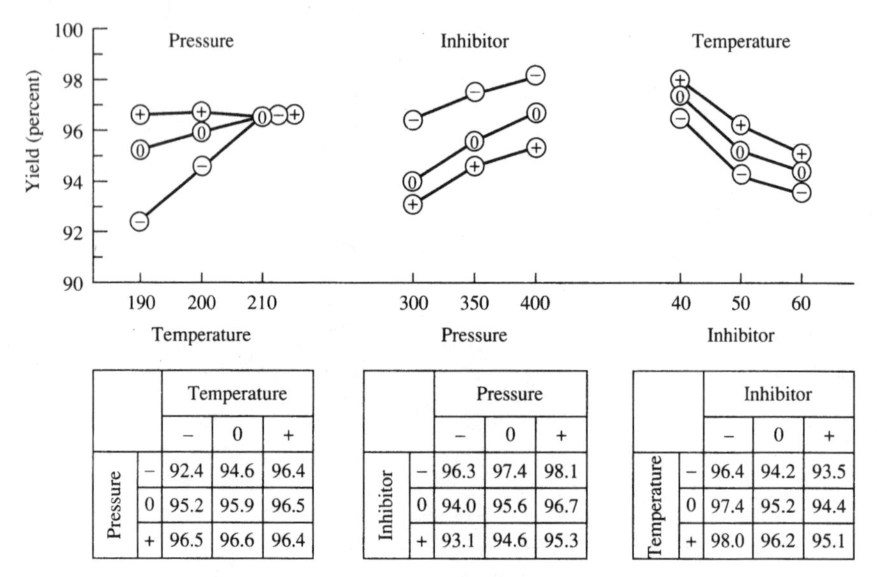


| Pressure | Temperature – | Temperature 0 | Temperature + |
|---|---|---|---|
| – | 92.4 | 94.6 | 96.4 |
| 0 | 95.2 | 95.9 | 96.5 |
| + | 96.5 | 96.6 | 96.4 |

| Inhibitor | Pressure – | Pressure 0 | Pressure + |
|---|---|---|---|
| – | 96.3 | 97.4 | 98.1 |
| 0 | 94.0 | 95.6 | 96.7 |
| + | 93.1 | 94.6 | 95.3 |

| Temperature | Inhibitor – | Inhibitor 0 | Inhibitor + |
|---|---|---|---|
| – | 96.4 | 94.2 | 93.5 |
| 0 | 97.4 | 95.2 | 94.4 |
| + | 98.0 | 96.2 | 95.1 |

**Figure 8.13** Response plots for two-factor interactions.

ture has no impact. Hence, yield can be increased slightly by setting temperature to the high level, but there is no increase if both temperature and pressure are increased. The increased temperature will also result in less variation in yield if pressure changes inadvertently.

Higher levels of inhibitor decrease the yield, but the decrease in yield as a result of inhibitor levels raised from 50 to 60 ppm is much less than that for inhibitor levels raised from 40 to 50 ppm. Decreasing the catalyst charge from the current level of 12 lb to 8 lb lowers the yield by about 3 percent. Increasing the catalyst charge does not increase the yield.

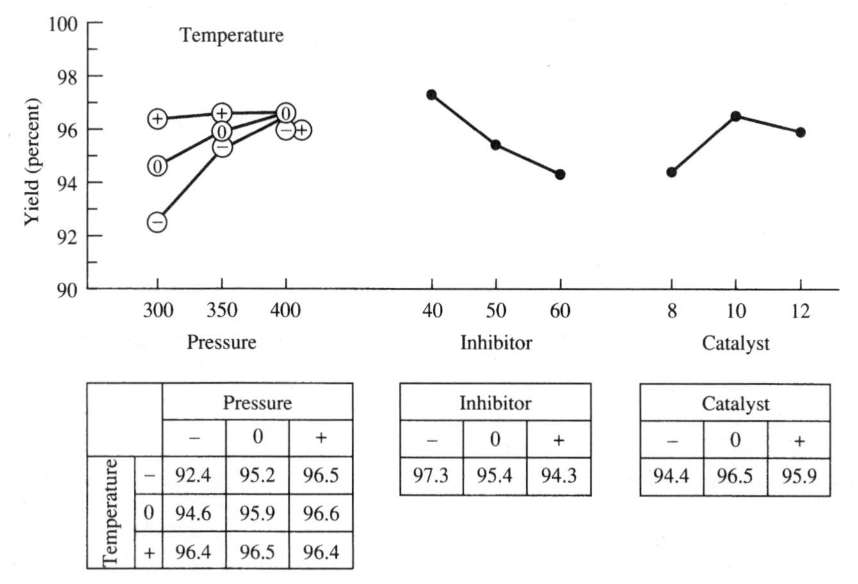


| | | Pressure | | |
|---|---|---|---|---|
| | | – | 0 | + |
| Temperature | – | 92.4 | 95.2 | 96.5 |
| | 0 | 94.6 | 95.9 | 96.6 |
| | + | 96.4 | 96.5 | 96.4 |

| Inhibitor | | |
|---|---|---|
| – | 0 | + |
| 97.3 | 95.4 | 94.3 |

| Catalyst | | |
|---|---|---|
| – | 0 | + |
| 94.4 | 96.5 | 95.9 |

**Figure 8.14** Response plots for important effects.

Response plots to summarize the important effects are shown in Fig. 8.14. Response plots are shown for the temperature/pressure interaction and the inhibitor and catalyst main effects. Note that the interaction plot for temperature and pressure is the same as in Fig. 8.13, but the pressure effect has been placed on the horizontal axis to get a second display of the interaction.

The process should be evaluated for improvement of yield after setting the temperature or pressure (or both if advantageous) at the high level in the study and reducing the level of the inhibitor.

This example illustrates the analysis of a three-level factorial design. In this case, previous studies had indicated that nuisance variables were not important. If this is not the case, some replication should be included in the design. The run chart and the response plots of main effects and two-factor interactions are the methods of analysis. Response plots summarizing the important effects should be prepared for presentation of results.

### Mixed two- and three-level factorial designs

Factorial designs can be run with some of the factors at two levels and other factors at three levels. Table 8.2 describes some of these designs that do not require an unreasonable number of tests. Response plots can be used to analyze the results of these designs in a manner similar to the last example.

**TABLE 8.2 Mixed Two- and Three-Level Factorial Designs**

| Design | Factors at two levels | Factors at three levels | Number of tests required |
|---|---|---|---|
| $2^1 3^1$ | 1 | 1 | 6 |
| $2^2 3^1$ | 2 | 1 | 12 |
| $2^3 3^1$ | 3 | 1 | 24 |
| $2^1 3^2$ | 1 | 2 | 18 |
| $2^2 3^2$ | 2 | 2 | 36 |

Section 8.4 discusses a special application of two-level factorial designs.

## 8.4 Experimental Design for Interchangeable Parts

Ott (1975) described the use of factorial designs in an experimental plan to study an assembled product that can be readily disassembled and reassembled. The plan uses the performance of the assembled product to determine the high and low levels of each factor in the study. Example 8.4 illustrates this design.

**Example 8.4** A manufacturer of automobiles developed a method of measuring noise in the engine. The method was calibrated to human responses. Engines that measured greater than 7.0 were considered noisy, and engines that measured less than 6.0 were considered quiet. Control charts on this measurement indicated that the engines currently being produced had greater noise levels than those that had been assembled in the previous month. The cause of the increase in noise was suspected to be found in the crank or the balancer since both of these components contained gears.

The following experimental plan was developed to isolate the source of the noise:

1. Eight engines were obtained for the study. Four of the engines were noisy (greater than 7.0), and four of the engines were quiet (less than 6.0).
2. The eight engines were disassembled into three components:
   - Balancer
   - Crankshaft
   - Rest of the engine (the block)

   The three components thus become the three factors in the study. Each factor was studied at two levels, quiet or noisy. The level was based on whether the component came from a noisy or quiet engine. For each component, four were labeled noisy and four were labeled quiet.

**TABLE 8.3 A $2^3$ Factorial Design for Noisy Engines**

| Test | Test order | Balancer | Crankshaft | Block | Measured noise |
|---|---|---|---|---|---|
| 1 | 3 | Noisy | Noisy | Noisy | 8.5 |
| 2 | 5 | Quiet | Noisy | Noisy | 9.0 |
| 3 | 1 | Noisy | Quiet | Noisy | 5.9 |
| 4 | 6 | Quiet | Quiet | Noisy | 5.2 |
| 5 | 2 | Noisy | Noisy | Quiet | 10.3 |
| 6 | 8 | Quiet | Noisy | Quiet | 10.5 |
| 7 | 4 | Noisy | Quiet | Quiet | 7.2 |
| 8 | 7 | Quiet | Quiet | Quiet | 5.7 |

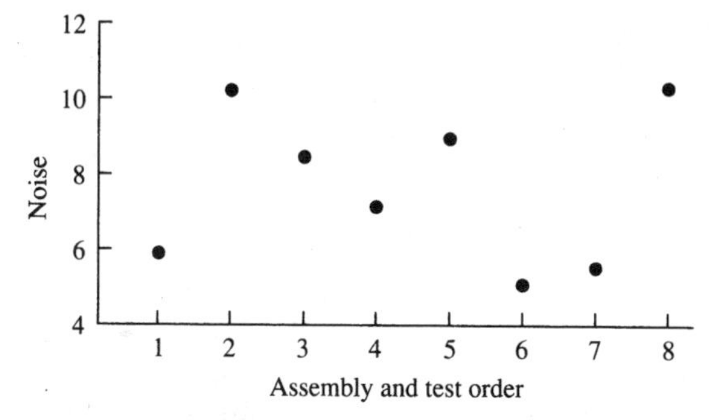


**Figure 8.15** Run chart of $2^3$ design for noisy engines.

3. Eight engines were reassembled using the $2^3$ factorial design shown in Table 8.3. Each of the three components of the engines was randomly selected from the noisy or quiet group according to the design. Each engine was then tested for noise. The results are shown in Table 8.3.

Figure 8.15 shows a run chart for the measured noise of the assembled engines. The run chart indicated no trends or special causes.

Figure 8.16 shows the analysis of the factor effects. The main effects and interactions were estimated from the design matrix in the usual manner (see Chap. 5).

A main effect in this study is the change in noise when going from a component labeled quiet to a component labeled noisy. An important interaction effect indicates that the main effect depends on whether one of the other components in the engine is labeled quiet or noisy.

The dot-frequency diagram of the effects shows that the effect of the balancer is more important than the other factors. None of the interactions appeared to be important. These results confirmed predictions made by the process engineers. The balancer was a new component to the engine.

A second study was done to understand what part of the balancer

| | | | Noisy (−) | | | Quiet (+) | | | |
|---|---|---|---|---|---|---|---|---|---|
| **Test** | **run order** | ***C*** | ***B*** | ***E*** | ***CB*** | ***CE*** | ***BE*** | **CBE** | **Measured noise** |
| 1 | 3 | − | − | − | + | + | + | − | 8.5 |
| 2 | 5 | + | − | − | − | − | + | + | 9.0 |
| 3 | 1 | − | + | − | − | + | − | + | 5.9 |
| 4 | 6 | + | + | − | + | − | − | − | 5.2 |
| 5 | 2 | − | − | + | + | − | − | + | 10.3 |
| 6 | 8 | + | − | + | − | + | − | − | 10.5 |
| 7 | 4 | − | + | + | − | − | + | − | 7.2 |
| 8 | 7 | + | + | + | + | + | + | + | 5.7 |
| Effects | | −0.38 | **3.58** | 1.28 | −.73 | −.28 | −.38 | −.13 | **Average** 7.79 |

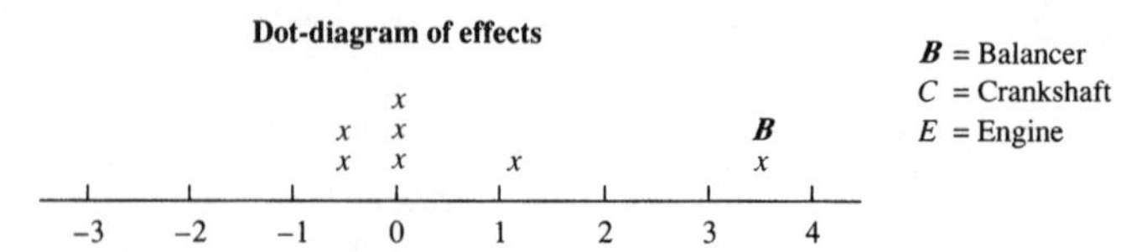


**Figure 8.16** Analysis of $2^3$ design for noisy engines.

was causing the noise. The consensus was that the noise was somehow related to the two gears in the balancer. Seven components of the balancer were identified that might contribute to the noise: the right and left gear, the right and left shaft, the housing, the cover, and the rotor. These seven components were then used as factors in a $2^{7-4}$ fractional factorial design.

The four balancers from the original noisy engines were disassembled, and each of the seven components (factors) was labeled noisy. Similarly, the four balancers from the quiet engines were disassembled, and the components labeled quiet. Eight balancers were then reassembled using a component specified as appropriate by the fractional factorial design. The eight balancers were then assembled in an engine for testing.

Table 8.4 shows the design matrix and the results of the study. The main effect of each factor is estimated by using the design matrix. Figure 8.17 shows a dot diagram of the effects. The two important effects are also shown on a square. As predicted, the two gears had the biggest effect on the noise. The effect of each type of gear on the noise was about the same. Figure 8.18 shows a response plot for the interaction of these two factors.

Follow-up studies focused on understanding the difference between noisy and quiet gears. The preparation of the materials in the gears turned out to be the cause of the noise.

**TABLE 8.4 A $2^{7-4}$ Fractional Factorial Design for Noise Study**

| Factors in study | Code | Factors in study | Code | |
|---|---|---|---|---|
| *RH* gear | *RG* | *RH* shaft | *RS* | Noisy (−) |
| Housing | *HO* | *LH* shaft | *LH* | Quiet (+) |
| *LH* gear | *LG* | Cover | *CO* | |
| *GE* rotor | *GR* | | | |

| Test | Run order | *RG* | *HO* | *LG* | *RS* | *LS* | *CO* | *GR* | Measured noise |
|---|---|---|---|---|---|---|---|---|---|
| 1 | 2 | − | − | − | + | + | + | − | 10.0 |
| 2 | 6 | + | − | − | − | − | + | + | 8.2 |
| 3 | 4 | − | + | − | − | + | − | + | 9.3 |
| 4 | 8 | + | + | − | + | − | − | − | 6.5 |
| 5 | 1 | − | − | + | + | − | − | + | 6.8 |
| 6 | 5 | + | − | + | − | + | − | − | 6.3 |
| 7 | 7 | − | + | + | − | + | − | − | 7.6 |
| 8 | 3 | + | + | + | + | + | + | + | 5.8 |
| | Effects | −1.73 | −.53 | −1.88 | −.58 | .58 | .68 | −.08 | Avg. = 7.56 |

**Dot-diagram of effects**

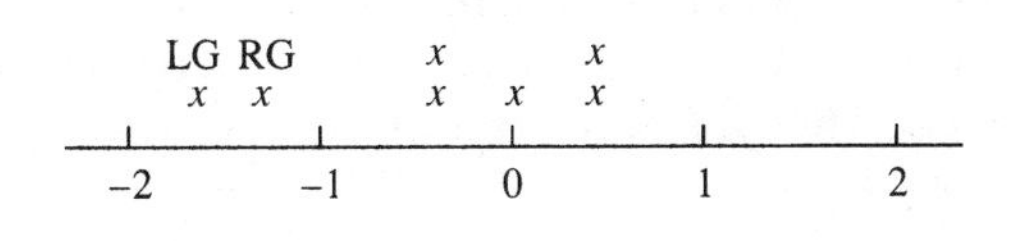


*LG* = *LH* gear
*RG* = *RH* gear

**Summary of important factors**

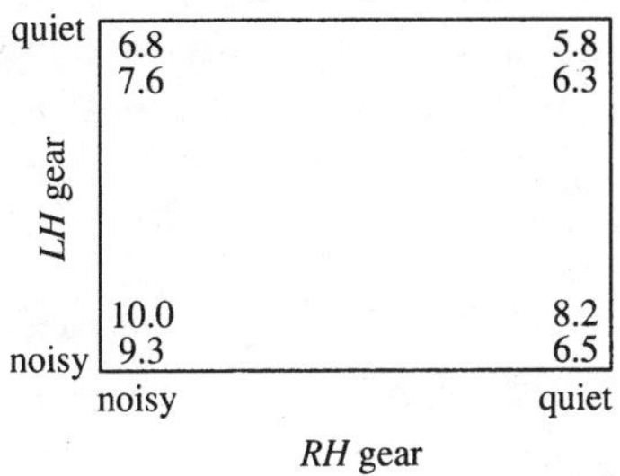


**Figure 8.17** Analysis of $2^{7-4}$ design.

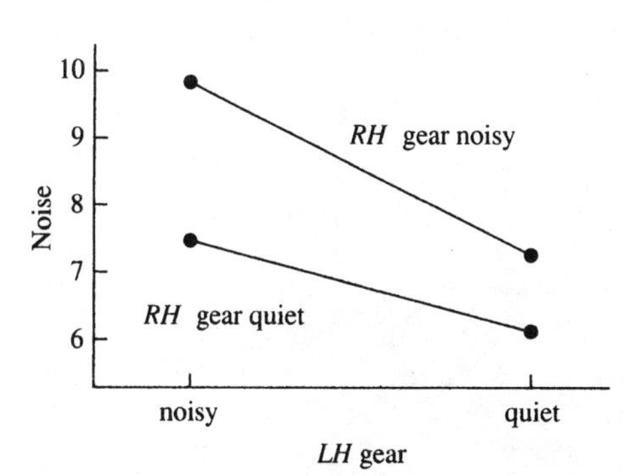


| *RH* gear | *LH* gear noisy | *LH* gear quiet |
|---|---|---|
| noisy | 9.7 | 7.2 |
| quiet | 7.4 | 6.1 |

**Figure 8.18** Response plot for important factors.

This example illustrates an application of two-level factorial designs that is effective and quick for improving the performance of products with multiple components. This type of study should be considered whenever an assembly of components is involved.

## 8.5 Experiments for Formulations or Mixtures

An experiment for a formulation or a mixture is a special type of study in which the factors are ingredients that are mixed together. The response variables are thought to depend on the relative proportions of the components rather than on the absolute amounts.

Examples of situations in which a mixture experiment is appropriate are a study of viscosity of oil blends with different concentrations of the additives, a study of the hardness of steel with different mixtures of alloys, and a test of taste preference of a drink with different levels of key ingredients.

In each of these cases, the levels of one factor cannot be changed independently of the levels of the other factors. The sum of the percentages of each factor in the mixture must be 100 percent. Therefore, a factorial design can be used for testing mixtures only in special circumstances. Other experimental patterns will be used for this type of experiment. Figure 8.19 compares the possible combinations of the

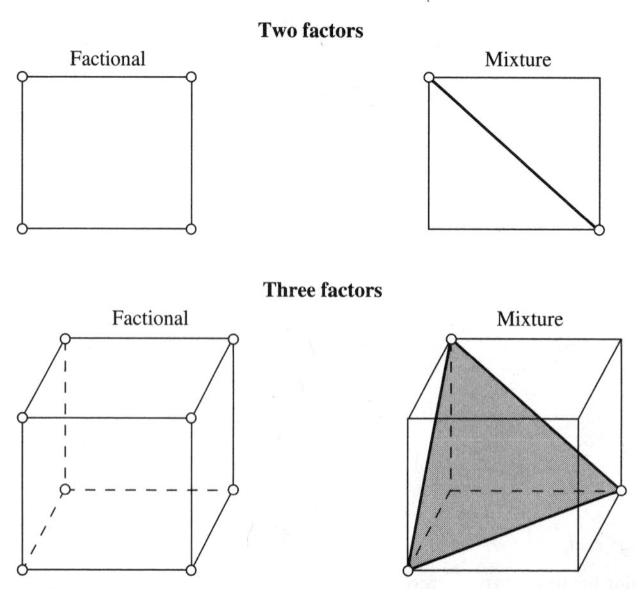


**Figure 8.19** Allowable combinations of a factorial design versus a mixture design.

factors for a factorial design to those for a mixture experiment. This section provides an introduction to two of these patterns: simplex and extreme vertices. For a more in-depth treatment of designing experiments for mixtures, see Cornell (1983, 1990).

### Using factorial designs for mixture experiments

In some situations factorial designs can be used to study formulations or mixtures even though the components must add to 100 percent. Some of these situations might be that

- One of the components makes up the majority (more than 80 percent) of the mixture.
- One of the components is an inert filler.
- One or more of the components are used to modify the properties of another component, and the ratio of the modifiers to the primary component is thought to be important.

In the first two situations, the components are changed independently, as they normally would be in a factorial design. The major component or the filler is then adjusted so that the percentages of all the components add to 100 percent. The components that are adjusted to make the sum 100 percent are called *slack components.* Table 8.5 contains a

**TABLE 8.5 Use of a $2^3$ Design for a Four-Component Mixture**

| Factor (component) | Levels (%) |
|---|---|
| *A* | 0–1 |
| *B* | 1–3 |
| *C* | 2–4 |
| *D* (slack) | Adjusted to 100% |

| Test | Factor *A* | *B* | *C* | *D* |
|---|---|---|---|---|
| 1 | 0 | 1 | 2 | 97 |
| 2 | 1 | 1 | 2 | 96 |
| 3 | 0 | 3 | 2 | 95 |
| 4 | 1 | 3 | 2 | 94 |
| 5 | 0 | 1 | 4 | 95 |
| 6 | 1 | 1 | 4 | 94 |
| 7 | 0 | 3 | 4 | 93 |
| 8 | 1 | 3 | 4 | 92 |

test for a four-component mixture with one of the components making up the majority of the mixture. The experiment is designed by changing three of the components according to a $2^3$ factorial design and adjusting the slack component to make the sum of the four components equal 100 percent.

The third situation mentioned above occurs when the ratios of components are important rather than the actual proportions of each in the mixture. This is frequently the case when some of the components are included to modify the properties of a primary component. Table 8.6 contains the pattern for an experiment to determine the formulation that achieved the desired properties of a rubber compound. Five components were studied; rubber and four components that were included in the formulation to modify the properties of the rubber. The amount of rubber was held constant. The levels of the four modifiers were designated as the ratio of the amount of the modifier to the amount of the rubber. A $2^{4-1}$ fractional factorial design was used to run the experiment.

In these examples the circumstances were such that factorial or fractional factorial patterns could be used to design a study for mixtures. The data obtained from these studies are analyzed in the usual way for factorial or fractional factorial patterns. There will be many other circumstances in which factorial patterns are not appropriate, and other patterns will be needed.

**TABLE 8.6 A $2^{4-1}$ Experiment to Study the Properties of Rubber Compounds**

| Factor | Levels (amount/amount of rubber) |
|---|---|
| 1 | .2–.3 |
| 2 | .05–.1 |
| 3 | .0–.02 |
| 4 | .02–.05 |

| Test | Factor 1 | Factor 2 | Factor 3 | Factor 4 |
|---|---|---|---|---|
| 1 | .20 | .05 | .00 | .02 |
| 2 | .30 | .05 | .00 | .05 |
| 3 | .20 | .10 | .00 | .05 |
| 4 | .30 | .10 | .00 | .02 |
| 5 | .20 | .05 | .02 | .05 |
| 6 | .30 | .05 | .02 | .02 |
| 7 | .20 | .10 | .02 | .02 |
| 8 | .30 | .10 | .02 | .05 |

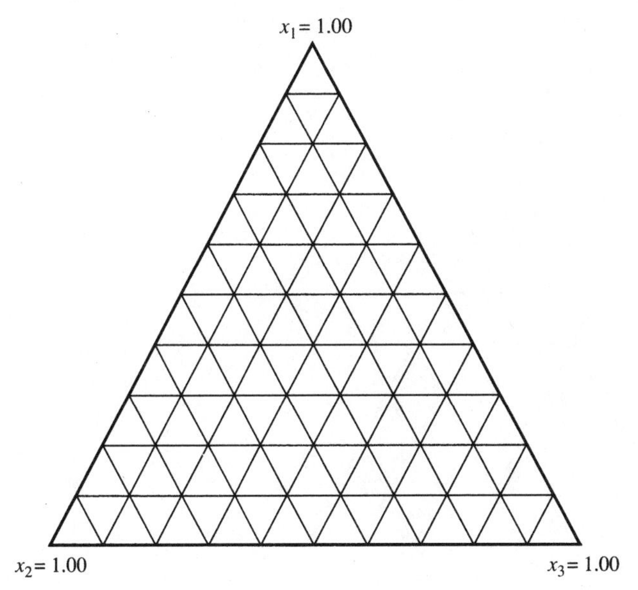


**Figure 8.20** Triangular graph paper.

### Simplex patterns when all components have no constraints

One of the situations in which factorial patterns cannot be used for mixture experiments occurs when all the components can be varied anywhere between 0 and 100 percent. In these situations, simplex experimental patterns are used. A simplex pattern for a three-component mixture is most easily displayed by using triangular graph paper. Figures 8.20 to 8.24 illustrate the scaling of triangular graph paper.

A simplex pattern for an experiment with three factors is developed by spreading the tests over the allowable experimental region. A simplex pattern using six tests is composed of the three pure mixtures and three mixtures with two components at 50 percent. This pattern is listed in Table 8.7. For simplex patterns with four or more factors, see Cornell (1990).

By adding the center point, all factors at one-third of the mixture, the interior of the experimental region can be studied. It is advisable to include a center point and to replicate the point to have a measure of the effect of nuisance variables. In addition to the center point, it is useful to include other points interior to the triangle if it is suspected that none of the factors will be excluded from the final mixture. Table 8.8 contains a simplex pattern using 10 tests, and Fig. 8.25 contains a display of the pattern on triangular graph paper.

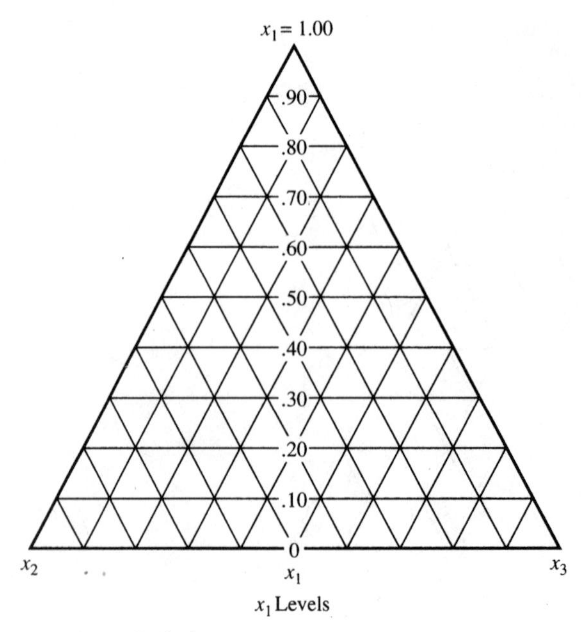


**Figure 8.21** Scale for component 1.

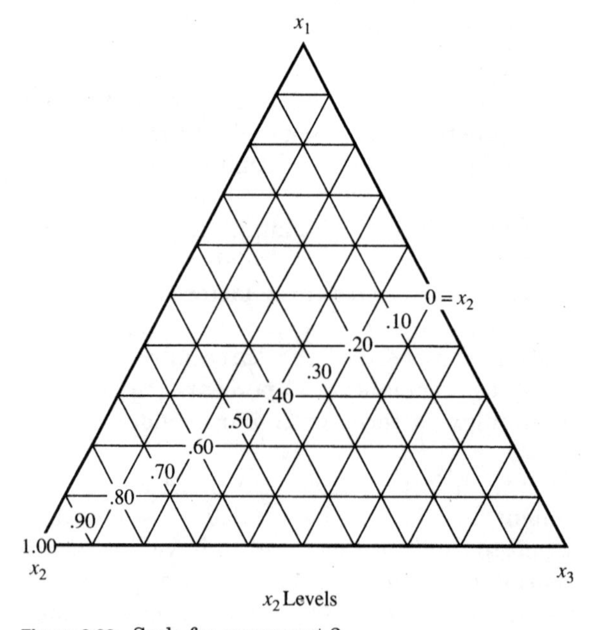


**Figure 8.22** Scale for component 2.

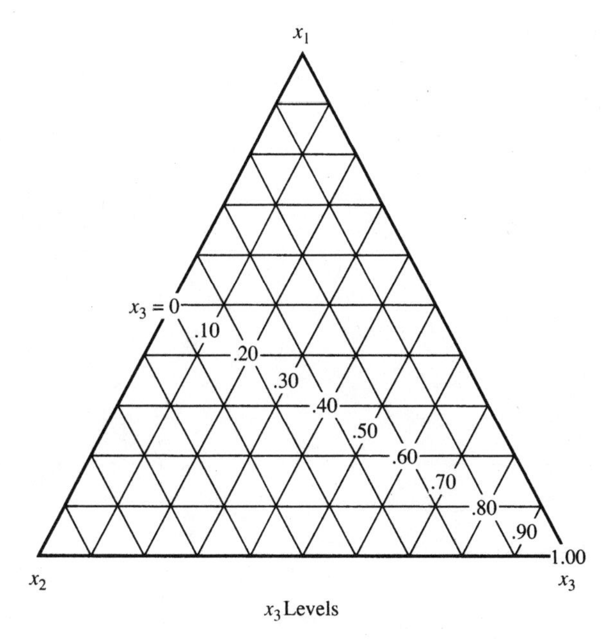


**Figure 8.23** Scale for component 3.

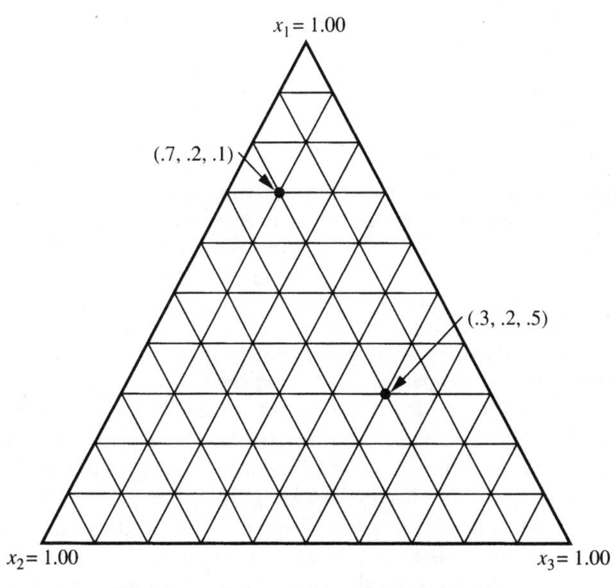


**Figure 8.24** Plotting points on triangular graph paper.

**TABLE 8.7 Simplex Pattern Using Six Tests**

| Test | Factor 1 | 2 | 3 |
|---|---|---|---|
| 1 | 100 | 0 | 0 |
| 2 | 0 | 100 | 0 |
| 3 | 0 | 0 | 100 |
| 4 | 50 | 50 | 0 |
| 5 | 50 | 0 | 50 |
| 6 | 0 | 50 | 50 |

**TABLE 8.8 Simplex Pattern Using 10 Tests**

| Test | Factor 1 | 2 | 3 |
|---|---|---|---|
| 1 | 100 | 0 | 0 |
| 2 | 0 | 100 | 0 |
| 3 | 0 | 0 | 100 |
| 4 | 50 | 50 | 0 |
| 5 | 50 | 0 | 50 |
| 6 | 0 | 50 | 50 |
| 7 | 33.3 | 33.3 | 33.3 |
| 8 | 66.7 | 16.6 | 16.6 |
| 9 | 16.6 | 66.7 | 16.6 |
| 10 | 16.6 | 16.6 | 66.7 |

**Example 8.5 Improving the Density of Ceiling Tile** One of the critical properties of ceiling tile is the density of the tile. The ceiling tiles perform best if the density of the tile is 0.9 lb/ft$^2$. One of the primary ingredients in ceiling tile is recycled material. There are three sources of the recycled material: *A, B,* and *C*. Source *A* is semifinished or finished tile that is scrapped in production. Source *B* is dust that is generated and collected at the sanding operation. Source *C* is sludge that is collected at the clarifier. Each of the different sources of recycled material has different properties that affect the density of the tile.

An experiment was performed to determine what percentage of each type of material should be used in the formulation for the tile. The total amount of recycled material in the formulation was kept approximately constant. The experiment consisted of a simplex design with 10 different mixtures. Two replications of each mixture were made. The run chart indicated that the effect of nuisance variables was small relative to the variation between different mixtures. Figure 8.26 contains the average density for each of the mixtures in the experiment.

The data indicate that density increases as the percentages of dust and sludge are increased. A mixture of 70 percent scrap, 15 percent sludge, and 15 percent dust results in a density of approximately 0.9 lb/ft$^2$. This mixture was tested in production over a period of a week, and the measure-

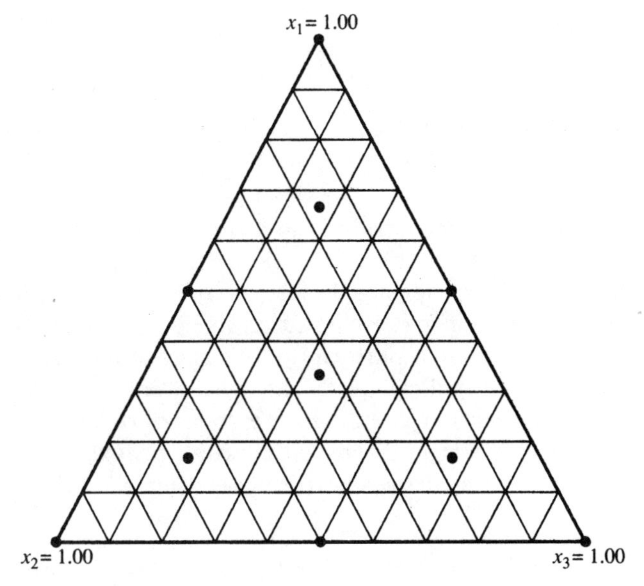


**Figure 8.25** Display of a simplex design using 10 tests.

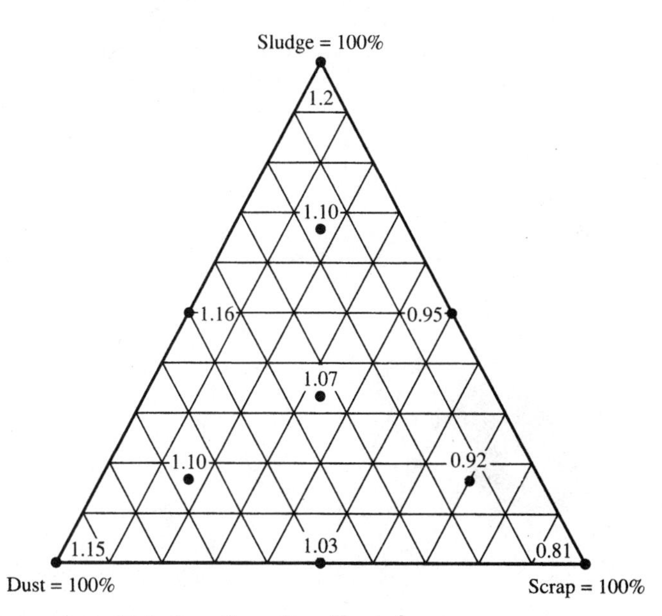


**Figure 8.26** Data from the ceiling tile study.

ments of density were plotted on a control chart to confirm the results of the experiment.

### Mixture experiments when the components are constrained

It is often the case that all the components cannot be varied from 0 to 100 percent. One or more may be constrained to a range, for example, 10 to 20 percent of the mixture. In these situations the simplex patterns discussed above are not appropriate. Before a pattern can be determined, the allowable region for experimentation (based on the constraints) must be determined. Then a pattern that consists of mixtures that cover the allowable region is developed. One type of pattern that is useful in this type of application is called an *extreme-vertices pattern.* The extreme-vertices design is discussed through an example. For more information on extreme-vertices patterns, see Cornell (1990).

**Example 8.6 Fruit Punch Study** Juices from watermelon, pineapple, and orange were combined to make a fruit punch. The proportions for the different fruits were constrained as follows:

$40\% < \text{watermelon} < 80\%$ $10\% < \text{pineapple} < 50\%$ $10\% < \text{orange} < 30\%$

Since there are three factors, the allowable region is obtained by drawing in the constraints on triangular graph paper. Figure 8.27 contains the allowable region for this example.

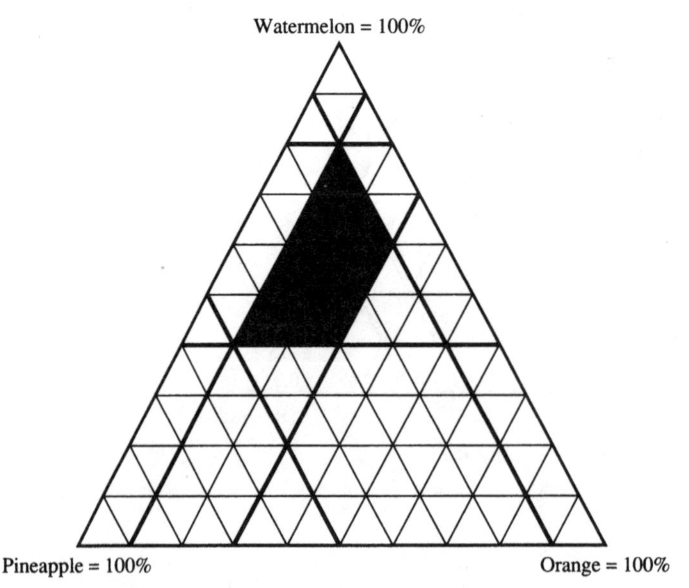


**Figure 8.27** Allowable region for the juice experiment.

Once the allowable region is determined, a pattern that covers this region is developed. In developing the region, first consideration should be given to the vertices. In this example there are four vertices, and all four were included in the study. If two or more vertices are close together, some can be left out. It is usually desirable to include the center of the region. The experimental pattern chosen for this experiment is contained in Table 8.9. Also to be considered for inclusion in the pattern are other interior points and points along the edges.

Six children were asked to rank the drinks from best to worst. A rank of 1 was given to the best drink, and a rank of 5 was given to the worst drink. The median ranks of the six children are displayed in Fig. 8.28. The data indicate that the children preferred the drinks with higher concentrations of orange and pineapple juice. It was decided to make the drink with a combination of 30 percent orange

**TABLE 8.9 Pattern for the Juice Experiment**

| Test | *W* | *P* | *O* |
|---|---|---|---|
| 1 | 40 | 50 | 10 |
| 2 | 80 | 10 | 10 |
| 3 | 60 | 10 | 30 |
| 4 | 40 | 30 | 30 |
| 5 | 55 | 25 | 20 |

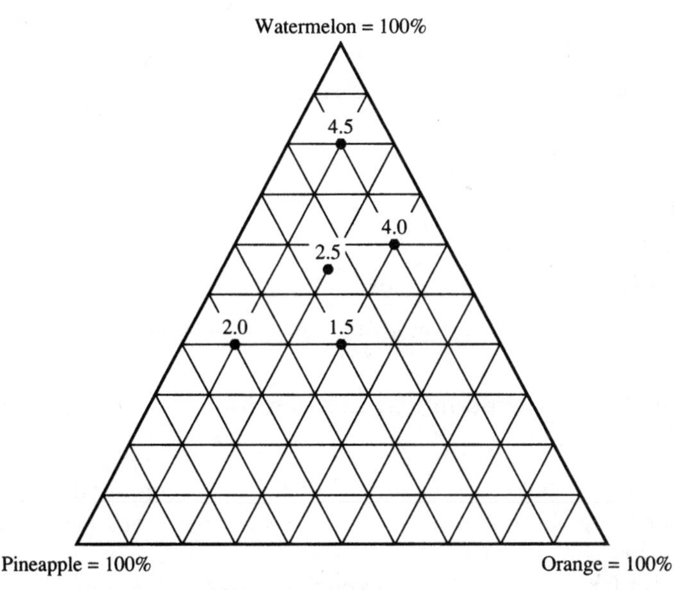


**Figure 8.28** Data from the juice experiment.

juice, 30 percent pineapple juice, and 40 percent watermelon juice, test number 4 with a 1.5 average response.

This section provides only a brief introduction to mixture experiments. For a more in-depth study, see Cornell (1983, 1990).

## 8.6 Evolutionary Operation

Evolutionary operation (EVOP) is a strategy developed by Box and Draper (1969) to run experiments on a process while in operation. The EVOP studies were designed to be run by operators on a full-scale manufacturing process while continuing to produce a product of satisfactory quality.

The important features of the EVOP strategy include the following:

1. A series of sequential experiments
2. Simple factorial designs with two or three factors that do not interrupt operations
3. Small changes to the factor levels so that radical changes in product quality or efficiency do not occur
4. A large number of runs required per factor combination
5. Simple graphical analysis

The concepts and methods presented in the first eight chapters of this book provide all the elements required to use an EVOP strategy. These elements include

1. The Model for Improvement with multiple PDSA cycles (Chap. 1)
2. The importance of stability of the response variables (App. A, Control Charts)
3. Factorial designs; selection of factor levels to meet different objectives (Chap. 5)
4. Replication (Chaps. 3. and 5)
5. Center points to evaluate nonlinearity of effects (Chap. 8)
6. Confirmation of new quality or efficiency levels (App. A, Control Charts)
7. Selection of the next experiment, i.e., continuous improvement (Chap. 1, the Model)

The Model for Improvement presented in Chap. 1 provides the road map to implementing the EVOP concept in an operating process. The model emphasizes the importance of the stability of the process as part of the current knowledge before experimenting on the process is begun. ("How do we know that a change is an improvement?") The four-step improvement cycle provides the structure to plan and conduct a series of sequential experiments on a process. After each experiment, first the current knowledge is updated, and then the next experiment is planned. Chapter 5 on factorial experiments provides the experimental designs to be used in a series of sequential studies.

Chapter 10 includes a case study that illustrates the application of the Model for Improvement to a manufacturing process. The sequential use of the PDSA cycle to guide improvement is illustrated in this case study.

## 8.7 Experimental Designs for Complex Systems

Most of the examples in this book have addressed improvement of a specific process or product. These have included machining, assembly, chemical, and electronic processes. In most of the examples, a small segment of the process was isolated for study. The variables (factors, response variables, and background variables) for that portion of the larger process (or system) were identified and studied in experiments. Sometimes it is desirable to run experiments on large systems before a detailed study of a particular part of the system. The initial study of the system should determine what part of the system to change to create the greatest impact on the overall system.

Table 8.10 describes some situations when this type of study might be of interest. In each of the cases given in the table, the objective is to make improvements in the overall system. Each system is made up of processes, steps, or stages that could be separately studied.

Table 8.11 shows some of the potential factors for the manufacturing example in Table 8.10. There are a total of 29 factors identified that could potentially affect the response of interest (visual defects). Experimenting with 29 factors is possible by using fractional factorial screening designs. (In Chap. 6, fractional factorial designs to include up to 16 factors were presented. The references given in Chap. 6 include designs with more than 16 factors.) An experiment of this size would require significant skill and resources. Another way to begin the investigation would be to choose one process step and run a series of fractional and full factorial studies on just that step. From the perspective of the overall system, experimenting with one step at a time is similar to the one-factor-at-a-time approach to experimentation that was criticized in Chap. 5.

Another approach is to view each step in the larger process as a potential factor. The levels of the factor could be either performing or not performing the step in the process. For steps that are essential to accomplish the basic purpose of the process, the levels could be defined by setting the variables in the process at extreme conditions. The complex process in Table 8.11 with 29 potential factors could thus be trans-

**TABLE 8.10 Examples of Experiments for Complex Systems**

**A. Manufacturing Process** *Objective:* Reduce visual defects (nicks, scratches, etc.) on parts
Process step 1. Incoming inspection
Process step 2. Sawing
Process step 3. Milling
Process step 4. Grinding
Process step 5. Measurement
Process step 6. Polishing and cleaning
Process step 7. Packaging

**B. Chemical Process** *Objective:* Maximize yield
Process stage 1. Mixing
Process stage 2. Reaction 1
Process stage 3. Blending
Process stage 4. Reaction 2
Process stage 5. Distillation
Process stage 6. Washing

**C. Medical Process** *Objective:* Reduce treatment time and improve outcomes
Subprocess 1. Collecting information
Subprocess 2. Triage
Subprocess 3. Examination
Subprocess 4. Laboratory tests
Subprocess 5. Prescription
Subprocess 6. Home treatment
Subprocess 7. Revisit

**D. Service Organization** *Objective:* Improve customer satisfaction
Process phase 1. Preparation for customer visit
Process phase 2. Initial interaction with customer
Process phase 3. Procedure 1
Process phase 4. Procedure 2
Process phase 5. Wrap-up
Process phase 6. Billing and collections
Process phase 7. Feedback

formed into a process with seven potential factors at two levels each. The screening designs in Chap. 6 could then be used to study the process. The initial experiment would screen out the steps that were not contributing to the response variables. Follow-up experiments could then identify the important factors in the steps that were contributing to the response. Example 8.7 illustrates this approach.

**Example 8.7 Visual Defects in a Manufacturing Process** Customer feedback had recently indicated that visual defects were impacting acceptance of a manufacturing company's product. There were many different theories about the causes of the defects, with some blame attributed to all seven major manufacturing steps. Rather than study each step, an experiment was designed to identify the steps contributing most of the problem. The processing steps and the potential factors in each step were listed in Table 8.11.

**TABLE 8.11 Variables in Complex Systems (Manufacturing Process)**

Objective: Reduce visual defects (nicks, scratches, etc.) on parts

| Process step | Number of factors | Potential factors for study |
|---|---|---|
| 1. Inspection | 5 | Supplier, grade of material, porosity, density, gage type |
| 2. Sawing | 5 | Feed rate, speed of blade, part position, blade type, coolant |
| 3. Milling | 5 | Rotating direction, mill type, operator, speed, vice pressure |
| 4. Grinding | 4 | Table speed, wheel speed, percent magnet, dress rate |
| 5. Measurement | 4 | Gage type, inspector, clean-up procedure, calibration |
| 6. Polishing | 3 | Time, type of polish, material |
| 7. Packaging | 3 | Type of padding, type of filler, size of package |
| | | Total potential factors = 29 |

The design selected was a $2^{7-4}$ fractional factorial design. Each of the seven steps in the manufacturing process was considered a factor. Two levels were identified for each step:

| Process step | Low level (−) | High level (+) |
|---|---|---|
| 1. Inspection | None | Normal inspection performed |
| 2. Sawing | Factors set to minimize speed/stress | Factors set at high speed/stress |
| 3. Milling | None | Standard operating conditions |
| 4. Grinding | Factors set to minimize speed/stress | Factors set at high speed/stress |
| 5. Measurement | None | Normal measurement procedures |
| 6. Polishing | None | Normal procedure |
| 7. Packaging | Minimum protection of parts | Maximum protection of parts |

It was essential that process steps 2 (sawing), 4 (grinding), and 7 (packaging) be done in order to produce and ship the basic part that the customer desired. A set of low and high settings was defined for the factors in each of these processes to represent extreme conditions. It was conceptually possible not to perform steps 1, 2, 5, and 6, so the two levels defined were (1) none (skip that step in the process) and (2) normal operating procedures.

| Test | Run Order | Inspection | Sawing | Milling | Grinding | Measure | Polishing | Package | Defects |
|---|---|---|---|---|---|---|---|---|---|
| | | SxG | IxG | IxM | IxS | IxM | SxM | MxG | |
| | | MxM | MxP | SxP | MxP | GxP | GxM | SxM | |
| | | PxP | MxP | GxP | MxP | SxP | IxP | IxP | |
| | – | None | Minimum | None | Minimum | None | None | Minimum | |
| | + | Normal | Maximum | SOP | Maximum | Normal | Normal | Maximum | |
| 1 | 4 | - | - | - | + | + | + | - | 3 |
| 2 | 8 | + | - | - | - | - | + | + | 16 |
| 3 | 3 | - | + | - | - | + | - | + | 33 |
| 4 | 6 | + | + | - | + | - | - | - | 18 |
| 5 | 7 | - | - | + | + | - | - | + | 14 |
| 6 | 1 | + | - | + | - | + | - | - | 4 |
| 7 | 5 | - | + | + | - | - | + | - | 18 |
| 8 | 2 | + | + | + | + | + | + | + | 31 |
| | Effect | 0.25 | 15.75 | -0.75 | -1.25 | 1.25 | -0.25 | 12.75 | |

Dot-diagram of effects on visual defects

**Figure 8.29** A $2^{7-4}$ fractional factorial design for complex systems (manufacturing process).

Figure 8.29 shows the results of this screening study, including a dot diagram for the effects. The analysis indicates that step 2 (sawing) and step 7 (packaging) are the biggest contributors to the visual defects. The high-speed saw process contributed about 16 additional defects above the low-speed process. Packaging with maximum protection of parts decreased the defects by an average of 13.

Follow-up studies were planned for both the saw process (study the five factors with a $2^{5-1}$ design) and the packaging process (study the three factors with a $2^3$ factorial design) to identify the important factors and interactions in each of these steps of the process. The initial screening study of the entire system resulted in a much more focused approach to study the individual factors in the process.

## 8.8 Summary

This chapter presented experimental patterns for special situations and applications. Cochran and Cox (1957) have described the designs with factors at more than two levels presented here, as well as many others. The key ideas in this chapter include these:

1. Some improvement studies require factorial designs at more than two levels.

2. Response plots are the primary tool to study nonlinear effects of factors.
3. Center points can be added to $2^k$ factorial designs to evaluate the importance of nonlinear effects.
4. Three-level factorials can be used in situations where factor interactions and nonlinear effects are expected.
5. Special experimental patterns may be required when the factors are components of a mixture.
6. The Model for Improvement, presented in Chap. 1, and the experimental patterns in other chapters provide a framework similar to EVOP for improvement of processes while in operation.
7. The concept of chunk variables can be used to study large, complex systems by using factorial designs.

## References

Box, G. E. P., and N. R. Draper (1969): *Evolutionary Operation: A Statistical Method for Process Improvement,* Wiley, New York.
Box, G. E. P., and N. R. Draper (1987): *Empirical Model-Building and Response Surfaces,* Wiley, New York.
Box, G. E. P., and K. B. Wilson (1951): "On the Experimental Attainment of Optimum Conditions," *Journal of the Royal Statistical Society B,* no. 13, pp. 1–45.
Cochran, W. G., and G. M. Cox (1957): *Experimental Design,* Wiley, New York.
Cornell, John A. (1983): *How to Run Mixture Experiments for Product Quality,* American Society for Quality Control, Milwaukee, Wis.
Cornell, John A. (1990): *Experiments with Mixtures: Designs, Models, and the Analysis of Mixture Data,* 2d ed., Wiley, New York.
Ott, Ellis R. (1975): *Process Quality Control,* McGraw-Hill, New York, chap. 6.

## Exercises

**8.1** Describe a situation where it would be desirable to include more than two levels of a factor in a study. Give an example for a qualitative factor and a quantitative factor.

**8.2** Refer to the $2 \times 2$ plasticizer example in Chap. 5. In addition to the eight runs in the study, two runs were made at the current process levels of 45 percent ingredient $X$ and 182°C. One run was made at the beginning of the study (reaction time = 6.5 h), and the second run was made at the end of the study (reaction time = 6 h). Analyze the example, including these two additional runs.

Are the conclusions of the study affected by the new data?

**8.3** Refer to the $2^3$ dye process example in Chap. 5. Four additional tests were made at a midlevel oxidation temperature and oven pressure. Two of the runs were done with material quality *A,* and two runs were done with material quality *B*. The following results were obtained:

| Run order | Material | Pressure | Temperature | Response shade |
|---|---|---|---|---|
| 1 | *A* | Midlevel | Midlevel | 191 |
| 5 | *B* | Midlevel | Midlevel | 207 |
| 8 | *A* | Midlevel | Midlevel | 189 |
| 12 | *B* | Midlevel | Midlevel | 210 |

Reanalyze this example, using the additional four runs. Plot the new runs on the cube. Estimate the nonlinear effect. Is the effect important relative to the factor effects? Plot the midlevel points on the response plots.

**8.4** Describe a potential study in which the concept of interchangeable parts (Sec. 8.4) would be appropriate. What is the response variable? What are potential factors? How would the levels of the factors be established?

**8.5** Describe a study to improve quality where it is appropriate to use a mixture design.

**8.6** An in-process scale was used in preparing additives for an industrial process. The improvement team had made significant improvements in the accuracy of the weighings done on the scale by standardizing operations and using control charts for calibration decisions. Environmental factors (wind and temperature) were identified as possible causes of weight variations. The scale was supposed to compensate for ambient temperature fluctuations, but special causes on the control chart indicated that sometimes temperature changes appeared to have an effect.

A study was designed to evaluate the effect of wind speed and ambient temperature on scale accuracy. The accuracy was measured by loading a standard weight on the scale and recording the difference from the standard value. The team conjectured that possible interactions and nonlinear effects of these factors made it difficult to understand their effects by using the control charts. A 3 × 3 factorial experiment was planned. Three levels of each factor were established from previous wind and temperature readings. The order of the tests was determined by the occurrence of the proper weather conditions. The standard weight was measured three times during each test. The following data were obtained:

| Test | | | | Measurement−standard (lb) | | |
|---|---|---|---|---|---|---|
| Test | Order | Wind | Temperature | Meas1 | Meas2 | Meas3 |
| 1 | 3 | Low | Low | .4 | −.8 | .6 |
| 2 | 8 | Low | Mid | −.7 | .5 | .3 |
| 3 | 4 | Low | High | 2.6 | 3.2 | 2.8 |
| 4 | 6 | Mid | Low | −1.0 | .8 | −.7 |
| 5 | 1 | Mid | Mid | −.5 | 1.3 | .6 |
| 6 | 5 | Mid | High | 3.6 | 2.5 | 3.5 |
| 7 | 2 | High | Low | 2.1 | −1.6 | −.8 |
| 8 | 7 | High | Mid | −1.3 | .5 | 1.6 |
| 9 | 9 | High | High | 1.5 | 4.3 | 2.6 |

Analyze the data from this study.

1. Prepare run charts of the average (to evaluate the bias of the weighing process) and the range (to evaluate the precision of the weighing process) of the weighings.
2. Prepare response plots for the average and range.
3. Do either of the environmental factors affect the accuracy of the scale?
4. What additional study should be done? What are recommendations for action?

**8.7** Plan an experiment for a large system that has multiple processes, subprocesses, stages, or steps.

Chapter

# 9

# New Product Design

## 9.1 Introduction

Today's technological innovations and changing markets make the improvement of existing products and the development of new products and services competitive necessities. Competition for an existing product will increase with time. New products now account for more than one-half of a company's sales (Cooper, 1993). These new products can be new to the company or new to the market. Also, they could include additions to existing product lines or new applications of existing products.

A study by the U.S. government (Office of Technology Assessment, 1990) identified several trends in product design. These trends include

1. Decentralization, with carefully managed division of responsibility among R&D and engineering groups
2. Simultaneous product and process development, where possible
3. Lower cost with no sacrifice of quality, coupled with substantial flexibility, greater reliance on purchased components and services
4. Greater reliance on contract engineering firms
5. Innovation (incremental as well as breakthrough)
6. More specialty products and services

Craig Barrett, chief operating officer of Intel Corporation, says, "You win the race by running faster. We're dedicated to obsoleting our own products before anyone else does. How quickly we get a product to market with features is what business is about."

A study by Cooper (1993) showed that out of every 7 new product ideas, 3 enter development, 1.5 are launched, and 1 succeeds. If one asks senior managers how effectively their organizations carry out new product design programs, you hear excuses for the past programs and little hope for dramatic improvement of current projects. Competitive forces of today demand even higher quality of products that must be brought to market faster than ever. These forces are at work across a wide range of industries.

The aim of this chapter is to link the methods of planned experimentation to the design of a new product (service or system).

As a foundation, organizations should consider adopting the following concepts in designing a new product:

- Up-front planning with increased learning in the design phase where the uncertainty is greater
- Functions (marketing, engineering, and operations) tightly integrated and coordinated
- Parallel, overlapping tasks instead of sequential or serial tasks
- Small, dedicated teams with an enlarged scope of jobs
- Focus on rapid learning by the teams

These concepts along with the methods of planned experimentation (and systems thinking) will help improve communication, accelerate learning, increase the leverage for higher-quality products, and reduce the time necessary to bring the new product to market.

A starting point in the design of a new product is to identify the needs of the customer and generate ideas about a possible product or service. This was illustrated by Deming's production, viewed as a system beginning with stage 0 (generation of ideas). Using stage 0 and the sequential building of knowledge and the Model for Improvement from Chap. 1 (develop, test, and implement in Fig. 1.7), four phases for a new product design follow:

*Phase 0:* Generate ideas.

*Phase 1:* Develop concepts and define product.

*Phase 2:* Test the product.

*Phase 3:* Produce the product.

Figure 9.1 illustrates the major activities in each of the four phases by each of the major functions (marketing, engineering, and operations).

The remaining sections of this chapter describe the activities in Fig. 9.1 and the use of the methods of planned experimentation in each phase. The applications presented in the figure are denoted by the shaded boxes that describe the activity. Lead functions are denoted by subheadings within each section of the chapter.

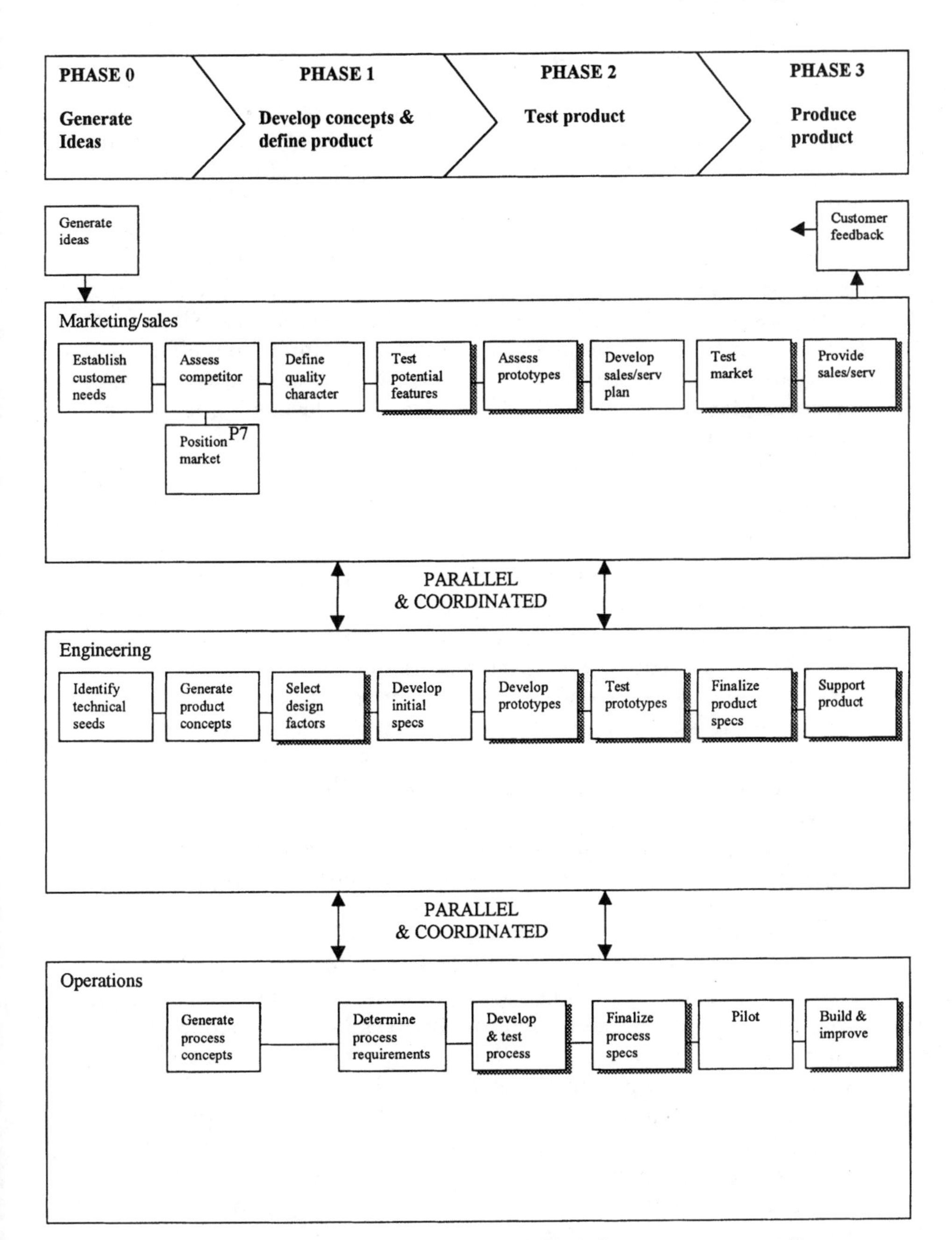


**Figure 9.1** Four-phase process for new product design. (Shaded areas represent applications of planned experimentation.)

## 9.2 Phase 0: Generate Ideas

Every successful product or service is based on satisfying a need of a customer or of society. Defining the need underlying a particular product or service is a major activity of this phase. How do customers use the product or service? Why is this product or service important to customers? What other product or service could be used instead?

Ideas about a product come from articulating the need. New ideas can come from (1) direct search of opportunities involving marketing, R&D, and engineering; (2) exploratory consumer studies; (3) technological information (patents and inventions); (4) individual effort; and (5) creative group methods. It is important that several concepts be considered for the product. (See References for developing ideas for a change in Chap. 1.)

### Marketing

Marketing may conduct surveys to obtain information from people about their feelings, beliefs, experiences, needs, expectations, or wants. Survey methods include personal interviews, group interviews, written questionnaires, or simply observation of customers' uses of current products (or the competitor's product). Recent marketing environments and trends should also be included in the analysis.

In planning a survey, make explicit what is to be learned from the survey. Questions must be asked in a way that the respondents can understand and answer. Test the questions with a small group of people who are representative of the anticipated respondents to the survey. Revise the questions based on this test. Segmentation of customer groups is determined by using the planned grouping tool of blocking of experimental units from Chap. 3.

Judgment samples are useful during the early stages of developing and testing a new product concept (analytic studies). Marketing books have recommended sampling strategies other than random sampling (probability samples). Dommermuth (1975) called these nonprobability samples *purposive samples*. Blankenship and Breen (1993) state, "Most samples chosen in marketing research are non-probability samples. A true probability sample, because of the stringent requirements, is far too expensive and too time-consuming for most uses." Johansson and Nonaka (1996) call this the "survey myopia" myth. They state, "Elaborate probability sampling designs are simply not necessary in marketing research."

Once needs to be satisfied are understood, quality can be defined by using a set of quality characteristics that then serve as response variables for experiments conducted during the later phases. This translation of customer needs to quality characteristics was facilitated by a diagram called the quality characteristic diagram, introduced in Chap. 1 (see Fig. 1.3). The needs of the customer are listed at the top. The first column of the diagram defines quality at a primary level in the language of the customer. Examples are *easy to service, easy to close, does not rattle, proper size, lasts a long time, comfortable to use, doesn't skip,* and *easy to read.* This primary level states how the product will give the customer what he or she wants.

The primary level of quality characteristics must be refined through more detailed steps, as illustrated by the additional columns (secondary and tertiary) of the diagram. Additional quality characteristics may be added by reviewing the dimensions of quality given in Chap. 1. The details of the quality characteristic should be measurable for comparison with competitive products. More details can be attained by subdividing a quality characteristic into two or more subcharacteristics.

A weighting factor of some type for each quality characteristic (e.g., most important to least important; or high, medium, and low) might provide useful information for potential tradeoffs later. Customer surveys may be necessary to justify the weightings.

Quality characteristics should be

- Continuous variables, as far as possible
- Measurable
- A family of measurements that provides a definition of quality sufficient for the product
- Specific enough to be useful for design of the product or process

The quality characteristic diagram defines quality and provides a communication link for marketing, engineering, and operations. It gives these departments major input into the design of the product.

**Example 9.1 Solenoid Valve** A manufacturer of components for automobile engines is designing a valve to turn a pollution control device on and off (see Example 5.3). How should the manufacturer define quality? What are the important quality characteristics? The quality characteristics selected are given by the quality characteristic diagram in Fig. 9.2.

**Example 9.2 Wallpaper** A manufacturer of wallpaper is redesigning an existing wallpaper product. The quality characteristic diagram is given in Fig. 9.3. Which quality characteristics are most important to you?

### Engineering

People in engineering (or R&D) must have a deep understanding of new technologies and new materials and an independent drive for creativity. Theirs must be an environment that allows risk taking and failure as a learning experience. The improvement cycle and planned

| Needs of the customer: Low pollution efficient engine | | | |
|---|---|---|---|
| **Quality characteristic**[1] | | | |
| **Primary**[2] | **Secondary**[3] | **Tertiary**[3] | **#** |
| performs well | airflow | airflow | 1 |
| | overload protection | overload protection | 2 |
| is reliable | time until replacement | time until replacement | 3 |
| | thermal stability | thermal stability | 4 |
| | impact resistance | impact resistance | 5 |
| can be serviced | easy to remove | time to remove | 6 |
| | | effort to remove | 7 |
| | easy to install | time to install | 8 |
| | | effort to install | 9 |

[1] Do not include design factors in this list. (*Test*: You should *not* be able to set the levels of these quality characteristics.)
[2] Express in the language of the customer.
[3] To add more detail, subdivide into two or more quality characteristics.

**Figure 9.2** Quality characteristic diagram for designing a solenoid valve.

experimentation will be important methods to maximize learning in this environment. Outcomes of this process include an increased understanding of new technologies and materials and concepts for new products based on these technologies.

Sometimes a new technology leads to the expansion of old concepts for new products. Miniaturization of the microchip is an example. New technology is generated by a combination of needs, concepts, and hardware. There must be sufficient activity addressing all major strategic needs of the customer.

Sometimes a new product or process idea does not require or warrant new technology. The capacity to generate technology is important, but the challenge is to bring the technology to the marketplace with lower cost and higher quality.

| Needs of the customer: Decorative walls | | | |
|---|---|---|---|
| Quality characteristics[1] | | | |
| Primary[2] | Secondary[3] | Tertiary[3] | # |
| looks good | consistent with fashion trends | | 1 |
| | attractive | | 2 |
| | lack of visible seams | | 3 |
| easy to clean | time to clean | | 4 |
| | effort to clean | | 5 |
| wears well | resistance to scratches | | 6 |
| | stain resistance | | 7 |
| | gloss retention | | 8 |

[1] Do not include design factors in this list. (*Test*: You should *not* be able to set the levels of these quality characteristics.)
[2] Express in the language of the customer.
[3] To add more detail, subdivide into two or more quality characteristics.

**Figure 9.3** Quality characteristic diagram for redesigning wallpaper.

What are the product possibilities? How well can they be manufactured? What is their potential in meeting the needs of the customer? The table given in Fig. 9.4 captures important information for new and current concepts. As a goal for the process of generating product concepts, the concepts selected must be capable of meeting the customer's needs through product possibilities that can be designed (with few parts) for ease of manufacture and assembly. Assessment of process concepts can be done by the table as well.

Identify several conceptual designs and the form or structure the product might take. Avoid selecting one concept without serious consideration of others. Screen product concepts while increasing current knowledge by running planned experiments to reduce the number of conceptual designs. An example of a planning form using a randomized block design for such selection of a product concept is given in Fig. 9.5.

| Product concept | Product form/ structure | Ease in manufacturing assembly | Needs addressed | Potential risk in meeting needs low/medium/high |
|---|---|---|---|---|
| (current) | | | | |
| (new) | | | | |

Figure 9.4 Assessment of product concepts.

**1. Objective:**
Select a product concept from available candidates.

**2. Background information:**
Information on each product concept that is in contention. (Figure 9.4)

**3. Experimental variables:**

A.

| Response variables | Measurement technique |
|---|---|
| 1. Quality characteristics | (may be subjective rankings) |
| 2. Manufacturability measures<br>• assembly time<br>• number of parts, steps<br>• manufacturing costs | |
| 3. Life in service | |

B.

| Factors under study | Levels |
|---|---|
| 1. Product concept | Current Concept 1 Concept 2 |

C.

| Background variables | Method of control |
|---|---|
| 1. Wide range of conditions (production and customer) | (consider chunk-type block variables) |

Figure 9.5 Example of a planning form for selecting a product concept.

## Operations

The generation of new ideas to be used for manufacturing a new product should follow the same process as engineering. Consideration should be given to modular design, number of assembly steps, assembly time, material cost and availability, and automation.

## 9.3 Phase 1: Develop Concepts and Define Product

### Marketing

In the introduction of Chap. 1, a question was posed: How are the best concepts or features in a new product selected for meeting customer needs from the many that are in contention? Marketing research is faced with the task of providing answers to that question. Methods used to help answer this question include sampling, surveys, and experimental design. Kano surveys (Kano, 1994) and conjoint analysis (*con*sider *joint*ly) are frequently mentioned as methods for testing multiple new product features. These methods are used with different customer groups that possess the need identified in phase 0.

The following example demonstrates the use of a conjoint analysis with a $2^{7-4}$ fractional factorial design.

**Example 9.2 (*Continued*) Conjoint Analysis on Potential Features for New Wallpaper** The marketing research group from a manufacturer of wallpaper has identified several potential features for a wallpaper product. The planning form is given in Fig. 9.6. The response variable is an average ranking (ordinal measurement) by the potential customers. Profile cards are prepared with each combination of factor combinations in the eight tests of the $2^{7-4}$ fractional factorial design. Pictures of the different combinations of colors and patterns of wallpaper were included on the profile cards.

An example of a profile card for condition 1 of the design matrix $(- - - + + + -)$ is given below:

| **Profile A:** current price | [Insert picture or show actual prototype with current pattern, standard installation instructions, no cleaning kit, new color family, vinyl coating, and gloss finish] |
|---|---|

Figure 9.7 shows the results of the study. The response variable is average rank with lower rank being better (most appealing). On the dot diagram, a negative effect means that moving from the minus to plus level increased the appeal by having that feature. Coating, color, and pattern were the most appealing features. Price had a positive effect, meaning that people were not willing to pay for additional features. An interaction of price with a feature could indicate a willingness to pay for that feature.

Based on the analysis and the current knowledge, the following conclusions are drawn:

- Price was important (not willing to pay for additional features).
- Vinyl coating was preferred.
- New color and patterns were preferred.

1. **Objective:**
Determine what features to offer in a product profile.

2. **Background information:**
The team has identified several potential features for the new wallpaper.

3. **Experimental variables:**

A.

| Response variables | Measurement technique |
|---|---|
| 1. Ranks (most appealing to least) | Average ranking of 8 product profile cards by 25 potential customers |

B.

| Factors under study | Levels | |
|---|---|---|
| 1. Pattern | current | new |
| 2. Installation instruction | standard | video |
| 3. Cleaning kit | no | yes |
| 4. Color family | old | new |
| 5. Coating | paper | vinyl |
| 6. Gloss finish | no | yes |
| 7. Price | current | 5% increase |

C.

| Background variables | Method of control |
|---|---|
| 1. Age of customer | Create 2 blocks |
| 2. Income of customer | Create 2 blocks |
| 3. All others | Hold constant |

4. **Replication:**
Each of the 8 profile cards will be ranked by 25 customers in each of the 4 blocks

5. **Methods of randomization:**
Shuffle the order of the 8 profile cards.

6. **Design Matrix:** (attach copy)
$2^{7-4}$ fractional factorial design with 8 runs.

7. **Data collection forms:** (attach copies)
Form in the customer packet.

8. **Planned methods of statistical analysis:**
Dot diagram and response plots for each block.

9. **Estimated cost, schedule, and other resource considerations:**
Total cost for this experiment is $10,000. Customers within each block will be participating in an evening session. All 4 sessions will be completed in 1 week.

**Figure 9.6** Planning form for conjoint analysis for wallpaper.

A methodology that brings marketing, engineering, and operations together is quality function deployment (QFD). QFD relates the factors for design to those quality characteristics identified in phase 0.

In its most basic form, QFD can be thought of as a matrix with the

Design matrix for a $2^{7-4}$ pattern

| Test | Run order | 1<br>24<br>35<br>67 | 2<br>14<br>36<br>57 | 3<br>15<br>26<br>47 | 4<br>12<br>56<br>37 | 5<br>13<br>46<br>27 | 6<br>23<br>45<br>17 | 7<br>34<br>25<br>16 | Average rank |
|---|---|---|---|---|---|---|---|---|---|
| 1 | 6 | - | - | - | + | + | + | - | 2.9 |
| 2 | 3 | + | - | - | - | - | + | + | 6.1 |
| 3 | 8 | - | + | - | - | + | - | + | 6.0 |
| 4 | 5 | + | + | - | + | - | - | - | 3.4 |
| 5 | 1 | - | - | + | + | - | - | + | 5.9 |
| 6 | 7 | + | - | + | - | + | - | - | 3.1 |
| 7 | 2 | - | + | + | - | - | + | - | 4.6 |
| 8 | 4 | + | + | + | + | + | + | + | 3.5 |
| Effects: | | -0.8 | 0.1 | -0.3 | -1.0 | -1.1 | -0.3 | 1.9 | |

Dot diagram of effects

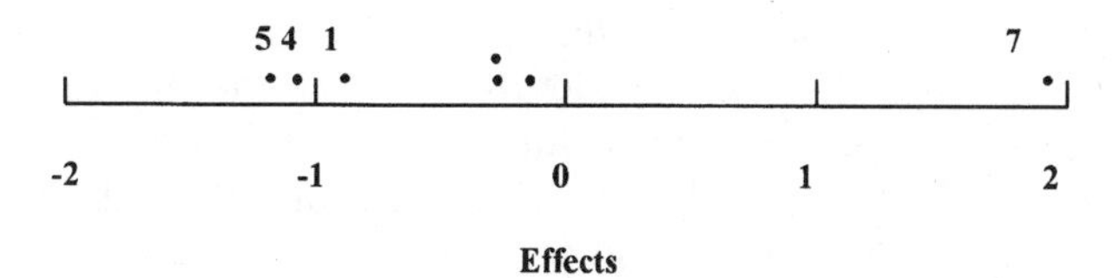


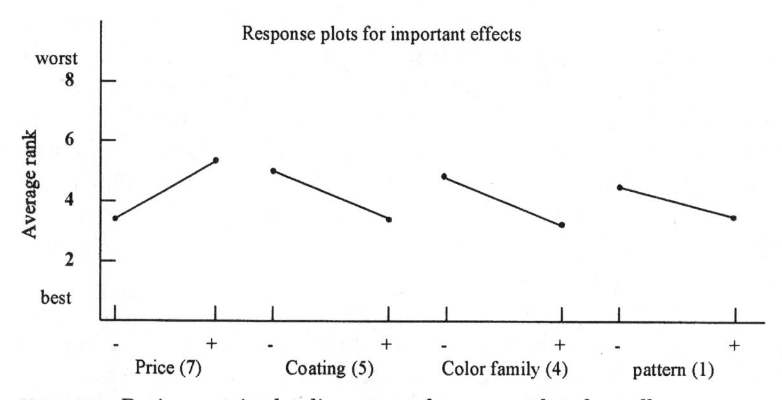


**Figure 9.7** Design matrix, dot diagram, and response plots for wallpaper.

rows representing the customers' needs (the "whats") and the columns representing the design aspects of a new product (the "hows"). At the intersections of the matrix, symbols are used to represent the degree of relationship between the rows and columns. Additional features may be added to the matrix. The relationships between the columns (design factors) can be added to the top of the columns. Analysis of competitors can be added to the right-hand column. Best settings or targets of the design factors can be added at the bottom of the columns.

QFD links the four major phases in the design of a new product. Constant interaction between marketing, research, engineering, manufacturing, and sales/service is needed to translate the needs of the customer to a new or improved product that will better meet those needs. QFD promotes this interaction and the breaking down of barriers between departments. See Hauser and Clausing (1988) and Akao (1990) for more discussion of QFD.

Multiple QFD matrices may be used to help plan the entire product design process from design characteristics to product specifications to process specifications. Each set of columns becomes the set of rows for the next-lower level relationship.

### Engineering

A diagram modifying the matrix of QFD is illustrated in Fig. 9.8. The QFD relation diagram gives the product designers a method for relating quality characteristics to factors that should be addressed in the design of a new product or component. This diagram helps define the current knowledge for a product and is analogous to a cause-and-effect diagram.

The extreme left-hand column of the QFD relation diagram lists the quality characteristics that have been identified using the quality characteristic diagram given in Fig. 1.3. Quality characteristics are used as response variables in planned experimentation to increase product knowledge. When there are multiple measurements of the quality characteristic at the same combination of levels of the factors, it is often useful to form several statistics. The overriding factor in selection of a quality characteristic statistic is its impact on customer loss and its relationship to customer need. Following is a discussion of some commonly used statistics:

**Average.** The average of the observations in the experiment reduces the magnitude of variation due to nuisance variables. The average is widely used as a response variable when replication is included in the design.

**Standard deviation.** The standard deviation of the observations in a replicated experiment represents the magnitude of the nuisance or background variables within the experimental pattern. Ranges could be used instead of standard deviation if the number of observations is small (less than 10) and constant for each factor combination.

**Average versus variation.** Both average and measures of variation such as standard deviations could be used as response variables (see

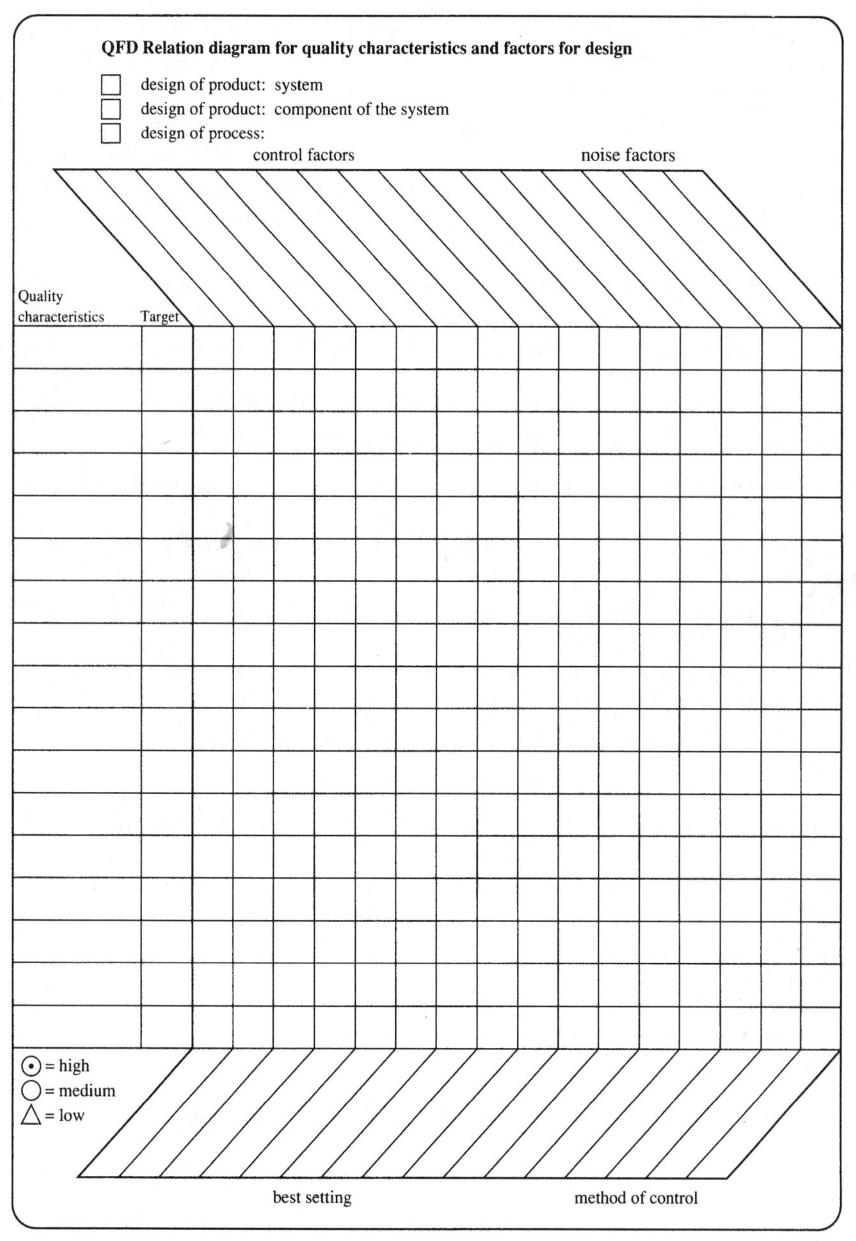


What planned experiments need to be run to test this diagram?

**Figure 9.8** QFD relation diagram.

Example 5.3). Choice of one depends on the relative impact of each on customer loss. The objective is to minimize the loss to the customer.

Taguchi (1987) uses a signal-to-noise ratio consisting of an average (signal) divided by some measure of variation (noise) as a response variable. This signal-to-noise ratio has no advantage, as a response variable, over looking at average and variation separately, and it is more difficult to interpret. [See Box (1988) for more discussion of signal-to-noise ratios.]

Once the statistic of the quality characteristic is defined, the target value for each statistic is determined. The target value is in the second column. There are three cases of the target value $t$:

1. Smaller is better ($t = 0$): wear, shrinkage, deterioration.
2. Bigger is better ($t =$ infinity): strength, life, fuel efficiency.
3. Target value is best ($t = t_0$): dimension, clearance, weight, viscosity.

This target is set at minimum loss to the customer. Figure 9.9 gives examples of needs, quality characteristics, selected statistics of quality characteristics, and target values from examples presented in earlier chapters of this book.

The remaining columns of the QFD relation diagram are factors that may have an effect on the quality characteristic. These factors are classified as control factors and noise factors and are defined as follows:

*Control factors:* Factors that can be assigned at specific levels.

| Need of the Customer | Quality characteristic | Statistic of the quality characteristics | Target values | Example in book |
|---|---|---|---|---|
| Increased tool-life | wear rate | slope of line | smaller | 4.3 |
| Labels on cans | labels loose or not | percent cans with loose labels | smaller | 4.5 |
| Desired viscosity | reaction time | average | 8 hr. | 5.1 |
| Color | shade | shade reading | 220 | 5.2 |
| Low-pollution efficient engine | air flow of solenoid | average,<br>standard deviations | .60 cfm<br>smaller | 5.3 |
| Smoothness of finished part | microfinish | average | smaller | 8.1 |

**Figure 9.9** Examples of quality characteristics and target.

These factors are set by those designing the product; they are not directly changed by the customer.

*Noise factors:* These factors can potentially affect the quality characteristic but cannot be controlled at the design phase. There are three types of noise factors (Taguchi, 1987): external factors (the environment in which the product is used or distributed), internal factors (product deterioration with age or use), and unit-to-unit factors (variations in the manufacturing process).

An example of control factors in designing, say, a new tennis racket would be the shape of the frame and the type of composite material selected for the frame. Examples of noise factors include temperature or humidity (external), wear or warp of the frame (internal), and the effect of variation in shape of different frames on accuracy of the hit (unit-to-unit).

These three sources of noise must be considered during the early phases of product design. Figure 9.10 illustrates how product performance variation can be reduced (based on Kackar, 1985).

The leverage for improvements during the early phases in the product cycle is many times greater than for improvements made during actual production of that product. When an engineer calls for changes early in product design, there is time to make improvements at the lowest cost. The majority of the production costs, including levels of scrap, are determined at the beginning during the design of the product and the design of the production process.

At the bottom of the QFD relation diagram is found the best setting for each of the control factors and method of control for each noise factor. These are determined by running planned experiments with selection of the experimental variables in the following way:

Response variables = quality characteristic

Factors under study = control factors and noise factors

| Process/noise: | External | Internal | Unit-to-unit |
|---|---|---|---|
| Design of product | high | high | medium |
| Design of production process | low | medium | high |
| Production | low | low | medium |
| Sales/service | medium | low | low |

**Figure 9.10** Leverage on noise factors to reduce variability.

Background variables = noise factors

Although the above relationships are the general rule, there will be some exceptions. For example, inner noise factors such as wear may be used as response variables. As more knowledge is gained from planned experiments, the QFD relation diagram should be updated for relationships between factors and quality characteristics, best settings for control factors, and method of control of noise factors.

Another powerful strategy for improving product at this early phase is robust design. Planned experiments are used to test the interactions between control factors and noise factors. The strategy is to take advantage of an interaction by setting a control factor at a level that desensitizes the noise factor. For example, suppose an engineer is considering different materials for brake pads to improve brake torque. The noise factor is the temperature of the pads during various driving conditions. Figure 9.11 displays the interaction plot of the control factor (pad material) and the noise factor (rotor temperature). The strategy for robust design would be to go with pad material *B*.

The overall strategy of robust design is to set control factors to desensitize the product or process to noise factors, and this requires a change from thinking about designing a product or process as a search for something that works. The leverage of achieving robustness upstream is given in Fig. 9.12.

The QFD relation diagram is useful in planning the types of experiments that are needed to decide the best settings for control factors to ensure that the quality characteristics will be close to the target (minimum loss) and have minimum sensitivity to the noise factor.

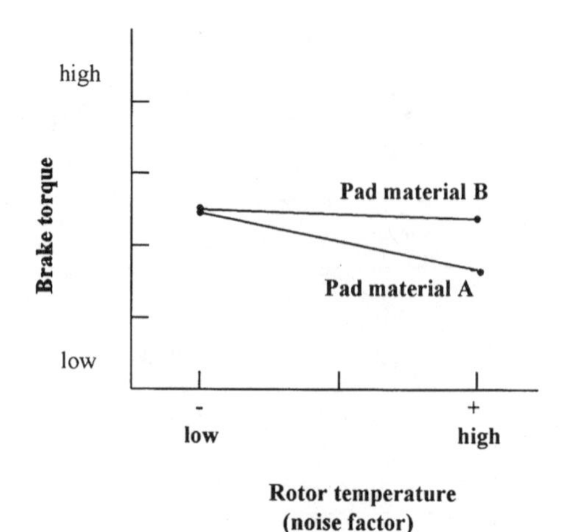


**Figure 9.11** Interaction of a control factor and a noise factor.

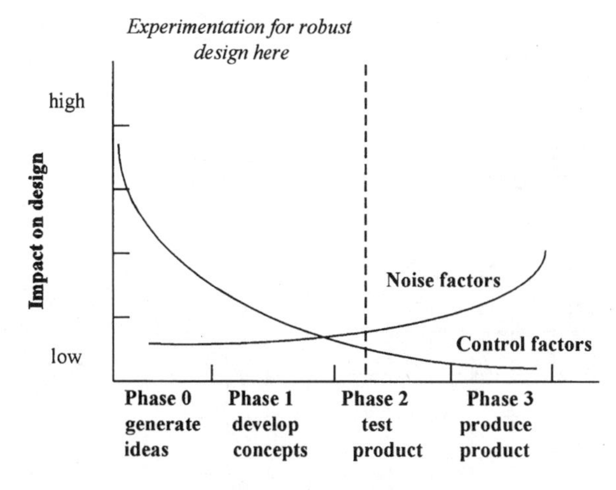


**Figure 9.12** Leverage for achieving robust design.

**Example 9.1 (*Continued*) QFD Relation Diagram of a Solenoid Valve** A planned experiment was run to determine the best settings for some important components in the design of a solenoid valve for a pollution control device on an automotive engine. This experiment is described in Example 5.3. Airflow was the quality characteristic chosen as the response variable in a particular experiment. Other experiments were run to study the other quality characteristics listed in Fig. 9.2.

The partially completed QFD relation diagram of the solenoid valve based on this planned experiment is given in Fig. 9.13. Best settings were chosen for two of the four control factors based on the response plots given in Fig. 5.22. The remaining two could be set to impact the other quality characteristics favorably.

**Example 9.2 (*Continued*) Setting of Control Factors for New Wallpaper** An engineering group from a manufacturer of wallpaper is identifying the factors for design of a vinyl-coated wallpaper product. Figure 9.14 provides the QFD relation diagram.

The group has a moderate level of knowledge that the new design will be perceived as more attractive. Previous tests have identified a seam problem because of shrinkage of the wallpaper (probably due to temperature or humidity during installation). When two sheets were put together, the seam tended to show. When the width of the seam was greater than 0.2 mil, this called attention to the seams and detracted from the appearance of the wall.

The group decided to determine the effect that installation factors have on seams and appearance. It was hoped that by setting control

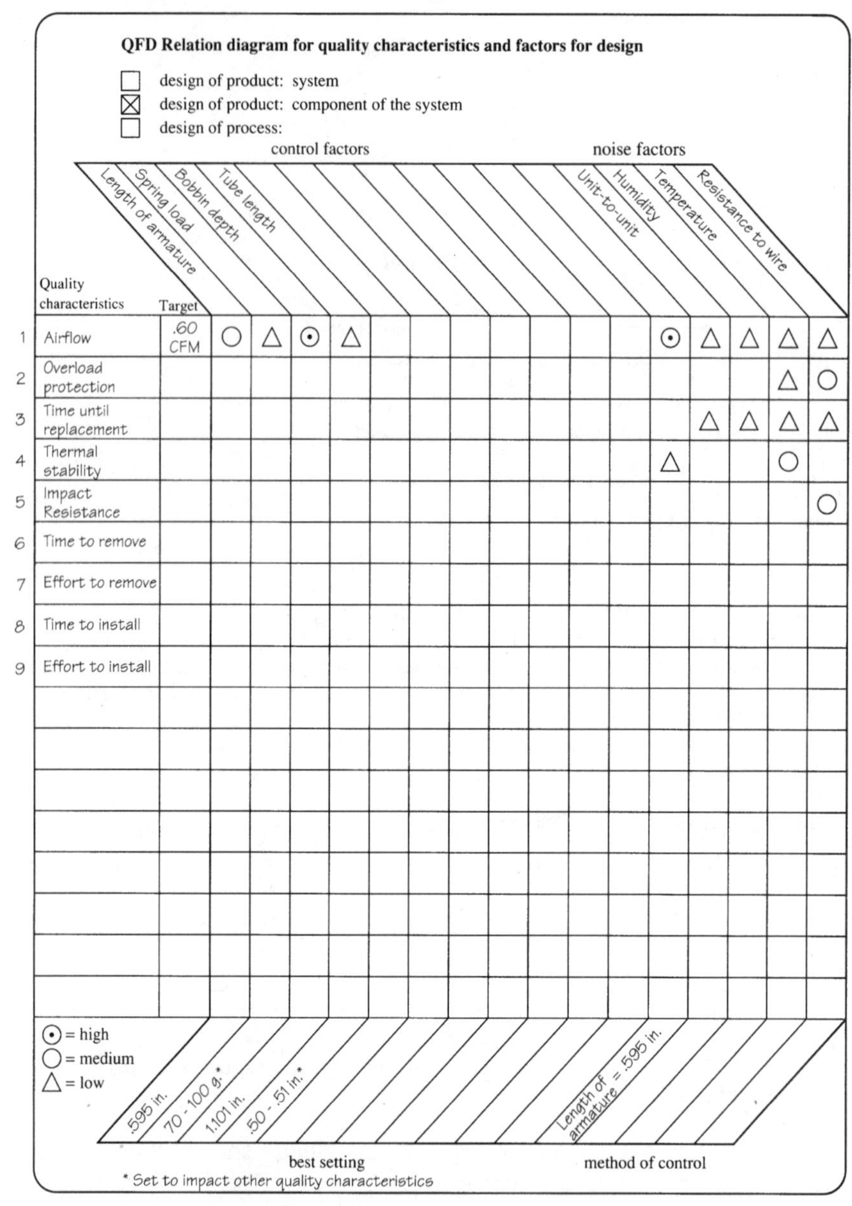


**Figure 9.13** QFD relation diagram for designing a solenoid valve (partially completed).

factors, noise factors would be reduced. Figure 9.15 contains the planning form for the study. Three control factors and two installation (noise) factors were studied. Four measurements were made on each seam, and averages and standard deviations were calculated. A large

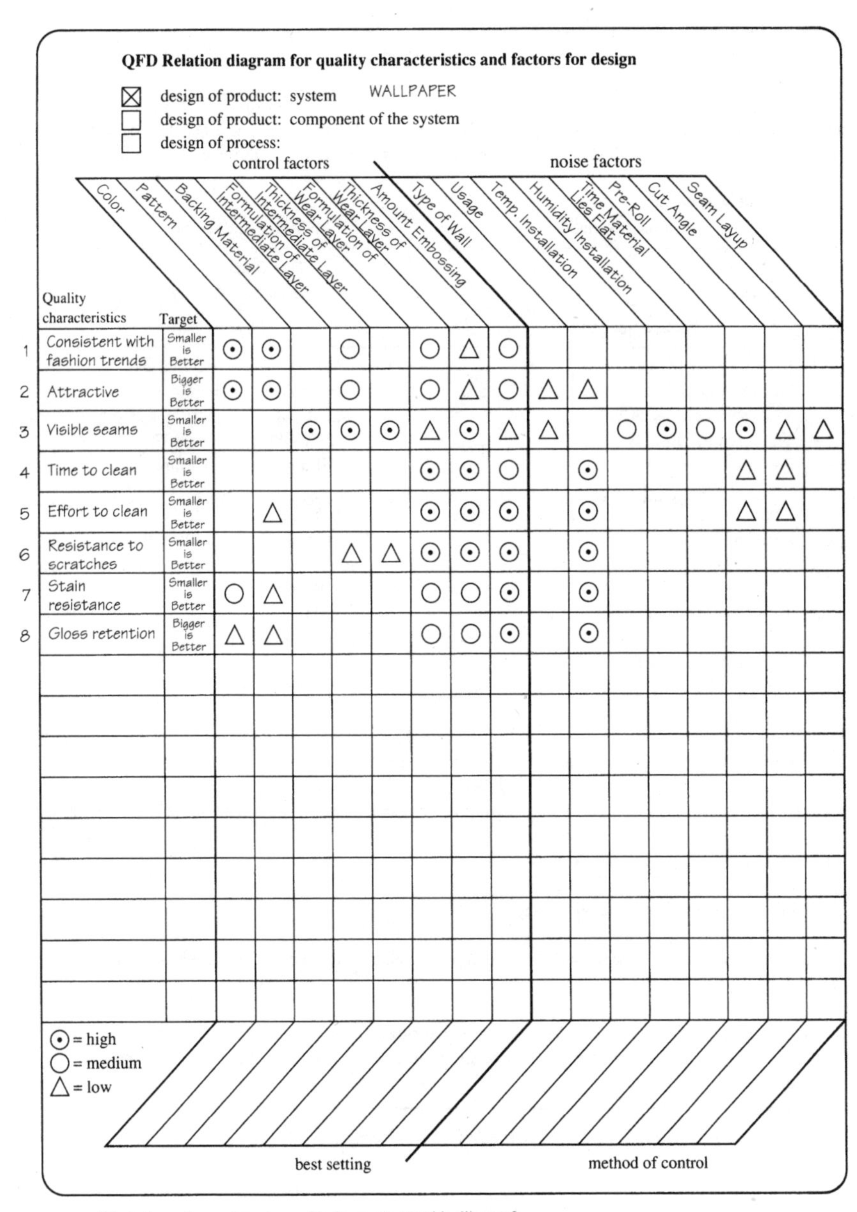


What planned experiments need to be run to test this diagram?

**Figure 9.14** QFD relation diagram for redesigning wallpaper.

standard deviation would show up as variation in seam width. A $2^{5-1}$ design was chosen because there is no confounding of two-factor interactions with main effect.

Appearance of the new design remained good throughout the 16

1. **Objective:** Find best settings of the control factors to minimize visible seams while preserving the appearance.

2. **Background information:** The redesign has a new color and pattern that has a beautiful appearance. There has been a problem with seams. Probable cause is temperature or humidity at the installation site.

3. **Experimental variables:**

| A. Response variables | Measurement technique | |
|---|---|---|
| 1. Visible seams | gage (mils) | |
| 2. Appearance | subjective scoring | |

| B. Factors under study | Levels | |
|---|---|---|
| 1. Backing material | Material *A* | Material *B* |
| 2. Formulation of intermediate layer | Formula 1 | Formula 2 |
| 3. Thickness of wear layer | 1.0 mils | 2.0 mils |
| 4. Temperature/humidity | 50°/40% | 90°/80% |
| 5. Pre-roll | 0 min. | 10 min. |

| C. Background variables | Method of control |
|---|---|
| 1. Installer | One person, standard instructions |
| 2. Type of adhesive | Use standard for all applications |
| 3. Subwall | Use plywood (porous material) |
| 4. Thickness intermediate layer | 4.0 mils |
| 5. Formulation of wear layer | Same as old design |
| 6. Time material lays flat | 4 minutes |
| 7. Cut angle | standard 90° |
| 8. Seam layup | set at low |

4. **Replication:** Four measurements for visible seams per piece.

5. **Methods of randomization:** Order of the 16 runs was randomized using a random permutation table.

6. **Design matrix:** $2^{5-1}$ factorial design.

7. **Data collection forms:** (not shown here)

8. **Planned methods of statistical analysis:** Compute average and standard deviation of four readings per piece.

9. **Estimated cost, schedule, and other resource considerations:** Evaluated in the test room with temperature and humidity controls. Four days are required to complete. Appearance scoring done by appearance team.

**Figure 9.15** Documentation of the wallpaper experiment.

runs of the experiment. Laboratory testing for scratches, stains, and gloss retention confirmed that no deterioration occurred for these quality characteristics.

Figure 9.16 contains the response plots for standard deviations of visible seams; they summarize the important results of the experiment. The first plot shows a strong interaction between a control fac-

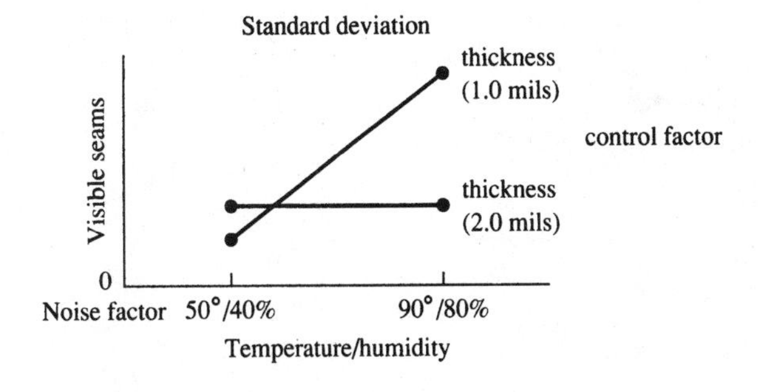


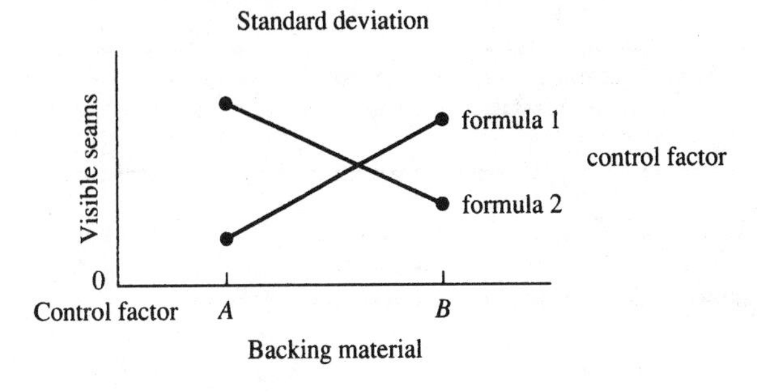


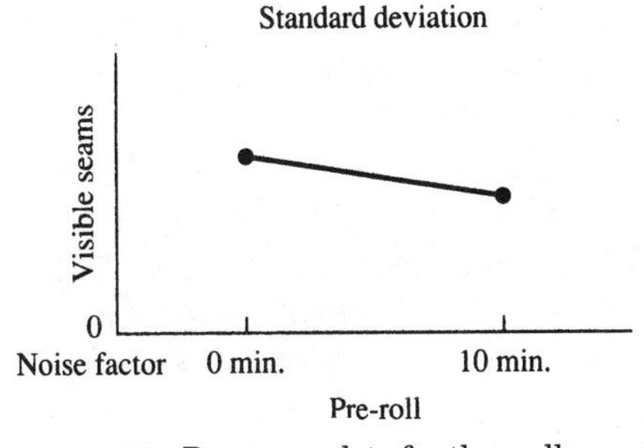


**Figure 9.16** Response plots for the wallpaper experiment.

tor and a noise factor. Thickness of wear level interacts with the temperature/humidity noise factor. The effect of this noise factor on visible seams is less with the 2.0-mil wear layer. The thinner wear level (1.0-mil) was affected by a high-temperature, high-humidity combination.

The second plot shows a strong interaction between the other two control factors. Backing material and the formulation of the intermediate layer interacted with respect to seams. Both backing materials are needed for different applications. Thus, material *A* is chosen with formula 1, and material *B* is chosen with formula 2.

The third plot demonstrates a strong relationship between an installation factor and visible seams. By laying out the roll for 10 minutes before installation, the seams will have less variation. This adds to the installation time.

Based on these analyses and the current knowledge, the following conclusions (and updates for Fig. 9.14) are drawn:

| Control factors | Best setting |
|---|---|
| Thickness of wear layer | 2.0 mils |
| Backing material | *A* with formula 1<br>*B* with formula 2 |

| Noise factors | Method of control |
|---|---|
| Pre-roll | Have installation instructions include a pre-roll at 10 min |
| Temperature/humidity | Desensitized by setting control factor, thickness of wear layer, at 2.0 mils |

**Example 9.3 Designing a New Course** A new course in planned experimentation is being designed. What are the needs of the customer and the corresponding quality characteristics that guide the design of this new course?

Figure 9.17 gives the quality characteristic diagram. This course should create an enjoyable learning experience for the students. It is also intended to enable them to apply the concepts to their jobs or hobbies. Fifteen quality characteristics were identified as important measures for this class.

What are the control factors in designing this new course? What are the noise factors? The QFD relation diagram is given in Fig. 9.18. Eleven control factors are identified based on concepts developed from various sources of adult education research. The control factor learning teams relates to many of the quality characteristics. The five noise factors may affect the learning.

How can the noise factors be desensitized or eliminated? Best settings of control factors are based on experimenting with related classes. The method of control for noise factors has proved successful in other classes.

What is the best setting (team size and makeup) for the control factor learning teams? A planned experiment for a prototype class can help answer this question. Figure 9.19 gives an example of the planning form for such an experiment.

This last example involves designing a new service rather than a new product. Inspection can prevent bad product from reaching the customer, but this is often not possible for a service. It is too late; the service to the customer has already been provided. Since this is the case in designing a class, achieving quality by design is even more important.

| Needs of the customer: Enjoy learning new concepts and be able to apply concepts to job or hobby. | | | |
|---|---|---|---|
| | Quality characteristic[1] | | |
| **Primary**[2] | **Secondary**[3] | **Tertiary**[3] | **#** |
| enjoy instruction | good teacher | pleasant person | 1 |
| | | good instructor | 2 |
| | good class | good textbook | 3 |
| | | good teaching process | 4 |
| enjoy learning | have fun in class | enjoy class | 5 |
| | | enjoy environment | 6 |
| | clear instruction | good teaching aids | 7 |
| | | good examples | 8 |
| | interesting | motivated | 9 |
| | | involved | 10 |
| recognize applications | problem clear | good charter | 11 |
| | formulate problem | document current knowledge | 12 |
| | | develop overall plan | 13 |
| know how to use methods successfully | use improvement cycle | use methods | 14 |
| | | improvements made | 15 |

[1] Do not include design factors in this list. (*Test*: You should *not* be able to set the levels of these quality characteristics.)
[2] Express in the language of the customer.
[3] To add more detail, subdivide into two or more quality characteristics.

**Figure 9.17** Quality characteristic diagram for designing a course in planned experimentation.

## Operations

As product designers of engineering are developing prototypes and initial product specifications, operations should determine process requirements to produce the product. Also, the strategy for robust design deployed by engineering will be applied by operations (manufacturing engineering) in the next phase.

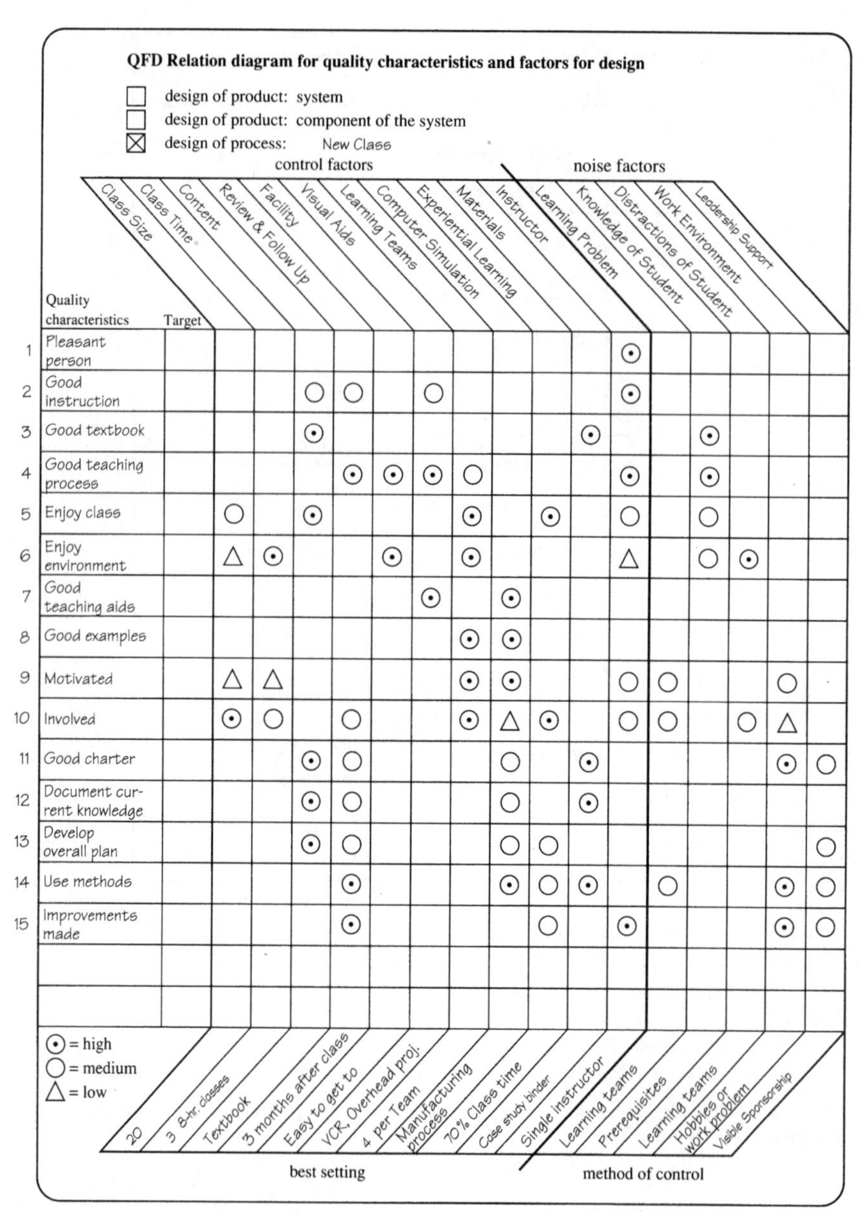


What planned experiments need to be run to test this diagram?

**Figure 9.18** QFD relation diagram for designing a course in planned experimentation.

**1. Objective:** Select size and makeup of a learning team for a planned experiment class.

**2. Background information:** Previous classroom experience with teams and a study of the literature on cooperative learning teams.

**3. Experimental variables:**

A.

| Response variables | Measurement technique |
|---|---|
| 1. Concepts learned | Test |
| 2. Enjoyed class | Interview |
| 3. Involved | Observation |

B.

| Factors under study | Levels | |
|---|---|---|
| 1. Size of team | 5 | 8 |
| 2. Makeup of team | Matched | Diverse |

C.

| Background variables | Method of control |
|---|---|
| 1. Learning disability | Measure |
| 2. Knowledge of student | Chunk-type block variable |
| 3. Distraction of student | Chunk-type block variable |

**Method of control:** Treat the 2nd and 3rd background variables as noise factors. Create four blocks made up of different treatment combinations of the background variables.

**4. Replication:** Four replications (four blocks as noise factors) will be made using the four combinations of the two factors under study.

**5. Methods of randomization:** Table of random permutations of four was used to assign the four treatment combinations.

**6. Design matrix:** $2^2$ design in four blocks.

**Figure 9.19** Documentation of an experiment to study learning teams.

## 9.4 Phase 2: Test

In the introduction of Chap. 1, another question was posed How is a new product designed to work under a wide range of conditions that will be encountered during actual production and use by the customer? The strategy of robust design continues as all three functions address the question in this phase.

### Marketing

With a better definition of the product, marketing should continue its testing with customer groups. This will help fine-tune the product to target customers. Experiments with focus groups could be run to assess the reaction to the current prototypes and some of the leading competitors' products as factors under study.

## Engineering

Once the product concepts have been selected and the product has been defined, this phase will test components and systems for the new product. Product specifications must be completed and prototypes tested.

The objectives of testing prototypes are

- To confirm that the criteria for function and performance are built into the design
- To detect likely causes of quality characteristic variation around targets under a variety of conditions
- To reevaluate costs.

Knowledge of reliability and accelerated testing as well as failure analysis on existing products is essential. The philosophy of testing prototypes should be to challenge the design rather than pamper it. The number of PDSA cycles used to test prototypes need not be large if sufficient knowledge is gained with each cycle.

**Example 9.1 (*Continued*) Testing Prototype Solenoid Valves** The quality characteristic of airflow in Example 5.3 is important in the design of a solenoid valve for a pollution control device on an automotive engine. The $2^4$ experiment (based on a moderate level of knowledge) showed that the control factor bobbin depth can be used to increase airflow. Armature length was recommended to be set at 0.595 in to minimize variation of airflow.

The best settings for a redesign of the solenoid are as follows:

| Control factors | Best setting |
|---|---|
| Bobbin depth | 1.101 in |
| Length of armature | 0.595 in |
| Spring load<br>Length of the tube | Set both based on other quality characteristics |

Thirty prototypes of the new design were developed. Will this design work well under the wide range of conditions that will be encountered during actual production and use by customers?

Figure 9.20 contains the planning form for this new experiment. The experiment chosen is a paired-comparison block design with the noise factors arranged in blocks (chunk variables). Four background variables that potentially affect the performance of the solenoid are pressure, humidity, temperature, and resistance. These variables are treated as noise factors arranged in blocks.

The blocks were chosen by judgment to create extreme conditions. This experiment will increase understanding of the performance of the new design with respect to potential problems. This allows for

**1. Objective:** Determine if a redesign of a solenoid valve holds up under various noise factors. This information will be used to determine if the current solenoid valve should be replaced by the new design.

**2. Background information:** Previous experiments using fractional factorial and factorial designs lead to redesigning the current valve. Noise factors are suspect in affecting variation of the current design.

**3. Experimental variables:**

| A. Response variables | Measurement technique | |
|---|---|---|
| 1. Average and standard deviation of flow (cfm) | Flow meter 65c | |

| B. Factors under study | Levels | |
|---|---|---|
| 1. Solenoid design | Current | Redesign |

| C. Background variables | Levels | |
|---|---|---|
| 1. Pressure (lb/sq. in.) | 30 | 45 |
| 2. Humidity (percent): | 80 | 95 |
| 3. Temperature (F) | 40 | 60 |
| 4. Resistance of wire (ohms) | 50 | 100 |

**Method of control:** Treat the four background variables as noise factors. Create five blocks made up of different treatment combinations of the background variables.

**4. Replication:** Five replications (blocks as noise factors) will be made using six current and six redesigned solenoids within each block.

**5. Methods of randomization:** Flip a coin to determine order (heads: current design; tails: redesign).

**6. Design matrix: (attach copy)** See Table 9.1.

**7. Data collection forms: (attach copy)** See Table 9.1.

**8. Planned methods of statistical analysis:** Compute the average and standard deviation for the six flow readings for current and redesigned solenoids. Plot run charts for average and standard deviation of flow and run charts for average and standard deviation of flow adjusted for block effect.

**9. Estimated cost, schedule, and other resource considerations:** Study can be completed during a 5-day period.

**Figure 9.20** Documentation of the solenoid experiment.

testing our current and new designs under the widely varying conditions that could be expected in the field.

Table 9.1 lists the design matrix and the data form for the study. The results of the study are given in Table 9.2. Run charts for average and standard deviation of flow and run charts for average and standard deviation of flow adjusted for block effect are given in Fig. 9.21.

Based on the analysis of the prototype study, these conclusions are drawn:

**TABLE 9.1 Design Matrix and Data Collection Form for Example 9.1**

| Noise factor | Block 1 | Block 2 | Block 3 | Block 4 | Block 5 |
|---|---|---|---|---|---|
| Pressure | 45 | 30 | 30 | 45 | 45 |
| Humidity | 80 | 80 | 95 | 80 | 95 |
| Temperature | 40 | 40 | 60 | 40 | 60 |
| Resistance | 50 | 100 | 50 | 100 | 100 |
| Response: Average and Standard Deviation of Six Solenoids for Flow (cfm) | | | | | |
| Current $c$ | (1) | (1) | (2) | (2) | (1) |
| Avg. | — | — | — | — | — |
| Std. dev. | — | — | — | — | — |
| Redesign $r$ | (2) | (2) | (1) | (1) | (2) |
| Avg. | — | — | — | — | — |
| Std. dev. | — | — | — | — | — |

*Note:* The (1) or (2) in parentheses indicates the order in which the values were tested.

**TABLE 9.2 Test Results for Example 9.1**

| Design | Block 1 | Block 2 | Block 3 | Block 4 | Block 5 | Average |
|---|---|---|---|---|---|---|
| Response: Average of Six Solenoids for Flow (cfm) | | | | | | |
| $c$ (current) | 0.59 | 0.54 | 0.64 | 0.60 | 0.60 | 0.594 |
| $r$ (redesigned) | 0.61 | 0.59 | 0.60 | 0.65 | 0.54 | 0.598 |
| Block average | 0.600 | 0.565 | 0.620 | 0.625 | 0.570 | |
| | | | | | Overall average | 0.596 |
| Response: Average of Six Solenoids for Flow (cfm) Adjusted for Block Effect (Average−Block Average + Overall Average) | | | | | | |
| $c$ | 0.586 | 0.571 | 0.616 | 0.571 | 0.626 | 0.594 |
| $r$ | 0.606 | 0.621 | 0.576 | 0.621 | 0.566 | 0.598 |
| Response: Standard Deviation of Six Solenoids for Flow (cfm) | | | | | | |
| $c$ | 0.067 | 0.068 | 0.040 | 0.052 | 0.041 | 0.054 |
| $r$ | 0.033 | 0.036 | 0.030 | 0.029 | 0.024 | 0.030 |
| Block average | 0.050 | 0.052 | 0.035 | 0.040 | 0.033 | |
| | | | | | Overall average | 0.042 |
| Response: Standard Deviation of Six Solenoids for Flow (cfm) Adjusted for Block Effect (Average−Block Average + Overall Average) | | | | | | |
| $c$ | 0.059 | 0.058 | 0.047 | 0.054 | 0.050 | 0.054 |
| $r$ | 0.025 | 0.026 | 0.037 | 0.031 | 0.033 | 0.030 |

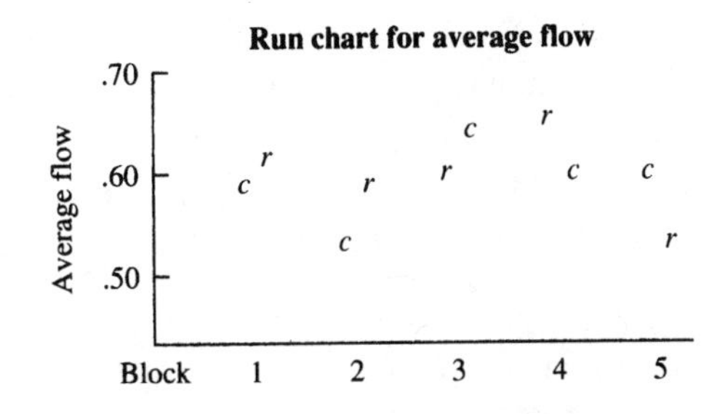


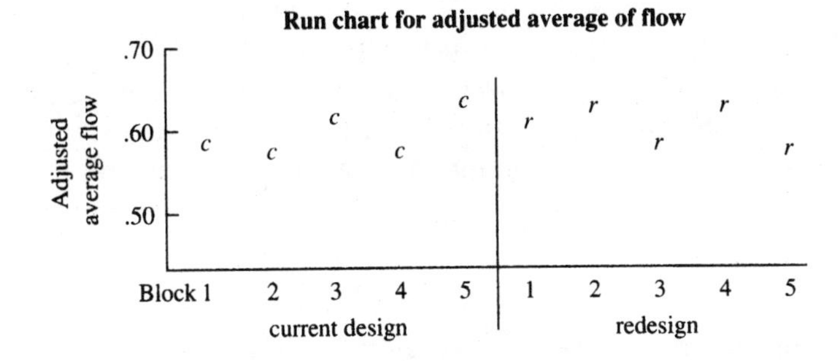


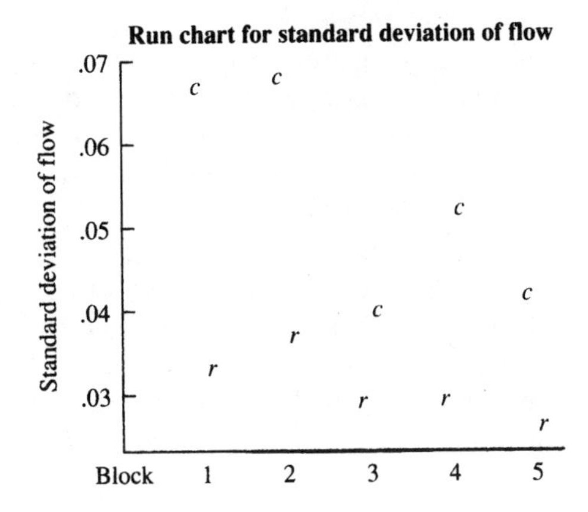


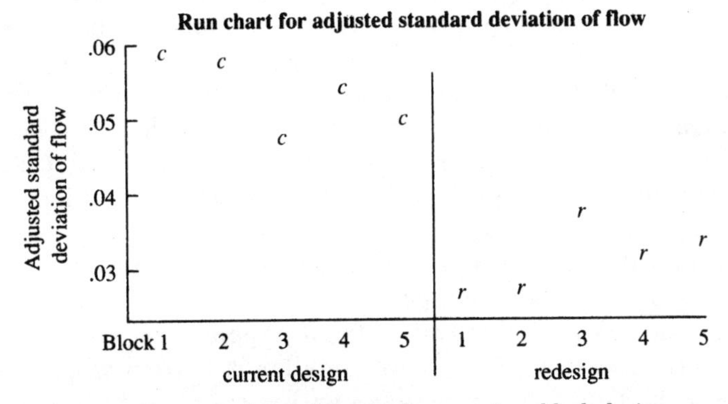


**Figure 9.21** Four run charts for paired-comparison block design.

1. Average flow for current (0.594) and redesign (0.598) are both close to the target value (0.600). The sensitivity (block effect) of both designs was similar for average flow.
2. Standard deviations for the redesign were smaller than those for the current design. The sensitivity (block effect) for the redesign was much better.

It was felt that these results will hold up in the future and that it would be best to accept the redesign and run a pilot of the redesigned solenoid.

An alternative design for Example 9.1 could use four background variables as factors and run a $2^{5-1}$ to test the interaction of the background variables with the solenoid design. Taguchi (1986, 1987) uses the "outer-array" approach to accommodate background variables as noise factors.

### Operations

Another question posed in Chap. 1 was: How are the best operating conditions chosen for a manufacturing process among the hundreds of choices? The strategy of robust design discussed for engineering in phase 1 applies directly to developing and testing the production process. The leverage for improvement of quality lies in reducing variation of the quality characteristics of the product due to unit-to-unit noise (variations in the manufacturing process).

The control factors for the product usually become the quality characteristics for the manufacturing process. For example, two important control factors for the solenoid in Example 9.1 are bobbin depth and length of armature. The targets are 1.101 in for bobbin depth and 0.595 in for length of armature. These two factors are control factors for the product design but become quality characteristics for the process design.

After the concepts for the production process are chosen (there are cases in which a new concept for production process may not be necessary), the next step is to select factors for designing the production process. The QFD relation diagram should be used again with the control factors and noise factors coming from the production process.

The aim of robust design in operations should be to set control factors that will improve process capability by reducing the effect of noise factors (highest leverage for unit-to-unit noise).

Examples from previous chapters will illustrate this strategy as it applies to designing the production process.

Example 4.5 showed that the control factor, application rate, gave the smallest percentage of loose labels (smaller is better) when the target was 14 oz/h. This factor interacted with a chunk variable (one or more background variables) used to define the block. Interaction was used to desensitize the product. A follow-up study would be a factorial design to study the background variables used to make up the chunk variable.

Example 5.2 has a shade of a dyed material as the quality characteristic. The target is at shade 220. The factorial design showed that running the process at high oxidation temperature makes the process less sensitive to variation in the material. The third factor, oven pressure, can be used to adjust the shade to 220.

Example 8.2 has fill weight as a quality characteristic. A number of factors that affected the average fill weight had been identified. A factorial design with a center point was run with fill weight variation to determine the presence of nonlinear effects. The results shown in Fig. 8.9 indicate a nonlinear response with line speed. If speed is kept below 350 cans per hour, the variation in fill weight will be less. Also, temperature and consistency interact. Keeping temperature above 200 minimizes the variation in fill weight due to varying consistency of the incoming ingredients.

Output of this phase in process design is process definition. One now needs standardization—things or methods that have already been found to be good (standard parts, units, or modules, and standard procedures).

Development of production capability and development of the product must be done in parallel. This ensures smooth and efficient transition to factory operations, for product and production design must be managed together.

Once the strategy of robust design is applied to designing the production process, the next step is to validate that the production process produces a good product. Specifications and tolerances help to define acceptable outcomes.

There are different approaches to the management of variability resulting from the production process. The aim should be to reduce the dependence on inspection to achieve quality. The purpose of inspection should be to improve the process and reduce costs. Mass inspection is costly and ineffective. The application of the Model for Improvement and the methods of control charting and planned experimentation provide the basis for the control plan. See App. A for more on inspection and the use of control charts.

Production capability must be developed to produce the parts and components, to assemble the product, and to have operating systems that are necessary for production and field operations. Building prototypes confirm the product. Often parts and components are handmade. Pilot runs with production tooling confirm the process. During a pilot run, product from the production process is studied for the first time. A pilot run also provides the opportunity to predict future production capability. The methods of control charting and determining the capability of a process (App. A) are useful here.

What are the important control factors? What are the noise factors that the product might encounter during production? Pilot runs provide the opportunity to establish the relationships between the control factors, noise factors, and quality characteristics as identified by the QFD relation diagram for the production process. Planned experimentation will increase the knowledge of these relationships. This will help establish the best operating conditions for the manufacturing process before production starts.

## 9.5 Phase 3: Produce Product

The major tasks in this phase for marketing are to provide sales and service and to establish a customer feedback system. The product designers need to support the product in production and the field. Operations is involved with producing the product and any problems associated with production.

All functions should be involved with any improvement activities associated with the product. Most of the examples in Chaps. 2 and 4 through 8 represent these types of improvement activities.

## 9.6 Summary

This chapter illustrated the importance of well-designed and well-executed experiments to aid in designing quality into new products (processes or systems). The development started with Deming's conception of production as a system and the emphasis on the design and redesign stage in matching products and services to a need.

A four-phase process in the design of a new product was presented. The key tasks of this process are to define quality, set targets, and design products or processes that are close to the targets under a wide range of conditions.

By integrating the concepts of quality function deployment, the ideas developed by Taguchi, and the methods of planned experimentation into the Model for Improvement, a strategy or road map for quality by design is created. Application of this strategy can

- Accelerate the evolution of new product cycles
- Reduce development costs
- Improve the transition from R&D to manufacturing
- Make the product or process robust against noise factors by selecting the proper level of control factors

This results in higher acceptance of the product and fewer warranty claims.

The role of each of the major functions (marketing, engineering, and operations) with respect to planned experimentation is illustrated below:

- Objective of experiment

  Marketing: Determine the best features that would satisfy the customers' needs and wants.

  Engineering: Select the best factors for design of product.

  Manufacturing: Select the best factors for design of the production process.

- Response variables

  Marketing: Preference ranking by potential customers.

  Engineering: Quality characteristics derived from customer needs.

  Manufacturing: Product specifications.

- Factors under study

  Marketing: New product features.

  Engineering: Factors for design of product.

  Manufacturing: Factors for design of production process.

Finally, applying these strategies of experimentation to the design of a new product provides the greatest leverage for improvement of quality.

## References

Akao, Yoji (1990): *Quality Function Deployment,* Productivity Press. Cambridge, MA.

Blankenship, A. B., and G. E. Breen (1993): *State of the Art Marketing Research,* American Marketing Association, Chicago.

Box, George (1988): "Signal-to-Noise Ratios, Performance Criteria and Transformations," *Technometrics,* vol. 30, no. 1, February.

Cooper, Robert (1993): *Winning at New Products,* Addison-Wesley, Reading, Mass.

Dommermuth, W. P. (1975): *The Use of Sampling in Marketing Research,* American Marketing Association, Chicago.

Hauser, John R., and Don Clausing (1988): "The House of Quality," *Harvard Business Review,* May–June.

Johansson, K. K., and I. Nonaka (1996): "Market Research the Japanese Way," *Harvard Business Review,* reprint 87303.

Kackar, R. (1985): "Off-Line Quality Control, Parameter Design and the Taguchi Method," *Journal of Quality Technology,* vol. 17, no. 4, October.

Kano, N. (1994): "Attractive Quality Creation," Paper presented at Convergence Conference, Dearborn, Mich., October 17, 1994.

Office of Technology Assessment (1990): *Worker Training: Competing in the New International Economy,* Government Printing Office, Washington, D.C.
Taguchi, Genichi (1986): *Introduction to Quality Engineering, Designing Quality into Products and Processes,* Unipub, White Plains, N.Y.
Taguchi, Genichi (1987): *System of Experimental Design,* vols. 1 and 2, Unipub, White Plains, N.Y.

## Exercises

**9.1** Study the effect of poor quality product design and production process design for products in an organization. (Use warranty claims and work backward.)

**9.2** Compare the processes in an organization with those in the evolution of a new product from Fig. 9.1.

**9.3** Identify the three sources of product performance variation (external, internal, and unit-to-unit) for the major products in an organization.

**9.4** Complete a QFD relationship diagram for a new or existing product.

**9.5** Determine how many planned experiments were carried out by product design engineers or manufacturing engineers in an organization in the last 2 years.

**9.6** *The paper box factory.* Design a paper box that stores change (quantity equal to the change in most people's pockets). Materials and technologies include

- Two pairs of scissors
- Two small staplers
- Four crayons
- A ream of $8\frac{1}{2}$" × 11" blank paper.

Appoint a product-planning team, a product design team, a production process design team, a manufacturing team, and a marketing team.

Use the four-phase process from Fig. 9.1 to develop your product and production processes. What are the customer's needs? (The box needs a removable lid.) What are the quality characteristics? What is your measurement process? What are the product control factors? Noise factors?

Manufacturing should produce 50 boxes. Is the process stable? Capable? Can the product be improved? How? Can the process be improved? How?

# Chapter 10

# Case Studies

The experimental patterns and the methods of analysis presented in the previous chapters form a system of sequential experimentation. The case studies presented in the next two sections of this chapter illustrate the application of this system to develop, test, and implement changes.

The first case study involves a redesign of an existing milling process for the manufacture of metallic bricks. Eight cycles are completed to increase the degree of belief in the prediction that the changes made to the process will improve the quality of bricks in the future. As new knowledge is gained, appropriate changes are made to the process.

The second case study involves a redesign of a wallpaper product. Eight cycles are completed to increase the degree of belief that a new wallpaper product will be an improvement over the current product.

## 10.1 Case Study 1: Improving a Milling Process

This case study is based on work in a small factory, Mid-State Brick Company. The factory contains multiple machining processes to make a simple product, a metallic brick. Figure 10.1 illustrates the production flow for the factory. The size and finish of the brick are important quality characteristics. Figure 10.2 contains the drawings and specifications for the brick.

Historically, Mid-State had shipped all the bricks that it produced. Minimal checks or analyses of the quality of the bricks had been done. Recent complaints from customers about product quality forced an assessment of quality in the plant. Initial checks at final inspection showed many parts out of specification for multiple dimensions and finish. The width of the brick and the finish on the sides were

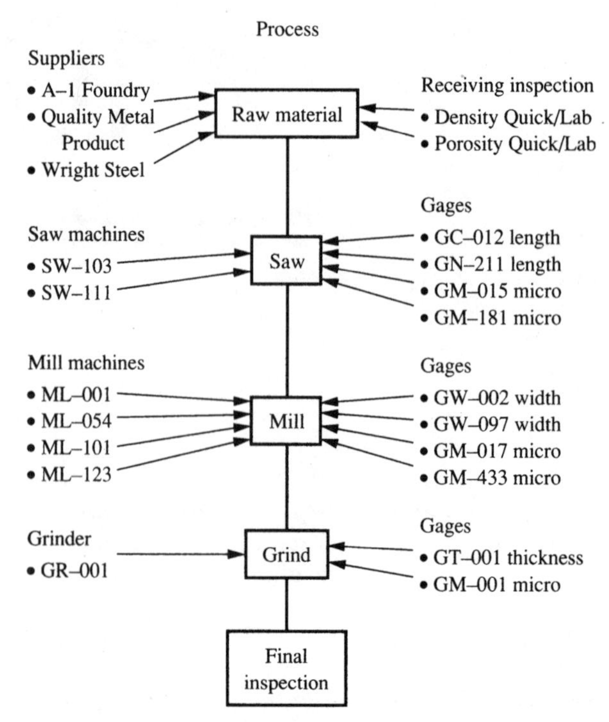


**Figure 10.1** Process flow for Mid-State Brick Company.

particular problems. A quality improvement team for the mill process was organized to address these issues.

The charter for the team is shown in Fig. 10.3. The team consisted of the factory engineer, the mill supervisor, two mill operators, and the contract maintenance person. After review of the charter, the team began focusing on the mill process. A schematic diagram of the mill operation and a flowchart of the machining process for each mill were prepared (Fig. 10.4). Next, a cause-and-effect diagram (Fig. 10.5) was prepared to summarize the possible causes of variation in width and microfinish.

Team discussions about their current knowledge of the mill process led to the following plan for the improvement effort:

*Cycle 1:* Document the current performance of the mill process and determine the capability of the process for width and microfinish. Determine the capability of the measurement process for width and microfinish.

*Cycle 2:* Identify the important sources of variation in the mill process, using nested designs.

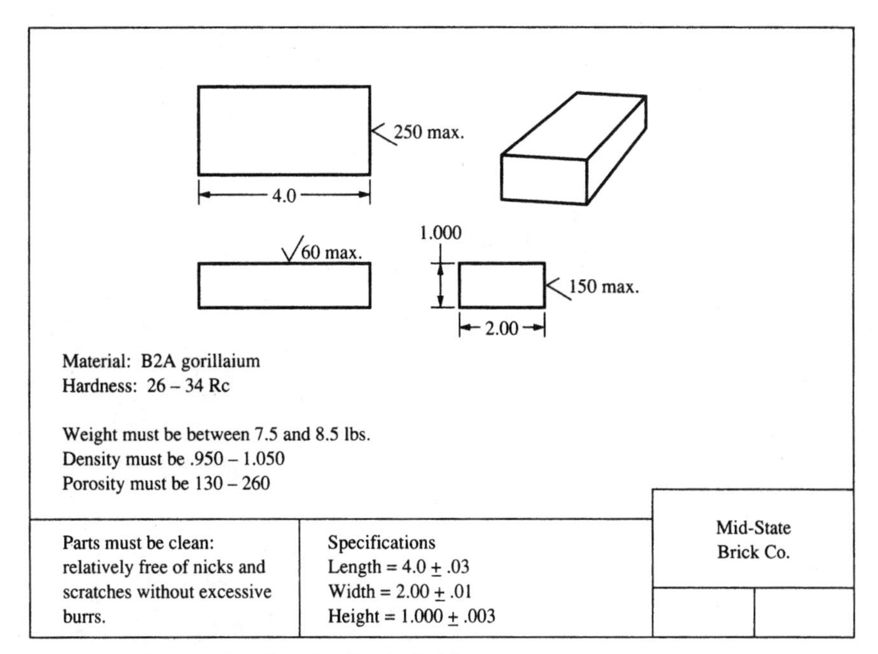


**Figure 10.2** Engineering drawing for the brick.

---

**General Description:**

Improve the quality of the bricks from the mill process. The focus should be on the width of the brick and the finish of the milled surface.

**Expected Results:**

1. A capable process: All bricks meet specifications for width and microfinish on the milled surface.
2. A better understanding of the factors that affect quality in the mill process.
3. Procedures for continuous monitoring and improvement of the mill process.
4. Understanding of improvement methods to apply to the other processes in the factory.

**Boundaries:**

All four mills in the process should be addressed by the team. Studies can be done on one mill and the results transferred to the other mills. The operators should be the primary resources for measurement and other data collection.

---

**Figure 10.3** Charter of quality improvement team for mill.

*Cycle 3:* Evaluate the different suppliers of mill cutters.

*Other cycles:* Conduct other studies as required to improve the quality of the bricks.

*Last cycle:* Document the performance of the process, using control charts.

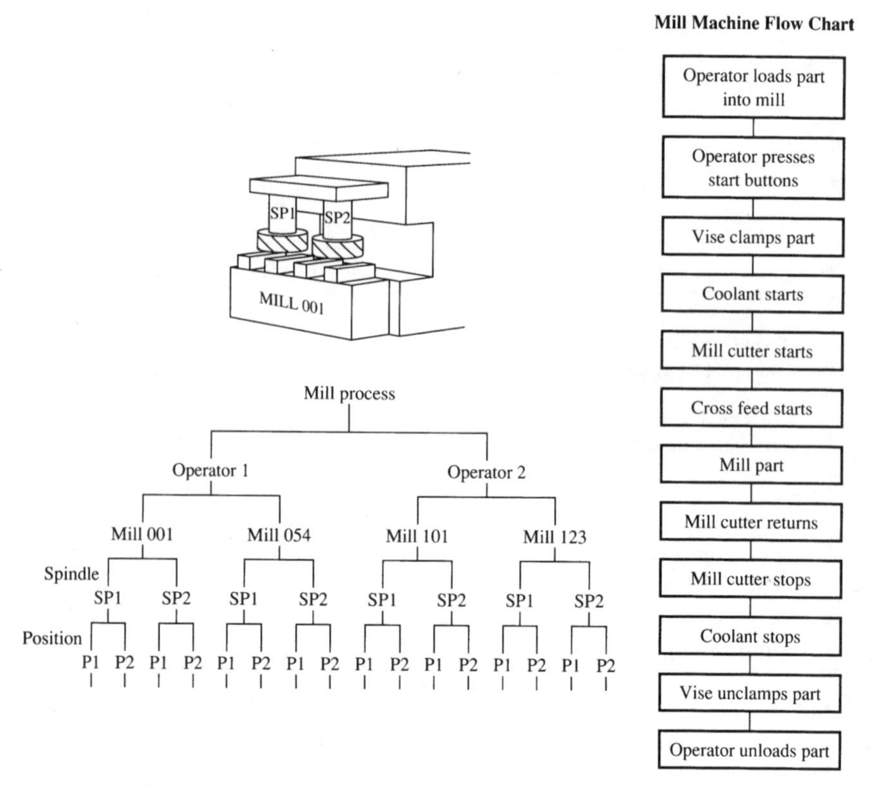


**Figure 10.4** Flowcharts of the mill process.

## First improvement cycle: Current performance of the mills

The objective of the first improvement cycle was twofold:

1. To learn about the current performance of the mill process and, if the process was stable, to determine the process capability for the width and microfinish
2. To learn about the quality of the gages used to measure the width and microfinish of the milled surface

Meeting these objectives requires the information needed to describe the current mill process. The team felt that the process might be unstable for both width and microfinish. Based on the results at final inspection, the team felt the process would not be capable for either quality characteristic.

Control charts were planned for the process. None of the team members had any ideas about the important sources of variation, so it was difficult to develop a rational subgrouping strategy. The team de-

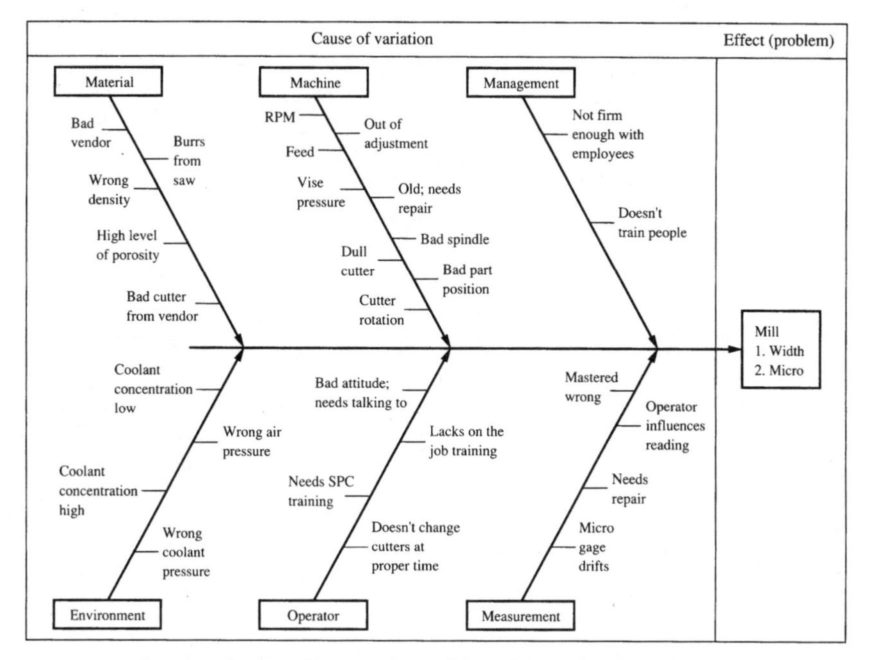


**Figure 10.5** Cause-and-effect diagram for width and microfinish.

cided to select one brick from each mill during every hour of operation for the next 2 days. The bricks from each mill would be measured for width and microfinish by the mill operator. Subgroups of four bricks would be formed from the bricks from each mill. The supervisor would gather the data and develop the control charts.

To document the quality of the measurement processes, the supervisor selected one part from the milled inventory each hour and measured it twice, using both the width gage (GW-002) and the microfinish gage (GM-015). In addition, a certified master brick (width = 2.00 in and microfinish = 150 μ) was measured using the gages each hour.

The team would meet at the end of the second day to review the control charts and complete the data analysis. Figure 10.6 contains the control chart for the width of the milled surface. No special causes were indicated by either the $\overline{X}$ or the $R$ chart. The averages tended to cluster around the centerline, indicating that an important source of variation might be included in the range. Some important differences in the mills were apparent. The supervisor did not keep track of which part came from which mill, so no further analysis was done with the data.

Although the clustering of the averages indicated that a systematic source of variation was probably present in the ranges, a capability

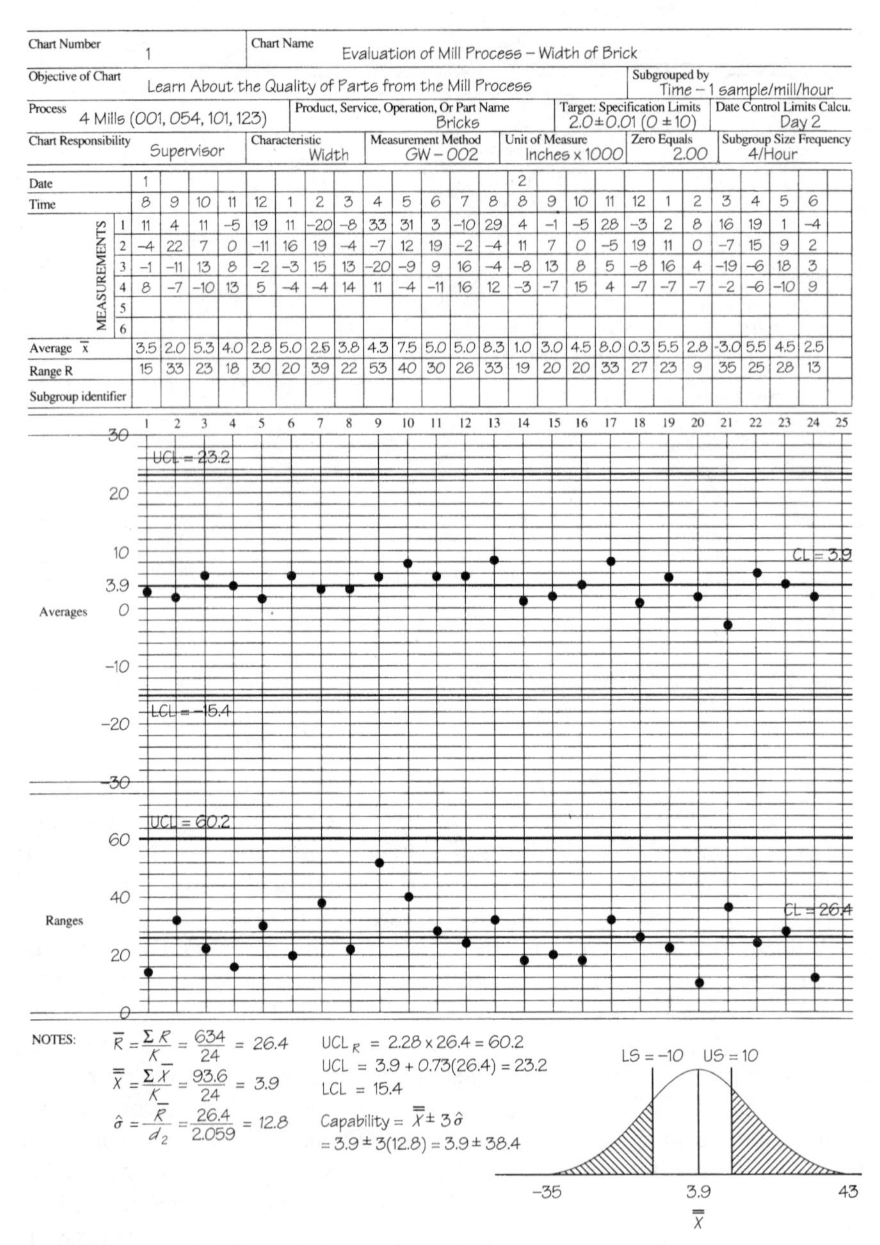

| Chart Number 1 | Chart Name Evaluation of Mill Process – Width of Brick | | |
|---|---|---|---|
| Objective of Chart Learn About the Quality of Parts from the Mill Process | | Subgrouped by Time – 1 sample/mill/hour | |
| Process 4 Mills (001, 054, 101, 123) | Product, Service, Operation, Or Part Name Bricks | Target: Specification Limits 2.0±0.01 (0 ± 10) | Date Control Limits Calcu. Day 2 |
| Chart Responsibility Supervisor | Characteristic Width | Measurement Method GW – 002 | Unit of Measure Inches x 1000 |
| Zero Equals 2.00 | Subgroup Size Frequency 4/Hour | | |

| | 1 | 2 | 3 | 4 | 5 | 6 | 7 | 8 | 9 | 10 | 11 | 12 |
|---|---|---|---|---|---|---|---|---|---|---|---|---|
| Date | 1 | | | | | | | | | | | |
| Time | 8 | 9 | 10 | 11 | 12 | 1 | 2 | 3 | 4 | 5 | 6 | 7 |
| Measurements 1 | 11 | 4 | 11 | –5 | 19 | 11 | –20 | –8 | 33 | 31 | 3 | –10 |
| 2 | –4 | 22 | 7 | 0 | –11 | 16 | 19 | –4 | –7 | 12 | 19 | –2 |
| 3 | –1 | –11 | 13 | 8 | –2 | –3 | 15 | 13 | –20 | –9 | 9 | 16 |
| 4 | 8 | –7 | –10 | 13 | 5 | –4 | –4 | 14 | 11 | –4 | –11 | 16 |
| 5 | | | | | | | | | | | | |
| 6 | | | | | | | | | | | | |
| Average $\bar{x}$ | 3.5 | 2.0 | 5.3 | 4.0 | 2.8 | 5.0 | 2.5 | 3.8 | 4.3 | 7.5 | 5.0 | 5.0 |
| Range R | 15 | 33 | 23 | 18 | 30 | 20 | 39 | 22 | 53 | 40 | 30 | 26 |
| Subgroup identifier | | | | | | | | | | | | |

| | 13 | 14 | 15 | 16 | 17 | 18 | 19 | 20 | 21 | 22 | 23 | 24 | 25 |
|---|---|---|---|---|---|---|---|---|---|---|---|---|---|
| Date | | 2 | | | | | | | | | | | |
| Time | 8 | 8 | 9 | 10 | 11 | 12 | 1 | 2 | 3 | 4 | 5 | 6 | |
| Measurements 1 | 29 | 4 | –1 | –5 | 28 | –3 | 2 | 8 | 16 | 19 | 1 | –4 | |
| 2 | –4 | 11 | 7 | 0 | –5 | 19 | 11 | 0 | –7 | 15 | 9 | 2 | |
| 3 | –4 | –8 | 13 | 8 | 5 | –8 | 16 | 4 | –19 | –6 | 18 | 3 | |
| 4 | 12 | –3 | –7 | 15 | 4 | –7 | –7 | –7 | –2 | –6 | –10 | 9 | |
| 5 | | | | | | | | | | | | | |
| 6 | | | | | | | | | | | | | |
| Average $\bar{x}$ | 8.3 | 1.0 | 3.0 | 4.5 | 8.0 | 0.3 | 5.5 | 2.8 | -3.0 | 5.5 | 4.5 | 2.5 | |
| Range R | 33 | 19 | 20 | 20 | 33 | 27 | 23 | 9 | 35 | 25 | 28 | 13 | |
| Subgroup identifier | | | | | | | | | | | | | |



**Figure 10.6** Initial width control chart for mill process.

analysis was done to give an indication of the quality of the parts being produced. The capability analysis of the process for width indicated that as many as one-half of the parts were either too narrow or too wide. The capability of the process ranged from −35 to +43 (in thousandths of an inch) from nominal. The tolerances for the width were ±10.

Figure 10.7 contains the control chart for the microfinish. The process was also stable for this quality characteristic. But the capability analysis indicated that only a small percentage of the bricks would have a microfinish less than the upper specification of 150 μ. The current process average of 166.6 μ would have to be reduced well below 150 μ in order to make all good bricks. The plant engineer thought the variation for microfinish was reasonably good but was very surprised to see how high the average was.

Figure 10.8 contains the control chart for precision (range chart for the two measurements of the same part) and bias (individual chart for the master part). Both charts were stable. The bias of the width measurement was +5 (in thousandths of an inch). The $\overline{X}$ chart for the measured part (Fig. 10.6) averaged about +4, so this bias in the measurement indicated the bricks being produced actually averaged about −1 below the nominal value (2.00 in).

The precision of the width measurement process was evaluated by using the range of the two measurements of the same brick. The average range was 2.4. Table 10.1 compares this variation to the variation of the process (from Fig. 10.6). The variation in the measurement process represents less than 4 percent of the total variation in the process. The actual variation from brick to brick represents most of the variation. Thus the GW-002 gage provides adequate precision to learn about the variation in the process. But the standard deviation of the measurement process ($\sigma = 2.13$) represents a significant portion of the tolerance for width (±10). The capability of the measurement process would be about ±6.

A similar analysis was done for the microfinish measurement process (GM-002). The bias was less than 1 μ, and the standard deviation for measurement was about 0.5 μ, which represents less than 1 percent of the process variation. Thus the microfinish gage was considered an unimportant source of variation for this process.

Based on the information obtained in this improvement cycle, the team decided to

1. Calibrate the width gage to remove the bias. The mill operators would run the master brick twice per day on both the width and micro gages and would plot the results on the master control charts developed in this cycle.

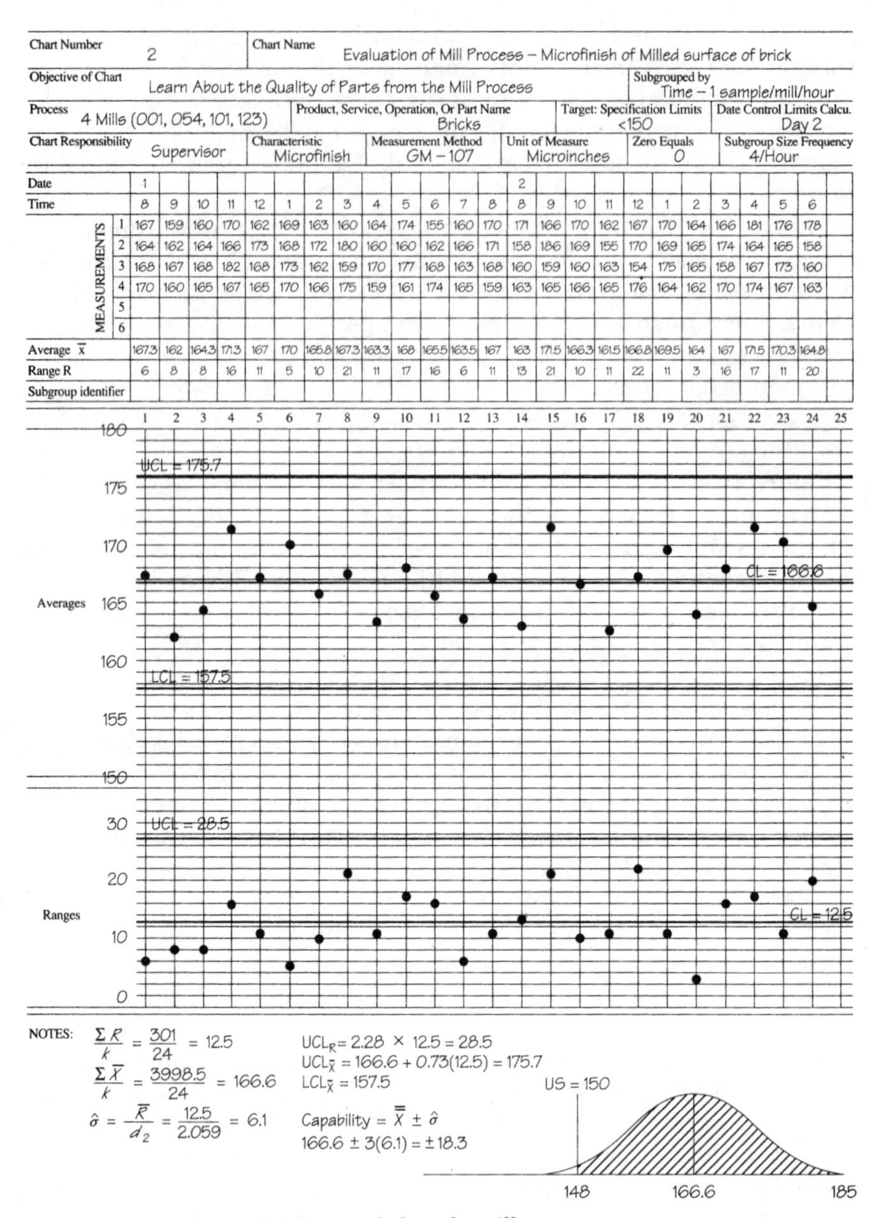

Chart Number: 2 | Chart Name: Evaluation of Mill Process – Microfinish of Milled surface of brick

Objective of Chart: Learn About the Quality of Parts from the Mill Process | Subgrouped by: Time – 1 sample/mill/hour

Process: 4 Mills (001, 054, 101, 123) | Product, Service, Operation, Or Part Name: Bricks | Target: Specification Limits: <150 | Date Control Limits Calcu.: Day 2

Chart Responsibility: Supervisor | Characteristic: Microfinish | Measurement Method: GM – 107 | Unit of Measure: Microinches | Zero Equals: 0 | Subgroup Size Frequency: 4/Hour

| | 1 | 2 | 3 | 4 | 5 | 6 | 7 | 8 | 9 | 10 | 11 | 12 | 13 | 14 | 15 | 16 | 17 | 18 | 19 | 20 | 21 | 22 | 23 | 24 | 25 |
|---|---|---|---|---|---|---|---|---|---|---|---|---|---|---|---|---|---|---|---|---|---|---|---|---|---|
| Date | 1 | | | | | | | | | | | | | 2 | | | | | | | | | | | |
| Time | 8 | 9 | 10 | 11 | 12 | 1 | 2 | 3 | 4 | 5 | 6 | 7 | 8 | 8 | 9 | 10 | 11 | 12 | 1 | 2 | 3 | 4 | 5 | 6 | |
| Measurements 1 | 167 | 159 | 160 | 170 | 162 | 169 | 163 | 160 | 164 | 174 | 155 | 160 | 170 | 171 | 166 | 170 | 162 | 167 | 170 | 164 | 166 | 181 | 176 | 178 | |
| 2 | 164 | 162 | 164 | 166 | 173 | 168 | 172 | 180 | 160 | 160 | 162 | 166 | 171 | 158 | 186 | 169 | 155 | 170 | 169 | 165 | 174 | 164 | 165 | 158 | |
| 3 | 168 | 167 | 168 | 182 | 168 | 173 | 162 | 159 | 170 | 177 | 168 | 163 | 168 | 160 | 159 | 160 | 163 | 154 | 175 | 165 | 158 | 167 | 173 | 160 | |
| 4 | 170 | 160 | 165 | 167 | 165 | 170 | 166 | 175 | 159 | 161 | 174 | 165 | 159 | 163 | 165 | 166 | 165 | 176 | 164 | 162 | 170 | 174 | 167 | 163 | |
| 5 | | | | | | | | | | | | | | | | | | | | | | | | | |
| 6 | | | | | | | | | | | | | | | | | | | | | | | | | |
| Average $\bar{x}$ | 167.3 | 162 | 164.3 | 171.3 | 167 | 170 | 165.8 | 167.3 | 163.3 | 168 | 165.5 | 163.5 | 167 | 163 | 171.5 | 166.3 | 161.5 | 166.8 | 169.5 | 164 | 167 | 171.5 | 170.3 | 164.8 | |
| Range R | 6 | 8 | 8 | 16 | 11 | 5 | 10 | 21 | 11 | 17 | 16 | 6 | 11 | 13 | 21 | 10 | 11 | 22 | 11 | 3 | 16 | 17 | 11 | 20 | |
| Subgroup identifier | | | | | | | | | | | | | | | | | | | | | | | | | |



NOTES:

$\frac{\Sigma R}{k} = \frac{301}{24} = 12.5$

$\frac{\Sigma \bar{X}}{k} = \frac{3998.5}{24} = 166.6$

$\hat{\sigma} = \frac{\bar{R}}{d_2} = \frac{12.5}{2.059} = 6.1$

$UCL_R = 2.28 \times 12.5 = 28.5$

$UCL_{\bar{X}} = 166.6 + 0.73(12.5) = 175.7$

$LCL_{\bar{X}} = 157.5$

Capability $= \bar{\bar{X}} \pm \hat{\sigma}$

$166.6 \pm 3(6.1) = \pm 18.3$

**Figure 10.7** Initial microfinish control chart for mill.

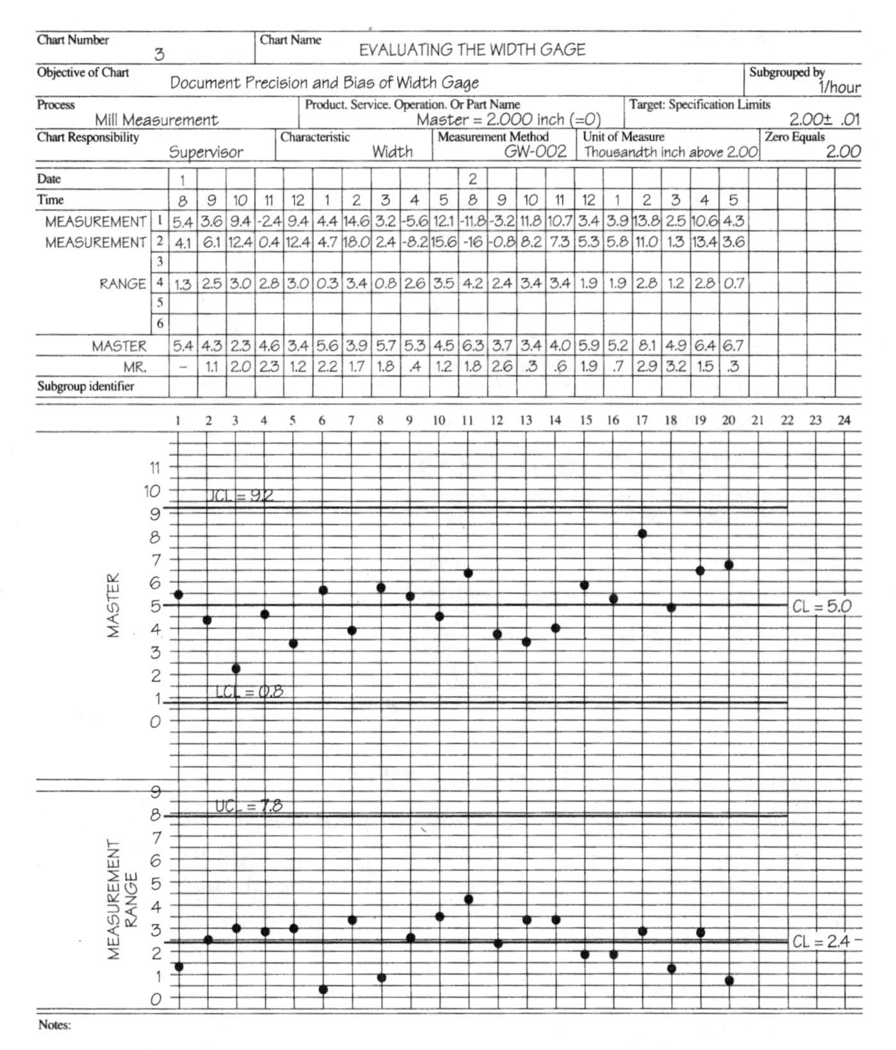
Chart Number: 3 | Chart Name: EVALUATING THE WIDTH GAGE

Objective of Chart: Document Precision and Bias of Width Gage | Subgrouped by: 1/hour

Process: Mill Measurement | Product. Service. Operation. Or Part Name: Master = 2.000 inch (=0) | Target: Specification Limits: 2.00± .01

Chart Responsibility: Supervisor | Characteristic: Width | Measurement Method: GW-002 | Unit of Measure: Thousandth inch above 2.00 | Zero Equals: 2.00

| | 1 | 2 | 3 | 4 | 5 | 6 | 7 | 8 | 9 | 10 | 11 | 12 | 13 | 14 | 15 | 16 | 17 | 18 | 19 | 20 | 21 | 22 | 23 | 24 |
|---|---|---|---|---|---|---|---|---|---|---|---|---|---|---|---|---|---|---|---|---|---|---|---|---|
| Date | 1 | | | | | | | | | | 2 | | | | | | | | | | | | | |
| Time | 8 | 9 | 10 | 11 | 12 | 1 | 2 | 3 | 4 | 5 | 8 | 9 | 10 | 11 | 12 | 1 | 2 | 3 | 4 | 5 | | | | |
| MEASUREMENT 1 | 5.4 | 3.6 | 9.4 | -2.4 | 9.4 | 4.4 | 14.6 | 3.2 | -5.6 | 12.1 | -11.8 | -3.2 | 11.8 | 10.7 | 3.4 | 3.9 | 13.8 | 2.5 | 10.6 | 4.3 | | | | |
| MEASUREMENT 2 | 4.1 | 6.1 | 12.4 | 0.4 | 12.4 | 4.7 | 18.0 | 2.4 | -8.2 | 15.6 | -16 | -0.8 | 8.2 | 7.3 | 5.3 | 5.8 | 11.0 | 1.3 | 13.4 | 3.6 | | | | |
| 3 | | | | | | | | | | | | | | | | | | | | | | | | |
| RANGE 4 | 1.3 | 2.5 | 3.0 | 2.8 | 3.0 | 0.3 | 3.4 | 0.8 | 2.6 | 3.5 | 4.2 | 2.4 | 3.4 | 3.4 | 1.9 | 1.9 | 2.8 | 1.2 | 2.8 | 0.7 | | | | |
| 5 | | | | | | | | | | | | | | | | | | | | | | | | |
| 6 | | | | | | | | | | | | | | | | | | | | | | | | |
| MASTER | 5.4 | 4.3 | 2.3 | 4.6 | 3.4 | 5.6 | 3.9 | 5.7 | 5.3 | 4.5 | 6.3 | 3.7 | 3.4 | 4.0 | 5.9 | 5.2 | 8.1 | 4.9 | 6.4 | 6.7 | | | | |
| MR. | – | 1.1 | 2.0 | 2.3 | 1.2 | 2.2 | 1.7 | 1.8 | .4 | 1.2 | 1.8 | 2.6 | .3 | .6 | 1.9 | .7 | 2.9 | 3.2 | 1.5 | .3 | | | | |
| Subgroup identifier | | | | | | | | | | | | | | | | | | | | | | | | |



**Figure 10.8** Control chart for width measurement process.

2. Notify management that most of the bricks being produced have a rougher finish than the microfinish specification allows and that about one-half the bricks are outside the width specifications.
3. Have the maintenance person investigate alternative width gages with better precision for future use.

Next, the team wanted to plan a study to learn about the important sources of variation in the width of the milled bricks.

**TABLE 10.1 Analysis of Precision of Width Measurement**

Variation for process

$$\bar{R} = 26.4 \qquad \sigma_p = \frac{\bar{R}}{d_2} = \frac{26.40}{2.059} = 12.8$$

Variation of measurement process

$$\bar{R}_m = 2.4 \qquad \sigma_m = \frac{\bar{R}_m}{d_2} = \frac{2.4}{1.128} = 2.1$$

Variation of product

$$\sigma_{\text{product}} = \sqrt{(\sigma_{\text{process}})^2 - (\sigma_{\text{measurement}})^2}$$

$$\sigma_{\text{product}} = \sqrt{(12.8)^2 - (2.1)^2} = 12.6$$

Summary of variation (units = 0.001 in)

| Source of variation | Standard deviation $\sigma$ | Variance component $\sigma^2$ | Percentage of variation |
|---|---|---|---|
| Product | 12.6 | 158.8 | 97.3 |
| Measurement | 2.1 | 4.4 | 2.7 |
| Total (process) | 12.8 | 163.2 | 100.0 |

### Second improvement cycle: Sources of variation

The objective of the second improvement cycle was to determine the important sources of variation in the mill process. Figure 10.4 indicated that the mill process consists of the following components:

1. Two operators
2. Two mills for each operator
3. Two spindles within each mill
4. Two positions within each spindle

The team felt that the operators and the spindles within the mills would be the most important sources of variation for width. For microfinish, the team thought that the positions within each spindle would contribute most of the variation. A nested study was planned to evaluate the four sources of variation in the process relative to the part-to-part variation within a spindle.

The experiment was performed the next afternoon by selecting three parts from each position within each spindle for each mill and measuring the parts for width and microfinish. Figure 10.9 contains dot-frequency diagrams for both the microfinish and the width data. For microfinish, none of the factors studied appears to be an important

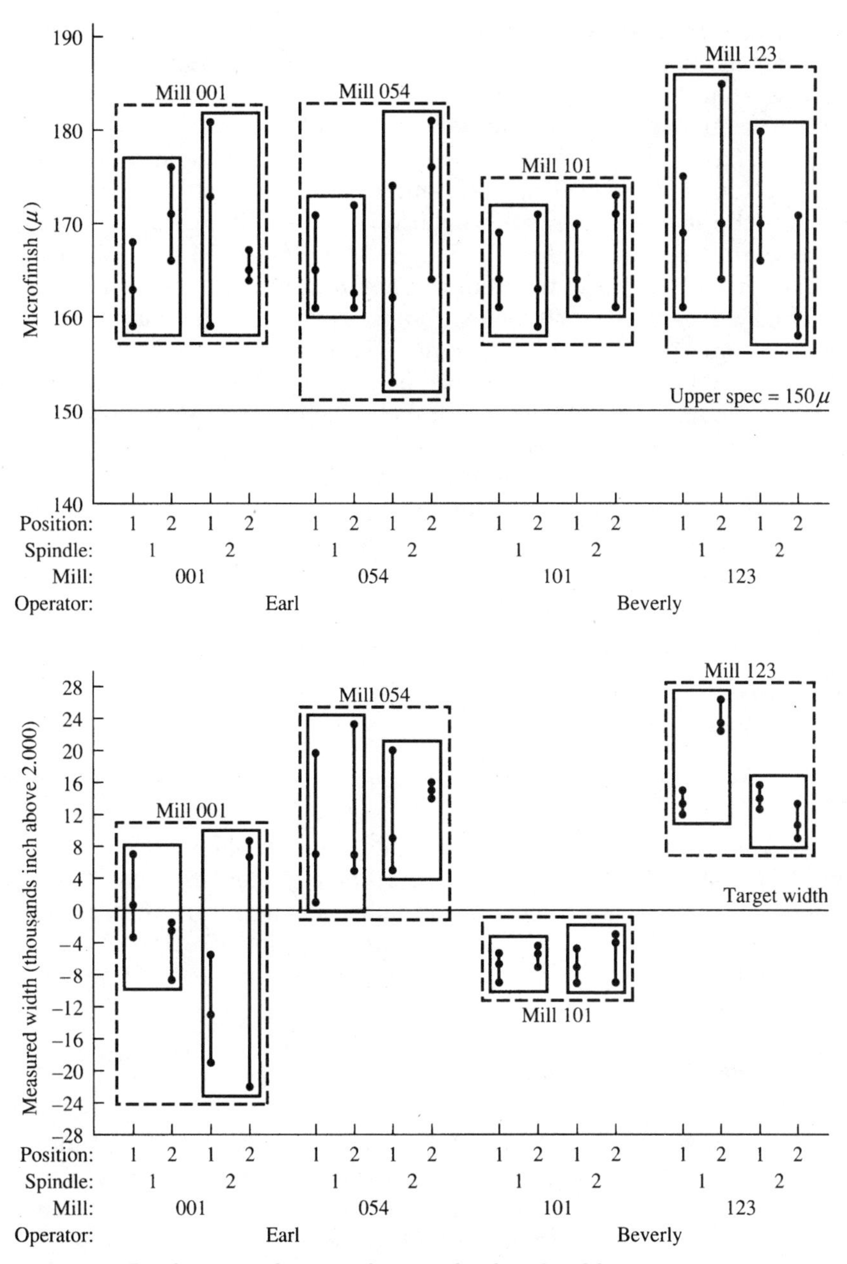


**Figure 10.9** Dot-frequency diagrams for microfinish and width.

source of variation. Most of the variation is attributable to the variation from part to part within a position. None of the parts in the study were less than the upper specification of 150 μ. The problems with microfinish seem to be common throughout the mill process.

The dot-frequency diagram for width contains information about the variation in the width measurements. Differences between the mills are the biggest source of variation. These differences inflated the range on the control chart during cycle 1. Mills 054 and 123 are producing parts with greater than the nominal width, while mill 101 is making all parts slightly below nominal width. The positions for spindle 1, mill 123 need to be aligned. The variation of parts from spindle 2, mill 001 and spindle 1, mill 054 seem to be greater than the variation of parts from the other spindles. The dot-frequency diagram was used to develop a repair/adjustment plan for the mills.

The actions taken in this cycle were primarily designed to reduce the variation in the width measurements. The contract maintenance person scheduled the following maintenance activities:

| Mill | Activity |
|---|---|
| 101 | Adjust up 0.006 in |
| 001 | Repair spindle 2 |
| 054 | Repair spindle 1 |
| | Adjust down 0.014 in |
| 123 | Rebuild spindle 1 |
| | Adjust down 0.012 in |

These activities were expected to take 2 days to complete.

While the maintenance activities were being completed, the team wanted to run studies on mill 101 to learn about factors that cause the microfinish to be so high. Also, plans were discussed to develop control charts that could be used to determine the need for future maintenance.

### Third improvement cycle: Evaluating mill cutter vendors

Mid-State Brick Company currently bought mill cutters from five different vendors. When a cutter broke, it was replaced with whatever brand was available. The operators thought that the type of cutter could impact both the microfinish and the width variation, but they disagreed on which brand gave the best results. A one-factor study was planned to evaluate the five types of cutters.

The study would be run on mill 101 (after adjustment) during the next morning. One at a time, each of the five cutters would be put on mill 101 during the morning. After about 10 parts were milled with the new cutter, three subgroups of four bricks each would be measured for both microfinish and width. The average microfinish and the range of the four width measurements would be analyzed to evaluate the cutters. At the end of the study, another cutter of the type that appeared to perform best would be installed and tested.

Figure 10.10 contains the results of the study in the third cycle. The Brite cutter had the best overall performance for both microfinish and width variation. When a second Brite cutter was tested, the results were similar. Even with the Brite cutter, though, the microfinish results averaged above 160 μ. The average range for width with the Brite cutter was about 0.004 in.

The primary action from this cycle was to use the Brite cutter for all further tests. The supervisor agreed to schedule a meeting with the sales representative of the Brite cutter and to obtain information for all the cutters from the purchasing group.

### Fourth improvement cycle: Screening process variables

For the fourth improvement cycle, the team designed a study to determine which process variables could be used to reduce the microfinish. Using the process flowchart (Fig. 10.4) and the cause-and-effect diagram (Fig. 10.5), the team identified seven variables it thought could affect the microfinish. Table 10.2 summarizes these potential factors.

Since team members believed that only two or three of the factors would have a substantial effect on the microfinish, they considered themselves to have a low level of knowledge. Thus a screening design (fractional factorial) was selected to eliminate the unimportant factors. The $2^{7-4}$ fractional factorial design was selected to accommodate the seven factors in Table 10.2.

Selecting the levels for each factor was a difficult task for the team. The plant engineer wanted to choose the levels near the extremes of the possible settings of the factors. Both operators were concerned about possible damage to the mills from running near the extremes for some of the factors. After discussion and debate, the team agreed on the following levels for the factors:

| | Levels | |
|---|---|---|
| Factor | – | + |
| 1. Mill rpm $R$ | 300 | 700 |
| 2. Feed rate $F$ | 4 | 10 |
| 3. Vise pressure $V$ | 200 | 600 |
| 4. Coolant pressure $P$ | 10 | 80 |
| 5. Coolant concentration $C$ | 5 | 25 |
| 6. Air pressure $A$ | 0 | 0.01 |
| 7. Movement $M$ | Conventional | Climb |

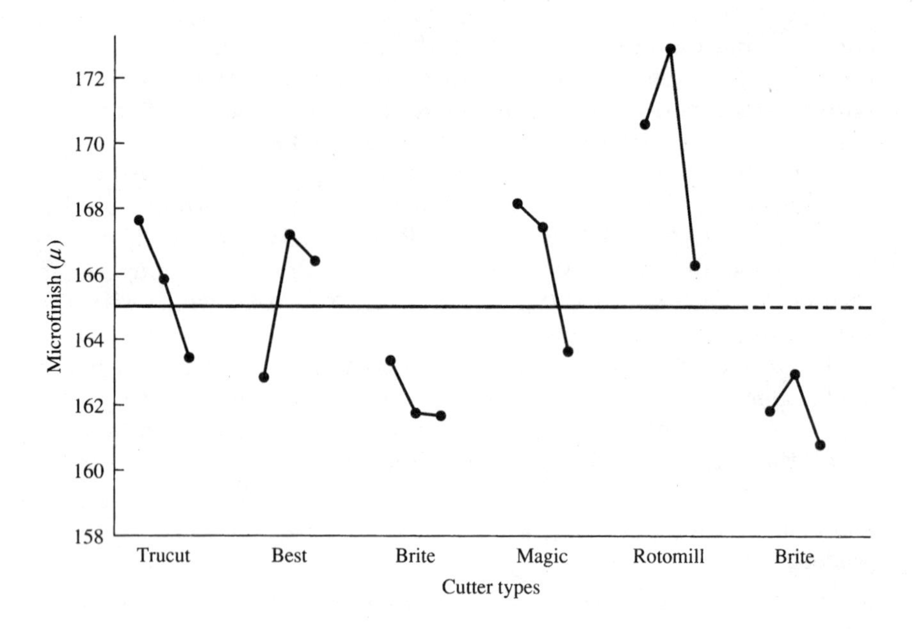


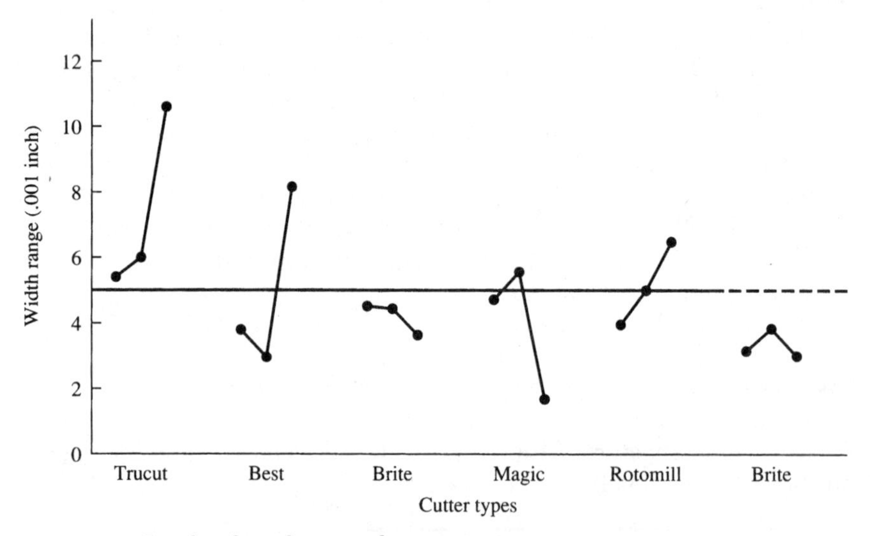


**Figure 10.10** Results of one-factor study on cutter types.

The study would be run on mill 101 using the Brite blade. After setting up the mill at the conditions in the design matrix, five parts would be run. Then two subgroups of four bricks each would be run and measured for microfinish and width. Although the focus of the study was on the microfinish, the team wanted to continue to evaluate the width variation. The average microfinish and the range of the width measurements would be the response variables for the study.

**TABLE 10.2 Possible Factors for the Mill Process**

| Factor | Description of factors | Range |
|---|---|---|
| Mill rpm | The revolutions per minute that the mill cutter turns. | (10–900) |
| Feed rate | The rate that the mill cutter travels through the part in inches per minute. | (0.00001–20) |
| Vise pressure | The amount of force that is clamping the part in pounds per square inch. | (1.5–950) |
| Coolant pressure | The pressure of the coolant flow on the part in pounds per square inch. | (2–100) |
| Coolant concentration | The amount of coolant concentration/ amount of water. This is stated in percent. | (0.00001–35) |
| Air pressure | The amount of air pressure blowing on the part while it is being milled. | (0–0.125) |
| Movement | The movement of the cutter into the part. | — |
| Conventional | Pushes the part into the vise during the cutting process. | — |
| Climb | Tries to pull the part out of the vise while it is cutting. | — |

The study was to be completed on the morning of the fifth day. During the fifth test, six cutters were broken. Figure 10.11 shows the results of the study for microfinish. Three factors were found to be important: mill rpm, feed rate, and coolant concentration. From the tests completed in this study, these three factors formed a full factorial design. The analysis of the cube in Fig. 10.11 indicated no interaction between the factors. Best microfinish could be achieved at high rpm, high coolant concentration, and low feed rate.

Figure 10.12 shows the results for width. Feed rate and possibly mill rpm were found to affect the width variation. The effect of changes in feed rate was greater at the low rpm setting. To minimize the width variation, low feed rates and high rpm settings were desirable. This is the same direction as the best conditions for microfinish.

The following actions were taken after this cycle:

1. Set the coolant concentration at 25 percent. The coolant was controlled through a central system that affected all the mills.
2. Set the feed rate at 5 in/min for each mill.
3. Set the mill rpm at 700 for each mill.

After maintenance was completed on the mills, the team planned to

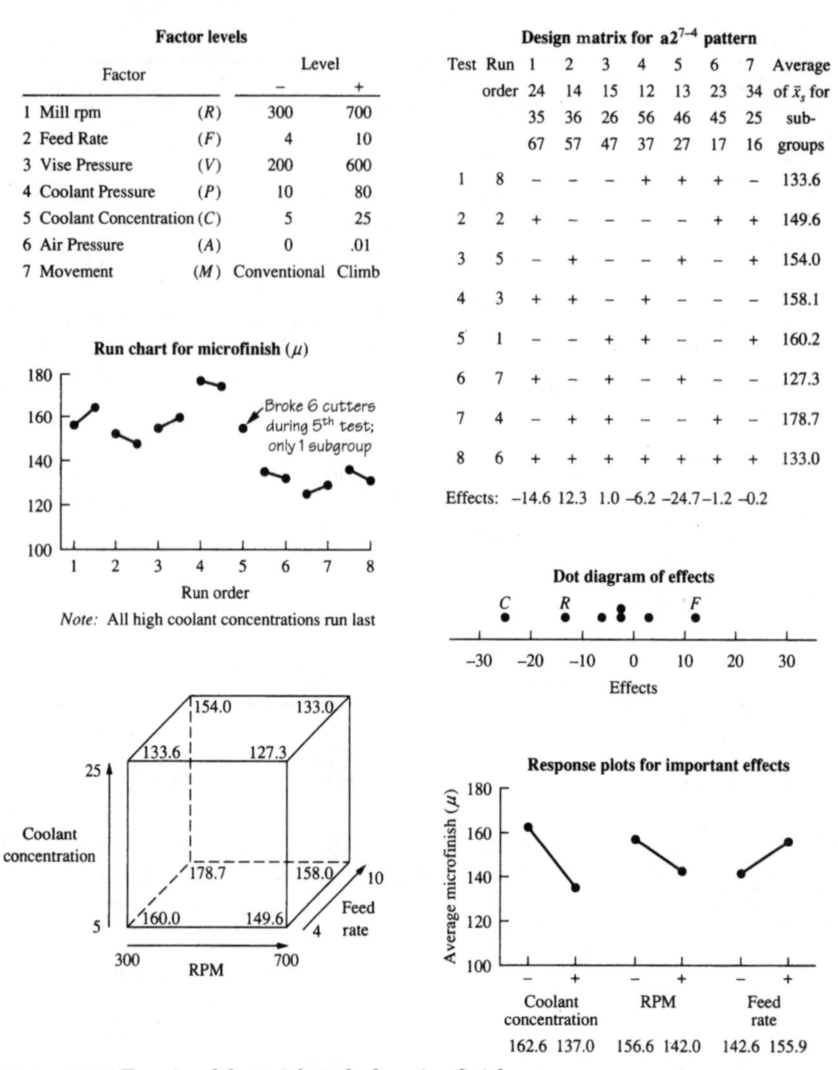

**Factor levels**

| Factor | | Level − | Level + |
|---|---|---|---|
| 1 Mill rpm | (R) | 300 | 700 |
| 2 Feed Rate | (F) | 4 | 10 |
| 3 Vise Pressure | (V) | 200 | 600 |
| 4 Coolant Pressure | (P) | 10 | 80 |
| 5 Coolant Concentration | (C) | 5 | 25 |
| 6 Air Pressure | (A) | 0 | .01 |
| 7 Movement | (M) | Conventional | Climb |

**Design matrix for a$2^{7-4}$ pattern**

| Test | Run order | 1 (24, 35, 67) | 2 (14, 36, 57) | 3 (15, 26, 47) | 4 (12, 56, 37) | 5 (13, 46, 27) | 6 (23, 45, 17) | 7 (34, 25, 16) | Average of $\bar{x}_s$ for sub-groups |
|---|---|---|---|---|---|---|---|---|---|
| 1 | 8 | − | − | − | + | + | + | − | 133.6 |
| 2 | 2 | + | − | − | − | − | + | + | 149.6 |
| 3 | 5 | − | + | − | − | + | − | + | 154.0 |
| 4 | 3 | + | + | − | + | − | − | − | 158.1 |
| 5 | 1 | − | − | + | + | − | − | + | 160.2 |
| 6 | 7 | + | − | + | − | + | − | − | 127.3 |
| 7 | 4 | − | + | + | − | − | + | − | 178.7 |
| 8 | 6 | + | + | + | + | + | + | + | 133.0 |
| Effects: | | −14.6 | 12.3 | 1.0 | −6.2 | −24.7 | −1.2 | −0.2 | |



**Figure 10.11** Fractional factorial study for microfinish.

determine the capability for the mill process. Plans were also made to evaluate a new width gage that the supervisor had obtained.

## Fifth improvement cycle: Evaluate effect of improvements

Control charts were planned to evaluate the status of the mill process after the maintenance and other process changes had been made. The subgroups were set up to allow an assessment of each of the mills. During the next day, every 30 min one of the mill operators selected a

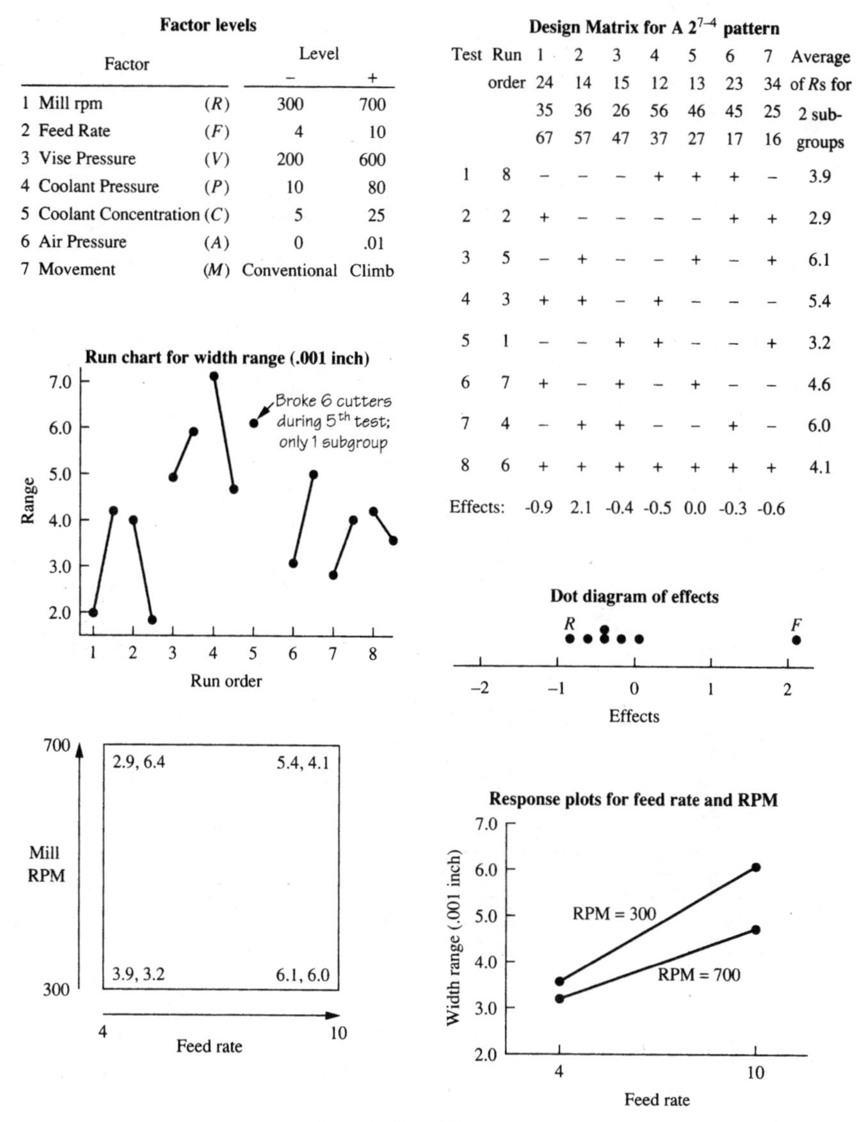

**Factor levels**

| Factor | | Level − | Level + |
|---|---|---|---|
| 1 Mill rpm | (*R*) | 300 | 700 |
| 2 Feed Rate | (*F*) | 4 | 10 |
| 3 Vise Pressure | (*V*) | 200 | 600 |
| 4 Coolant Pressure | (*P*) | 10 | 80 |
| 5 Coolant Concentration | (*C*) | 5 | 25 |
| 6 Air Pressure | (*A*) | 0 | .01 |
| 7 Movement | (*M*) | Conventional | Climb |

**Design Matrix for A $2^{7-4}$ pattern**

| Test | Run order | 1<br>24<br>35<br>67 | 2<br>14<br>36<br>57 | 3<br>15<br>26<br>47 | 4<br>12<br>56<br>37 | 5<br>13<br>46<br>27 | 6<br>23<br>45<br>17 | 7<br>34<br>25<br>16 | Average of *R*s for 2 subgroups |
|---|---|---|---|---|---|---|---|---|---|
| 1 | 8 | − | − | − | + | + | + | − | 3.9 |
| 2 | 2 | + | − | − | − | − | + | + | 2.9 |
| 3 | 5 | − | + | − | − | + | − | + | 6.1 |
| 4 | 3 | + | + | − | + | − | − | − | 5.4 |
| 5 | 1 | − | − | + | + | − | − | + | 3.2 |
| 6 | 7 | + | − | + | − | + | − | − | 4.6 |
| 7 | 4 | − | + | + | − | − | + | − | 6.0 |
| 8 | 6 | + | + | + | + | + | + | + | 4.1 |
| Effects: | | -0.9 | 2.1 | -0.4 | -0.5 | 0.0 | -0.3 | -0.6 | |



**Figure 10.12** Fractional factorial study for width.

subgroup of four bricks from one of the mills. The operators coordinated the sampling so that the four mills were rotated. The supervisor conducted a study of the new width gage following the design used in cycle 1.

Figure 10.13 shows the control charts for the width measurements. Both the average chart and the range chart indicated special causes. The measurements from mill 001 were consistently high. This mill

| Chart Number | Chart Name | | |
|---|---|---|---|
| 4 | Evaluation Mill Process-width | | |
| Objective of Chart | | | Subgrouped by |
| Learn about current capability for width after maintenance and improvements | | | Mill/Time |
| Process | Product. Service. Operation. Or Part Name | Target: Specification Limits | |
| Mill-each subgroup from a mill | Bricks | | |

| Chart Responsibility | Characteristic | Measurement Method | Unit of Measure | Zero Equals |
|---|---|---|---|---|
| Operators | Width | GW-097 | .001 inch | 2.000 |

| | 1 | 2 | 3 | 4 | 5 | 6 | 7 | 8 | 9 | 10 | 11 | 12 | 13 | 14 | 15 | 16 | 17 | 18 | 19 | 20 |
|---|---|---|---|---|---|---|---|---|---|---|---|---|---|---|---|---|---|---|---|---|
| Date | 6 | | | | | | | | | | | | | | | | | | | |
| Time | 8 | | 9 | | 10 | | 11 | | 12 | | 1 | | 2 | | 3 | | 4 | | 5 | |
| Measurements 1 | 4.9 | 2.2 | -0.7 | 0.8 | 7.4 | -2.0 | 0.0 | -0.5 | 8.7 | 0.1 | -2.8 | -3.0 | 5.4 | 0.3 | 3.5 | -2.0 | 4.5 | 4.0 | -1.9 | -2.1 |
| 2 | 10.4 | 1.4 | -2.4 | 9.4 | 6.4 | -1.1 | -4.8 | 15.6 | 4.9 | -1.7 | -1.7 | 4.3 | 6.7 | -0.7 | 1.6 | 13.2 | 7.6 | -0.4 | 1.6 | 19.7 |
| 3 | -2.4 | -1.0 | -0.8 | -2.6 | 10.3 | -2.7 | 5.4 | -4.4 | 10.5 | 2.7 | 2.0 | 2.0 | 8.3 | 2.2 | -2.9 | 1.3 | -7.8 | -0.2 | 1.2 | -2.3 |
| 4 | 4.8 | -2.7 | -2.2 | -2.0 | 1.1 | 0.6 | -2.9 | 0.6 | -1.3 | 0.4 | 3.2 | -6.7 | 15.5 | 1.1 | -2.8 | -1.9 | 0.5 | -1.1 | 1.8 | -6.2 |
| 5 | | | | | | | | | | | | | | | | | | | | |
| 6 | | | | | | | | | | | | | | | | | | | | |
| Average | 4.4 | 0.0 | -1.5 | 1.4 | 6.3 | -1.3 | -0.5 | 2.8 | 5.7 | 0.4 | 0.2 | -0.8 | 9.0 | 0.7 | -0.2 | 2.6 | 1.2 | 0.6 | 0.7 | 2.3 |
| Range P. | 12.8 | 4.9 | 1.7 | 12 | 9.2 | 3.3 | 10.2 | 20 | 11.9 | 4.4 | 6.0 | 11 | 10.1 | 2.9 | 6.3 | 15.2 | 15.5 | 5.2 | 3.7 | 25.9 |
| Mill | 001 | 054 | 101 | 123 | 001 | 054 | 101 | 123 | 001 | 054 | 101 | 123 | 001 | 054 | 101 | 123 | 001 | 054 | 101 | 123 |

Averages

Ranges

NOTES: $\bar{\bar{X}} = 1.7$ UCL = 8.7 LCL = −5.3

$\bar{R} = 9.6$ $UCL_R = 21.9$

**Figure 10.13** Control chart for width after improvements.

had been repaired after the third cycle, and was probably in need of adjustment. The other mills appeared to be averaging near the nominal width. There was an indication of a special cause on the range chart for mill 123. All five of the ranges for this mill were above the centerline. Spindle 1 on this mill had been rebuilt after the third cycle.

Figure 10.14 shows the control charts for the microfinish measurements. The process was stable during day 6. A capability analysis shows great improvement—a very small percentage of the bricks are expected to be greater than the specification of 150 μ.

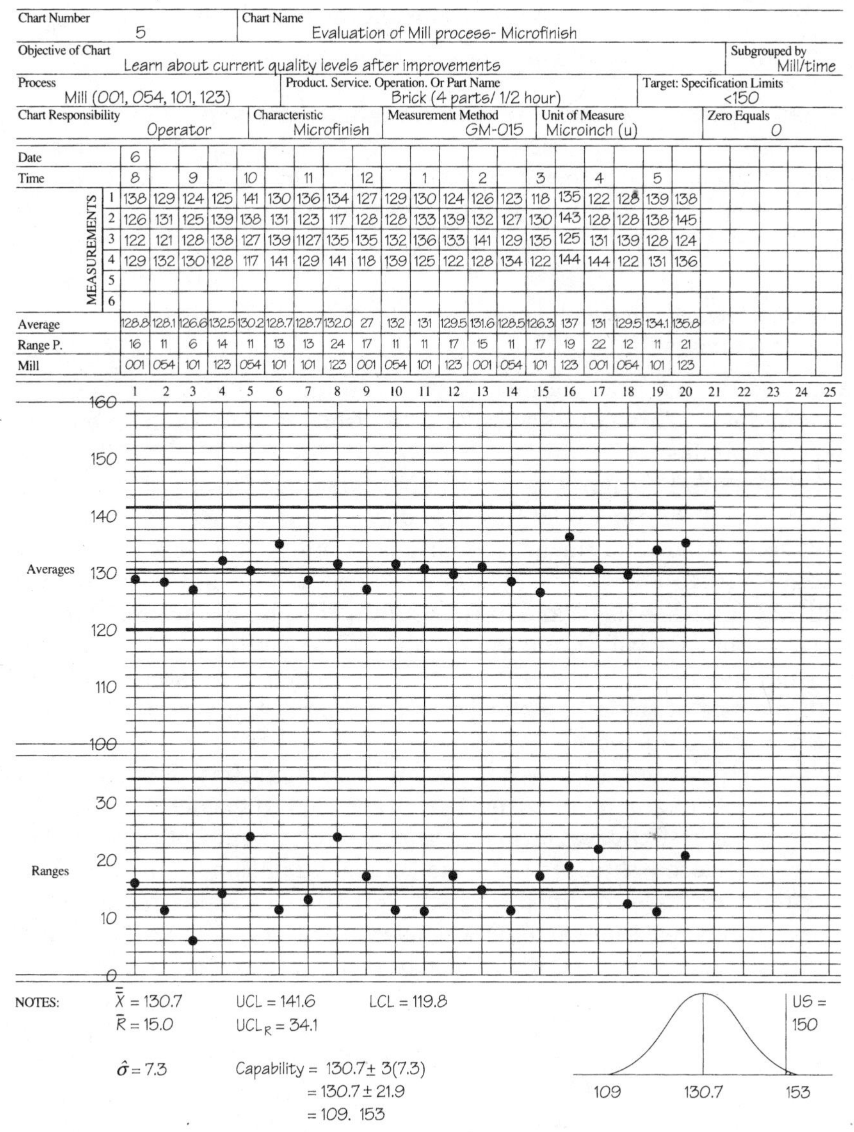

| Chart Number: 5 | Chart Name: Evaluation of Mill process- Microfinish | |
|---|---|---|
| Objective of Chart: Learn about current quality levels after improvements | | Subgrouped by: Mill/time |
| Process: Mill (001, 054, 101, 123) | Product. Service. Operation. Or Part Name: Brick (4 parts/ 1/2 hour) | Target: Specification Limits: <150 |

| Chart Responsibility | Characteristic | Measurement Method | Unit of Measure | Zero Equals |
|---|---|---|---|---|
| Operator | Microfinish | GM-015 | Microinch (u) | 0 |

| | 1 | 2 | 3 | 4 | 5 | 6 | 7 | 8 | 9 | 10 | 11 | 12 | 13 | 14 | 15 | 16 | 17 | 18 | 19 | 20 |
|---|---|---|---|---|---|---|---|---|---|---|---|---|---|---|---|---|---|---|---|---|
| Date | 6 | | | | | | | | | | | | | | | | | | | |
| Time | 8 | | 9 | | 10 | | 11 | | 12 | | 1 | | 2 | | 3 | | 4 | | 5 | |
| Measurements 1 | 138 | 129 | 124 | 125 | 141 | 130 | 136 | 134 | 127 | 129 | 130 | 124 | 126 | 123 | 118 | 135 | 122 | 128 | 139 | 138 |
| 2 | 126 | 131 | 125 | 139 | 138 | 131 | 123 | 117 | 128 | 128 | 133 | 139 | 132 | 127 | 130 | 143 | 128 | 128 | 138 | 145 |
| 3 | 122 | 121 | 128 | 138 | 127 | 139 | 1127 | 135 | 135 | 132 | 136 | 133 | 141 | 129 | 135 | 125 | 131 | 139 | 128 | 124 |
| 4 | 129 | 132 | 130 | 128 | 117 | 141 | 129 | 141 | 118 | 139 | 125 | 122 | 128 | 134 | 122 | 144 | 144 | 122 | 131 | 136 |
| 5 | | | | | | | | | | | | | | | | | | | | |
| 6 | | | | | | | | | | | | | | | | | | | | |
| Average | 128.8 | 128.1 | 126.6 | 132.5 | 130.2 | 128.7 | 128.7 | 132.0 | 27 | 132 | 131 | 129.5 | 131.6 | 128.5 | 126.3 | 137 | 131 | 129.5 | 134.1 | 135.8 |
| Range P. | 16 | 11 | 6 | 14 | 11 | 13 | 13 | 24 | 17 | 11 | 11 | 17 | 15 | 11 | 17 | 19 | 22 | 12 | 11 | 21 |
| Mill | 001 | 054 | 101 | 123 | 054 | 101 | 101 | 123 | 001 | 054 | 101 | 123 | 001 | 054 | 101 | 123 | 001 | 054 | 101 | 123 |



**Figure 10.14** Control chart for microfinish after improvements.

The study of the width gage (GW-097) showed no bias and better precision ($\sigma$ = 1.3 in thousandths of an inch) than the previous gage. Since the range chart for the width was not in control, this could not be compared to the process variation.

The following actions were taken after this cycle:

1. Adjust mill 001 down by 5.5. thousandths of an inch.
2. Adopt the new width gage for all further measurements.

Plans for the next cycle were to run a nested design on mill 123 to isolate the variation problem and then to schedule appropriate maintenance to correct the problems. Also, the team planned to conduct a study to optimize the important factors identified during the fourth cycle.

### Sixth improvement cycle: Evaluating important factors

A nested design was planned to evaluate mill 123 during the first hour of day 7. Three parts were selected from each position on each of the two spindles.

A $2^3$ factorial design was used to optimize the factors identified as important in the fractional factorial design in the fourth cycle. Levels were selected around the best conditions from the previous study.

Figure 10.15 shows the results of the nested design on width for mill 123. Figures 10.16 and 10.17 contain the summaries for the factorial studies for width and microfinish. The results of the nested design for mill 123 indicated that position 1 on spindle 1 was out of line with the rest of the mill. After the rebuilding of this position, the entire mill was adjusted up 0.003 in.

The factorial design was run on mills 101 and 054. For each test condition, a subgroup of four parts was selected from each of the mills. The analysis of the variation in the measurements of width (Fig. 10.16) indicated that mill 101 had more variation than mill 054. Maintenance was scheduled to check mill 101 for needed repairs. The average range for the two mills was used as the response variable. The analysis of the effects indicated that the feed rate and the rpm coolant

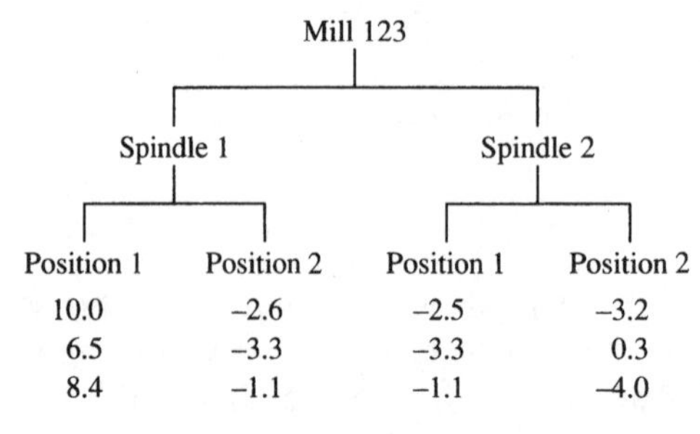


**Figure 10.15** Nested study for mill 123.

**Three factor levels for width**

| Factor | Level – | Level + |
|---|---|---|
| 1 Mill rpm | 500 | 800 |
| 2 Feed Rate | 2 | 8 |
| 3 Coolant Concentration ($C$) | 20 | 30 |

**Design Matrix: three factor factorial study of width** / **Width range**

| Test | Run order | 1 | 2 | 3 | 12 | 13 | 23 | 123 | Mill 101 | Mill 054 | Average |
|---|---|---|---|---|---|---|---|---|---|---|---|
| 1 | 4 | – | – | – | + | + | + | – | 5.3 | 1.2 | 3.3 |
| 2 | 2 | + | – | – | – | – | + | + | 2.7 | 1.5 | 2.1 |
| 3 | 3 | – | + | – | – | + | – | + | 5.8 | 2.6 | 4.2 |
| 4 | 1 | + | + | – | + | – | – | – | 4.5 | 1.5 | 3.0 |
| 5 | 7 | – | – | + | + | – | – | + | 3.4 | 2.9 | 3.2 |
| 6 | 5 | + | – | + | – | + | – | – | 4.6 | 3.2 | 3.9 |
| 7 | 8 | – | + | + | – | – | + | – | 4.8 | 2.3 | 3.6 |
| 8 | 6 | + | + | + | + | + | + | + | 6.8 | 4.8 | 5.8 |
| Effects: | | .13 | 1.03 | .98 | .38 | 1.3 | .13 | .38 | | | |

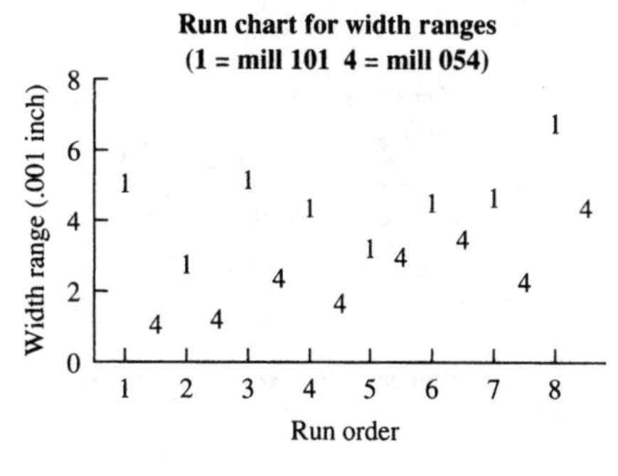


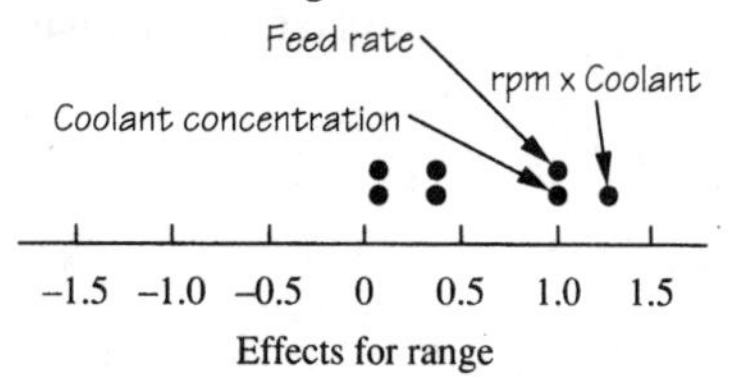

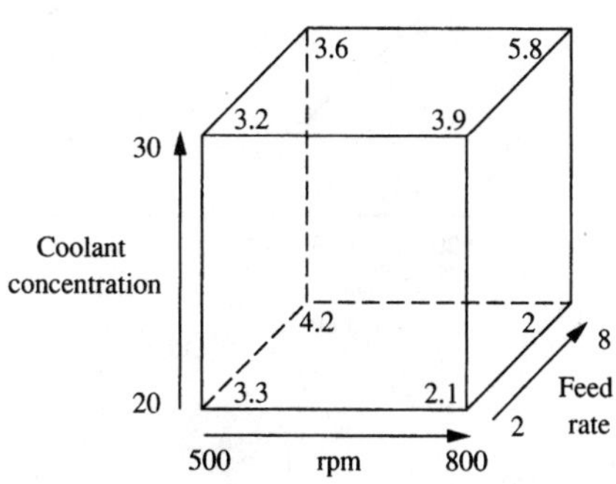


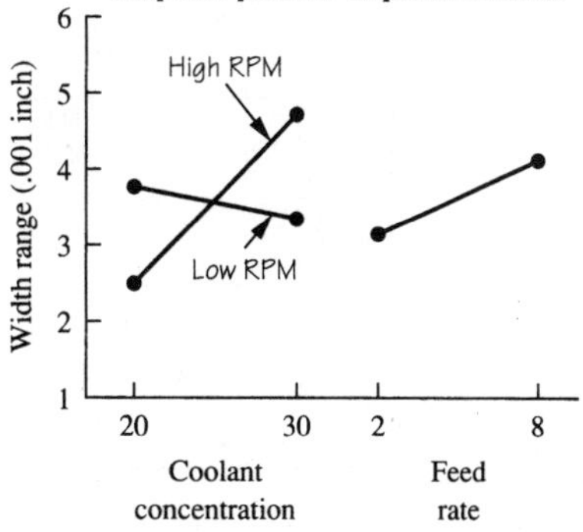


**Figure 10.16** Factorial design for width.

**Three factor levels for width**

| Factor | Level – | Level + |
|---|---|---|
| 1 Mill rpm | 500 | 800 |
| 2 Feed Rate | 2 | 8 |
| 5 Coolant Concentration ($C$) | 20 | 30 |

**Design Matrix: three factor factorial study of width** / **Microfinish average ($\mu$)**

| Test | Run order | 1 | 2 | 3 | 12 | 13 | 23 | 123 | Mill 101 | Mill 054 | Average |
|---|---|---|---|---|---|---|---|---|---|---|---|
| 1 | 4 | – | – | – | + | + | + | – | 132.3 | 130.5 | 131.4 |
| 2 | 2 | + | – | – | – | – | + | + | 130.8 | 134.0 | 132.4 |
| 3 | 3 | – | + | – | – | + | – | + | 142.9 | 140.2 | 141.6 |
| 4 | 1 | + | + | – | + | – | – | – | 133.0 | 137.0 | 135.0 |
| 5 | 7 | – | – | + | + | – | – | + | 123.7 | 128.1 | 125.9 |
| 6 | 5 | + | – | + | – | + | – | – | 118.8 | 121.5 | 120.2 |
| 7 | 8 | – | + | + | – | – | + | – | 132.2 | 131.8 | 132.0 |
| 8 | 6 | + | + | + | + | + | + | + | 131.4 | 129.9 | 130.7 |
| Effects: | | –3.2 | 7.4 | –8.0 | –.8 | –.4 | 1.0 | 3.1 | | | |

**Run chart for microfinish range**
**(1 = Mill 101 4 = Mill 054)**

Average microfinish (μ) vs. Run order

| Run order | 1 | 2 | 3 | 4 | 5 | 6 | 7 | 8 |
|---|---|---|---|---|---|---|---|---|

**Dot diagram of effects**

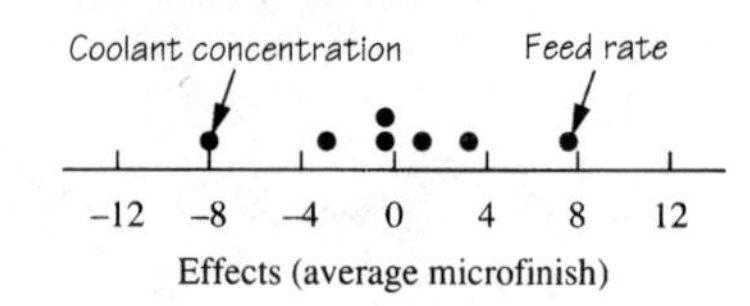


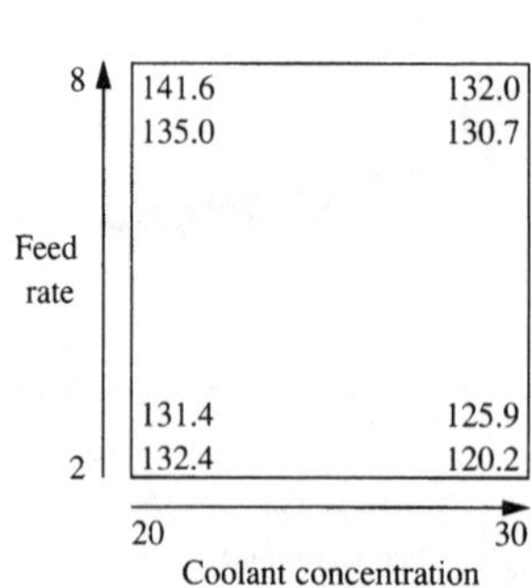


**Response plots for important effects**

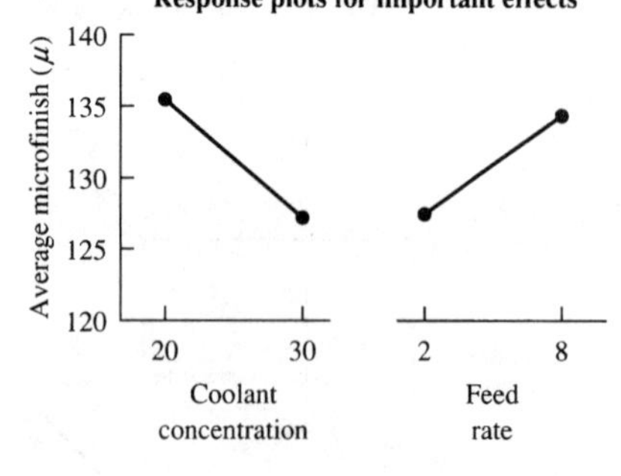


**Figure 10.17** Factorial design for microfinish.

concentration interaction were the most important effects. Low coolant levels were desirable when running at high rpm, while high coolant levels were preferred when running at low rpm. The variation in width increased as the feed rate was increased from 2 to 8 in/min.

The analysis of the microfinish confirmed the effects found in the earlier fractional factorial design for coolant concentration and feed rate. The increase in coolant above 25 percent showed a continuing decrease in microfinish. The effect of feed rate was the same as that of the previous study. If kept in the range of 500 to 800, rpm did not have an important effect on microfinish.

The following actions were taken as a result of this cycle:

1. Increase the coolant concentration to 35 percent. This should further improve the microfinish.
2. Set the rpm to 500. This should help the width variation while running at the high coolant level.

The team next planned to study the effect of costs at different feed rate levels. Since the feed rate directly affected the productivity of the mill process, the cost to the mill should be considered when the feed rate is set.

#### Seventh improvement cycle: Determining optimum levels

The supervisor asked the plant accountant to develop cost estimates in the mill process for running at feed rates from 2 to 10 in/min. The team planned a one-factor study to evaluate the variation of the width measurements and the microfinish average. One subgroup would be taken from each mill at feed rates of 2, 4, 6, 8, and 10 in/min.

The results of the study are presented in Table 10.3. The analysis of the data is shown in Fig. 10.18.

Based on this analysis, the team decided to set the feed rate at 5 in/min. This should give a cost to mill of about $2.50 per brick and should produce bricks that meet the quality specifications for both microfinish and width. The team felt that a final cycle could be run to demonstrate the performance of the process and to test a control chart plan for continuous monitoring of the mill process.

#### Eighth improvement cycle: Confirmation of improvements

After the feed rate was set to 5 in/min, process conditions were checked for each of the mills. Control charts were planned to be developed on the next day for both microfinish and width. Subgroups of

**TABLE 10.3 One-Factor Study to Evaluate Effect of Feed Rate**

| Feed rate (in/min) | Mill 001 | Mill 054 | Mill 101 | Mill 123 | Average | Cost per brick |
|---|---|---|---|---|---|---|
| | Width Variation (Range for Subgroup of 4; Unit = 0.001 in) | | | | | |
| 2 | 5.6 | 4.2 | 1.7 | 2.8 | 3.60 | $5.59 |
| 3 | | | | | | 4.00 |
| 4 | 3.9 | 1.6 | 1.3 | 3.7 | 2.63 | 3.14 |
| 6 | 3.6 | 6.0 | 1.7 | 5.8 | 4.28 | 2.29 |
| 8 | 6.6 | 5.9 | 2.0 | 5.6 | 5.03 | 1.90 |
| 10 | 3.3 | 4.6 | 3.9 | 7.5 | 4.83 | 1.66 |
| | Microfinish (Average for Subgroup of 4; Unit = microinches) | | | | | |
| 2 | 118.9 | 125.6 | 128.6 | 120.1 | 123.3 | $5.59 |
| 3 | | | | | | 4.00 |
| 4 | 126.7 | 129.8 | 130.1 | 128.7 | 128.8 | 3.14 |
| 6 | 127.2 | 130.1 | 128.4 | 131.4 | 129.3 | 2.29 |
| 8 | 136.2 | 135.1 | 132.8 | 138.3 | 135.6 | 1.90 |
| 10 | 134.6 | 134.9 | 137.2 | 140.4 | 136.8 | 1.66 |

four bricks would be selected from one of the mills every 30 min. One brick would be selected from each of the four positions on the mill. The position and mill would be recorded on the chart for each brick. Thus, if a special cause were found on the chart, the particular problem spindle or position could be identified. The operators would each take responsibility for one of the charts.

Figures 10.19 (microfinish) and 10.20 (width) show the control charts developed during day 9. Both charts indicate that the process is stable. Capability analyses for each of the quality characteristics indicated that all bricks milled are expected to meet the specifications.

#### Final actions of the team

The team reviewed its charter (Fig. 10.3) and believed it had successfully accomplished all the expected results. The operators would continue to use the control charts developed during the last (eighth) improvement cycle to monitor the mill process. Since the process was operating so well, the frequency of subgroups was reduced to one per hour. This would allow a check on each mill twice during a normal 8-hour shift.

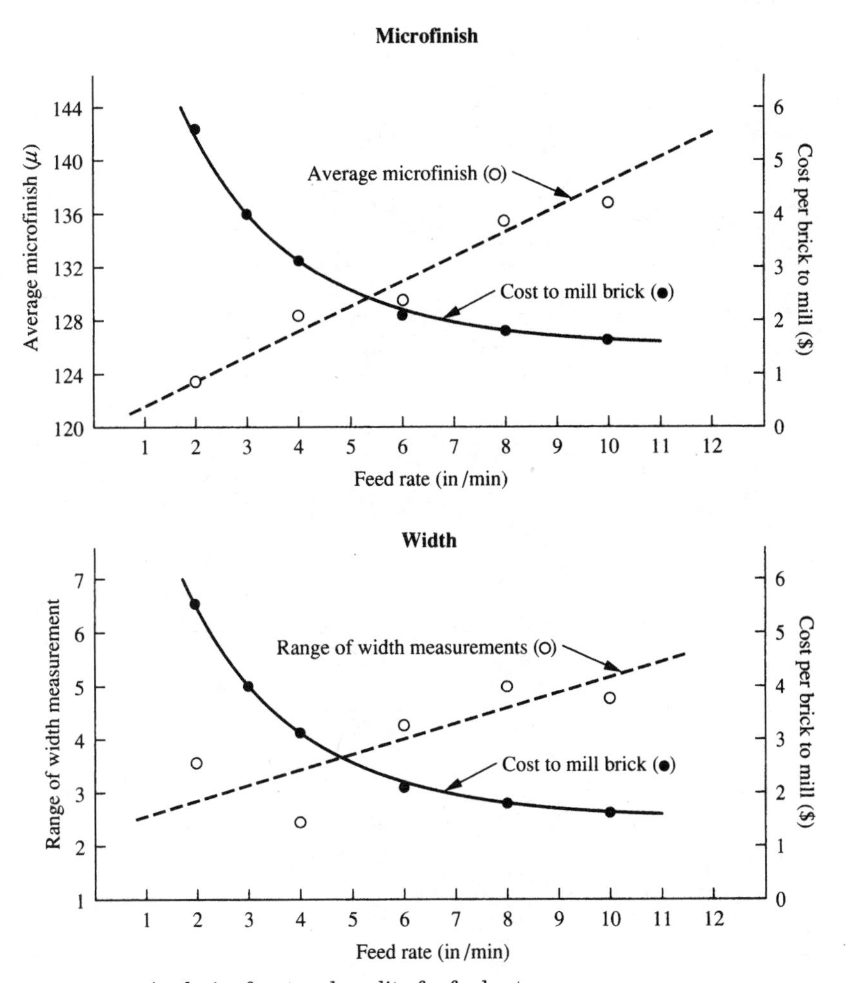


**Figure 10.18** Analysis of cost and quality for feed rate.

Team members were looking forward to sharing their experiences in improving the mill process with the other operations in the plant. A meeting was scheduled to present their work to the other employees of Mid-State Brick. After this meeting, two new quality improvement teams would be formed to work on the saw and grinder processes.

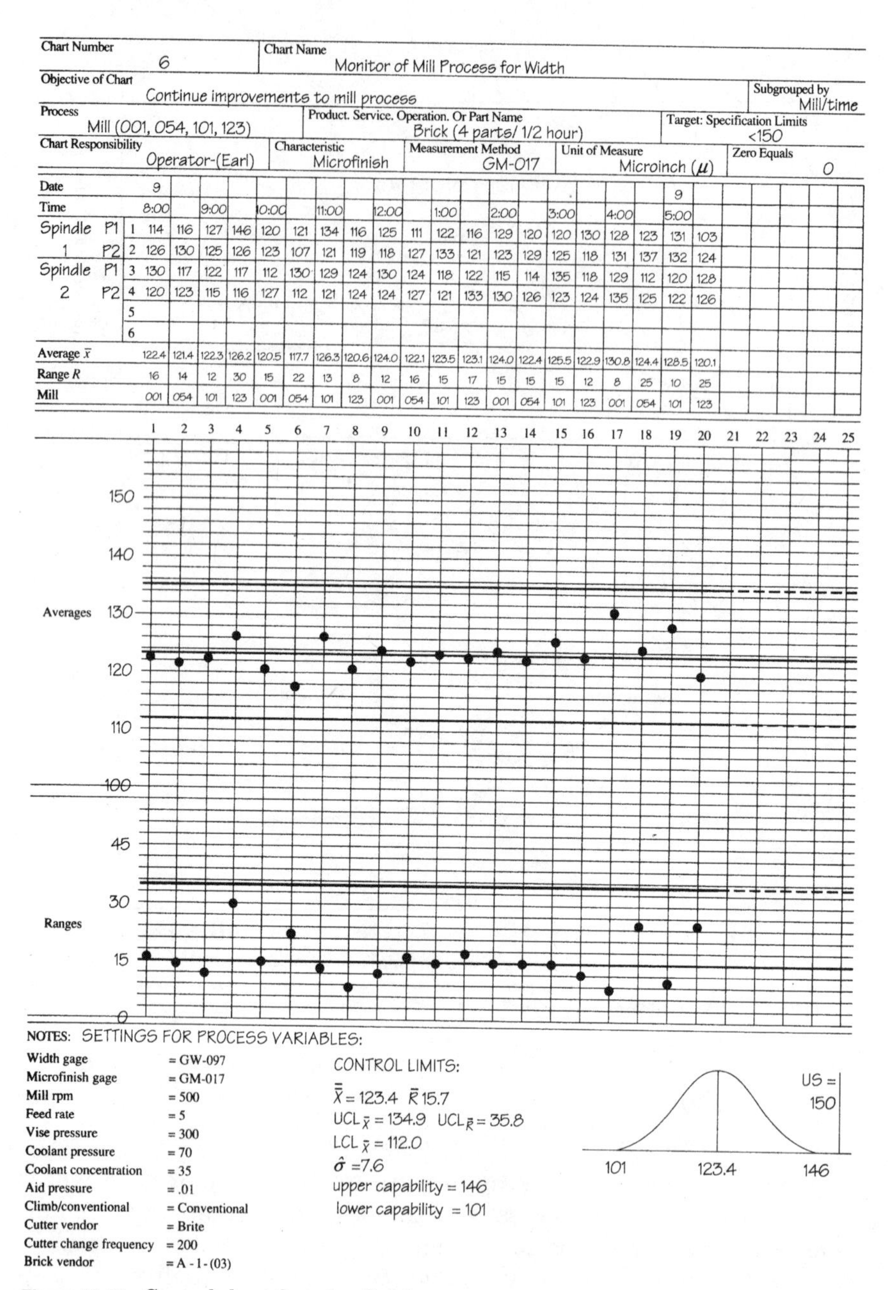

Chart Number: 6

Chart Name: Monitor of Mill Process for Width

Objective of Chart: Continue improvements to mill process

Subgrouped by: Mill/time

Process: Mill (001, 054, 101, 123)

Product. Service. Operation. Or Part Name: Brick (4 parts/ 1/2 hour)

Target: Specification Limits: <150

Chart Responsibility: Operator-(Earl)

Characteristic: Microfinish

Measurement Method: GM-017

Unit of Measure: Microinch ($\mu$)

Zero Equals: 0

| | | 1 | 2 | 3 | 4 | 5 | 6 | 7 | 8 | 9 | 10 | 11 | 12 | 13 | 14 | 15 | 16 | 17 | 18 | 19 | 20 |
|---|---|---|---|---|---|---|---|---|---|---|---|---|---|---|---|---|---|---|---|---|---|
| Date | | 9 | | | | | | | | | | | | | | | | | | 9 | |
| Time | | 8:00 | | 9:00 | | 10:00 | | 11:00 | | 12:00 | | 1:00 | | 2:00 | | 3:00 | | 4:00 | | 5:00 | |
| Spindle 1 P1 | 1 | 114 | 116 | 127 | 146 | 120 | 121 | 134 | 116 | 125 | 111 | 122 | 116 | 129 | 120 | 120 | 130 | 128 | 123 | 131 | 103 |
| Spindle 1 P2 | 2 | 126 | 130 | 125 | 126 | 123 | 107 | 121 | 119 | 118 | 127 | 133 | 121 | 123 | 129 | 125 | 118 | 131 | 137 | 132 | 124 |
| Spindle 2 P1 | 3 | 130 | 117 | 122 | 117 | 112 | 130 | 129 | 124 | 130 | 124 | 118 | 122 | 115 | 114 | 135 | 118 | 129 | 112 | 120 | 128 |
| Spindle 2 P2 | 4 | 120 | 123 | 115 | 116 | 127 | 112 | 121 | 124 | 124 | 127 | 121 | 133 | 130 | 126 | 123 | 124 | 135 | 125 | 122 | 126 |
| | 5 | | | | | | | | | | | | | | | | | | | | |
| | 6 | | | | | | | | | | | | | | | | | | | | |
| Average $\bar{x}$ | | 122.4 | 121.4 | 122.3 | 126.2 | 120.5 | 117.7 | 126.3 | 120.6 | 124.0 | 122.1 | 123.5 | 123.1 | 124.0 | 122.4 | 125.5 | 122.9 | 130.8 | 124.4 | 128.5 | 120.1 |
| Range $R$ | | 16 | 14 | 12 | 30 | 15 | 22 | 13 | 8 | 12 | 16 | 15 | 17 | 15 | 15 | 15 | 12 | 8 | 25 | 10 | 25 |
| Mill | | 001 | 054 | 101 | 123 | 001 | 054 | 101 | 123 | 001 | 054 | 101 | 123 | 001 | 054 | 101 | 123 | 001 | 054 | 101 | 123 |



NOTES: SETTINGS FOR PROCESS VARIABLES:

| | |
|---|---|
| Width gage | = GW-097 |
| Microfinish gage | = GM-017 |
| Mill rpm | = 500 |
| Feed rate | = 5 |
| Vise pressure | = 300 |
| Coolant pressure | = 70 |
| Coolant concentration | = 35 |
| Aid pressure | = .01 |
| Climb/conventional | = Conventional |
| Cutter vendor | = Brite |
| Cutter change frequency | = 200 |
| Brick vendor | = A - 1 - (03) |

CONTROL LIMITS:

$\bar{\bar{X}} = 123.4$ $\bar{R}$ 15.7

$UCL_{\bar{X}} = 134.9$ $UCL_{\bar{R}} = 35.8$

$LCL_{\bar{X}} = 112.0$

$\hat{\sigma} = 7.6$

upper capability = 146

lower capability = 101

**Figure 10.19** Control chart for microfinish.

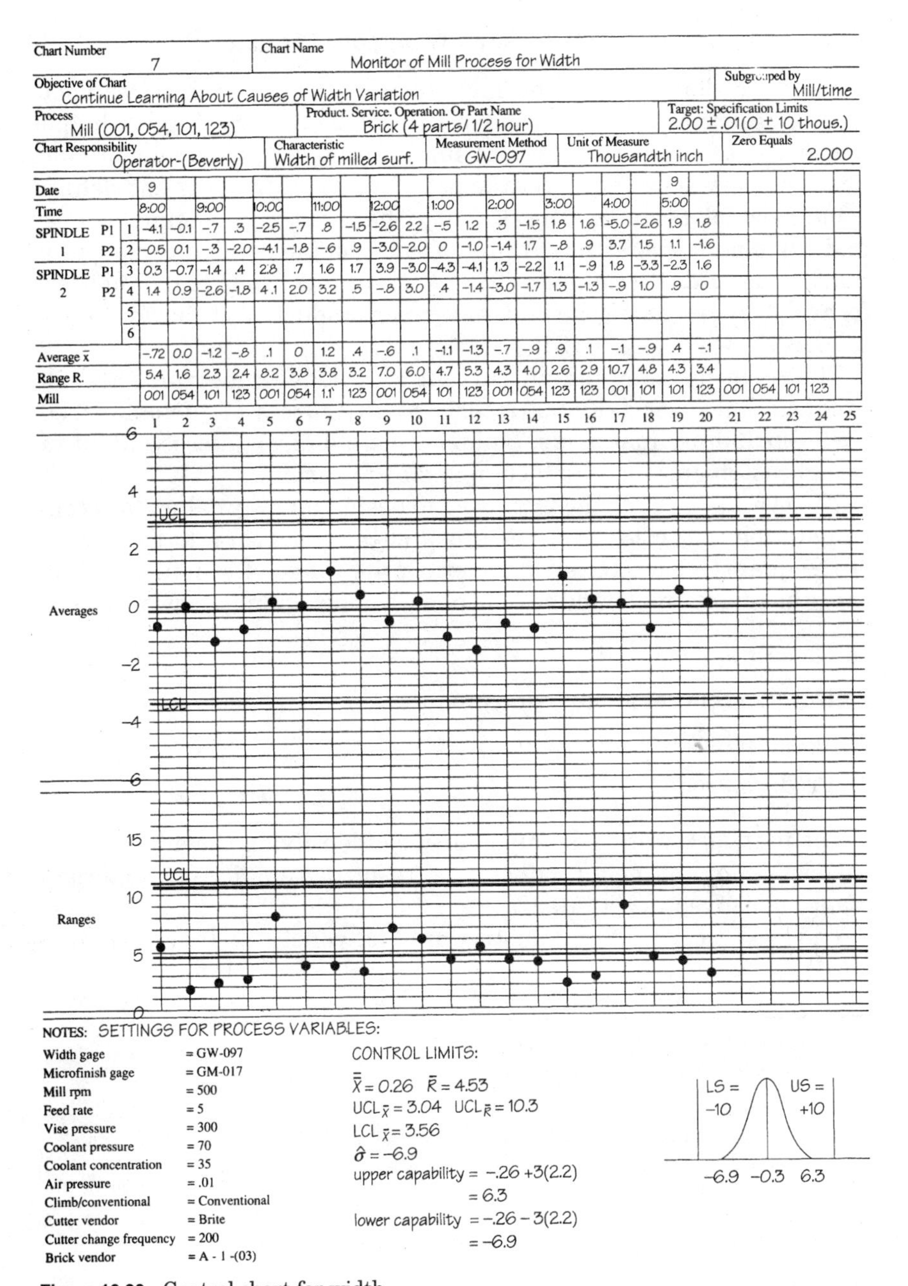

Chart Number: 7

Chart Name: Monitor of Mill Process for Width

Objective of Chart: Continue Learning About Causes of Width Variation

Subgrouped by: Mill/time

Process: Mill (001, 054, 101, 123)

Product. Service. Operation. Or Part Name: Brick (4 parts/ 1/2 hour)

Target: Specification Limits: 2.00 ± .01(0 ± 10 thous.)

Chart Responsibility: Operator-(Beverly)

Characteristic: Width of milled surf.

Measurement Method: GW-097

Unit of Measure: Thousandth inch

Zero Equals: 2.000

| | | | 1 | 2 | 3 | 4 | 5 | 6 | 7 | 8 | 9 | 10 | 11 | 12 | 13 |
|---|---|---|---|---|---|---|---|---|---|---|---|---|---|---|---|
| Date | | | 9 | | | | | | | | | | | | |
| Time | | | 8:00 | | 9:00 | | 10:00 | | 11:00 | | 12:00 | | 1:00 | | 2:00 |
| SPINDLE 1 | P1 | 1 | -4.1 | -0.1 | -.7 | .3 | -2.5 | -.7 | .8 | -1.5 | -2.6 | 2.2 | -.5 | 1.2 | .3 |
| | P2 | 2 | -0.5 | 0.1 | -.3 | -2.0 | -4.1 | -1.8 | -.6 | .9 | -3.0 | -2.0 | 0 | -1.0 | -1.4 |
| SPINDLE 2 | P1 | 3 | 0.3 | -0.7 | -1.4 | .4 | 2.8 | .7 | 1.6 | 1.7 | 3.9 | -3.0 | -4.3 | -4.1 | 1.3 |
| | P2 | 4 | 1.4 | 0.9 | -2.6 | -1.8 | 4.1 | 2.0 | 3.2 | .5 | -.8 | 3.0 | .4 | -1.4 | -3.0 |
| | | 5 | | | | | | | | | | | | | |
| | | 6 | | | | | | | | | | | | | |
| Average x̄ | | | -.72 | 0.0 | -1.2 | -.8 | .1 | 0 | 1.2 | .4 | -.6 | .1 | -1.1 | -1.3 | -.7 |
| Range R. | | | 5.4 | 1.6 | 2.3 | 2.4 | 8.2 | 3.8 | 3.8 | 3.2 | 7.0 | 6.0 | 4.7 | 5.3 | 4.3 |
| Mill | | | 001 | 054 | 101 | 123 | 001 | 054 | 1.1` | 123 | 001 | 054 | 101 | 123 | 001 |

| | | | 14 | 15 | 16 | 17 | 18 | 19 | 20 | 21 | 22 | 23 | 24 | 25 |
|---|---|---|---|---|---|---|---|---|---|---|---|---|---|---|
| Date | | | | | | | | 9 | | | | | | |
| Time | | | | 3:00 | | 4:00 | | 5:00 | | | | | | |
| SPINDLE 1 | P1 | 1 | -1.5 | 1.8 | 1.6 | -5.0 | -2.6 | 1.9 | 1.8 | | | | | |
| | P2 | 2 | 1.7 | -.8 | .9 | 3.7 | 1.5 | 1.1 | -1.6 | | | | | |
| SPINDLE 2 | P1 | 3 | -2.2 | 1.1 | -.9 | 1.8 | -3.3 | -2.3 | 1.6 | | | | | |
| | P2 | 4 | -1.7 | 1.3 | -1.3 | -.9 | 1.0 | .9 | 0 | | | | | |
| | | 5 | | | | | | | | | | | | |
| | | 6 | | | | | | | | | | | | |
| Average x̄ | | | -.9 | .9 | .1 | -.1 | -.9 | .4 | -.1 | | | | | |
| Range R. | | | 4.0 | 2.6 | 2.9 | 10.7 | 4.8 | 4.3 | 3.4 | | | | | |
| Mill | | | 054 | 123 | 123 | 001 | 101 | 101 | 123 | 001 | 054 | 101 | 123 | |



NOTES: SETTINGS FOR PROCESS VARIABLES:

| | |
|---|---|
| Width gage | = GW-097 |
| Microfinish gage | = GM-017 |
| Mill rpm | = 500 |
| Feed rate | = 5 |
| Vise pressure | = 300 |
| Coolant pressure | = 70 |
| Coolant concentration | = 35 |
| Air pressure | = .01 |
| Climb/conventional | = Conventional |
| Cutter vendor | = Brite |
| Cutter change frequency | = 200 |
| Brick vendor | = A - 1 -(03) |

CONTROL LIMITS:

$\bar{\bar{X}} = 0.26$ $\bar{R} = 4.53$

$UCL_{\bar{X}} = 3.04$ $UCL_{\bar{R}} = 10.3$

$LCL_{\bar{X}} = 3.56$

$\hat{\sigma} = -6.9$

upper capability = $-.26 + 3(2.2)$
$= 6.3$

lower capability = $-.26 - 3(2.2)$
$= -6.9$



**Figure 10.20** Control chart for width.

## 10.2 Case Study 2: Redesign a Wallpaper Product

A company that designs and manufactures wall coverings for home and commercial use has decided to replace an existing wallpaper product with a more up-to-date line. Although the current product has been very successful over the past 3 years, marketing had defined some new colors and patterns that it believed would improve sales and customer satisfaction. A product development team composed of marketing, R&D, engineering, and manufacturing personnel was formed to redesign the current wallpaper product. The charter for the team is shown in Fig. 10.21.

The team documented all current knowledge concerning the old product. Data were summarized from all sources of customer feedback—including final users, builders, installers, retailers, and all internal customers from production and engineering.

The overall task for the team was to design a new product to accommodate the new color and patterns and that was close to the targets of the quality characteristics under a wide range of conditions. To do this, the team completed the four phases of designing a new product:

1. Generate ideas.
2. Develop concepts and define product.
3. Test product.
4. Produce product.

The first task was to define quality by identifying the quality characteristics for the new product. The resulting quality characteristic diagram is given in Fig. 10.22*a*.

The next task was to generate product concepts. Team members

**General description:**

Redesign the current wallpaper product.

**Expected results:**

1. Improved appearance of the pattern and color choices.
2. Other quality characteristics will be at least as good as the old product.
3. Easy to install.
4. It will wear well and be easy to clean.
5. Manufacturing costs should not be increased.

**Boundaries:**

1. Stay within the R&D budget allocated for the project.
2. Be able to use existing manufacturing process.
3. Product must be available to retailers by next spring.

**Figure 10.21** Charter of product development team for redesign of a wallpaper product.

| Needs of the customer: Decorative walls | | | |
|---|---|---|---|
| Quality characteristic [1] | | | |
| Primary [2] | Secondary [3] | Tertiary [3] | # |
| looks good | consistent with fashion trends | | 1 |
| | attractive | | 2 |
| | lack of visible seams | | 3 |
| easy to clean | time to clean | | 4 |
| | effort to clean | | 5 |
| wears well | resistance to scratches | | 6 |
| | stain resistance | | 7 |
| | gloss retention | | 8 |

[1] Do not include design factors in this list. (*Test*: You should *not* be able to set the levels of these quality characterisitics.)
[2] Express in the language of the customer.
[3] To add more detail, subdivide into two or more quality characteristics.

**Figure 10.22*a*** Quality characteristic diagram for redesigning wallpaper.

listed all products that related to the charter. They included products from competitors as well as their own. R&D had developed some new formulations based on a different chemistry that resulted in brighter colors and more distinctive patterns.

The concepts that were tested related to composition and thickness of the three layers of the wallpaper. The layering of the wallpaper is illustrated in Fig. 10.22*b*.

The team made a subjective assessment of product concepts. The summary table is given in Fig. 10.23.

### First improvement cycle: Potential features

The team identified several potential features for the new wallpaper and decided to run a conjoint analysis to test these features. The planning form is given in Fig. 10.24. The response variable is an average ranking (ordinal measurement) by the potential customers. Profile

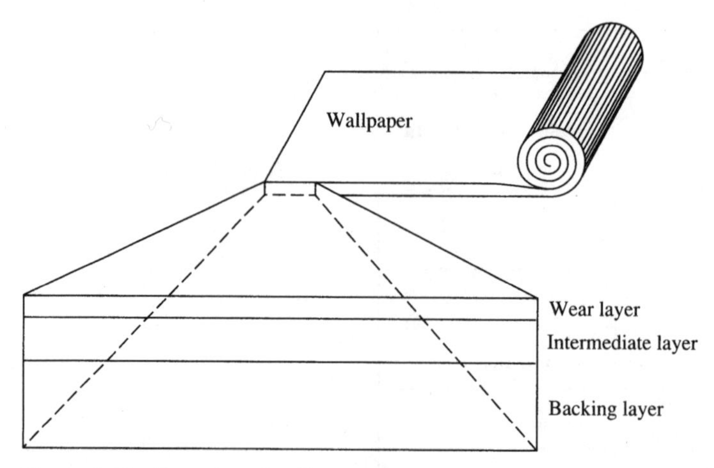


**Figure 10.22*b*** Layering of wallpaper.

| Attributes | Product concept<br>Current product | <br>Best competitor | New concepts<br>1 | <br>2 | <br>3 |
|---|---|---|---|---|---|
| Product form | Wallpaper | Wallpaper | Handmade prototypes | | |
| Ease in manufacturing | yes | yes | yes | yes | yes |
| Needs addressed: | | | | | |
| 1. Consistent with trends | 3 | 4 | 5 | 5 | 5 |
| 2. Attractive | 2 | 3 | 5 | 4 | 5 |
| 3. Lack of visible seams | 5 | 4 | 4 | 3 | 4 |
| 4. Time to clean | 4 | 3 | 5 | 4 | 3 |
| 5. Effort to clean | 3 | 2 | 4 | 3 | 3 |
| 6. Resistance to scratches | 4 | 3 | 5 | 4 | 4 |
| 7. Stain resistance | 3 | 3 | 5 | 4 | 4 |
| 8. Gloss retention | 3 | 3 | 5 | 5 | 4 |

Scale: 5 = greatly exceeds needs
4 = exceeds needs
3 = meets needs
2 = partially meets needs
1 = fails to meet needs

**Figure 10.23** Assessment of product concepts.

1. **Objective:**
   Determine what features to offer in a product profile.

2. **Background information:**
   The team has identified several potential features for the new wallpaper.

3. **Experimental variables:**

A.

| Response variables | Measurement technique |
|---|---|
| 1. Ranks (most appealing to least) | Average ranking of 8 product profile cards by 25 potential customers |

B.

| Factors under study | Levels | |
|---|---|---|
| 1. Pattern | current | new |
| 2. Installation instruction | standard | video |
| 3. Cleaning kit | no | yes |
| 4. Color family | old | new |
| 5. Coating | paper | vinyl |
| 6. Gloss finish | no | yes |
| 7. Price | current | 5% increase |

C.

| Background variables | Method of control |
|---|---|
| 1. Age of customer | Create 2 blocks |
| 2. Income of customer | Create 2 blocks |
| 3. All others | Hold constant |

4. **Replication:**
   Each of the 8 profile cards will be ranked by 25 customers in each of the 4 blocks.

5. **Methods of randomization:**
   Shuffle the order of the 8 profile cards.

6. **Design Matrix:** (attach copy)
   $2^{7-4}$ fractional factorial design with 8 runs.

7. **Data collection forms:** (attach copies)
   Form in the customer packet.

8. **Planned methods of statistical analysis:**
   Dot diagram and response plots for each block.

9. **Estimated cost, schedule, and other resource considerations:**
   Total cost for this experiment is $10,000. Customers within each block will be participating in an evening session. All 4 sessions will be completed in 1 week.

**Figure 10.24** Planning form for conjoint analysis for wallpaper.

cards are prepared with each of the eight tests of the $2^{7-4}$ fractional factorial design.

Pictures of the different combinations of colors and patterns of wallpaper were included on the profile cards. An example of a profile card for condition 1 of the design matrix (− − − + + + −) is given below:

| Profile A: current price | [Insert picture or show actual prototype with current pattern, standard installation instructions, no cleaning kit, new color family, vinyl coating, and gloss finish. |
|---|---|

Figure 10.25 shows the results of the study. Based on the analysis and the current knowledge, the following conclusions are drawn:

- Price was important (not willing to pay for additional features).
- Vinyl coating was preferred.
- New color and patterns were preferred.

#### Second improvement cycle: Preference test

The team members developed a concept product (concept 1) based on the results of the previous study. They decided to run a study with a twofold objective:

1. Determine if the appearance of new concept 1 is preferred over the old product.
2. Evaluate the installation of a product based on the new concept.

The plan was to use a judgment sample of eight panelists. These people covered the range of potential customers. Five wallpaper samples of old and new design were prepared based on color and pattern combinations. Sufficient material was made to evaluate the installation of the new product (versus the old) at a test room in the R&D laboratory. Sample presentation was done randomly for each panelist. Panelists could not collaborate with one another.

The resultant data are given in Fig. 10.26. The results were impressive. The new concept was clearly preferred over the old. The team discussed the results with the panelists. The panelists who preferred the old to the new on samples 1, 2, and 3 based their preference on the new samples having too wide a gap at the seams. This resulted in the seams being visible. A return trip to the laboratory indicated that the first three samples of the new design did have visible seams. The problem is illustrated here:

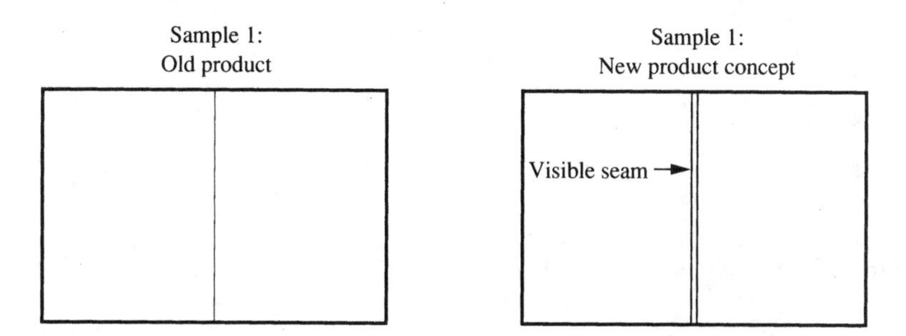


Previous testing in R&D had not identified a seam problem. They believed humidity during installation could be a factor contributing to the seam problem. They have no objective measurement for the gap of seams.

The team decided any further testing of new product for causes of visible seams would require a gage. Engineering agreed and developed a gage.

#### Third improvement cycle: Test of new gage

The next activity of the team was to evaluate the variability of the new gage with respect to the gap of seams. A three-factor nested experiment was selected to evaluate the following components:

1. Samples
2. Position within each sample
3. Measurement within each position

The planning form is given in Fig. 10.27. The five samples are the samples of the new product from the previous preference study. Three positions per sample will be measured to determine the unevenness of seams. Three measurements will be made per position by different operators. The design matrix is given in Fig. 10.28.

Figure 10.29 shows a run chart of the data. No obvious trends or other special causes are seen. The dot-frequency diagram is given in Fig. 10.30. The dot-frequency diagram indicates that most of the variation was attributable to sample differences and, to a lesser extent, to position within sample. The gage results were very repeatable. Variation between positions within sample is easily seen. The first three samples had the greatest problem with seams, which confirmed

Design matrix for a $2^{7-4}$ pattern

| Test | Run order | 1<br>24<br>35<br>67 | 2<br>14<br>36<br>57 | 3<br>15<br>26<br>47 | 4<br>12<br>56<br>37 | 5<br>13<br>46<br>27 | 6<br>23<br>45<br>17 | 7<br>34<br>25<br>16 | Average rank |
|---|---|---|---|---|---|---|---|---|---|
| 1 | 6 | - | - | - | + | + | + | - | 2.9 |
| 2 | 3 | + | - | - | - | - | + | + | 6.1 |
| 3 | 8 | - | + | - | - | + | - | + | 6.0 |
| 4 | 5 | + | + | - | + | - | - | - | 3.4 |
| 5 | 1 | - | - | + | + | - | - | + | 5.9 |
| 6 | 7 | + | - | + | - | + | - | - | 3.1 |
| 7 | 2 | - | + | + | - | - | + | - | 4.6 |
| 8 | 4 | + | + | + | + | + | + | + | 3.5 |
| Effects: | | -0.8 | 0.1 | -0.3 | -1.0 | -1.1 | -0.3 | 1.9 | |

Dot diagram of effects

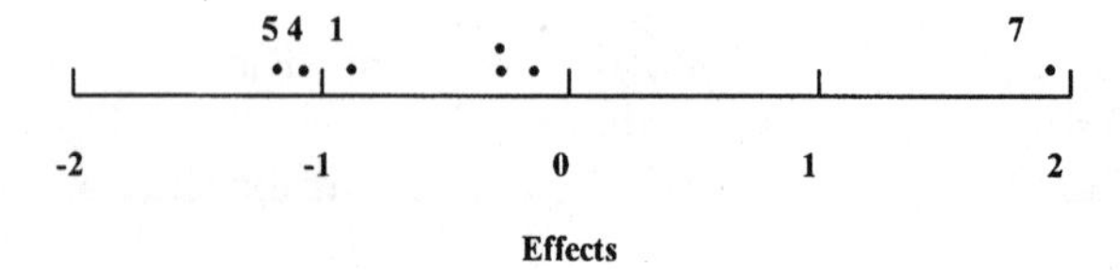


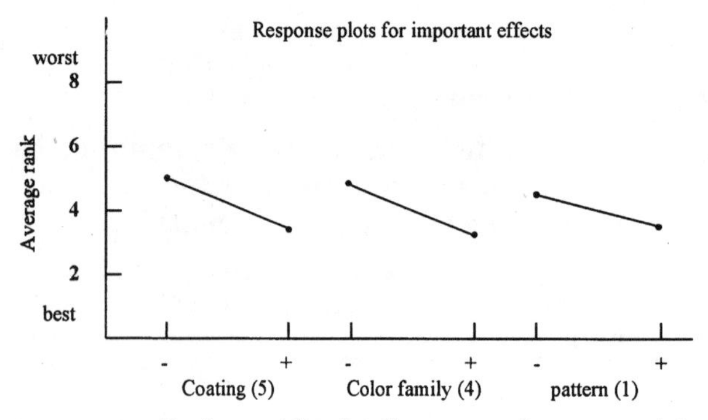


**Figure 10.25** Design matrix, dot diagram, and response plots for wallpaper.

the previous preference study. (*Note:* The gage measurement scale is 0.1 mil and 0 = deviation from target.)

The team decided to accept the gage for additional product development work and to take measurements from four different positions within a sample and to calculate averages and standard deviations for any future studies. The team then proceeded to design the product based on the selection of product concept.

| Panelist | Samples (color/pattern) 1 | 2 | 3 | 4 | 5 |
|---|---|---|---|---|---|
| 1 | N | N | N | N | N |
| 2 | N | N | O | N | N |
| 3 | O | N | N | N | N |
| 4 | N | O | N | N | N |
| 5 | N | N | N | N | N |
| 6 | N | N | N | N | N |
| 7 | O | O | O | N | N |
| 8 | N | N | N | N | N |

N = preferred new product
O = preferred old product

**Figure 10.26** Overall preference test for new versus old product.

---

**1. Objective:**
Evaluate the variability of the new gage with respect to seams.

**2. Background information:**
This measurement system is new.

**3. Experimental variables:**

A.

| Response variables | Measurement technique |
|---|---|
| 1. Visible seams (.1 mils) | New gage. |

B.

| Factors under study | Levels |
|---|---|
| 1. Samples (new product) | Five samples from last study. |
| 2. Position within sample | Three positions per sample made along the seam. |
| 3. Measurement within position | Three measurements made with three operators. |

C.

| Background variables | Method of control |
|---|---|
| 1. Installer | One person. |
| 2. Type of adhesive | Use standard adhesive. |
| 3. Calibration | Checked at the beginning and end of study. |

**4. Replication:**
The study has to be completed in one day. Only one replication of the experimental pattern will be done.

**5. Methods of randomization:**
The five samples and three positions per sample were randomly ordered.

**6. Design matrix:** (see Figure 10.28)

**7. Data collection forms:** (not shown here)

**8. Planned methods of statistical analysis:**
Run chart and dot-frequency diagram.

**9. Estimated cost, schedule, and other resource considerations:**
Study can be completed in one day using the laboratory testing room.

---

**Figure 10.27** Form for documentation of a planned experiment for cycle 3.

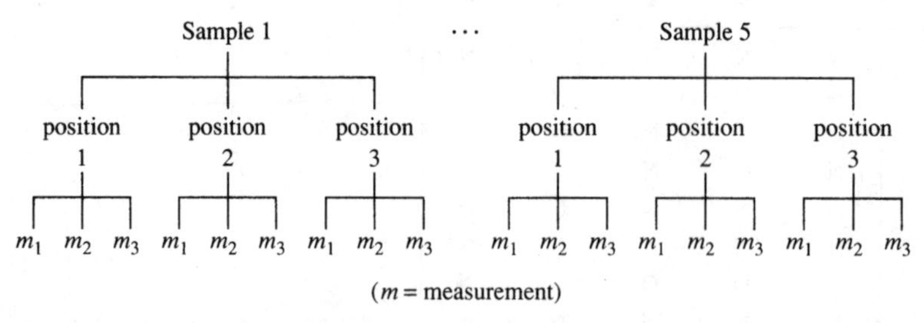


**Figure 10.28** Three-factor nested study design matrix (experimental pattern).

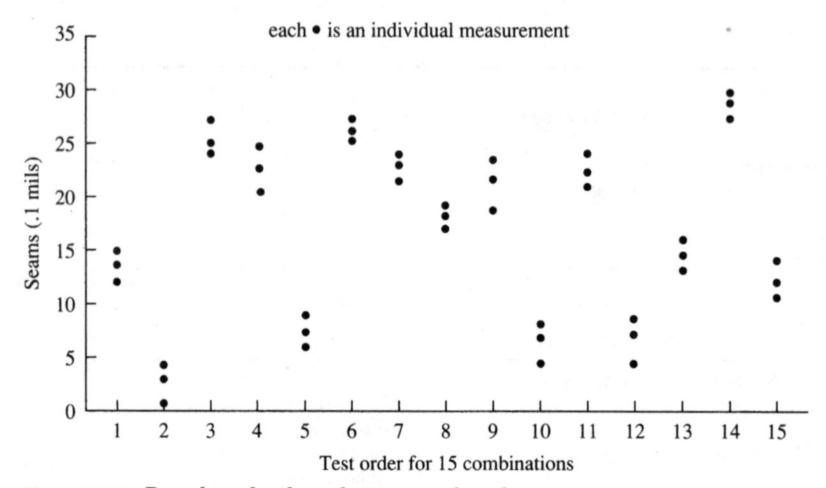


**Figure 10.29** Run chart for three-factor nested study.

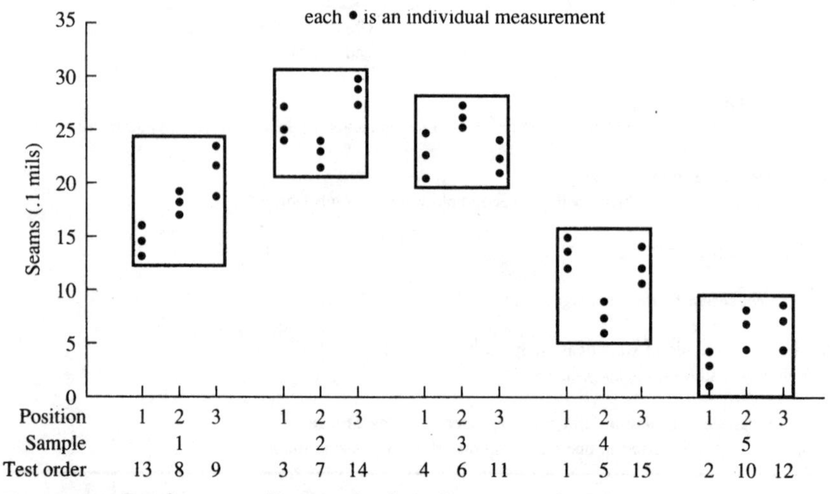


**Figure 10.30** Dot-frequency diagram for three-factor nested study.

Documentation of current knowledge of the new product is displayed as a QFD relation diagram in Fig. 10.31. (*Note:* This diagram is analogous to the cause-and-effect diagram used to define current knowledge of an existing process.) Eight control factors represent design parameters that can be set to specific levels. Those thought to

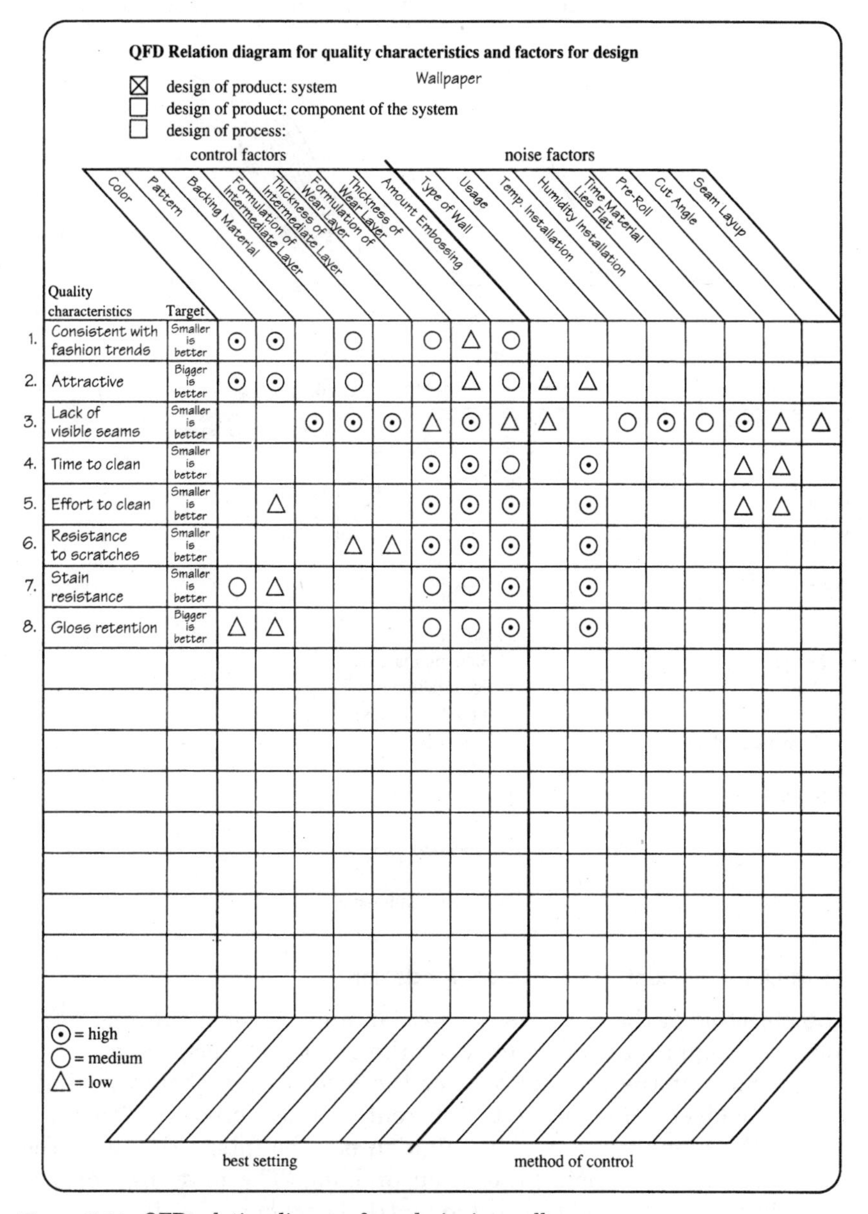


| | Quality characteristics | Target | Color | Pattern | Backing Material | Formulation of Intermediate Layer | Thickness of Intermediate Layer | Formulation of Wear Layer | Thickness of Wear Layer | Amount Embossing | Type of Wall | Usage | Temp. Installation | Humidity Installation | Time Material Lies Flat | Pre-Roll | Cut Angle | Seam Layup |
|---|---|---|---|---|---|---|---|---|---|---|---|---|---|---|---|---|---|---|
| 1. | Consistent with fashion trends | Smaller is better | ⊙ | ⊙ | | ○ | | ○ | △ | ○ | | | | | | | | |
| 2. | Attractive | Bigger is better | ⊙ | ⊙ | | ○ | | ○ | △ | ○ | △ | △ | | | | | | |
| 3. | Lack of visible seams | Smaller is better | | | ⊙ | ⊙ | ⊙ | △ | ⊙ | △ | △ | | ○ | ⊙ | ○ | ⊙ | △ | △ |
| 4. | Time to clean | Smaller is better | | | | | | ⊙ | ⊙ | ○ | | ⊙ | | | | △ | △ | |
| 5. | Effort to clean | Smaller is better | | △ | | | | ⊙ | ⊙ | ⊙ | | ⊙ | | | | △ | △ | |
| 6. | Resistance to scratches | Smaller is better | | | | △ | △ | ⊙ | ⊙ | ⊙ | | ⊙ | | | | | | |
| 7. | Stain resistance | Smaller is better | ○ | △ | | | | ○ | ○ | ⊙ | | ⊙ | | | | | | |
| 8. | Gloss retention | Bigger is better | △ | △ | | | | ○ | ○ | ⊙ | | ⊙ | | | | | | |

**Figure 10.31** QFD relation diagram for redesigning wallpaper.

have a strong relationship to the quality characteristics can be adjusted to move the quality characteristics closer to target.

Of the eight noise factors identified in Fig. 10.31 as affecting the quality characteristics, most relate to visible seams. The team developed an overall plan to test the relationships in the QFD relation diagram, but in this case study only the work on visible seams and appearance will be described. The identified noise and control factors are illustrated below.

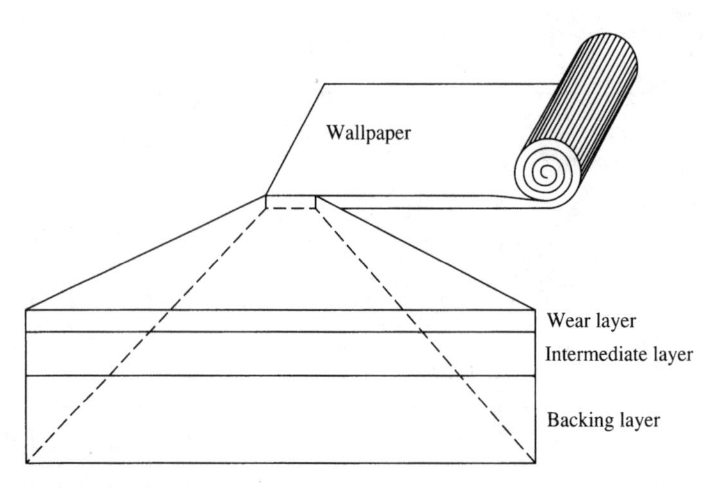


Noise factors
1. Type of wall
2. Usage
3. Temperature
4. Humidity
5. Pre-roll
6. Time material lays flat
7. Cut angle
8. Seam layup

Control factors
1. Color
2. Pattern
3. Backing material
4. Formulation of intermediate layer
5. Thickness of intermediate layer
6. Formulation of wear layer
7. Thickness of wear layer
8. Amount embossing

### Fourth improvement cycle: Factors for design

The objective of the next cycle is to plan an experiment to select factors for design that have an effect on the quality characteristics.

The plan for the cycle is to vary the formulation of intermediate and wear layers to simulate the current and new product. The belief is that the humidity during installation may be interacting with the new product and causing the seam problem. The planning form is given in Fig. 10.32, and the design matrix is given in Fig. 10.33.

**1. Objective:**
Determine the effect control factors and noise factors have on a redesign for wallpaper.

**2. Background information:**
The redesign has a new color and pattern that has a beautiful appearance. There is a need to reduce the magnitude of visible seams.

**3. Experimental variables:**

| A. | Response variables | Measurement technique | |
|---|---|---|---|
| 1. | Visible seams (average, standard deviation) | new gage (.1 mils) | |

| B. | Factors under study | Levels | |
|---|---|---|---|
| 1. | Formulation | old | new |
| 2. | Type of wall | wood | concrete |
| 3. | Temperature | 50° | 90° |
| 4. | Humidity | 40% | 80% |
| 5. | Time material lays flat | 2 min. | 4 min. |
| 6. | Pre-roll | 0 min. | 10 min. |
| 7. | Cut angle | 45° | 90° |
| 8. | Seam layup | low | high |

| C. | Background variables | Method of control |
|---|---|---|
| 1. | Installer | One person, standard instructions |
| 2. | Type of adhesive | Use standard for all applications |
| 3. | Thickness of intermediate layer | 4.0 mils |
| 4. | Thickness of wear layer | 1.0 mils |
| 5. | Backing material | Material *A* (same as old) |

**4. Replication:**
Measurements for visible seams at four positions per sheet.

**5. Methods of randomization:**
Order of the 16 runs was randomized using a random permutation table.

**6. Design matrix:**
$2^{8-4}$ fractional factorial design (see Figure 10.33).

**7. Data collection forms:** (see margin in Figure 10.33)

**8. Planned methods of statistical analysis:**
Compute average and standard deviation of four readings per sheet.

**9. Estimated cost, schedule, and other resources:**
Evaluated in the test room with temperature and humidity controls. Six days are required to complete.

**Figure 10.32** Documentation of the wallpaper experiment for cycle 4.

The run charts for the averages and standard deviations of visible seams at each of the 16 combinations are shown in Fig. 10.34. No obvious patterns were present. Three of the points on the average chart obviously indicate special causes. Further analysis will disclose that those points were due to changing the factors under study. The effects

| Test | Run order | 1 | 2 | 3 | 4 | 5 | 6 | 7 | 8 | 12 | 13 | 14 | 15 | 16 | 17 | 24 | Response | |
|---|---|---|---|---|---|---|---|---|---|---|---|---|---|---|---|---|---|---|
| | | | | | | | | | | 37 | 27 | 36 | 26 | 25 | 23 | 35 | | |
| | | | | | | | | | | 56 | 46 | 57 | 47 | 34 | 45 | 67 | | Standard |
| | | F | | T | H | | P | | | 48 | 58 | 28 | 38 | 78 | 68 | 18 | Average | deviation |
| 1 | 8 | − | − | − | + | + | + | − | + | + | + | − | − | − | + | − | 0 | 4 |
| 2 | 15 | + | − | − | − | − | + | + | + | − | − | − | − | + | + | + | 11 | 11 |
| 3 | 16 | − | + | − | − | + | − | + | + | − | + | + | − | + | − | − | 4 | 5 |
| 4 | 6 | + | + | − | + | − | − | − | + | + | − | + | − | − | − | + | 4 | 5 |
| 5 | 3 | − | − | + | + | − | − | + | + | + | − | − | + | + | − | − | 9 | 12 |
| 6 | 12 | + | − | + | − | + | − | − | + | − | + | − | + | − | − | + | 42 | 18 |
| 7 | 9 | − | + | + | − | − | + | − | + | − | − | + | + | − | + | − | 5 | 5 |
| 8 | 7 | + | + | + | + | + | + | + | + | + | + | + | + | + | + | + | 77 | 14 |
| 9 | 1 | + | + | + | − | − | − | + | − | + | + | − | − | − | + | − | 18 | 11 |
| 10 | 4 | − | + | + | + | + | − | − | − | − | − | − | − | + | + | + | 21 | 10 |
| 11 | 10 | + | − | + | + | − | + | − | − | − | + | + | − | + | − | − | 80 | 25 |
| 12 | 13 | − | − | + | − | + | + | + | − | + | − | + | − | − | − | + | 12 | 9 |
| 13 | 11 | + | + | − | − | + | + | − | − | + | − | − | + | + | − | − | 9 | 5 |
| 14 | 2 | − | + | − | + | − | + | + | − | − | + | − | + | − | − | + | 2 | 4 |
| 15 | 14 | + | − | − | + | + | − | + | − | − | − | + | + | − | + | − | 1 | 8 |
| 16 | 5 | − | − | − | − | − | − | − | − | + | + | + | + | + | + | + | 1 | 2 |
| Divisor = 8 | | | | | | | | | | | | | | | | | | |
| | Average | 23.5 | −2.0 | 29 | 11.5 | 4.5 | 12.0 | −3.5 | 2.0 | −4.5 | 19.0 | 10.4 | −0.5 | 16 | −3.5 | 5.5 | | |
| | Standard deviation | 5.8 | −3.8 | 7.5 | 2.0 | −0.2 | 0.8 | 0.1 | 0 | −3.0 | 2.2 | −0.2 | −1.5 | 2.5 | −2.8 | −0.2 | | |

**Figure 10.33** $2^{8-4}$ design matrix for the wallpaper study.

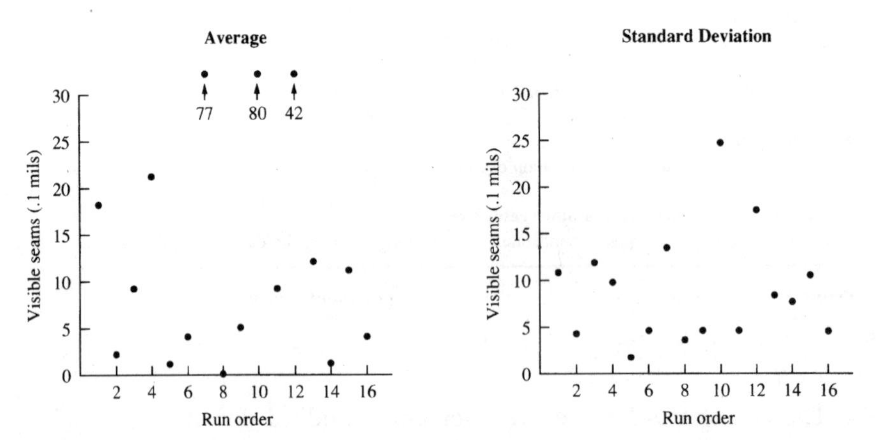


**Figure 10.34** Run charts for the wallpaper study.

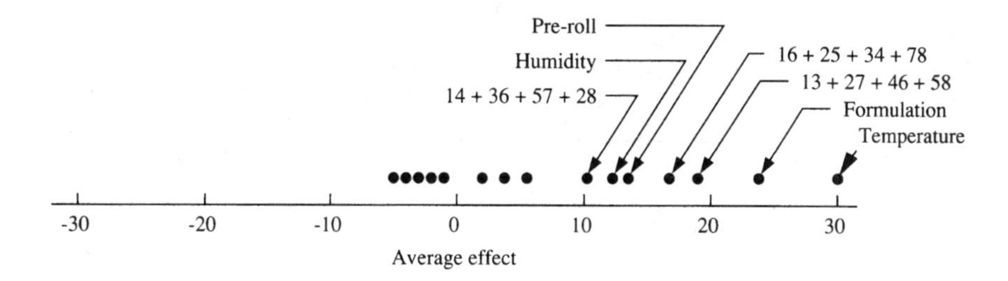


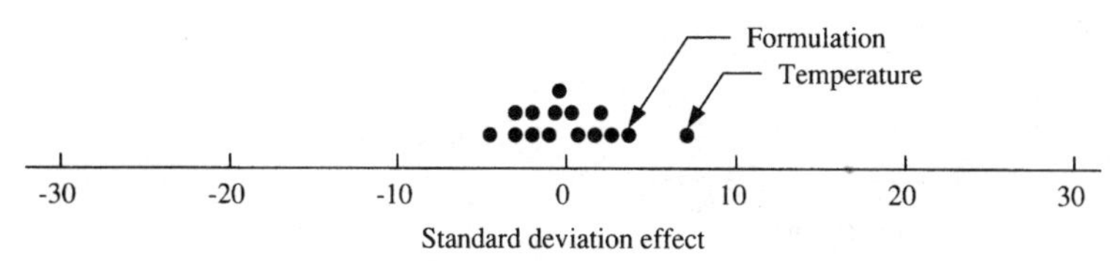


**Figure 10.35** Dot diagrams for the wallpaper study.

of the factors computed from the design matrix were included in Fig. 10.33. The dot diagrams are illustrated in Fig. 10.35.

One control factor, formulation, and three noise factors—pre-roll, humidity, and temperature—were found to be important. Temperature was a real surprise. It was thought that humidity was the problem.

Analysis of the two cubes from Fig. 10.36 (see also Fig. 10.37) as a full factorial design indicates a possible interaction between formulation and temperature for averages. These interactions accounted for the three high points on the run chart for averages in Fig. 10.34.

The prediction of the team was that the new concept will have installation problems with seams. Claims could be high! The plan will be to run another cycle with a follow-up study using a new run of material and to learn more about the interactions.

#### Fifth improvement cycle: Follow-up study

The objective of this study was to use a new run of material and to determine which interactions persist. A three-factor experiment was run with formulation (old and new), pre-roll, and temperature/humidity as a chunk variable with the low level at both low temperature and low humidity and the high level at both high temperature and high humidity. It was believed that these combinations would create the most extreme conditions as noise factors in the field.

The planning form for the study is given in Fig. 10.38. Visible seams and appearance are the response variables. Appearance will be scored subjectively by an appearance team. A $2^3$ factorial design, given in Fig. 10.39, was chosen so that all interactions could be studied.

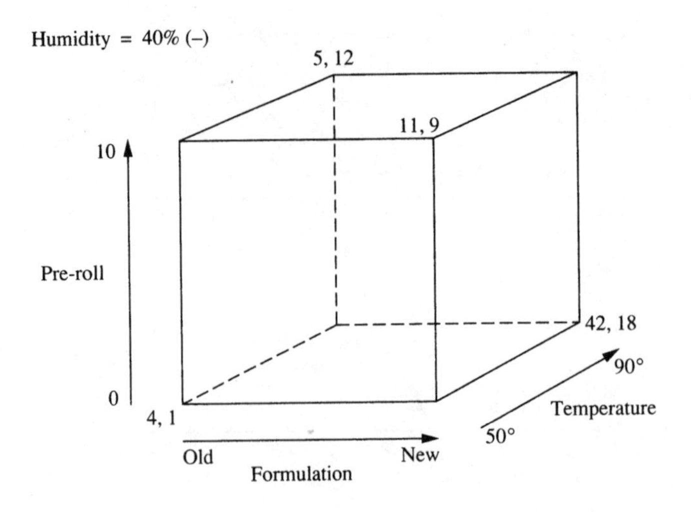


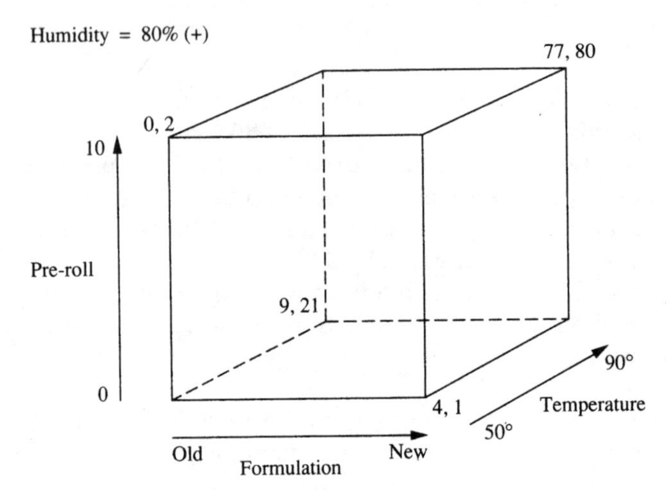


**Figure 10.36** Cube for the wallpaper study ($2^{8-4}$ design, any three factors form a full factorial design).

The run charts for the averages and standard deviations of visible seams at each of the eight combinations are contained in Fig. 10.40. No obvious patterns were present. The effects of the factors computed from the design matrix are included in Fig. 10.39.

The dot diagrams for the average and standard deviation of visible seams are contained in Fig. 10.41. Results are similar to those of the previous experiment and the dot diagram in Fig. 10.35. Formulation interacted with the temperature/humidity combination to affect the average of visible seams.

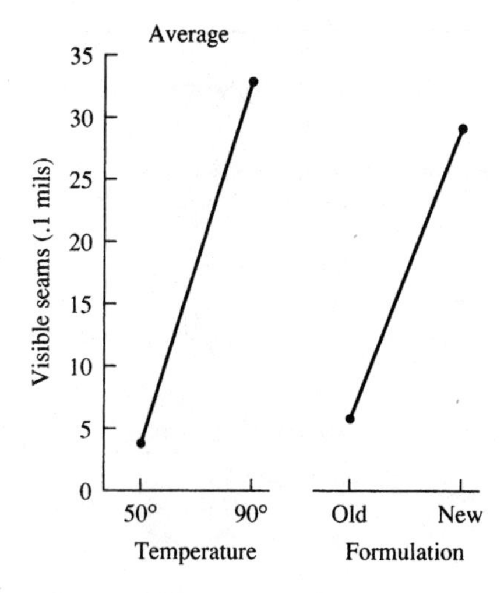


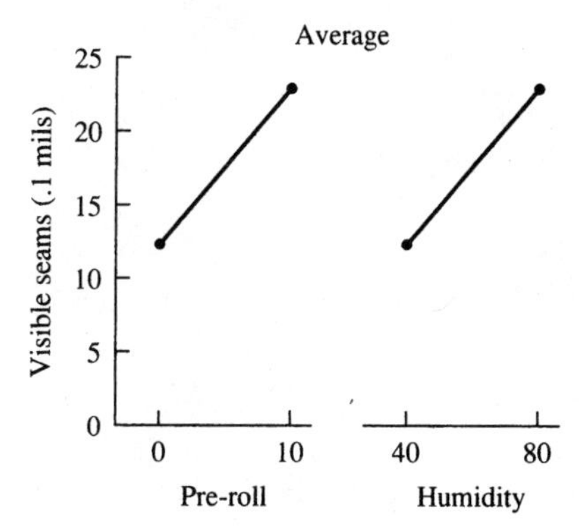


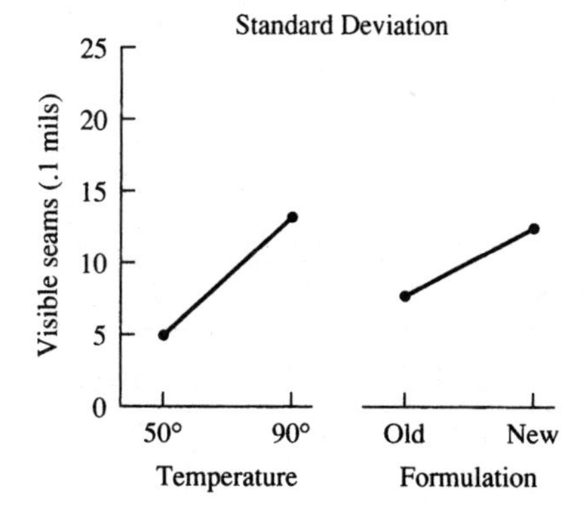


**Figure 10.37** Response plots for important effects.

**1. Objective:**
Determine the effect of control factors and noise factors on a new run of material for wallpaper.

**2. Background information:**
The redesign has a new color and pattern that has a beautiful appearance. There is a need to reduce the magnitude of visible seams. Previous study indicated three noise factors (temperature, pre-roll, and humidity) may contribute to the seam problem.

**3. Experimental variables:**

| A. Response variables | Measurement technique |
|---|---|
| 1. Visible seams (average, standard deviation) | new gage (.1 mils) |
| 2. Appearance | subjective scoring |

| B. Factors under study | Levels | |
|---|---|---|
| 1. Formulation | old | new |
| 3. Temperature/Humidity | 50°/40% | 90°/80% |
| 3. Pre-roll | 0 min. | 10 min. |

| C. Background variables | Method of control |
|---|---|
| 1. Installer | One person, standard instructions |
| 2. Type of adhesive | Use standard for all applications |
| 3. Thickness of intermediate layer. | 4.0 mils |
| 4. Thickness of wear layer | 1.0 mils |
| 5. Backing material | Material *A* (same as old) |
| 6. Type of wall | Wood |
| 7. Time material lays flat | 4 hours |
| 8. Cut angle | 90° |
| 9. Seam layup | Low |

**4. Replication:**
Four measurements for visible seams per sheet.

**5. Methods of randomization:**
Order of the eight runs was randomized using a random permutation table.

**6. Design matrix:** $2^3$ factorial design

**7. Data collection forms:** (see margin in Figure 10.39)

**8. Planned methods of statistical analysis:**
Run charts, dot diagram, cube, and response plots.

**9. Estimated cost, schedule, and other resources:**
Evaluated in the test room with temperature and humidity controls. One day is required to complete. Appearance scoring done by appearance team.

**Figure 10.38** Documentation of the wallpaper experiment for cycle 5.

| Test | Run order | 1 F | 2 T/H | 3 P | 12 | 13 | 23 | 123 | Response Average | Response Standard Deviation |
|---|---|---|---|---|---|---|---|---|---|---|
| 1 | 3 | − | − | − | + | + | + | − | 5 | 8 |
| 2 | 4 | + | − | − | − | − | + | + | 5 | 8 |
| 3 | 1 | − | + | − | − | + | − | + | 10 | 14 |
| 4 | 7 | + | + | − | + | − | − | − | 32 | 16 |
| 5 | 6 | − | − | + | + | − | − | + | 3 | 2 |
| 6 | 2 | + | − | + | − | + | − | − | 10 | 6 |
| 7 | 5 | − | + | + | − | − | + | − | 13 | 8 |
| 8 | 8 | + | + | + | + | + | + | + | 40 | 15 |
| Divisor = 4 | | | | | | | | | | |
| Average effect | | 14.0 | 18.0 | 3.5 | 10.5 | 3.0 | 2.0 | −0.5 | | |
| Standard Deviation | | 3.2 | 7.2 | −3.8 | 1.2 | 2.2 | 0.2 | 0.2 | | |

**Figure 10.39** $2^3$ design matrix for the wallpaper study.

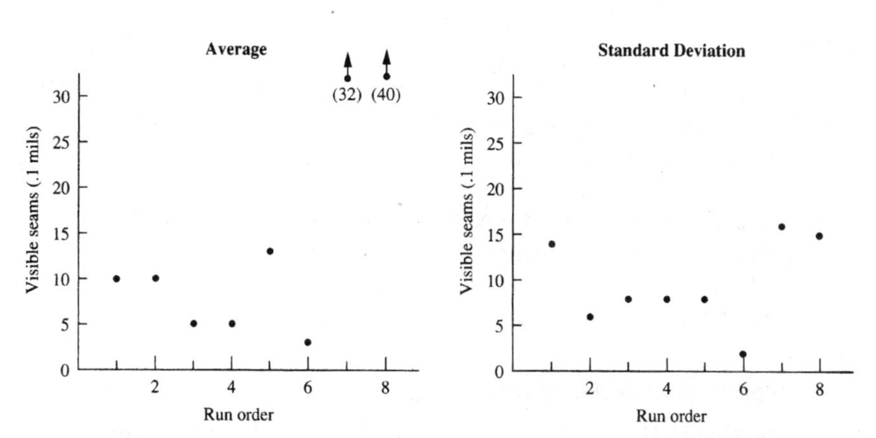


**Figure 10.40** Run charts for the wallpaper study.

Pre-roll showed a larger effect on standard deviations of visible seams. Pre-roll had an effect for the average in cycle 4 that was not confirmed in this study. The standard-deviation response plot showed fewer visible seams with a 10-min pre-roll waiting period. The team attributed this to the new run of material. This factor will be watched in future studies.

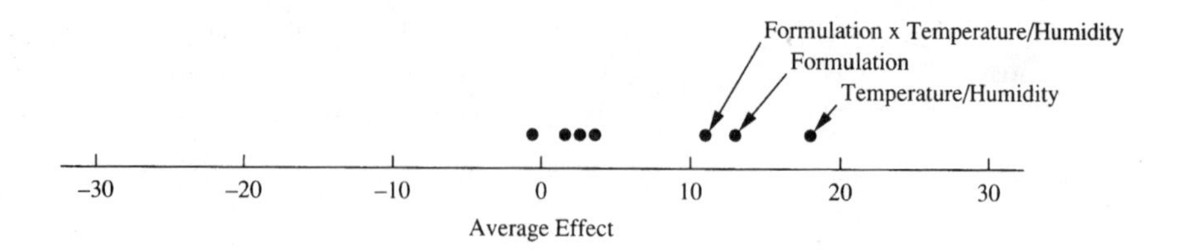


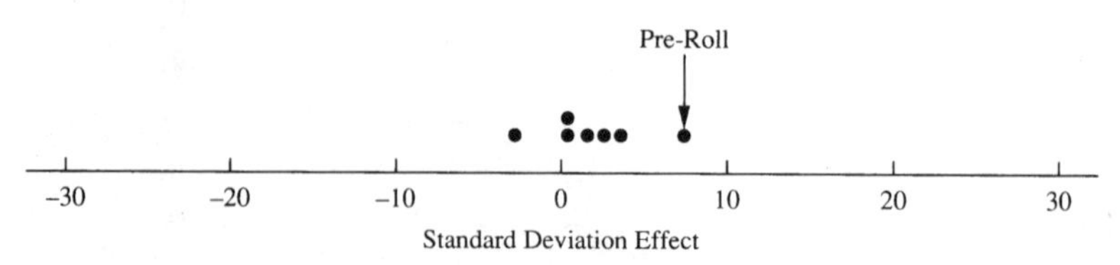


**Figure 10.41** Dot diagram for the wallpaper study.

Analysis of the cube in Fig. 10.42*a* revealed the interaction of formulation and temperature/humidity. Response plots for the important factors are given in Fig. 10.42*b*.

The result of cycle 5 was a confirmation of cycle 4. The team still had a problem with seams. Cycle 5, however, showed that taking a pre-roll step of allowing 10 min for the roll to lie flat prior to installation helped some with the seams. The team therefore decided that pre-roll could be controlled as a noise factor by including this finding in the instructions given to the installer.

Because the appearance team had preferred the new product to the old in all other quality characteristics, the product development team decided to study the effect of some of the control factors on variations in the noise factors identified during installation of the wallpaper.

### Sixth improvement cycle: Back to design factors

The objective of the next cycle was to determine the effect that installation factors had on seams and appearance in a new prototype formulation. The hope was to reduce the effect of noise factors by changing the levels of three control factors:

| Factors | Old prototype | New prototype |
|---|---|---|
| Wear layer | 1.0 mil | 2.0 mils |
| Intermediate layer | Formula 1 | Formula 2 |
| Backing layer | Material A | Material B |

Hopefully, the new settings for these three control factors would desensitize the product to the noise factor, temperature/humidity, and the need for a 10-min pre-roll.

Figure 10.43 contains the planning form for the study. Three control factors and two installation (noise) factors are studied. Four measurements were made on seams for each sheet, and averages and standard deviations were calculated. A $2^{5-1}$ design was chosen because there was no confounding of the two-factor interactions.

At completion of the experiment, prototypes were tested for appearance. It was found that appearance of the new design remained good throughout the 16 runs. Laboratory testing for scratches, stains, and gloss retention confirmed the robustness of the new prototype.

Figure 10.44 contains the response plots for standard deviations of visible seams that summarize the important results of the experiment. The first plot shows a strong interaction between thickness of wear layer, a control factor, and temperature/humidity, a noise factor. The effect of this noise factor on visible seams is less with the 2.0-mil wear layer. The thinner wear level (1.0 mil) was affected by the high-temperature/high-humidity combination.

The second plot shows a strong interaction between the other two control factors. Backing material and formulation of the intermediate layer interacted with respect to seams. Because both backing materials are needed for different applications, material *A* is chosen with formula 1 and material *B* is chosen with formula 2.

The third plot demonstrates a strong relationship between an installation factor and visible seams. By laying out the roll for 10 min before installation, the seams will have less variation. This adds to the installation time.

Based on these analyses and the team's current knowledge, the following conclusions are drawn:

| Control factors | Best setting |
|---|---|
| Thickness of wear layer | 2.0 mils |
| Backing material *A* | Formula 1 |
| Backing material *B* | Formula 2 |

| Noise factors | Method of control |
|---|---|
| Pre-roll | Have installation instructions include a 10-min pre-roll step. |
| Temperature/humidity | Desensitized by setting control factor, *thickness of wear layer,* at 2.0 mils. |

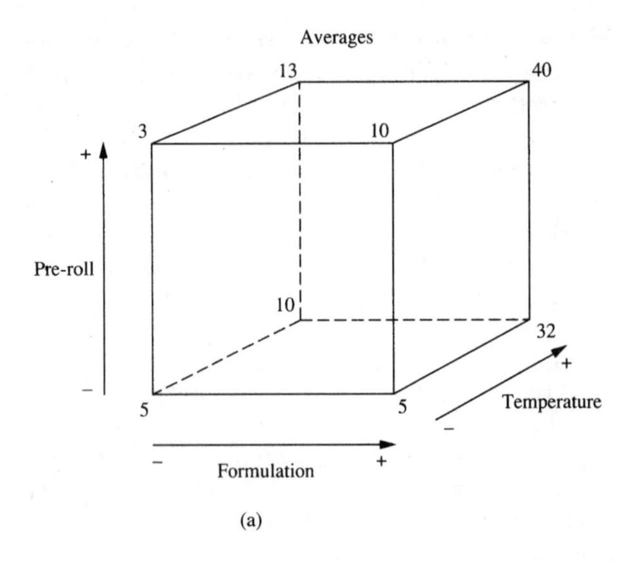


(a)

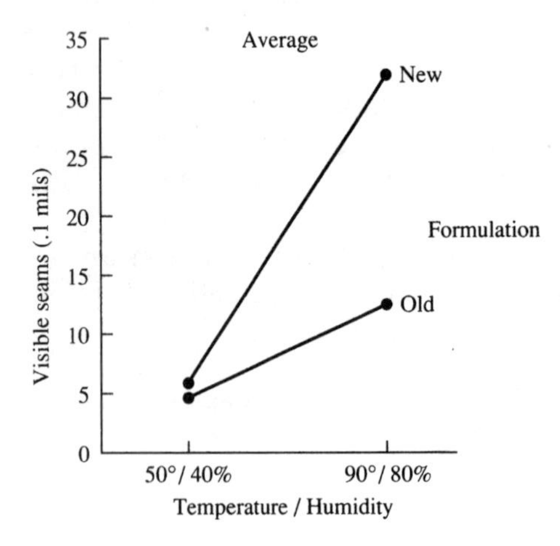


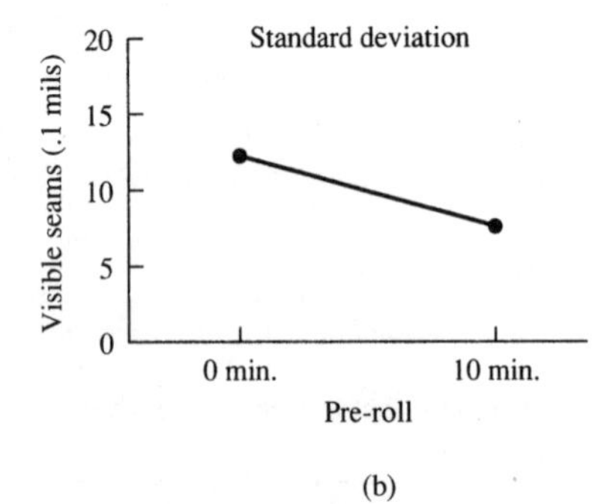


(b)

**Figure 10.42** (*a*) Cube plot and (*b*) response plots for the wallpaper experiment.

**1. Objective:**

Determine the effect of installation factors on a redesign for wallpaper by changing three control factors.

**2. Background information:**

The redesign has a new color and pattern that has a beautiful appearance. There has been a problem with seams. Probable cause is high temperature or high humidity at the installation site.

**3. Experimental variables:**

A.

| Response variables | Measurement technique |
|---|---|
| 1. Visible seams (average, standard deviation) | new gage (.1 mils) |
| 2. Appearance | subjective scoring |

B.

| Factors under study | Levels | |
|---|---|---|
| 1. Backing material | Material *A* | Material *B* |
| 2. Formulation of intermediate layer | Formula 1 | Formula 2 |
| 3. Thickness of wear layer | 1.0 mils | 2.0 mils |
| 4. Temperature/humidity | 50°/40% | 90°/80% |
| 5. Pre-roll | 0 min. | 10 min. |

C.

| Background variables | Method of control |
|---|---|
| 1. Installer | One person, standard instructions |
| 2. Type of adhesive | Use standard for all applications |
| 3. Wall | Use coarse material |
| 4. Thickness of intermediate layer | 4.0 mils |
| 5. Formulation of wear layer | Same as old design |
| 6. Time material lays flat | 4 hours |
| 7. Cut angle | 90° |
| 8. Seam layup | low |

**4. Replication:**

Four measurements for visible seams per sheet.

**5. Methods of randomization:**

Order of the 16 runs was randomized using a random permutation table.

**6. Design matrix:** $2^{5-1}$ fractional factorial design.

**7. Data collection forms:** (not shown here)

**8. Planned methods of statistical analysis:**

Run charts, dot diagram, cube, and response plots.

**9. Estimated cost, schedule, and other resources:**

Evaluated in the test room with temperature and humidity controls. Four days are required to complete. Appearance scoring done by appearance team.

**Figure 10.43** Documentation of the wallpaper experiment for cycle 6.

Updates to the QFD relation diagram are given in Fig. 10.45.

The team next had to design the production process. No new equipment was needed for the production process. Backing material *B* was new and required a new supplier. The increase in wear thickness to 2.0 mils increased costs. This increase was offset by the lower cost of backing material *B*.

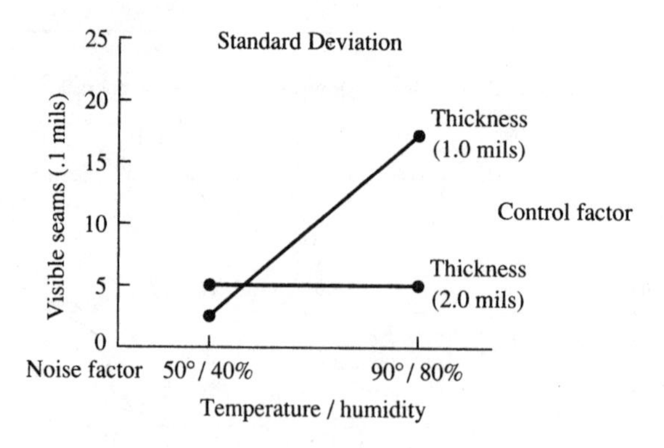


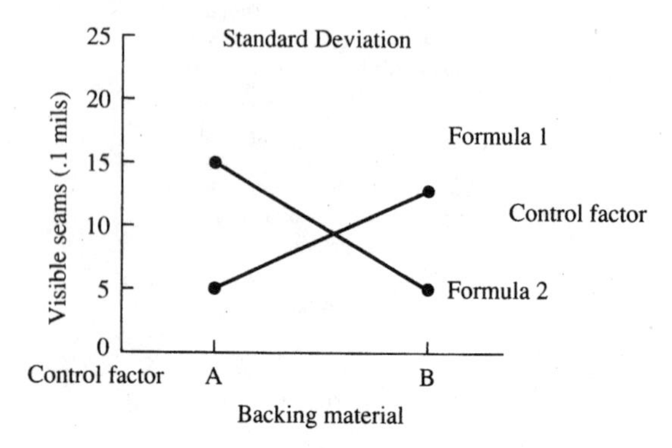


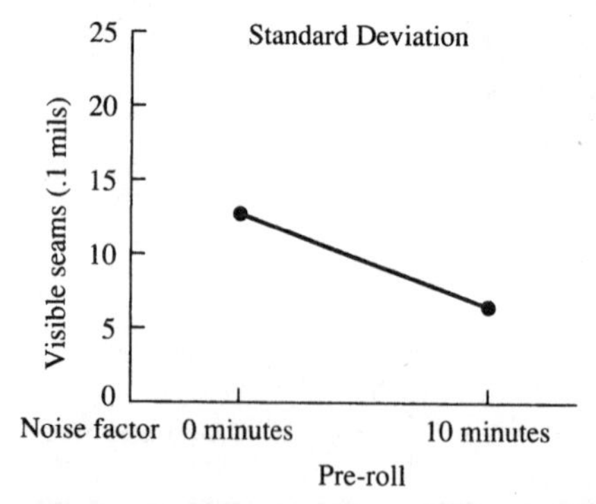


**Figure 10.44** Response plots for the wallpaper experiment.

Training based on the change to the production process was conducted for all those involved. Short production runs indicated no new production problems. The changes made in the redesign of the product did not result in any difficulties for the production process.

The team had decided to proceed with a pilot run of samples for each

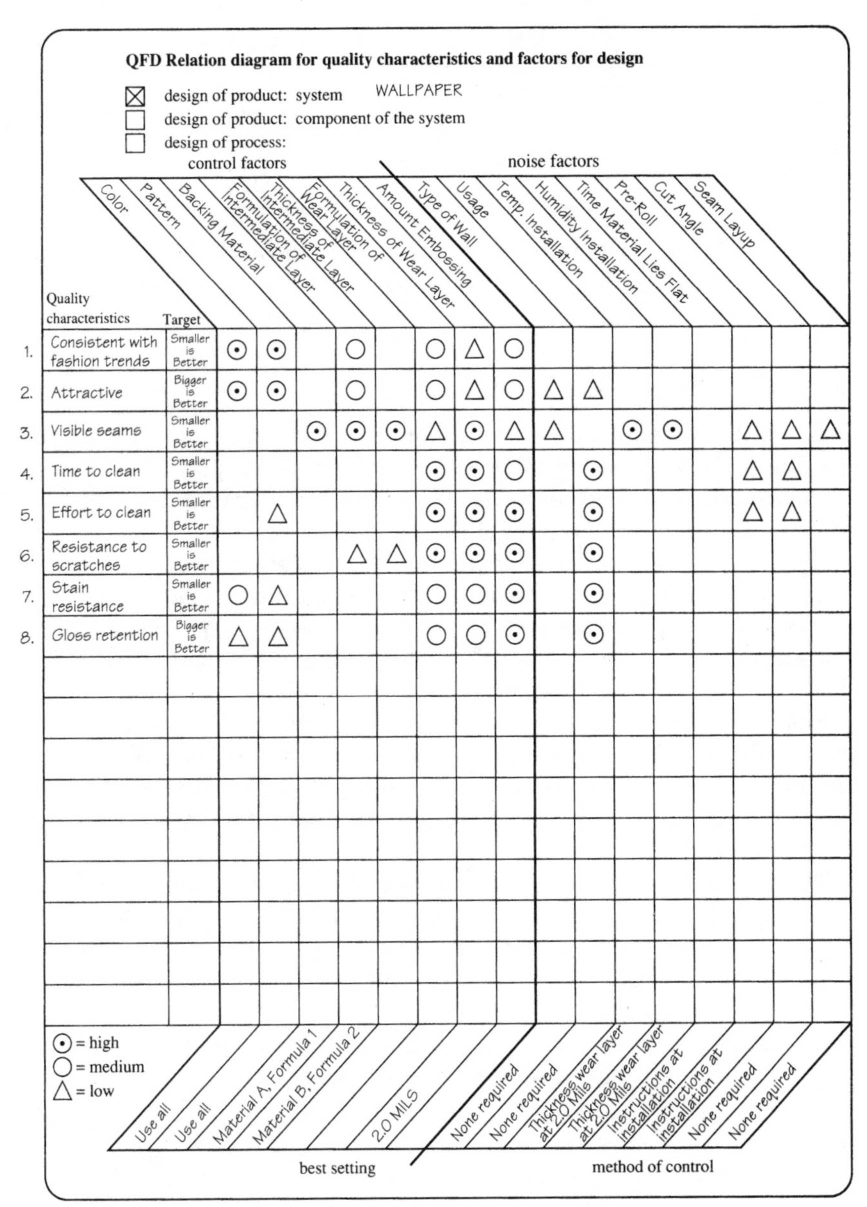


What planned experiments need to be run to test this diagram? $2^{8-4}$ cycle 3, $2^{3}$ cycle 4, $2^{5-1}$ cycle 5

**Figure 10.45** Updated QFD relation diagram for redesigning wallpaper.

of the five basic color/pattern product types evaluated in cycle 2 (see Fig. 10.26). Formulation for each type of color/pattern was as follows:

| Product type (color/pattern) | 1 | 2 | 3 | 4 | 5 |
|---|---|---|---|---|---|
| Backing material | *A* | *A* | *A* | *B* | *B* |
| Formula | 1 | 1 | 1 | 2 | 2 |

Backing material *A* was chosen for the first three product types because of the successful use of that material for other current product types similar to them.

### Seventh improvement cycle: Field test

The objective of this study was to field-test the pilot run of wallpaper. All five product types from the pilot run were evaluated along with the old product. A field test was chosen over the temperature and humidity environment of the R&D laboratory.

This cycle must increase understanding of performance of the new design with respect to potential problems. Including the old product with the pilot product of the new design under widely varying conditions should increase the team's degree of belief about its future predictions of product performance.

Field sites were used to create a blocking factor. The blocks chosen were judged likely to create extreme conditions. Figure 10.46 gives the planning form for the study. Important results of the study are given in Fig. 10.47. A run chart for standard deviation of visible seams is given in Fig. 10.48.

The run chart in Fig. 10.48 confirmed the results of the sixth cycle. Standard deviations for visible seams are 14 or less. The pilot run of five product types compared favorably with the old product types. The team predicted that claims against the redesigned product will not exceed claims against the old product.

For all five product types, appearance exceeded the old product at all four field sites. The action of the team was a decision to start production of the redesigned product.

### Eighth improvement cycle: Capability

During cycle 7, production continued to run the pilot for 4 days to determine capability of the processes. The objective of the eighth cycle was to learn about the current performance of the production processes before production of the redesigned product.

Figure 10.49 contains control charts for thickness of the wear layer. Five samples of wallpaper were taken every 2 h and measured for thickness. Control limits were calculated at the end of 20 subgroups. A spe-

**1. Objective:**
Determine the performance of a redesign of a wallpaper product under various field conditions. This information will be used to determine if the current product should be replaced by the new design.

**2. Background information:**
Previous experiments using fractional factorial and factorial designs lead to changes in the current design. Noise factors during installation are affecting the variation in visible seams.

**3. Experimental variables:**

A.

| Response variables | Measurement technique |
|---|---|
| 1. Visible seams (average, standard deviation) | gage (.1 mils) |
| 2. Appearance | subjective scoring |

B.

| Factors under study | Levels |
|---|---|
| 1. Product types | 1 2 3 4 5 6 = old product |

C.

| Background variables | Method of control |
|---|---|
| 1. Type of adhesive | Use standard for all applications |
| 2. Time material lays flat | 2 hours |
| 3. Cut angle | Standard at 90° |
| 4. Seam layup | Set at low |
| 5. Pre-roll | Instructions (10 min.) |
| 6. Thickness of wear layer | Measure thickness |

**Method of control for other background variables:** Create four blocks (field sites) made up of different treatment combinations of the background variables.

| Background factor | Block 1 | Block 2 | Block 3 | Block 4 |
|---|---|---|---|---|
| 1. Field site | N.E. | N.W. | S.E. | S.W. |
| 2. Humidity | middle | low | high | low |
| 3. Temperature | middle | low | high | high |
| 4. Wall | wood | old | cement | old |

**4. Replication:**
Four measurements for visible seams per sheet.

**5. Methods of randomization:**
Order of the six levels of product types within block is randomized using a random permutation table.

**6. Design matrix: (attach copy)** Not shown.

**7. Data collection forms: (attach copy)** See Figure 10.47.

**8. Planned methods of statistical analysis:**
Compute the average and standard deviation of the four measurements. Plot run charts for the six sample types adjusted for block effect.

**Figure 10.46** Documentation of the wallpaper experiment for cycle 7.

cial cause occurred at start-up on July 21. A check of viscosity indicated too high a reading. Viscosity was brought down, and the pilot continued.

Process capability is a prediction of individual measurements of thickness of wear layer. The action from cycle 6 was to increase wear layer from 1.0 to 2.0 mils. This would help desensitize the possible problem of

Response: standard deviation of visible seams (run order in parentheses)

| Product | Block 1 | Block 2 | Block 3 | Block 4 |
|---|---|---|---|---|
| 1 | (3) 2 | (2) 4 | (6) 12 | (3) 5 |
| 2 | (4) 7 | (4) 6 | (1) 9 | (1) 7 |
| 3 | (6) 2 | (5) 12 | (3) 12 | (5) 1 |
| 4 | (1) 9 | (1) 6 | (2) 14 | (2) 10 |
| 5 | (2) 7 | (3) 4 | (4) 10 | (4) 8 |
| 6 = old | (5) 9 | (6) 10 | (5) 12 | (6) 11 |

**Figure 10.47** Test results for wallpaper study.

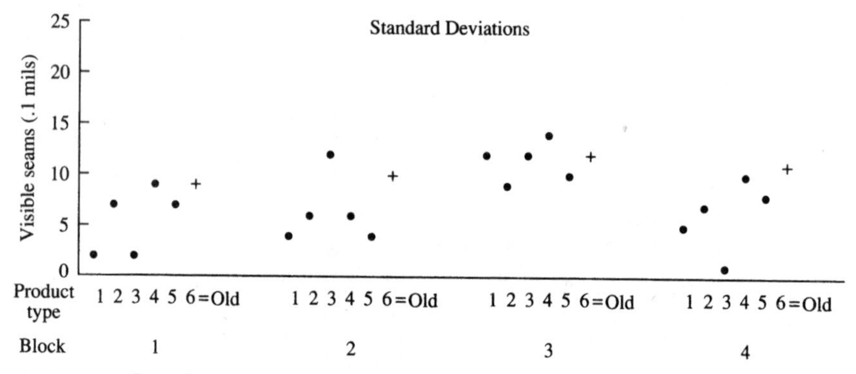


**Figure 10.48** Run chart for block design of wallpaper study, cycle 7.

high temperature and high humidity that could occur during installation. The capability from this cycle was from 1.25 to 2.87 mils. The average was 2.06 mils. This was slightly higher than the target of 2.0 mils. The thickness of the wear layer from samples involved in the field test of cycle 7 ranged from 1.4 to 2.5 mils.

Based on the knowledge gained from this cycle, the following actions were taken:

1. Target for thickness of wear layer should continue at 2.0 mils (increases production costs).
2. Process engineering should do a study to establish the relationship of viscosity to thickness of wear layer.
3. Monitor viscosity with control charts. Take action on special causes. Identify common causes of variation.
4. Discuss results of pilot with product development team.

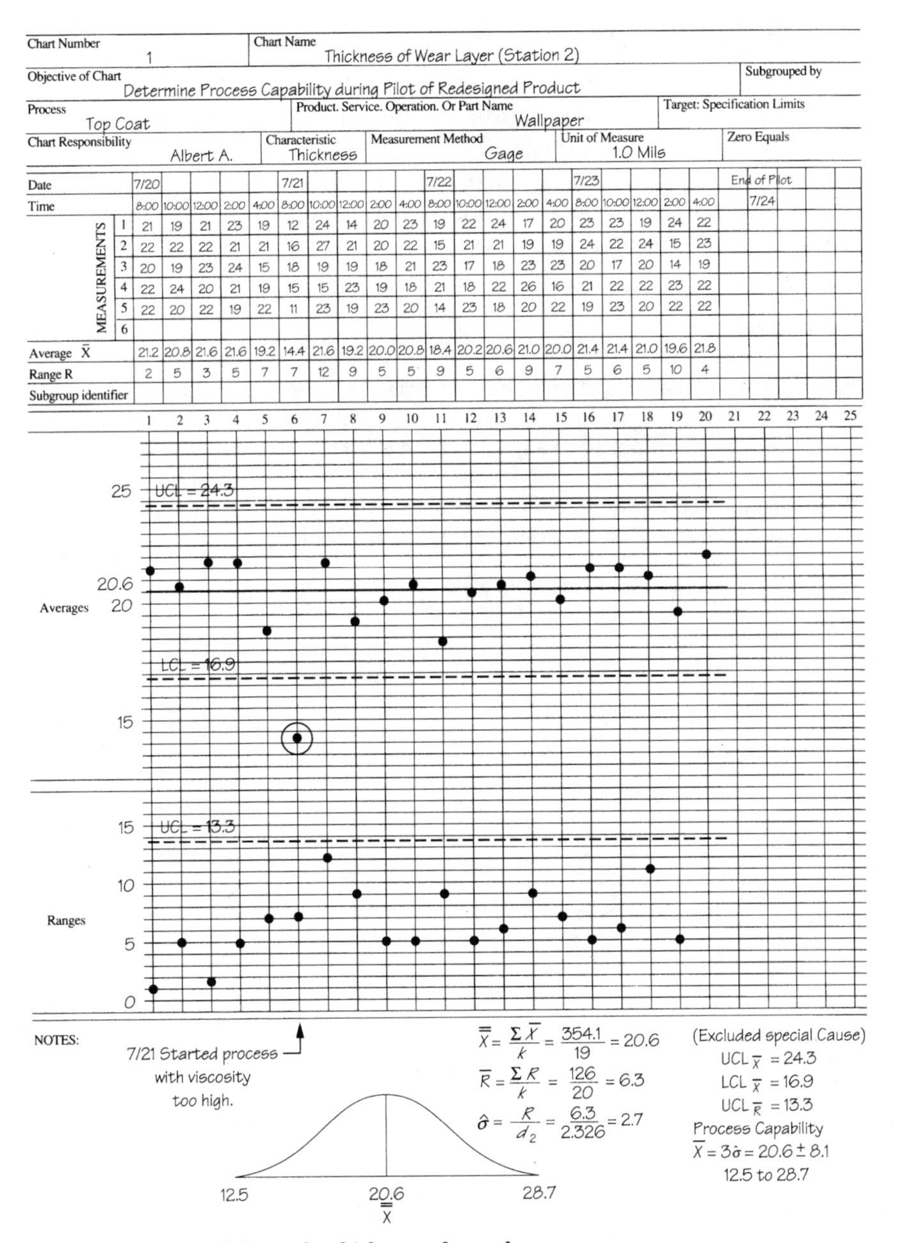

| Chart Number | Chart Name | Subgrouped by |
|---|---|---|
| 1 | Thickness of Wear Layer (Station 2) | |

| Objective of Chart |
|---|
| Determine Process Capability during Pilot of Redesigned Product |

| Process | Product. Service. Operation. Or Part Name | Target: Specification Limits |
|---|---|---|
| Top Coat | Wallpaper | |

| Chart Responsibility | Characteristic | Measurement Method | Unit of Measure | Zero Equals |
|---|---|---|---|---|
| Albert A. | Thickness | Gage | 1.0 Mils | |

| | 1 | 2 | 3 | 4 | 5 | 6 | 7 | 8 | 9 | 10 | 11 | 12 | 13 | 14 | 15 | 16 | 17 | 18 | 19 | 20 | 21–22 |
|---|---|---|---|---|---|---|---|---|---|---|---|---|---|---|---|---|---|---|---|---|---|
| Date | 7/20 | | | | | 7/21 | | | | | 7/22 | | | | | 7/23 | | | | | End of Pilot |
| Time | 8:00 | 10:00 | 12:00 | 2:00 | 4:00 | 8:00 | 10:00 | 12:00 | 2:00 | 4:00 | 8:00 | 10:00 | 12:00 | 2:00 | 4:00 | 8:00 | 10:00 | 12:00 | 2:00 | 4:00 | 7/24 |
| Measurements 1 | 21 | 19 | 21 | 23 | 19 | 12 | 24 | 14 | 20 | 23 | 19 | 22 | 24 | 17 | 20 | 23 | 23 | 19 | 24 | 22 | |
| 2 | 22 | 22 | 22 | 21 | 21 | 16 | 27 | 21 | 20 | 22 | 15 | 21 | 21 | 19 | 19 | 24 | 22 | 24 | 15 | 23 | |
| 3 | 20 | 19 | 23 | 24 | 15 | 18 | 19 | 19 | 18 | 21 | 23 | 17 | 18 | 23 | 23 | 20 | 17 | 20 | 14 | 19 | |
| 4 | 22 | 24 | 20 | 21 | 19 | 15 | 15 | 23 | 19 | 18 | 21 | 18 | 22 | 26 | 16 | 21 | 22 | 22 | 23 | 22 | |
| 5 | 22 | 20 | 22 | 19 | 22 | 11 | 23 | 19 | 23 | 20 | 14 | 23 | 18 | 20 | 22 | 19 | 23 | 20 | 22 | 22 | |
| 6 | | | | | | | | | | | | | | | | | | | | | |
| Average $\bar{X}$ | 21.2 | 20.8 | 21.6 | 21.6 | 19.2 | 14.4 | 21.6 | 19.2 | 20.0 | 20.8 | 18.4 | 20.2 | 20.6 | 21.0 | 20.0 | 21.4 | 21.4 | 21.0 | 19.6 | 21.8 | |
| Range R | 2 | 5 | 3 | 5 | 7 | 7 | 12 | 9 | 5 | 5 | 9 | 5 | 6 | 9 | 7 | 5 | 6 | 5 | 10 | 4 | |
| Subgroup identifier | | | | | | | | | | | | | | | | | | | | | |



**Figure 10.49** Control charts for thickness of wear layer.

### Final actions of the product development team

The team reviewed the charter (Fig. 10.21). Based on their current knowledge, the expected results can be achieved. The team was able to stay within the boundaries of the charter.

The team recommended proceeding with production of the redesigned wallpaper. Recommendation was approved, and production started the next week.

## Exercises

**10.1** Review each case study. Discuss possible alternative approaches to the design and analysis of data. Discuss why both cases were analytic studies. Were there any enumerative aspects of the two case studies?

**10.2** Study case studies involved with improving quality. Compare and discuss approaches used to those presented in this book.

**10.3** Create your own case study for improving quality from an area of interest to you.

# Appendix A

# Improvement Using Control Charts

## A.1 Introduction

This appendix describes the use of Shewhart control charts in improvement. As discussed in Chap. 2, simple run charts of the key measures of interest often provide an adequate analysis and presentation of the data from the test. Control charts offer a higher level of sophistication which may be warranted in some studies.

Walter Shewhart (1931) introduced the concept that variation in a quality characteristic can be attributed to two types of causes:

*Common causes of variation.* Causes that are inherent in the process over time, that affect everyone working in the process, and that affect all outcomes of the process.

*Special causes of variation.* Causes that are not part of the process all the time or that do not affect everyone, but arise because of specific circumstances.

A process whose outcomes are affected only by common causes is called a *stable process* or one that is in a state of statistical control. A stable process implies only that the variation is predictable within bounds. A process whose outcomes are affected by both common causes and special causes is called an *unstable process.* For an unstable process, the variation from one time period to the next is unpredictable. As special causes are identified and removed, the process becomes stable.

The Shewhart control chart provides an operational definition of these concepts. The control chart is a statistical tool used to distinguish variation in a process due to common causes and variation due to special causes. Dr. Walter A. Shewhart is credited with developing the control chart. In 1931 Shewhart published *Economic Control of Quality of Manufactured Product,* which discussed the theory and application of control charts.

The control chart method has general applicability throughout an organization. Top managers can use a control chart to study variation in sales, supervisors can use the tool to assign responsibility for improvement of a process, administrative personnel can use it to identify opportunities for improvement, and operators can use a control chart to determine when to adjust a process.

The control chart method is often confused with the use of specifications and tolerances. Figure A.1 contrasts these two very different ap-

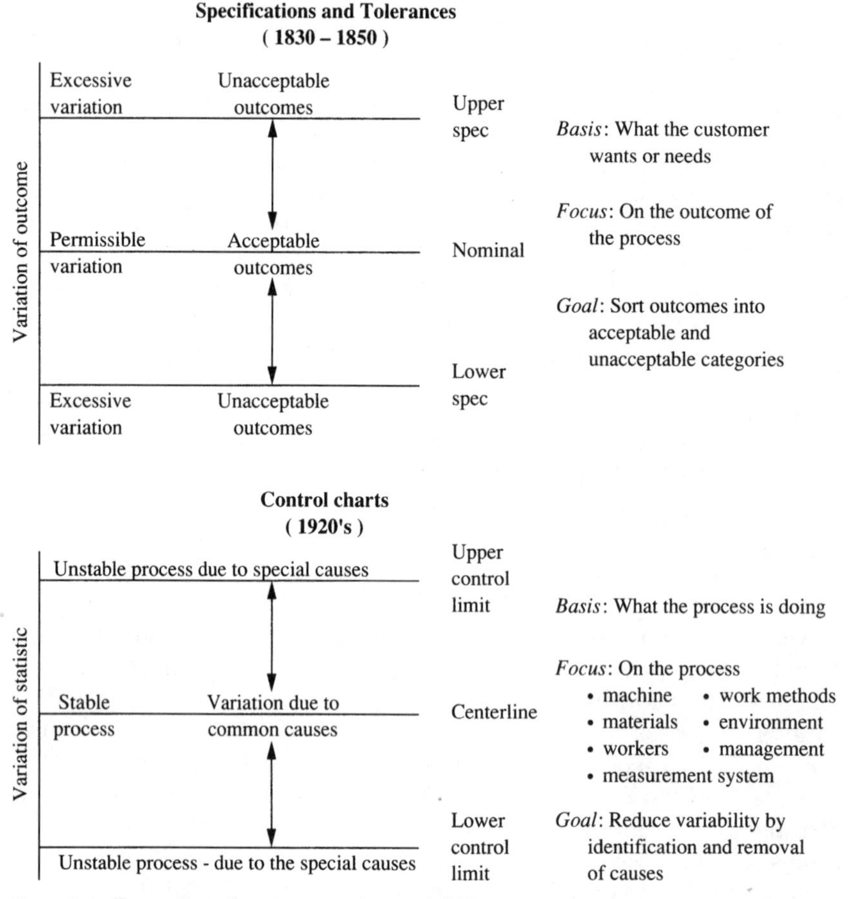


**Figure A.1** Two approaches to managing variability.

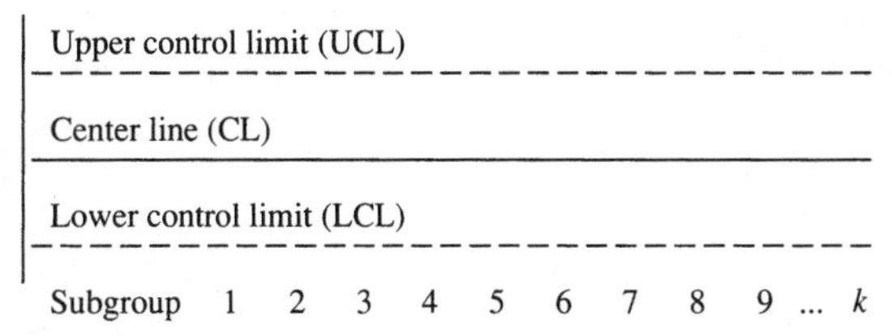


**Figure A.2** Illustration of the form of a control chart.

proaches to the management of variability. The control chart consists of three lines and points plotted on a graph. Figure A.2 illustrates the form of a typical control chart.

The construction of a control chart typically involves

- Plotting the data or some summary statistic of a subgroup of the data in a run order (time is the most common order)
- Determining some measure of the central tendency of the data (such as the average)
- Determining some measure of the common-cause variation of the data
- Calculating a centerline and upper and lower control limits
- Plotting the statistic on the chart, with the horizontal axis representing the subgroup number and the vertical axis being a scale for the statistic.

When 20 to 30 subgroups have been plotted on the chart, the control limits can be computed. The control limits bound the variation of the statistic that is due to common causes. Formulas for control limits have been developed for all common types of control charts. Although the limits are based on statistical theory, Deming (1986, p. 334) states that control limits should not be associated with any calculation of probability.

## A.2 Control Charts for Individual Measurements

One of the most useful types of control charts is the control chart for individual measurements, or the *X* chart. This control chart is a simple extension of the run chart. The control chart for individuals is useful when

- The data consist of some type of measure (see later discussion of continuous data).

- There is no rational way to organize the data into subgroups (see later section on subgrouping and stratification with control charts).
- Measures of performance of the process can be obtained only infrequently.
- The variation at any one time (within a subgroup) is insignificant relative to that between subgroups.

Examples of situations and data in which a control chart for individuals can be useful include batch processes, accounting data, maintenance records, shipment data, yields, efficiencies, sales, costs, and forecast or budget variances. Often the frequency of data collection cannot be controlled for these situations and types of data.

Instrument readings such as temperatures, flows, pressures, etc. often have minimal variation at any one time, but will change over time. The study of tool wear is another example of insignificant short-term variation relative to variation over time. Control charts of the individual measurements can often be useful in these cases.

These are some advantages of the control chart for individuals (compared to other types of charts for continuous data):

- The chart is an extension of the familiar run chart.
- No calculations of statistics is required for plotting on the chart.
- Plotting is done each time a measurement is made, providing fast feedback.
- Charts for multiple measures of performance can be stacked for presentation purposes.
- The capability of a process can be evaluated directly from the control limits on the chart.

The $X$ chart is somewhat less sensitive in its ability to detect the presence of a special cause than the other types of charts for continuous data. Besides this reduced sensitivity, there are some other disadvantages to using an $X$ chart to study variation in data:

- Since each individual measure is plotted on the chart, there is no opportunity to focus on different sources of variation through subgrouping.
- All sources of variation are combined on one chart, sometimes making identification of the important sources of variation difficult.
- The $X$ chart is sensitive to a nonsymmetric distribution of data and may require data transformation to be used effectively.

To develop an $X$ chart, 20 to 30 measurements are required to characterize the process of interest. The symbol for the number of measures used to calculate control limits is $k$. The individual measurements are plotted on the $X$ chart, and the average of the individual measurements is used for the centerline of the chart. The *moving ranges* of consecutive measurements are used to estimate the variation of the process and to develop control limits for the $X$ chart.

The moving range is calculated by pairing consecutive measurements. The range is calculated for each set of two measurements by subtracting the low value from the high value. Each individual measurement is considered twice in the calculation of the moving ranges. Since a "previous" measurement is not available for the first measurement in the set, only $k-1$ moving ranges can be calculated. The average of the moving ranges $\overline{\text{MR}}$ is used for control limit calculations. Figure A.3 contains a form used to calculate the appropriate control limits. The steps for developing a control chart for individuals are also shown on the form in Fig. A.3.

Since the $X$ chart of individual measurements contains all the information available in the data, it is not necessary to plot the moving ranges. However, plotting the moving ranges will make an increase or decrease in the variation of the process more obvious than that on the $X$ chart. If the moving ranges are plotted, the additional control chart rules (discussed later in this appendix) should not be used to evaluate for special causes.

An example of an $X$ chart concerns a chemical product that is shipped in hopper cars with a sample taken from each car during loading. Laboratory tests are made on each sample for product certification, and this test result becomes one dot on the chart. The cars are loaded from storage bins that are filled on an intermittent basis from a process unit. Laboratory results for the concentration of an additive for the last 25 cars loaded are used to develop control charts for the product shipped.

The control chart for the additive is shown in Fig. A.4, and the calculation sheet is shown in Fig. A.5. As can be seen in the two figures, the moving ranges for car numbers 14 and 15 are greater than the moving-range upper control limit of 12.8. These two values are removed, and the average moving range is recalculated. The revised $\overline{\text{MR}} = 2.86$ was used to calculate the control limits for the $X$ chart. The control chart indicates there is a special cause present for car 14. Note that the 236-ppm concentration for car 14 is associated with the two moving ranges that were above the upper control limit.

## A.3 Subgrouping

The concept of subgrouping is one of the most important components of the control chart method. The other types of control charts require

NAME ______________________ DATE ______________

PROCESS ______________________ SAMPLE DESCRIPTION ______________

NUMBER OF SUBGROUPS (k) ____________ BETWEEN (dates) ______ — ______

---

$\bar{X} = \frac{\Sigma X}{k} = $ ________ = ______   $\overline{MR} = \frac{\Sigma MR}{k-1} = $ ________ = ______

**X-CHART**

$UCL = \bar{X} + (2.66 * \overline{MR})$

$UCL = \quad + (2.66 * \quad )$

$UCL = \quad +$

$UCL = $ ____________

$LCL = \bar{X} - (2.66 * \overline{MR})$

$LCL = \quad - (2.66 * \quad )$

$LCL = \quad -$

$LCL = $ ____________

**MR CHART**

$UCL_{MR} = 3.27 * \overline{MR}$

$UCL_{MR} = 3.27 *$

$UCL_{MR} = $ ________

Recalculate $\overline{MR}$ after removing MR's greater than $UCL_{MR}$

$\overline{MR}^* = \sum MR \ / \ k - ?$

$\overline{MR}^* = $ ______ / ______

$\overline{MR}^* = $ ________

---

**CALCULATION OF CONTROL LIMITS**

The steps for developing a control chart for individuals are the following:

1. Calculate the k-1 moving ranges and the average of the moving ranges ($\overline{MR}$).
2. Calculate the upper control limit for the moving range using $UCL_{MR} = 3.27 * \overline{MR}$.
3. Remove any moving range bigger than the $UCL_{MR}$ and recalculate the average moving range ($\overline{MR}$).◊
4. Calculate the average of the k measurements ($\bar{X}$).
5. Calculate the control limits for the X-chart using:

   $UCL = \bar{X} + (2.66 * \overline{MR})$

   $LCL = \bar{X} - (2.66 * \overline{MR})$

6. Calculate and draw a scale such that the control limits "enclose" the inner 50% of the charting area.
7. Plot the k = 20 to 30 measurements on an X-chart.
8. Draw the centerline ($\bar{X}$) and the control limits on the X-chart.

◊ Note: This recalculation should be done only once.

**Figure A.3** An $X$ chart calculation form.

some type of subgrouping strategy. Shewhart's principle is to organize (classify, stratify, group, etc.) data from the process in a way that ensures the greatest similarity among the data in each subgroup and the greatest difference among the data in different subgroups. The aim of rational subgrouping is to include only common causes of vari-

Chart: 133 | Chart name: Additive at hopper car loading
Objective: Study variation and identify special causes | Subgrouped by: Hopper car
Process: Hopper car loading | Product: Q100 through Q209 | Target: 225 +/- 25 | date: 9/93
Chart responsibility: Lab post #2 | Characteristic: Additive | Measurement method: GC | Unit: PPM | Zero = 0

| Car | 1 | 2 | 3 | 4 | 5 | 6 | 7 | 8 | 9 | 10 | 11 | 12 | 13 | 14 | 15 | 16 | 17 | 18 | 19 | 20 | 21 | 22 | 23 | 24 | 25 | | | | | |
|---|---|---|---|---|---|---|---|---|---|---|---|---|---|---|---|---|---|---|---|---|---|---|---|---|---|---|---|---|---|---|
| Time | | | | | | | | | | | | | | | | | | | | | | | | | | | | | | |
| Measure | 215 | 218 | 222 | 217 | 216 | 214 | 219 | 221 | 216 | 220 | 218 | 218 | 221 | 236 | 222 | 221 | 216 | 218 | 223 | 217 | 218 | 221 | 220 | 219 | 215 | | | | | |
| | | | | | | | | | | | | | | | | | | | | | | | | | | | | | | |
| MR | - | 3 | 4 | 5 | 1 | 2 | 5 | 2 | 5 | 4 | 2 | 0 | 3 | 15 | 14 | 1 | 5 | 2 | 5 | 6 | 1 | 3 | 1 | 1 | 4 | | | | | |

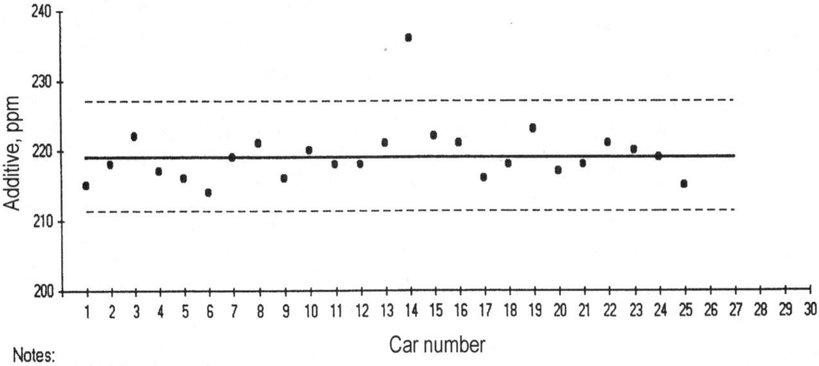

Additive, ppm

Car number

Notes:

**Figure A.4** Individual control chart for additive.

ation within a subgroup, with all special causes of variation occurring between subgroups.

The most common method to obtain rational subgroups is to hold time "constant" within a subgroup. Only data taken at the same time (or in some selected time period) are included in a subgroup. Data from different time periods will be in different subgroups. This use of time as the basis of subgrouping allows the detection of causes of variation related to time.

Statistics calculated from the data in the subgroups are usually plotted in order of time. The subgroups can also be ordered by other factors, such as supplier, shift, operator, or part position, to investigate the importance of these factors.

As an example of subgrouping, consider a study planned to reduce late payments. Historical data from the accounting files would be used to study the variation in late payments. The statistic, percentage of late payments, can be used to summarize the data in each subgroup. What is a good way to subgroup the historical data on late payments? The data could be grouped by billing month, by receiving month, by major account, by product line, or by account manager. Knowledge or theories about the process should be used to develop rational subgroups. Some combination of time (either receiving or billing month) and one or more of the other process variables would be a reasonable way to develop the first control chart.

NAME Additive DATE 9/93

PROCESS Hopper Car Loading SAMPLE DESCRIPTION Composite

NUMBER OF SUBGROUPS (k) 25 BETWEEN (dates) Car 1 - 25

---

$\bar{X} = \frac{\Sigma x}{k} = \frac{5481}{25} = 219.2$ $\overline{MR} = \frac{\Sigma MR}{k-1} = \frac{94}{24} = 3.92$

---

**X-CHART**

$UCL = \bar{X} + (2.66 * \overline{MR})$

$UCL = 219.2 + (2.66 * 2.95)$

$UCL = 219.2 + 7.8$

$UCL = 227.0$

$LCL = X - (2.66 * \overline{MR})$

$LCL = 219.2 - (2.66 * 2.95)$

$LCL = 219.2 - 7.8$

$LCL = 211.4$

**MR CHART**

$UCL_{MR} = 3.27 * \overline{MR}$

$UCL_{MR} = 3.27 * 3.92$

$UCL_{MR} = 12.8$

Recalculate $\overline{MR}$ after removing MR's greater than $UCL_{MR}$

$\overline{MR}^* = \Sigma MR \ / \ k - ?$

$\overline{MR}^* = 65 \ / \ 22$

$\overline{MR}^* = 2.95$

---

**Figure A.5** Calculation form for individual control chart for additive.

## A.4 Interpretation of a Control Chart

The control chart provides a basis for taking action to improve a process. A process is considered to be stable when there is a random distribution of the plotted points within the control limits. If there are points outside the limits, or if the distribution of points within the limits is not random, then the process is considered to be unstable, and action should be taken to uncover the special causes of variation.

Figure A.6 lists four rules that are recommended for general use with control charts. Special circumstances may warrant use of some of the additional tests given by Nelson (1984). Deming (1986, p. 319) emphasizes that it is necessary to state in advance what rules to apply. Revision of the control limits should be done only when the existing limits are no longer appropriate. This is the case when improvements have been made to the process and the improvements result in special causes on the control chart. Control limits should then be calculated for the new process.

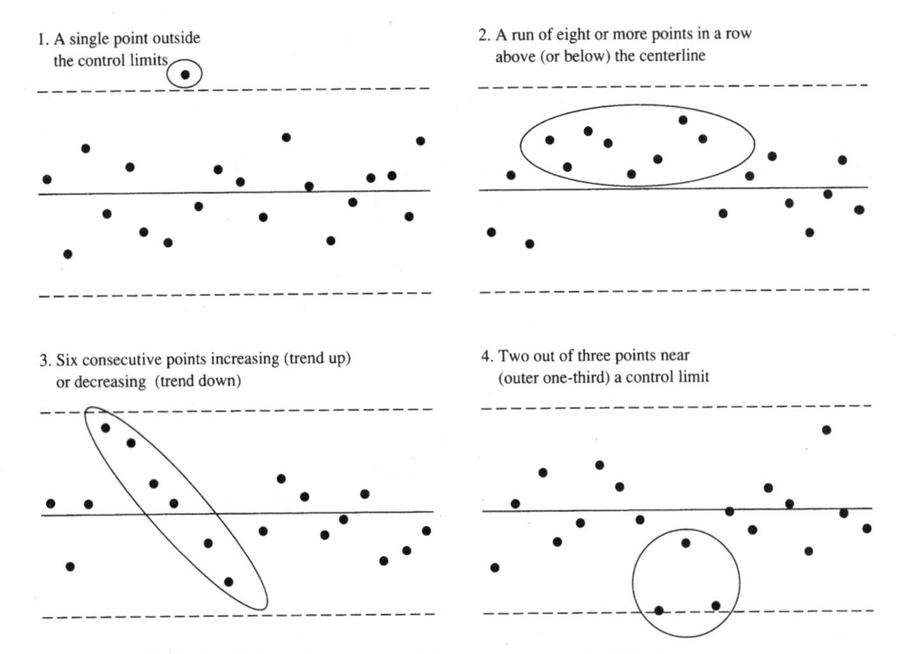


**Figure A.6** Rules for determining a special cause on a control chart.

## A.5 Types of Control Charts

There are many different types of control charts besides the basic $X$ chart. This section will discuss the various types. Examples will be given for the $\overline{X}$ and $R$ charts, the $P$ chart, the $C$ chart, and the $U$ chart.

Selection of the control chart to use in a particular application primarily depends on the type of data. The different types of data can be classified into three categories:

1. Classification data
2. Count data
3. Continuous data

The first two types are called *attribute data*; the third type is also called *variable data* or continuous data. For classification data, the quality characteristic is recorded in one of two classes. Examples of classes are conforming units and nonconforming units, go and no go, and good and bad. To obtain count data, the number of incidences of a particular type is recorded, such as number of mistakes, number of accidents, or number of sales leads. For continuous data, a measured numerical value of the quality characteristic is recorded, such as a dimension, physical attribute, cost, or time.

In general, data should be collected as continuous data whenever possible, since learning then requires many fewer measurements than with the use of attribute classifications or counts. The control charts for continuous data thus require fewer measurements in each subgroup than the attribute control chart. Typical subgroup sizes for continuous data charts range from 1 to 10, whereas subgroup sizes for attribute charts range from 30 to 1000.

Figure A.7 contains a summary of frequently used charts and the types of data to which they apply. When continuous data can be put in

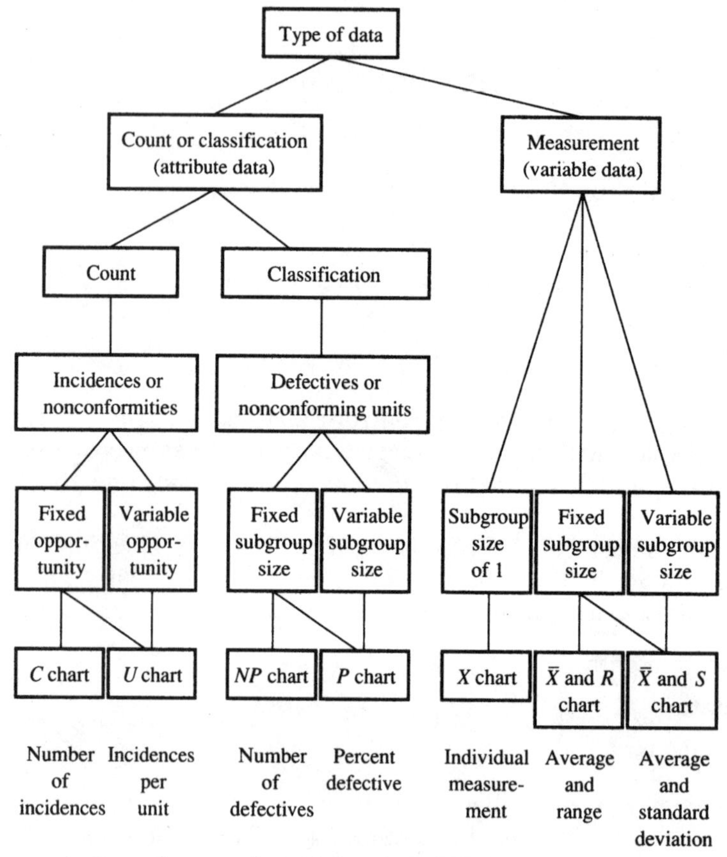


**Figure A.7** Selection of a particular type of control chart.

rational subgroups, the $\overline{X}$ and $R$ chart is the most frequently used. The $\overline{X}$ and $S$ chart is required when different subgroup sizes are necessary. The $S$ chart is also preferred if very large (greater than 10) subgroups are desirable. As discussed earlier, the $X$ chart is used when the data are not organized in subgroups.

### $\overline{X}$ and R charts

For the $\overline{X}$ and $R$ chart, the statistics used are the average of the measurements in each subgroup and the range (largest measurement minus the smallest measurement) in each subgroup. The average is designated as $\overline{X}$ and the range as $R$. Two control charts are required: the $\overline{X}$ chart for the averages and the $R$ chart for the ranges. Two to ten measurements are included in each subgroup, with subgroup sizes of 3 to 6 being most common.

Figure A.8 shows a worksheet that can be used to calculate limits for an $\overline{X}$ and $R$ chart. Figure A.9 shows an example of an $\overline{X}$ and $R$ chart.

If the subgroup size is variable (i.e., the number of measurements changes from subgroup to subgroup), the standard deviation, another descriptive statistic for variation, is used in place of the range. The $\overline{X}$ and $S$ chart is used in this case, where $S$ is the symbol for the standard deviation of the subgroup values. For many applications with continuous data, the subgroup size can be held constant. Since the calculation and presentation (variable limits) are more complex, the $S$ chart thus has limited use (except for applications that are on a computer).

### Attribute control charts

When sample units are classified into two categories (conforming and nonconforming units), the $P$ chart ($P$ = percentage nonconforming) is appropriate. The $P$ chart can be used with either a fixed or a variable subgroup size (so it is never necessary to use the $NP$ chart). For an $NP$ chart the number of nonconforming units is plotted rather than a percentage. The $NP$ chart is appropriate only for a fixed subgroup size. Subgroup sizes for $P$ charts typically range from 30 to 1000. Figure A.10 is a worksheet for calculating limits for a $P$ chart. Figure A.11 shows an example of a $P$ chart for both fixed and variable subgroup sizes.

For count data, the $C$ chart (number of incidences) is appropriate when the opportunity for occurrence is relatively constant (within 20 percent of the average) among subgroups. The $U$ chart (incidences per unit) is required for count data when the opportunity for occurrence is variable among subgroups. The subgroup size is replaced by the concept of *area of opportunity* for $C$ and $U$ charts. Figure A.12 is a worksheet for a $C$ chart or $U$ chart. Figure A.13 shows examples of both a $C$ chart and a $U$ chart.

## $\overline{X}$ and $R$ Control Chart Calculation Form

Name ______________________ Date ______________________
Process ______________________ Sample description ______________________
Number of subgroups ($k$) ________ Between (dates) ________ – ________
Number of samples or measurements per subgroup ($n$) ______________________

$\overline{\overline{X}} = \frac{\Sigma \overline{X}}{k} =$ ________ = ________

$\overline{R} = \frac{\Sigma \overline{R}}{k} =$ ________ = ________

| $\overline{X}$ Chart | | | | | | R Chart | | | | |
|---|---|---|---|---|---|---|---|---|---|---|
| UCL | = | $\overline{\overline{X}}$ | + ( $A_2$ | * | $\overline{R}$ ) | UCL | = | $D_4$ | * | $\overline{R}$ |
| UCL | = | | + ( | * | ) | UCL | = | | * | |
| UCL | = | | + | | | UCL | = | ________ | | |
| UCL | = | ________ | | | | | | | | |
| LCL | = | $\overline{\overline{X}}$ | – ( $A_2$ | * | $\overline{R}$ ) | LCL | = | $D_3$ | * | $\overline{R}$ |
| LCL | = | | – ( | * | ) | LCL | = | | * | |
| LCL | = | | – | | | LCL | = | ________ | | |
| LCL | = | ________ | | | | | | | | |

**Factors for Control Limits**

| $n$ | $A_2$ | $D_3$ | $D_4$ | $d_2$ |
|---|---|---|---|---|
| *1 | 2.66 | — | 3.27 | 1.128 |
| 2 | 1.88 | — | 3.27 | 1.128 |
| 3 | 1.02 | — | 2.57 | 1.693 |
| 4 | 0.73 | — | 2.28 | 2.059 |
| 5 | 0.58 | — | 2.11 | 2.326 |
| 6 | 0.48 | — | 2.00 | 2.534 |
| 7 | 0.42 | 0.08 | 1.92 | 2.704 |
| 8 | 0.37 | 0.14 | 1.86 | 2.847 |
| 9 | 0.34 | 0.18 | 1.82 | 2.970 |
| 10 | 0.31 | 0.22 | 1.78 | 3.087 |

*Use moving range of 2 for determining $R$

**Process Capability**

If the process is in statistical control, the standard deviation is:

$\hat{\sigma} = \overline{R} \;/\; d_2$

$\hat{\sigma} = \quad /$

$\hat{\sigma} =$ ________

The process capability is:

$\overline{\overline{X}} - 3 * \hat{\sigma}$ to $\overline{\overline{X}} + 3 * \hat{\sigma}$

– to +

________ to ________

Figure A.8 Worksheet for $\overline{X}$ and $R$ charts.

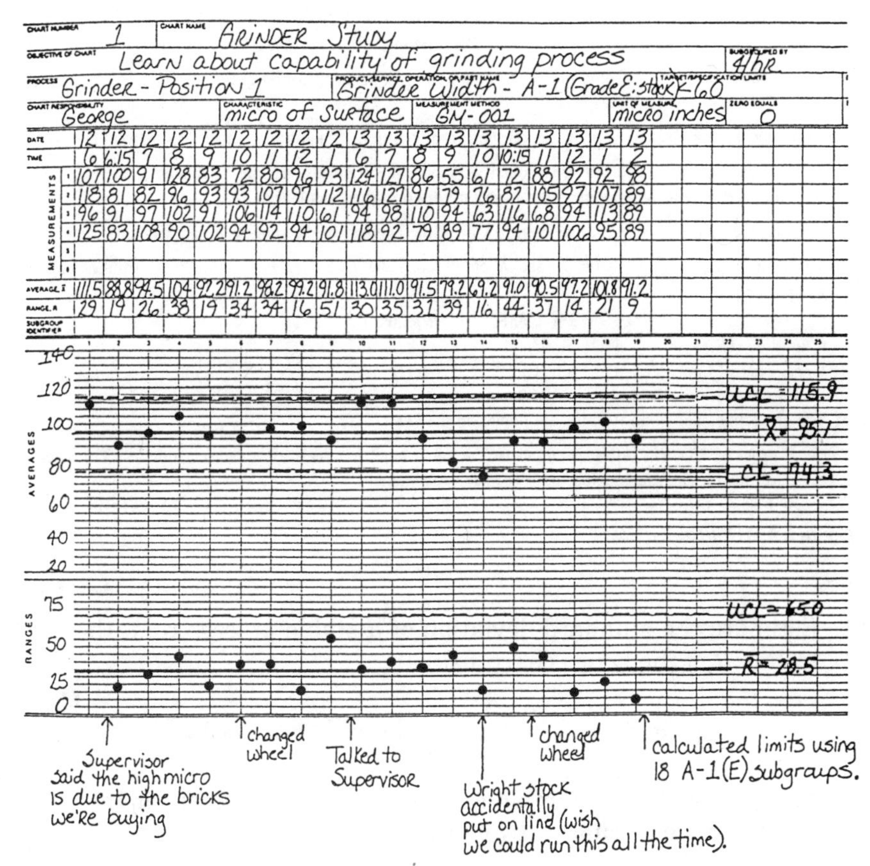


**Figure A.9** Example and discussion of an $\bar{X}$ and $R$ chart.

There are other types of control charts, such as the median control chart and the cumulative sum chart, used for special applications. The seven types of charts shown in Fig. A.7 are the most common Shewhart control charts found in practice.

## A.6 Capability of a Process

If a process is found to be stable for a particular quality characteristic, the process capability can be determined. Process capability is a prediction of the individual outcomes or measurements of a characteristic of the process.

For continuous data, the process capability can be compared to customer specifications to determine if the process is capable of meeting them. The following procedure can be used to determine process capability for continuous data:

**Discussion**

This is a control chart for the microfinish (smoothness) of a product from a grinding process. All samples are taken from position 1 on the grinder.

*Subgrouping:* Four consecutive parts are selected from position 1 each hour to form a subgroup ($n = 4$).

*Steps to computer limits:* After 19 subgroups (two days), the control limits are calculated. Since subgroup 14 was from a different material supplier, it was not used in the calculations ($k = 18$).

1. The average of the 18 subgroup averages ($\overline{X}$) and the average of the ranges ($\overline{R}$) are calculated.
2. The control limits for both the $\overline{X}$ and $R$ charts are calculated (see Figure A.8 for formulas and factors):

| | **$\overline{X}$ Chart** | ***R* chart** |
|---|---|---|
| Centerline (CL) | 95.1 | 28.5 |
| Upper control limit (UCL) | 115.9 | 65.0 |
| Lower control limit (LCL) | 74.3 | none |

3. The centerlines and control limits are drawn on the chart.

*Status of process:* Unstable—the special cause for subgroup 14 is due to the material supplier. The process is stable for the 18 subgroups using the A-1 material vendor.

**Figure A.9** *(Continued)*

1. Estimate the standard deviation of a stable process from the centerline $R$ of the range chart, using

$$\sigma = \frac{R}{d_2} \qquad (d_2 \text{ is tabulated in Fig. A.8})$$

2. The practical minimum and maximum of the process for the characteristic are calculated by using

$$\text{Minimum} = \text{X} - 3\sigma$$

$$\text{Maximum} = \text{X} + 3\sigma$$

3. A process is considered *capable* if the interval from the practical minimum to the practical maximum falls within the customer specifications.

For attribute data, the centerline of the appropriate control chart for a stable process is usually used to express process capability. For a $P$ chart, 100 percent $- P$ is sometimes presented as the capability. Calculation can be done by using the control limit formula to express a range of expected values for a given sample size or area of opportunity.

## A.7 Planning a Control Chart

The effective use of control charts requires careful planning to develop and maintain the chart. Many attempts to use control charts have not

---

$d$ = Nonconforming sample units per subgroup
$n$ = Number of sample units per subgroup
$k$ = Number of subgroups
$p$ = Percent nonconforming units = $100 * d/n$

---

Control limits when subgroup size ($n$) is constant:

$$\overline{p} = \frac{\Sigma p}{k} = \_\_\_\_\_\_ = \_\_\_\_\_\_ \quad \text{(centerline)}$$

$$\hat{\sigma}_p = \sqrt{\frac{\overline{p}*(100-\overline{p})}{n}} = \sqrt{\frac{\quad *(100- \quad)}{\quad}} = \_\_\_\_\_\_$$

UCL = $\overline{p}$ + (3* $\hat{\sigma}_p$ )  LCL = $\overline{p}$ − (3* $\hat{\sigma}_p$ )

UCL = ___ + (3*___)  LCL = ___ − (3*___)

UCL = ___ + _____  LCL = ___ − _____

UCL = _____  LCL = _____

---

Control limits when subgroup size ($n$) is variable:

$$\overline{p} = \frac{\Sigma d}{\Sigma n} * 100 = \_\_\_\_\_\_ * 100 = \_\_\_\_\_\_ \quad \text{(centerline)}$$

$$\hat{\sigma}_p = \frac{\sqrt{\overline{p}(100-\overline{p})}}{\sqrt{n}} = \frac{\sqrt{\quad *(100- \quad)}}{\sqrt{n}} = \frac{\_\_\_\_\_\_}{\sqrt{n}}$$

UCL = $\overline{p}$ + (3 * $\hat{\sigma}_p$ )  LCL = $\overline{p}$ − (3 * $\hat{\sigma}_p$ )

UCL = ___ + (3 * ___ / $\sqrt{n}$)  LCL = ___ − (3 * ___ / $\sqrt{n}$)

UCL = ___ + (___ / $\sqrt{n}$)  LCL = ___ − (___ / $\sqrt{n}$)

| | | | | | |
|---|---|---|---|---|---|
| $n$: | ___ | ___ | ___ | ___ | ___ |
| $\sqrt{n}$: | ___ | ___ | ___ | ___ | ___ |
| $3 * \hat{\sigma}_p$: | ___ | ___ | ___ | ___ | ___ |
| UCL: | ___ | ___ | ___ | ___ | ___ |
| LCL: | ___ | ___ | ___ | ___ | ___ |

---

**Figure A.10** Worksheet for $P$ chart calculations.

ABSENTEEISM

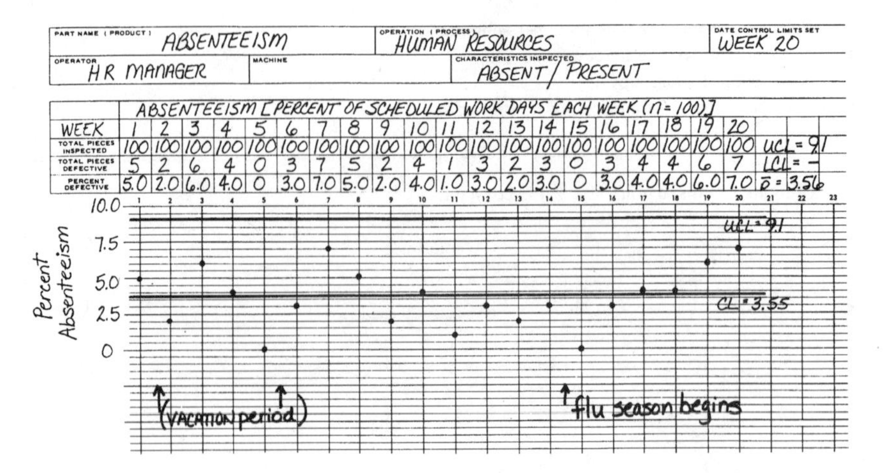

PART NAME (PRODUCT): ABSENTEEISM | OPERATION (PROCESS): HUMAN RESOURCES | DATE CONTROL LIMITS SET: WEEK 20

OPERATOR: HR MANAGER | MACHINE: | CHARACTERISTICS INSPECTED: ABSENT / PRESENT

ABSENTEEISM [PERCENT OF SCHEDULED WORK DAYS EACH WEEK (n = 100)]

| WEEK | 1 | 2 | 3 | 4 | 5 | 6 | 7 | 8 | 9 | 10 | 11 | 12 | 13 | 14 | 15 | 16 | 17 | 18 | 19 | 20 | |
|---|---|---|---|---|---|---|---|---|---|---|---|---|---|---|---|---|---|---|---|---|---|
| TOTAL PIECES INSPECTED | 100 | 100 | 100 | 100 | 100 | 100 | 100 | 100 | 100 | 100 | 100 | 100 | 100 | 100 | 100 | 100 | 100 | 100 | 100 | 100 | UCL = 9.1 |
| TOTAL PIECES DEFECTIVE | 5 | 2 | 6 | 4 | 0 | 3 | 7 | 5 | 2 | 4 | 1 | 3 | 2 | 3 | 0 | 3 | 4 | 4 | 6 | 7 | LCL = – |
| PERCENT DEFECTIVE | 5.0 | 2.0 | 6.0 | 4.0 | 0 | 3.0 | 7.0 | 5.0 | 2.0 | 4.0 | 1.0 | 3.0 | 2.0 | 3.0 | 0 | 3.0 | 4.0 | 4.0 | 6.0 | 7.0 | $\bar{p}$ = 3.56 |



DELIVERIES NOT ON TIME

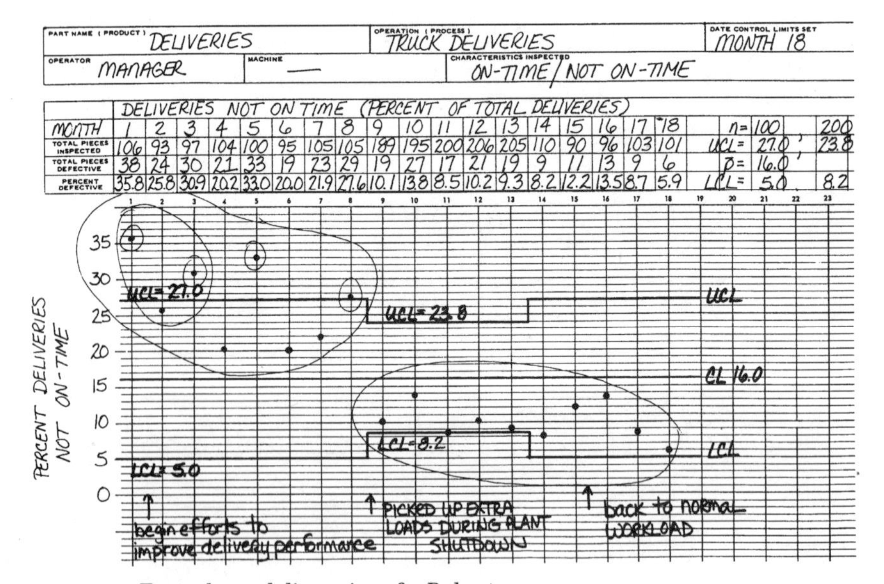

PART NAME (PRODUCT): DELIVERIES | OPERATION (PROCESS): TRUCK DELIVERIES | DATE CONTROL LIMITS SET: MONTH 18

OPERATOR: MANAGER | MACHINE: — | CHARACTERISTICS INSPECTED: ON-TIME / NOT ON-TIME

DELIVERIES NOT ON TIME (PERCENT OF TOTAL DELIVERIES)

| MONTH | 1 | 2 | 3 | 4 | 5 | 6 | 7 | 8 | 9 | 10 | 11 | 12 | 13 | 14 | 15 | 16 | 17 | 18 | | | |
|---|---|---|---|---|---|---|---|---|---|---|---|---|---|---|---|---|---|---|---|---|---|
| TOTAL PIECES INSPECTED | 106 | 93 | 97 | 104 | 100 | 95 | 105 | 105 | 189 | 195 | 200 | 206 | 205 | 110 | 90 | 96 | 103 | 101 | n = | 100 | 200 |
| TOTAL PIECES DEFECTIVE | 38 | 24 | 30 | 21 | 33 | 19 | 23 | 29 | 19 | 27 | 17 | 21 | 19 | 9 | 11 | 13 | 9 | 6 | UCL = | 27.0 | 23.8 |
| PERCENT DEFECTIVE | 35.8 | 25.8 | 30.9 | 20.2 | 33.0 | 20.0 | 21.9 | 27.6 | 10.1 | 13.8 | 8.5 | 10.2 | 9.3 | 8.2 | 12.2 | 13.5 | 8.7 | 5.9 | $\bar{p}$ = | 16.0 | |
| | | | | | | | | | | | | | | | | | | | LCL = | 5.0 | 8.2 |



**Figure A.11** Examples and discussion of a *P* chart.

## Discussion

**I. Absenteeism**. There are 20 employees in the department, each scheduled to work five days per week. Thus, there are 100 total work days scheduled each week. Each day all 20 employees are each classified as "present" or "absent" (using an operational definition of "absent"). Each week the percent absent ($P$) is calculated and plotted.

*Subgrouping:* The 20 employees and the five days of the week are combined to form a subgroup of $n = 100$ each week.

*Steps to compute limits:* After five months (20 weeks), the control limits are calculated. Thus, $k = 20$.

1. The average of the 20 subgroup percentages ($\overline{P}$) is calculated.
2. The control limits for the $P$ chart are calculated (see Figure A.10 for formulas):

   Centerline (CL) = 3.55%
   Upper control limit (UCL) = 9.1
   Lower control limit (LCL) = none

*Status of process:* Stable—there is a possible trend developing beginning at week 15.

**II. Deliveries not on time.** A carrier keeps track of delivery performance for each of its shippers. Each month the total number of deliveries and the number of deliveries not arriving on time are recorded. (An operational definition of "on time" was jointly developed with the carrier and shippers.) The percent not on time is calculated and plotted each month.

*Subgrouping:* All deliveries within a particular month for each shipper are grouped together to form a subgroup. Since the number of deliveries varies, $n$ is variable.

*Steps to compute limits:* After 18 months, the control limits are calculated. Thus, $k = 18$.

1. The weighted average of the percentages ($\overline{P}$) is calculated by dividing the total number of deliveries not on time by the total number of deliveries.
2. The control limits for the $P$ chart are calculated for $n = 100$ and $n = 200$ (Figure A.10 contains the formulas):

   Centerline (CL) = 16.0%
   Upper control limit (UCL) = 27.0($n$ = 100), 23.8($n$ = 200)
   Lower control limit (LCL) = 5.0($n$ = 100), 8.2($n$ = 200)

*Status of process:* Unstable—there are two distinct periods of performance on the chart. An important improvement occurred after the ninth month.

**Figure A.11** *(Continued)*

been successful because of lack of planning and preparation. Figure A.14 shows a planning form that can be used to develop a control chart.

### Objective of the chart

Every control chart should be associated with one or more specific objectives. The objective might be to improve the yield of the process, to identify and remove special causes from a process, or to establish statistical control so that the capability of the process can be determined. The objectives should be summarized on the control chart form. After a period of time, once the objective has been met, the control chart should be discontinued or a new objective developed.

**Control Chart Calculation Form ($C$ and $U$ Charts)**

$C$ Chart Control Limits (area of opportunity constant)

$c$ = Number of incidences per subgroup
$k$ = Number of subgroups
**Note:** The subgroup size is defined by the "area of opportunity" for incidences and must be constant.

$\bar{c} = \frac{\Sigma c}{k}$ = ________ = ________ (centerline)

UCL = $\bar{c} + (3 * \sqrt{\bar{c}})$ LCL = $\bar{c} - (3 * \sqrt{\bar{c}})$

UCL = ____ + (3*____) LCL = ____ − (3*____)

UCL = ____ + ______ LCL = ____ − ______

UCL = ______ LCL = ______

$U$ Chart Control Limits (area of opportunity may vary)

$c$ = Number of incidences per subgroup
$n$ = Number of standard area of opportunities in a subgroup ($n$ may vary)
$u$ = Incidences per standard area of opportunity = $c/n$
$k$ = Number of subgroups
**Note:** The standard area of opportunity will be defined by the people planning the control chart in units such as worker-hours, miles driven, per 10 invoices, etc.

$\bar{u} = \frac{\Sigma c}{\Sigma n}$ = ________ = ________ (centerline)

UCL = $\bar{u} + (3 * \sqrt{\bar{u}}) / \sqrt{n}$ LCL = $\bar{u} - (3 * \sqrt{\bar{u}}) / \sqrt{n}$

UCL = ____ + (3*____) / $\sqrt{n}$ LCL = ____ − (3*____) / $\sqrt{n}$

UCL = ____ + (______ / $\sqrt{n}$) LCL = ____ − (______ / $\sqrt{n}$)

| | | | | | |
|---|---|---|---|---|---|
| $n$: | ____ | ____ | ____ | ____ | ____ |
| $\sqrt{n}$: | ____ | ____ | ____ | ____ | ____ |
| UCL: | ____ | ____ | ____ | ____ | ____ |
| LCL: | ____ | ____ | ____ | ____ | ____ |

**Figure A.12** Worksheet for $C$ chart or $U$ chart.

MISTAKES IN DELIVERIES (C CHART)

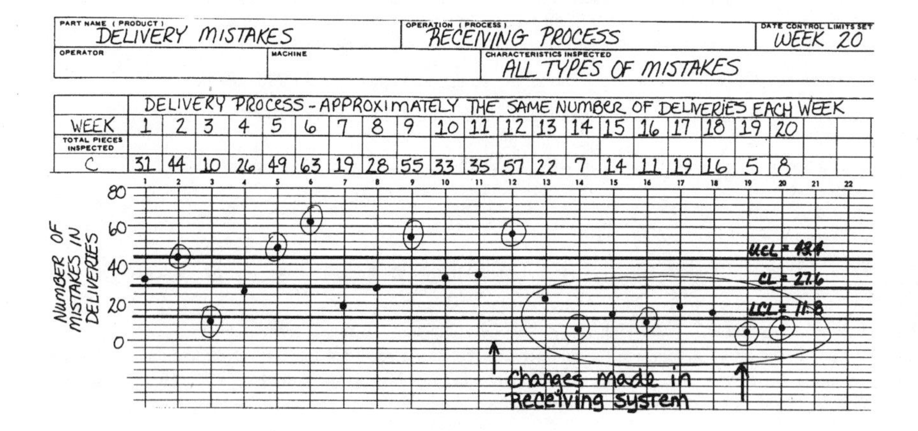


TRAFFIC ACCIDENTS (U CHART)

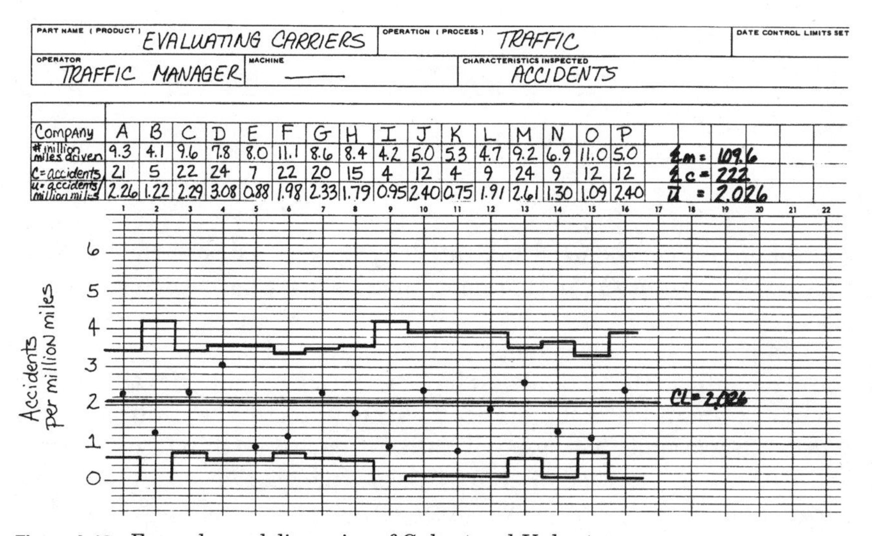


**Figure A.13** Examples and discussion of $C$ chart and $U$ chart.

## Sampling, measurement, and subgrouping

There are a number of measurement and sampling issues that must be resolved prior to beginning a control chart. The most useful control charts will monitor measures of process performance rather than quality characteristics of the outcome of the process. Initially

**I. Number of mistakes in deliveries.** There are many different types of mistakes that can occur in the delivery/receiving process. The total of all types of mistakes ($c$) is recorded each week and plotted on a $C$ chart. There are approximately the same number of deliveries (about 500) each week.

*Subgrouping:* All deliveries during the week are grouped together to form a subgroup.

*Steps to compute limits:* After 20 weeks, the control limits are calculated ($k = 20$).

1. The average of the 20 subgroup $\bar{c}$ is calculated.
2. The control limits for the $C$ chart are calculated (see Figure A.12 for formulas):

| | |
|---|---|
| Centerline (CL) | 27.6 |
| Upper control limit (UCL) | 43.4 |
| Lower control limit (LCL) | 11.8 |

*Status of process:* Unstable—there are special causes during the first 12 weeks. Improvement occurred after the 12th week.

**II. Traffic accidents.** The driving records of the company's 16 major carriers are evaluated each year. The number of miles driven and the number of accidents is recorded for each of the carriers. A $U$ chart is prepared to compare the carriers' records since the number of miles driven varies. The statistic $u$ is computed by dividing the number of accidents by the number of million miles driven.

*Subgrouping:* A subgroup is constructed from all the deliveries from a given carrier for the past year. The 16 major carriers form 16 subgroups.

*Steps to compute limits:*

1. The average $\bar{u}$ is calculated by dividing the total number of accidents from all carriers by the total number of million miles driven.
2. The control limits for the $U$ chart are calculated (see Figure A.12 for formulas):

| | |
|---|---|
| Centerline (CL) | 2.026 |
| Upper control limit (UCL) | Different for each carrier |
| Lower control limit (LCL) | Different for each carrier |

*Status of process:* Stable—no special causes, no important differences in the carriers' driving records.

**Figure A.13** *(Continued)*

control charts can be used to evaluate characteristics of the outcome that are important to the customer, such as pH, color, weight, yield, or size.

Then process variables (such as temperature, reaction time, or voltage) that affect these quality characteristics can be identified. The variables should also be charted as close to the beginning of the process as possible. The type of data for each variable to be charted will determine the type of chart to use. Information about the variability and stability of the measurement system to be used should be documented. If the variability is not known, an effort to develop that information should be planned.

Important sampling issues for control charts include the point (location) of sampling, the frequency of sampling, and the strategy for subgrouping measurements. The concept of subgrouping (previously discussed) is one of the most important components of the control chart method. After selecting a method of subgrouping, the user of the control chart should be able to state which sources of variation in the

1. **Objective of the chart:**
2. **Sampling, measurement, and subgrouping**
   *Variable(s) to be charted:*
   *Method of measurement:*
   *Magnitude of measurement variation:*
   *Point (location) of sampling:*
   *Strategy for subgrouping:*
   *Frequency of subgroups:*
3. **Most likely special causes:**
4. **Notes required:**

| Note | Responsibility |
|---|---|
| | |

5. **Reaction plan for out-of-control points (attach copy):**
6. **Administration:**

| Task | Responsibility |
|---|---|
| Making measurements | |
| Recording data on charts | |
| Computing statistics | |
| Plotting statistics | |
| Extending/changing control limits | |
| Filing | |

7. **Schedule for analysis:**

**Figure A.14** Planning form for a control chart.

process will act to produce variation within subgroups and which sources will affect variation between subgroups. The specific objective of the control chart will often help determine the strategy for subgrouping of the data. For example, if the objective is to evaluate differences between raw material suppliers, then only material from a single supplier should be included in data within a subgroup.

The frequency of obtaining subgroups should depend on the objective of the chart (i.e., how quickly the objective needs to be obtained). Another consideration in selecting the frequency is how often potential causes act to produce changes in the process. In practice, limitations on the availability of measurements often set the frequency.

### Most likely special causes and notes required

The documentation of process information and activities is the most important part of many control charts. This documentation includes changes in the process, identification of special causes, investigations of special causes, and other relevant process data. Information from flowcharts and cause-and-effect diagrams should be used to identify particular notes that should be recorded. Responsibility for recording this critical information should be clearly stated.

### Reaction plan for special causes

A reaction plan for special causes revealed on the chart should be established. Often a checklist of items to evaluate or a flowchart of the steps to follow is useful. The reaction plan should specify the transfer of responsibility for identification of the special cause if it cannot be done at the local level.

As an example, a reaction plan for a control chart in a laboratory to monitor a measurement system might be as follows:

1. Run the quality control standard.
2. Notify operations of a potential problem.
3. Review the logbook for any recent instrument changes.
4. Prepare a new quality control standard and test it.
5. Replace the column in the instrument.
6. Notify the supervisor and call instrument repair.
7. Document the results of these investigations on the control chart.

### Administration of a control chart

There are a number of administrative duties required to maintain an effective control chart. Responsibility for timely measurement, recording data, calculating statistics, and plotting the statistics on the chart must be delineated. Proper revision and extension of control limits are an important consideration.

Revision of the control limits should be done only when the existing limits are no longer appropriate. There are four circumstances under which the original control limits should be recalculated:

1. When the initial control chart has special causes and there is a desire to use the calculated limits for analysis of data to be collected in the future. In this case, control limits should be recalculated after the data associated with the special causes are removed.
2. When "trial" control limits have been calculated with fewer than 20 to 30 subgroups (*Note:* trial limits should not be calculated with fewer than 12 subgroups). In this case, the limits should be recalculated when 20 to 30 subgroups become available.
3. When improvements have been made to the process and the im-

provements result in special causes on the control chart. Control limits should then be calculated for the new process.

4. When the control chart remains out of control for an extended period (20 or more subgroups) and approaches to identify and remove the special cause(s) have been exhausted. Control limits should be recalculated to determine if the process has stabilized at a different operating level.

The date on which the control limits were last calculated should be a part of the ongoing record for the control chart.

The form used to record the data and to plot the control chart is another important consideration. The form should allow for a continuing record and not have to be restarted every day or week. The control chart form should include space to document the important decisions and process information from the planning form. The recorded data should include the time and place and the person making the measurements as well as the results of the measurements.

The scale on the charts should be established to give a clear visual interpretation of the variation in the process. As a guide in selecting the scale, when the control limits are centered on the chart, about one-half of the scale should be included inside the control limits.

### Schedule for analysis

A schedule for analysis should be established for every active control chart. The frequency of analysis may vary for different levels of management. For example, the quality improvement team may meet to analyze the chart once per week, the department manager may meet with the team once per month to review the chart, and the production vice president might review the chart with the department manager at the end of each quarter.

Figure A.15 shows an example of a completed planning form for a control chart maintained by an accounting group.

## A.8 Summary

The most powerful concept associated with control charts is their universal applicability. Although Shewhart focused his initial work on manufacturing processes, the concepts of common and special causes and of stable and unstable processes have applications in many areas, including administrative and service activities and management and supervision.

Figure A.16 indicates the roles of workers and managers guided by a control chart to improve a process. All levels of an organization and all departments in the organization should have knowledge of the

1. **Objective of the chart:** To reduce the number of returned invoices that have to be billed again
2. **Sampling, measurement, and subgrouping**
   *Variables to be charted:* Percent of invoices returned that are not paid
   *Method of measurement:* Accounting supervisor records number of invoices sent each week and number returned unpaid
   *Magnitude of measurement variation:* Complete, accurate counts can be made; totals can be validated
   *Point (location) of sampling:* Master list and returns that cross the supervisor's desk
   *Strategy for subgrouping:* Subgroup will be all invoices mailed in a given week (historically, 35–90 invoices)
   *Frequency of subgroups:* One per week—100% of invoices for that week
3. **Most likely special causes:** New customers, price changes, computer program updates, new employees in the accounting department
4. **Notes required:**

| Note | Responsibility |
|---|---|
| Number of new customers each week | Supervisor |
| New employees | Supervisor |
| Changes in computer program | Systems |

5. **Reaction plan for out-of-control points** (attach copy)**:**
   1. Supervisor will call meeting of department to discuss all special causes.
6. **Administration:**

| Task | Responsibility |
|---|---|
| Making measurements | Supervisor |
| Recording data on charts | Supervisor |
| Computing statistics | Supervisor |
| Plotting statistics | Supervisor |
| Extending/changing control limits | Department QI team |
| Filing | Supervisor |

7. **Schedule for analysis:** QI team review once per month; supervisor sends copies to marketing and customer service group quarterly

**Figure A.15** Example of a completed control chart planning form.

concepts of common and special causes of variation and the use of control charts to differentiate these causes.

This chapter has presented an overview of the control chart method, including

- Construction of a control chart (the concept of subgrouping)
- Interpretation of a control chart (rules for determining special causes)
- Types of control charts
- Calculation sheets for control limits
- The concept of process capability
- Planning of a control chart

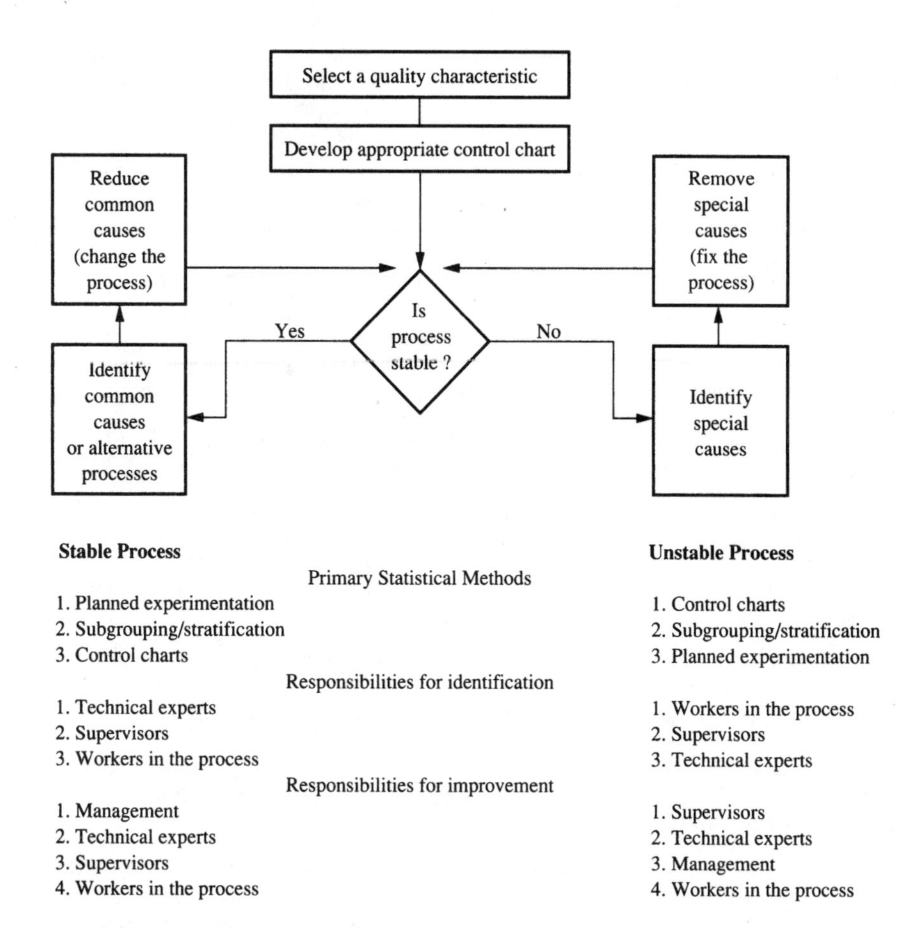


**Figure A.16** Methods and responsibilities for improvement.

The use of control charts in planned experimentation is presented in other chapters of this book.

## References


American Society for Quality Control (1985): Z1.1-1985, *Guide for Quality Control Charts*; Z1.2-1985, *Control Chart Method of Analyzing Data*; Z1.3-1985, *Control Chart Method of Controlling Quality during Production,* American Society for Quality Control, Milwaukee, Wis.

Deming, W. Edwards (1986): *Out of the Crisis,* Massachusetts Institute of Technology, Center for Advanced Engineering Study, Cambridge, Mass.

Grant, Eugene L., and Richard S. Leavenworth (1980): *Statistical Quality Control,* 5th ed., McGraw-Hill, New York.

Nelson, Lloyd S. (1984): "The Shewhart Control Chart—Tests for Special Causes," *Journal of Quality Technology,* vol. 16, no. 4, pp. 237–239, October.

Shewhart, Walter A. (1931): *Economic Control of Quality of Manufactured Product,* American Society for Quality Control, Milwaukee, Wis. (reprinted in 1980).

Shewhart, Walter A. (1939): *Statistical Method from the Viewpoint of Quality Control,* W. E. Deming (ed.), Graduate School, U.S. Department of Agriculture, Washington, D.C.

## Exercises

**A.1** For each of the following situations, list three likely common causes of variation and three likely special causes of variation.

| Process | Quality characteristic |
|---|---|
| (*a*) Service in a commercial bank | Time waiting in line |
| (*b*) Batter in baseball | Number of hits per game |
| (*c*) Plastic extruding process | Shrinkage of parts |
| (*d*) Food service in a restaurant | Temperature of the food |
| (*e*) Delivery of mail | Deliveries to wrong address |
| (*f*) Machining process | Size of a drilled hole |
| (*g*) Teaching a class | Students' scores on a test |
| (*h*) Driving a car | Fuel consumption (mpg) |

**A.2** Under what conditions should specifications and tolerances be used instead of a control chart?

**A.3** Consider a potential control chart for each of the following situations. Describe the variable to be charted, the type of control chart, and the method of obtaining subgroups (including subgroup frequency and subgroup size).

(*a*) The owner of a new car is interested in understanding the variation of fuel consumption. She uses the car for business and averages about 700 mi/week. The EPA rating on the car is 20 mi/gal.

(*b*) A housekeeper is concerned about the family's high electric bill. He thinks that one important cause is leaving the lights on in the house when they are not being used. To understand this problem, he decided to collect some data. Each day for a month, at 9:00 a.m. and again at 5:00 p.m., he went through the house and counted the number of lights on with no one in the room.

(*c*) A machine shop is concerned about variation in dimensions of the parts it makes. Customer tolerances are $\pm 0.003$ in for most of the critical dimensions. There have been a number of parts returned recently because of the size of the drilled holes. The drilling operation consists of three machines, each with four stations. An operator can drill about 45 parts per hour.

(*d*) A saleswoman wants to evaluate the effect of two potential new marketing strategies on sales closures. She typically calls on 20 to 30 prospective buyers each day. Historically, about one-third of her prospects buy her product after a sales call.

(*e*) The typist in a department has to work overtime because of changes to materials he has typed. He wants to collect some data on the extent of this problem to present at the department's monthly quality meeting. He types a total of 15 to 30 reports per day for six department personnel.

(*f*) A runner is interested in progress in improving his times in the mile run. He trains 4 days per week and has a time trial or a race once per week.

(*g*) The president of an office supply company decided to evaluate the delivery performance of her suppliers. During a period of 3 months, she evaluated each order from her 30 largest suppliers and recorded the date requested and the actual date of delivery. The number of deliveries from these suppliers ranged from 8 to 24.

**A.4** Complete a control chart planning form for the following scenario:

Some of the customers of a specialty chemical have been asking for lower levels of a particular impurity in the product. Although the yield varies widely from batch to batch, the operators felt that attempts to lower the impurity level through temperature and pressure controls resulted in lower yield. Frequent catalyst additions also confused the interpretation of the results.

The chemical process consisted of two parallel continuous reactors and three sequential distillation towers. The primary feedstock was obtained by pipeline from three suppliers. The impurity was measured by gas chromotography (GC), which had good precision for the levels of interest. The GC was in the unit so that samples could be run as frequently as every 20 min. Yield was measured by a material balance, which was done after each shift (every 8 hours). The material balance was thought to be accurate to within 1 percent.

The operators in the unit had recently had statistical process control (SPC) training and were eager to use control charts to learn about the process. The marketing manager for the product asked to be kept abreast of progress. He wanted to respond to the customers' requests within 2 months.

**A.5** The personnel director at a trucking company has been keeping track of absenteeism of her employees for about 4 months. Each week she records the total number of absent days (100 possible). The weekly absenteeism standard for the company is 10 percent. After looking at the data for the first 5 months, the director is pleased: Her 20 employees' absenteeism has never been over 10 percent on any week. Develop a control chart for the data given here, and answer the following questions:

(*a*) Is the company's absenteeism in statistical control?

(*b*) Should the personnel director be concerned about the absenteeism record?

(*c*) What does the control chart show that was missed by simple inspection of the data?

| week | 1 | 2 | 3 | 4 | 5 | 6 | 7 | 8 | 9 | 10 | 11 | 12 | 13 | 14 | 15 | 16 | 17 | 18 |
|---|---|---|---|---|---|---|---|---|---|---|---|---|---|---|---|---|---|---|
| days missed: | 5 | 2 | 3 | 7 | 2 | 4 | 6 | 4 | 2 | 4 | 2 | 3 | 2 | 4 | 5 | 7 | 7 | 8 |

**A.6** The accounting department is instituting process improvement and has been studying causes of delays, rework, and excess overtime. The department's quality control charts indicate that a large number of invoices have to be manually handled (requiring extra phone calls, rerouting documents, and other types of rework) because of mistakes or incomplete information on purchase orders. The head of the accounting department has asked the manager of the purchasing department to investigate this problem.

The purchasing manager decided to select some of the prepared orders for each purchasing agent over a one-week period and review them for completeness and errors. Sixty documents were randomly selected from the work of each of the 20 agents and were reviewed. Orders with one or more mistakes were identified. The following data were obtained at the end of the week.

| Agent | Number of orders with mistakes | Agent | Number of orders with mistakes |
|---|---|---|---|
| Dan | 5 | Bart | 4 |
| Hank | 0 | Linda | 1 |
| Ann | 2 | Judy | 3 |
| Bill | 6 | Helen | 9 |
| Mary | 7 | Larry | 8 |
| Dave | 5 | Ron | 4 |
| Fred | 8 | John | 5 |
| Sue | 4 | Mark | 0 |
| Chris | 0 | Emma | 6 |
| Tom | 5 | Tina | 8 |

(*a*) What can the purchasing manager learn from these results?

(*b*) Which are responsible for causing the mistakes, special or common causes?

(*c*) Which agents should be selected for special consideration?

(*d*) What feedback should the manager give to the head of the accounting department?

(*e*) Should the purchasing manager continue to collect data? If so, how should he analyze the data?

(*f*) What other types of data would be useful?

**A.7** Approximately 200 deliveries are recorded by the receiving department each week. Each delivery includes accompanying paperwork. There are many different mistakes that can occur with these deliveries, including wrong materials, damage, overage materials, shortages, late arrivals, and errors in the paperwork. Data are collected for a period of 20 weeks to examine the extent of these quality problems.

A clipboard is placed in the delivery area, and all personnel are asked to record the various problems in one of three categories:

- Critical errors (complete failure of the delivery)
- Major errors (problems that delay the delivery)
- Minor errors (errors that do not slow down the delivery)

The following data were obtained from this study.

| Week | Critical errors | Major errors | Minor errors |
|---|---|---|---|
| 1 | 4 | 8 | 31 |
| 2 | 2 | 12 | 44 |
| 3 | 8 | 6 | 10 |
| 4 | 3 | 4 | 26 |
| 5 | 13 | 11 | 49 |

| Week | Critical errors | Major errors | Minor errors |
|---|---|---|---|
| 6 | 6 | 7 | 63 |
| 7 | 7 | 16 | 19 |
| 8 | 4 | 9 | 28 |
| 9 | 0 | 5 | 55 |
| 10 | 5 | 10 | 33 |
| 11 | 3 | 14 | 35 |
| 12 | 4 | 8 | 57 |
| 13 | 7 | 4 | 22 |
| 14 | 1 | 10 | 7 |
| 15 | 5 | 13 | 14 |
| 16 | 2 | 7 | 11 |
| 17 | 14 | 9 | 19 |
| 18 | 4 | 15 | 16 |
| 19 | 7 | 6 | 5 |
| 20 | 6 | 10 | 8 |
| Total | 105 | 184 | 552 |

(*a*) Prepare appropriate control charts for these data.

(*b*) What can you learn about the delivery/receiving process?

(*c*) What are some possible reasons for the special causes?

(*d*) What other types of data would be useful in improving this process?

**A.8** The safety standard for the trucking industry's accident record is no more than 2 accidents per 1 million mi. The U.S. Department of Transportation gathered accident data for the 14 largest companies for the most recent year to evaluate their safety performance. The data are given below. Prepare an appropriate control chart, and answer the following questions.

| Company | Miles driven (millions) | Accidents |
|---|---|---|
| A | 9.3 | 21 |
| B | 4.1 | 5 |
| C | 9.6 | 22 |
| D | 7.8 | 24 |
| E | 8.0 | 17 |
| F | 11.1 | 22 |
| G | 8.6 | 8 |
| H | 8.4 | 15 |
| I | 4.2 | 5 |
| J | 5.0 | 16 |
| K | 5.3 | 6 |
| L | 4.7 | 11 |
| M | 9.2 | 20 |
| N | 6.9 | 8 |
| Total | 102.2 | 200 |

(*a*) Do the accidents come from a stable process?

(*b*) What does the control chart tell us about the accident records of the 14 trucking companies?

(*c*) Given the current system, are the companies capable of having no more than 2 accidents per 1 million mi?

**A.9** A critical contaminant is regularly monitored during continuous production of a chemical. Samples are selected three times (at somewhat regular intervals) during each 8-hour shift. The following data were obtained during the last week.

| | Concentration of contaminant (ppm) | | |
|---|---|---|---|
| | A shift | B shift | C shift |
| Monday | 18 | 17 | 18 |
| | 17 | 20 | 18 |
| | 17 | 18 | 16 |
| Tuesday | 19 | 16 | 18 |
| | 19 | 17 | 19 |
| | 18 | 16 | 19 |
| Wednesday | 17 | 16 | 20 |
| | 17 | 18 | 19 |
| | 17 | 19 | 18 |
| Thursday | 17 | 21 | 21 |
| | 18 | 22 | 20 |
| | 16 | 22 | 21 |
| Friday | 24 | 21 | 22 |
| | 22 | 21 | 22 |
| | 23 | 22 | 22 |
| Saturday | 20 | 20 | 23 |
| | 21 | 22 | 22 |
| | 23 | 22 | 22 |
| Sunday | 21 | 21 | 22 |
| | 20 | 21 | 22 |
| | 22 | 21 | 20 |

(*a*) Was the process stable during this week?

(*b*) What is the capability of the process for this contaminant?

**A.10** The materials handling department depends on forecasts from the various outlying plants that receive chemical products. The forecasts are required for planning local production, scheduling transportation equipment, and allocating various resources. Currently, forecasts made 1 week in advance are used, but this is not enough lead time for many of the activities and resources that must be scheduled.

Forecasts are currently made 1, 2, 3, and 4 weeks in advance. The 1-week forecast is thought to be reasonably reliable. If the 2-week forecast were reliable, the materials handling department's job could be done much more efficiently. The following data show the 2-week forecasts (stated in thousands of pounds) made by the three largest receiving plants for the past 20 weeks. The actual production use is also shown.

| | Plant A | | Plant B | | Plant C | |
|---|---|---|---|---|---|---|
| Week | Forecast | Actual | Forecast | Actual | Forecast | Actual |
| 1 | 350 | 330 | 200 | 210 | 250 | 240 |
| 2 | 420 | 430 | 220 | 205 | 300 | 300 |
| 3 | 310 | 300 | 230 | 215 | 130 | 120 |
| 4 | 340 | 345 | 190 | 200 | 210 | 200 |
| 5 | 320 | 345 | 200 | 200 | 220 | 215 |
| 6 | 240 | 245 | 210 | 200 | 210 | 190 |
| 7 | 200 | 210 | 210 | 205 | 230 | 215 |
| 8 | 300 | 320 | 190 | 200 | 240 | 215 |
| 9 | 310 | 330 | 210 | 220 | 160 | 150 |
| 10 | 320 | 340 | 200 | 195 | 340 | 335 |
| 11 | 320 | 350 | 180 | 185 | 250 | 245 |
| 12 | 400 | 385 | 180 | 200 | 340 | 320 |
| 13 | 400 | 405 | 180 | 240 | 220 | 215 |
| 14 | 410 | 405 | 220 | 225 | 230 | 235 |
| 15 | 430 | 440 | 220 | 215 | 320 | 310 |
| 16 | 330 | 320 | 220 | 220 | 320 | 315 |
| 17 | 310 | 315 | 210 | 200 | 230 | 215 |
| 18 | 240 | 240 | 190 | 195 | 160 | 145 |
| 19 | 210 | 205 | 190 | 185 | 240 | 230 |
| 20 | 330 | 320 | 200 | 205 | 130 | 120 |

(*a*) Is the forecasting process in statistical control for each of the three plants?

(*b*) What is the current capability of the three forecasting processes?

(*c*) What recommendations would you make to improve these processes?

**A.11** Approximately 40 invoices are prepared by the accounting department each day in one of the product line divisions. Some of the invoices are returned by customers for corrections and adjustments. Recently, more and more customers have complained to their sales contacts about "incorrect invoices." The accounting manager set up a data collection program to study this problem.

All invoices that required adjustment and the reason for adjustment were recorded and grouped by the day the invoice was prepared. Three-day groupings of invoices were studied (approximately 120 invoices). The following data were obtained during the first 2 months.

| First month | | | Second month | | |
|---|---|---|---|---|---|
| Days | Number of invoices | Number requiring adjustment | Days | Number of invoices | Number requiring adjustment |
| 1–3 | 125 | 20 | 1–3 | 122 | 21 |
| 4–6 | 120 | 16 | 4–6 | 110 | 11 |
| 7–9 | 118 | 9 | 7–10 | 125 | 24 |

| First month | | | Second month | | |
|---|---|---|---|---|---|
| Days | Number of invoices | Number requiring adjustment | Days | Number of invoices | Number requiring adjustment |
| 10–12 | 123 | 23 | 11–13 | 122 | 20 |
| 14–16 | 130 | 12 | 14–17 | 116 | 17 |
| 17–19 | 127 | 14 | 18–20 | 117 | 7 |
| 20–22 | 121 | 18 | 22–24 | 118 | 23 |
| 23–26 | 114 | 23 | 25–27 | 123 | 15 |
| 27–30 | 119 | 19 | 28–30 | 128 | 20 |
| Total | 1,097 | 154 | | 1,081 | 158 |

(*a*) Is the process stable?

(*b*) What predictions can be made for future periods?

Process improvement projects were begun in the accounting department. Flowcharts were prepared for the various processes in the invoicing system. A Pareto analysis on the causes of the adjustments was done and used to prioritize the projects. The following data on invoice adjustments were obtained 3 months after the improvement projects were begun:

| Sixth month | | | Seventh month | | |
|---|---|---|---|---|---|
| Days | Number of invoices | Number requiring adjustment | Days | Number of invoices | Number requiring adjustment |
| 1–3 | 115 | 3 | 1–3 | 126 | 18 |
| 4–6 | 120 | 8 | 4–7 | 110 | 13 |
| 7–9 | 128 | 6 | 8–9 | 114 | 16 |
| 10–13 | 113 | 4 | 11–13 | 122 | 14 |
| 14–16 | 120 | 9 | 14–17 | 127 | 3 |
| 17–19 | 117 | 5 | 18–20 | 110 | 0 |
| 21–23 | 125 | 13 | 22–24 | 112 | 4 |
| 24–26 | 122 | 22 | 25–28 | 125 | 8 |
| 28–30 | 129 | 12 | 29–31 | 120 | 6 |
| Total | 1,089 | 82 | | 1,066 | 82 |

(*c*) Have the improvement projects affected the number of required adjustments?

(*d*) What should be the next step to improve the process?

Appendix

# B

# Evaluating Measurement Systems

## B.1 Introduction

*"In any program of control we must start with observed data; yet data may be either good, bad, or indifferent. Of what value is the theory of control if the observed data going into that theory are bad? This is the question raised again and again by the practical man."*
DR. W. A. SHEWHART (1931, P. 376)

*"...statistical control of the process of measurement is vital; otherwise there is no meaningful measurement."*
DR. W. E. DEMING (1986, P. 332)

The identification of relevant quality characteristics is a critical step in improvement. Quality characteristics include characteristics of an outcome of a process, measures of performance of a process, needs of a customer, requirements for inputs to a process, etc. Once relevant quality characteristics have been selected, some method must be used to obtain measures of these characteristics. Understanding and improving the process for measurement play an important role in the improvement of processes and products.

The selected quality characteristics provide "windows" through which we are able to observe the performance of a process or the quality of a product. If those windows do not provide predictable, consistent views, intelligent decisions about actions to be taken cannot be made. As the saying goes, a person with one watch knows what time it is. A person with two watches is never sure.

When a serious effort begins for improvement, it is not long before the people involved discover the important role of the measurement process. Practitioners of improvement discover what Shewhart and Deming learned years ago: A measurement process must be in a state of statistical control for a useful method of measurement to exist. When there is an effort to improve, the measurement process is often a leading candidate for improvement.

There is a close tie between measurement and operational definitions (Deming, 1986, chap. 9). An operational definition is a definition that gives communicable meaning to a concept by specifying how the concept is applied within a particular set of circumstances. An important component of an operational definition is the statement of the measurement process used.

In studying variation of quality characteristics of a process or product, variation due to the measurement process plays a unique role. Figure B.1*a* shows a schematic for a typical cause-and-effect diagram. Variation of the quality characteristic due to measurement is included as one of the major categories on the diagram. Figure B.1*b* shows an alternative form of the cause-and-effect diagram that emphasizes the special role of measurement variation.

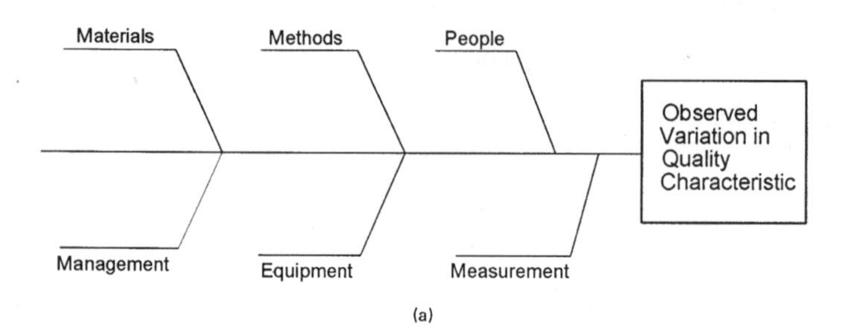


(a)

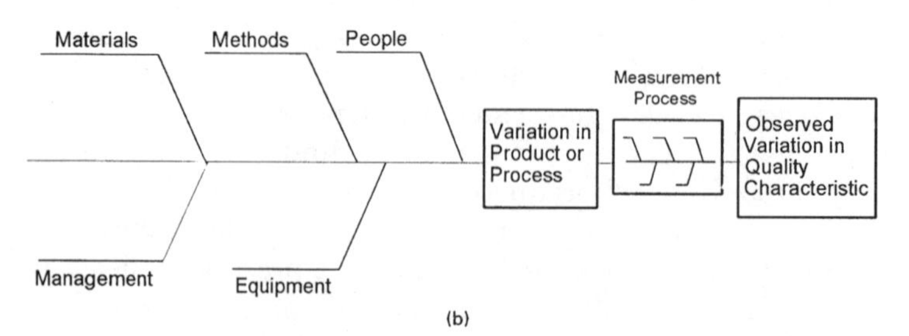


(b)

**Figure B.1** Measurement on a cause-and-effect diagram. (*a*) Measurement as one of the major categories of variation; (*b*) measurement as a process to observe variation in a product or process.

From Fig. B.1*b*, one can see that when measurements of a quality characteristic are made, the variation in the measured value can come from two sources:

1. The variation of the product or process being measured (due to material, equipment, method, etc.)
2. The variation contributed by the measurement process

If the measurement process is supplying more variation than the product or process being measured, then it will be difficult to learn from the data about the causes of variation in the product process. We must understand the capability of the measurement system in order to advance knowledge of the cause-and-effect system in our work processes and products.

Shewhart (1939, chaps. 3 and 4) contains further theoretical discussion of measurement and its role in improvement.

## B.2 Studying Measurement Processes

When systems for measurement are studied, it is useful to think of the "process" of measurement. Measurement systems consist of standard units of measure (i.e., feet for length, hours for time, early or late for time, etc.) and procedures for producing values in terms of these units of measure. The procedures may include physical instruments such as a gas chromatograph, a speedometer, or a caliper. The procedures may also be subjective determinations made by people using one or more of the senses.

There are many characteristics of a measurement process that could be of interest. These include

| | |
|---|---|
| Accuracy | Reproducibility |
| Bias | Robustness (ruggedness) |
| Cost | Sensitivity |
| Limit of detection | Simplicity |
| Linearity | Speed |
| Precision | Stability |
| Reliability | Validity |
| Repeatability | |

Definitions for these terms vary, so operational definitions must be established for each application. The following definitions for five important characteristics are from American National Standards Institute (ANSI) and American Society for Quality Control (ASQC) documents (1987b):

**Accuracy** The extent to which the measured value of a quantity agrees with the accepted value for that quantity.

It is important to understand that there is no "true" value for a quantity. Accepted values (or accepted standards) may be legal values, consensus values, agreement values, or values obtained from a standard (or reference) method. Accepted standard values will always be subject to modification or obsolescence. The amount of accuracy required for a test method is dependent on how the results of the measurement are to be used. A measurement process with an acceptable level of accuracy is called *accurate*.

**Precision** The degree of agreement among independent measurements of a quantity under specified conditions. Precision refers to the ability of a measurement process to reproduce its own outcome. Precision is usually defined in terms of the criteria for repeatability and reproducibility (see below). A measurement process with an acceptable level of precision is called **precise.**

**Bias** The difference between the average of a series of measurements and the accepted standard value. Bias is also called *systematic error*. A measurement system with an acceptable level of bias is called **unbiased.**

Note that accuracy is the combination of precision and bias. To be accurate, a measurement process must have *both* a useful level of precision and bias. Figure B.2 illustrates the meaning of these three terms.

Only one of the four measurement processes illustrated in Fig. B.2 has a "useful" level of precision and bias and can thus be considered accurate.

The precision of a measurement process will depend on the sources of variation which are included in the determination of precision. The terms *repeatability* and *reproducibility* are used to define two extreme conditions [*note:* Mandel (1971) gives formal definitions of repeatability and reproducibility in terms of statistical properties of test results]:

**Repeatability** Variation in measurements obtained under the best conditions. This often means measurements taken with one instrument, by one operator, in the same time period.

**Reproducibility** Variation in measurements obtained by measuring the same item under normal operating conditions. These conditions

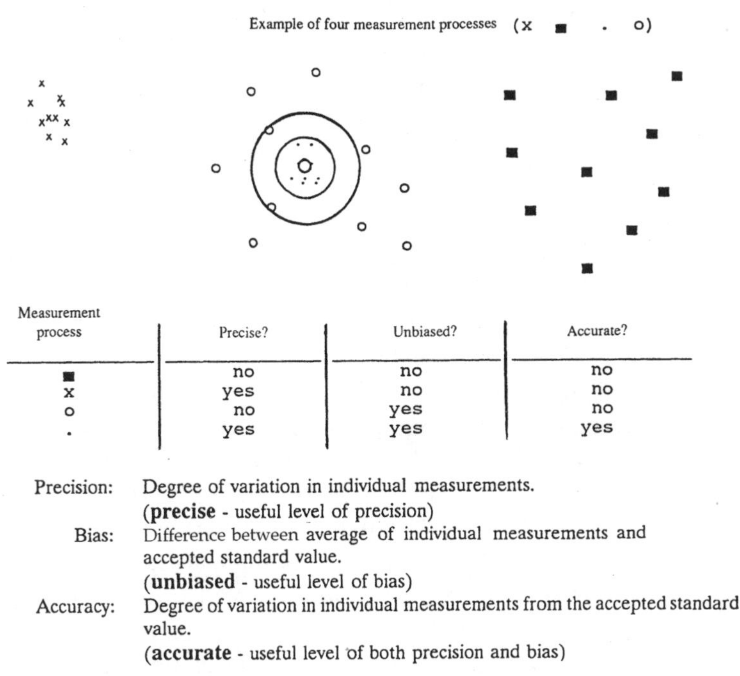


| Measurement process | Precise? | Unbiased? | Accurate? |
|---|---|---|---|
| ■ | no | no | no |
| x | yes | no | no |
| o | no | yes | no |
| . | yes | yes | yes |

Precision: Degree of variation in individual measurements.
(**precise** - useful level of precision)

Bias: Difference between average of individual measurements and accepted standard value.
(**unbiased** - useful level of bias)

Accuracy: Degree of variation in individual measurements from the accepted standard value.
(**accurate** - useful level of both precision and bias)

**Figure B.2** Illustration of precision, bias, and accuracy.

would reflect the normal work situation: different operators, environmental conditions, setup methods and calibration, etc.

The stability of a measurement process has the same meaning as for a manufacturing or service process. The measurement process must be in a state of statistical control before the bias or precision has any meaning. *The stability of the bias is much more important than the actual magnitude of the bias, since a known bias can be corrected through calibration.* (See discussion later.)

The definitions of the other characteristics used to describe the measurement process vary greatly from application to application. Definitions of terms from the references are included at the end of this appendix.

## B.3 Data from a Measurement Process

The outcome of a measurement process is called a *measurement* and will be in the form of some type of data. These data can be grouped into different types. Four useful categories of data from a measurement process are

1. *Continuous data.* Measurement is obtained by using some type of instrument with a continuous numeric scale (either interval or ratio). Examples are a dimension of a machined part, the viscosity of a liquid, the weight of a person, and the time to complete a task. Continuous data include data on an interval scale (where zero has no specific meaning, e.g., temperature) and data on a ratio scale (zero means absence of the property of interest, e.g., length).
2. *Count data.* Measurements are counts of a particular characteristic of interest. Examples are the number of black specks in a sample, the number of errors on a page, the number of accidents in a month, the number of times a task is completed, and counts of inventory.
3. *Rank data.* Measurements are on an ordinal scale (ordered values). Examples include inspection of a product (conforming or nonconforming), severity of a visual defect, preferred taste or smell, color scores, comparison of functions of competing products, and results of a medical diagnosis (critical, serious, or stable condition).
4. *Classification data.* Measurements are categories or classes of the characteristic of interest (also called *qualitative data,* or data on a nominal scale). Examples include results of a medical diagnosis (cold, measles, or flu), determination of religious preference (Catholic, Protestant, other), classification of a product (type A or type B).

Measurements can be transformed from one type of data to another, but usually only in one direction. Continuous or count data can usually be converted to rank or classification data, but not the other way around.

Data from measurement processes can also be grouped in other ways. The different kinds of control charts (see App. A) are based on two groupings of types of data (American National Standards Institute, 1987a):

1. *Attribute data.* They note the presence or absence of some characteristic or attribute in each of the units in the group under consideration, and count how many units do (or do not) possess the attribute, or how many such events occur in the unit, group, or area.
2. *Variable data.* They measure and record the numerical magnitude of a quality characteristic for each of the units in the group under consideration.

Attribute data could include classification, count, and rank data. Variable data refer primarily to continuous data, but rank data are often analyzed by using a variable control chart (realizing that the arithmetic functions may not be valid). Otherwise the ranks can be

| Category of Data | Control Chart Grouping | | |
|---|---|---|---|
| | Continuous Data (Variable Data) | Attribute Data | |
| | | Count | Classification |
| Continuous (ratio or interval scale) | $\bar{X}$, R Chart<br>X Chart | | |
| Count (ratio scale) | X Chart | C Chart | |
| Rank (ordinal scale) | X Chart<br>Median chart | | P Chart |
| Classification (nominal scale) | | | P Chart |

**Figure B.3** Control charts for types of measurement data. (*Note:* See App. A for information on the different types of control charts.)

converted to classification data and analyzed using attribute control charts. Figure B.3 summarizes the types of control charts appropriate for the two types of groupings of data.

### Measurement discrimination and rounding of numbers

Another issue that often arises in the use of a measurement process is the rounding off of quantitative data when the data are recorded. This becomes an important issue in improvement because many methods of improvement are based on studying the variation in data. Rounding numbers can affect the observed variation. Rounding of data becomes important whenever the smallest unit of measurement exceeds the process standard deviation.

The paper by Propst (1989–1990) suggests that the smallest increment that can be read using the measurement system should not exceed one-tenth of the total process variation. Holmes and Mergen (1991–1992) use information theory methodology to evaluate the importance of the increment of measure (or unit of calibration). They conclude that a gage should have units which are smaller than 30 percent of the process standard deviation.

Some general guidelines can be followed to ensure that the way the measurements are being recorded gives adequate discrimination:

1. In general, record data to as many places as the measurement process will allow. The data can then be rounded for different uses as needed. When rounding of the data is necessary, record the data to at least one more decimal than the order of magnitude of the process standard deviation (for example, $\sigma = 0.03$, round the data to thousandths).

2. When data using variable control charts are studied, the data should include at least 10 different *possible values* between the lowest and highest recorded values.
3. Wheeler and Lyday (1989) show that the range chart gives the best indication of inadequate measurement discrimination. They give the following rule based on the range chart for continuous data:

   When there are less than five possible values for the range within the control limits or more than one-fourth of the calculated ranges are zero, then the measurements have been made with inadequate discrimination.

## B.4 Monitoring and Improving a Measurement Process

Improvement of a measurement process can proceed in the same manner as for any other process. Figure B.4 shows a flow diagram for evaluating and improving a measurement process. The remainder of this appendix describes methods for monitoring and improving measurement processes. The case study at the end of this appendix follows the flow diagram and uses the appropriate methods.

When measurement processes are studied, a number of concepts and methods particular to measurement become important. The use of standards, reference materials, or check standards is a part of most measurement processes. The concept of calibration is also important. Working definitions for each of these terms are included at the end of this appendix. Formal definitions can be found in the ANSI standard (1987b). The types of standards or reference materials can vary greatly for different measurement systems. For chemical measurement processes, liquid standards of certified purity are common. For measuring parts, certified weights and dimensional items are used.

Calibration is a critical component of the measurement process. The bias of the measurement is controlled by the calibration procedures. These procedures will include a variety of methods depending on the particular measurement process. The *American National Standard for Calibration Systems* (1987b) contains guidelines for developing calibration systems for a measurement process. There are a number of statistical issues in calibration procedures. Hunter (1981) discusses these issues.

Control charts can play an important role in the calibration of many types of measurement equipment [see Schumacher (1983), Propst (1989–1990), and McNeese and Klein (1991–1992) for further examples of use of control charts with measurement processes]. The control chart can be used to determine the frequency or need for recalibration. Figure B.5 is an example of such a control chart for a gas

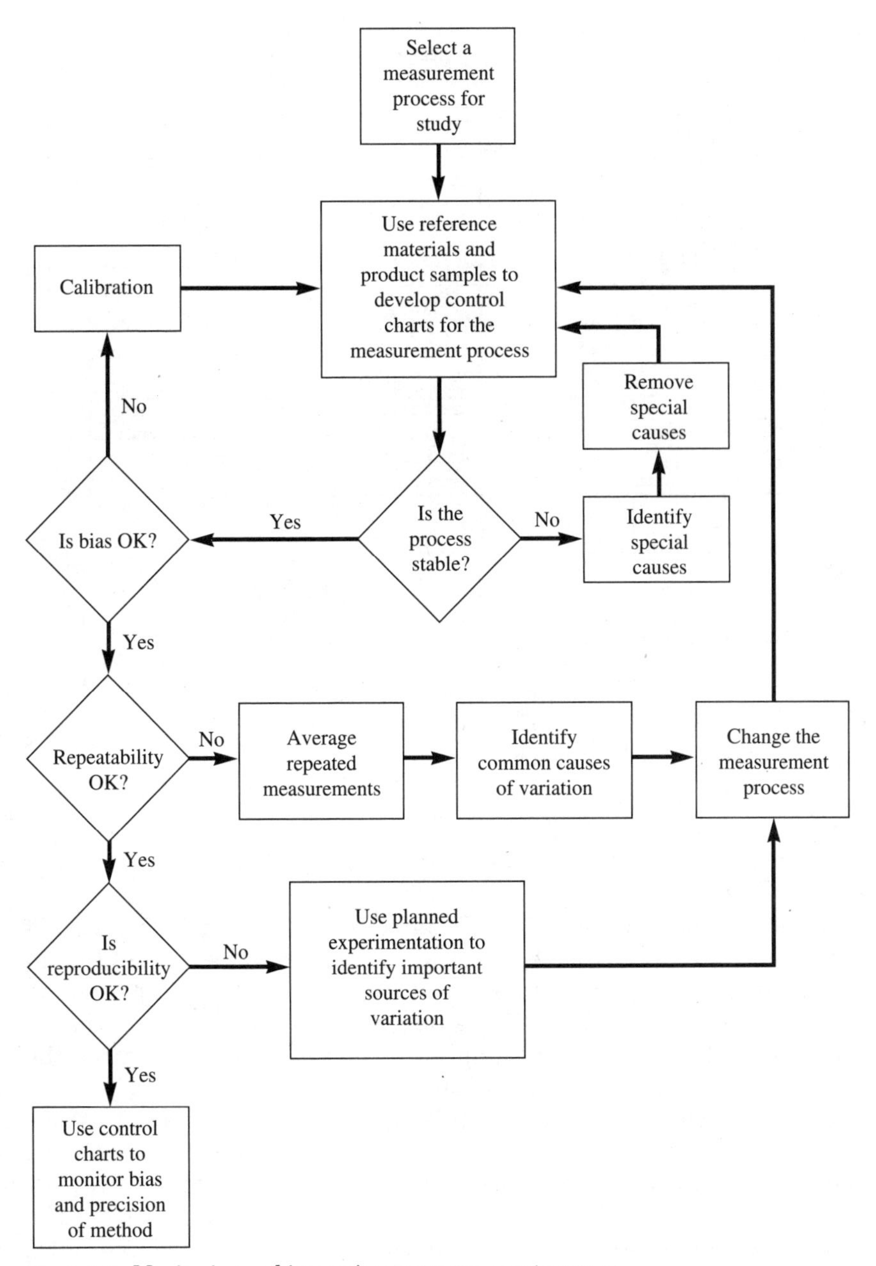


**Figure B.4** Monitoring and improving a measurement process.

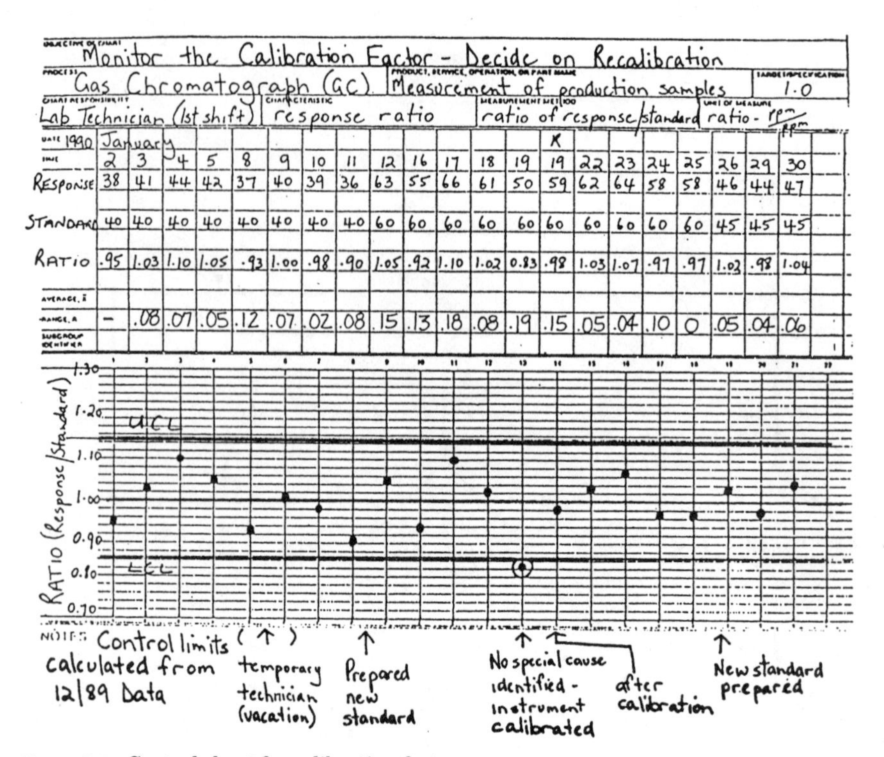


Figure B.5 Control chart for calibration factor.

chromatograph system used to measure low levels of chemical impurities.

In the example in Fig. B.5, each day that the instrument is used, a reference material is injected into the instrument. The ratio of the instrument response to the standard value is calculated and plotted on an individual control chart. If the response ratio is in control, the instrument is used to analyze samples. If the ratio is out of control, the instrument is recalibrated according to the calibration procedure.

Figure B.6, adapted from the ANSI standard (1987b), illustrates how these various types of standards can be used to monitor and control the accuracy of a measurement process. Check standards are evaluated at regular intervals using the normal measurement process. The established values for the check standard are validated using a reference material. The properties of the reference material are established and monitored using the reference method or reference standard.

Standards and reference materials are primarily used to study the bias of a measurement process. To study precision, it is not necessary to have a standard. Repeated measurements of the same material,

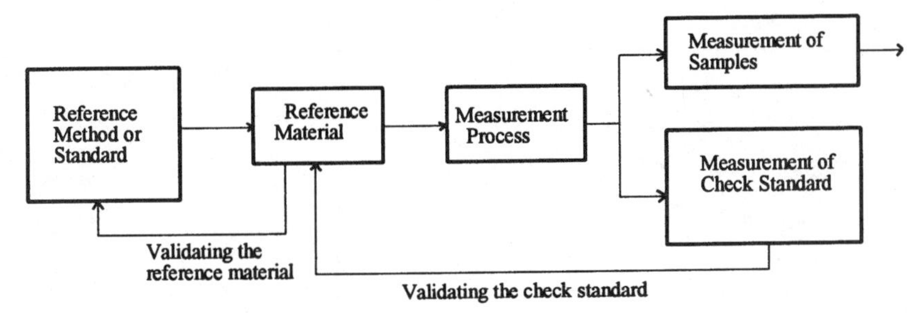


**Figure B.6** Use of standards with a measurement process.

sample, or situation can be used to study precision without knowing the accepted value of the element being measured.

In studying precision, the homogeneity of the material being evaluated becomes an important issue. This is especially true when the measurement process is destructive. An example would be a pull test that stretches a material until it breaks and records the force required to break the material. If the measured material is destroyed, it is not possible to repeat the test on the exact same material. The next section discusses procedures to study precision of a measurement process.

### Evaluating precision of a measurement process

Whenever precision of a measurement process is evaluated, it is important to understand which sources of variation in the measurement process will be included in the evaluation. When the sources of variation are minimized, the focus is on the repeatability of some aspect of the measurement process (often the instrument). When multiple sources of variation are studied (reproducibility), the focus is on the overall measurement process.

The procedures to study the precision of a measurement process depend on the type of measurement data. For continuous data, $\overline{X}$ and $R$ control charts can be used to study precision. For the other types of data, special procedures are required.

### Precision for classification data

Precision for classification data can be evaluated by repeating the measurement process on the same samples or objects of measurement. If the measurement relies heavily on a person's judgment, the repeat measures should be made in a blind (unknown to the observer) manner. If two different observers or inspectors measure the same objects, the reproducibility of the measurement process can be determined.

Table B.1 summarizes the results of a study to evaluate the precision of a classification measurement based on the examination of

**TABLE B.1 Precision for Classification Data**

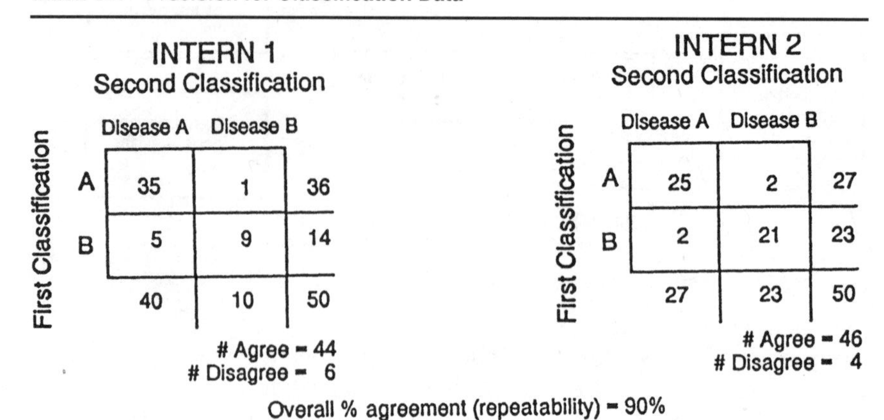

INTERN 1

| First Classification \ Second Classification | Disease A | Disease B | |
|---|---|---|---|
| A | 35 | 1 | 36 |
| B | 5 | 9 | 14 |
| | 40 | 10 | 50 |

\# Agree = 44
\# Disagree = 6

INTERN 2

| First Classification \ Second Classification | Disease A | Disease B | |
|---|---|---|---|
| A | 25 | 2 | 27 |
| B | 2 | 21 | 23 |
| | 27 | 23 | 50 |

\# Agree = 46
\# Disagree = 4

Overall % agreement (repeatability) = 90%

| INTERN 1 \ INTERN 2 | Disease A | Disease B | |
|---|---|---|---|
| A | 51 | 25 | 76 |
| B | 3 | 21 | 24 |
| | 54 | 46 | 100 |

\# Agree = 72
\# Disagree = 28
Reproducibility = 72%

X-rays by two interns. Fifty X-rays were selected for the study. About two-thirds of the X-rays were from patients later determined to have disease *A*, and the remainder were from patients with disease *B*. The X-rays were coded and independently evaluated by the two interns. The X-rays were then reordered and again submitted to each of the interns. (The same X-rays were presented as a second group of 50 X-rays.)

The repeatability of the measurement process is evaluated by comparing the first classification to the second classification for each intern. A very precise measurement method would show good agreement between the two evaluations. There were 44 (out of 50) agreements for intern 1 and 46 agreements for intern 2. The overall repeatability of the measurement method could be described as "90 percent agreement on repeated evaluations."

The reproducibility of the measurement process in this example is determined by comparing the results of the two interns on the 100

X-rays (two sets of the same 50 X-rays). They agreed on 72 of the evaluations and disagreed on 28. The reproducibility of the measurement method could be described as "72 percent agreement between interns on the same X-rays."

Two-way tables comparing repeated uses of the measurement method will generally be the best way to evaluate and describe the precision of a measurement method that gives classification data. Rank data with two levels (e.g., conforming and nonconforming parts) could be analyzed in a similar manner.

### Precision for rank data

Six different technicians were used to rank the quality of manufactured materials by visual observation. The study was done to evaluate the overall precision of the measurement process. Five samples of materials were selected from the manufacturing process for the study. Each technician was asked to rank the five samples from low to high quality. They repeated this process on three different days. Table B.2 contains the results of this study.

Figure B.7 shows a plot of the rankings for each of the samples. While there is variation in the ranks for each sample, the plot shows a tendency for each sample to be ranked either low or high. To summarize the precision of this measurement method, the median and range of the individual measurements (rank) were computed for each sample as determined by each technician. The median range over all the samples was used to quantify the repeatability for each technician (see Table B.2). Technician 5 could not repeat the rankings as well as the other technicians. The median of the median ranges was used to quantify the overall repeatability of the method.

Reproducibility for each sample was quantified by calculating the range of the median ranks for all the technicians. This measure of reproducibility varied from 1 to 3 for the five samples. When technician 5 was removed, the measure varied from 1 to 2.

In conducting precision studies for measurement systems that produce rank data, a graphical analysis (similar to Fig. B.7) should always be prepared. The interpretation of this plot will help in understanding the precision of the test method. Care should be taken when doing arithmetic on the ranks. Medians are generally preferred over averages for rank data since medians do not require arithmetic operations.

### Precision for continuous data

Control charts can be used to study the precision of a test method that produces continuous or variable data. A set of $\overline{X}$ and $R$ control charts for a process can be modified to include an evaluation of the

**TABLE B.2 Study of Precision of Rank Measurement**

| Technician | Day | Sample A | B | C | D | E |
|---|---|---|---|---|---|---|
| 1 | 1 | 2 | 3 | 4 | 1 | 5 |
| | 2 | 1 | 4 | 5 | 2 | 3 |
| | 3 | 2 | 3 | 4 | 1 | 5 |
| 2 | 1 | 1 | 3 | 5 | 2 | 4 |
| | 2 | 1 | 3 | 5 | 2 | 4 |
| | 3 | 1 | 2 | 4 | 3 | 5 |
| 3 | 1 | 1 | 3 | 5 | 2 | 4 |
| | 2 | 1 | 4 | 5 | 2 | 3 |
| | 3 | 2 | 4 | 5 | 1 | 3 |
| 4 | 1 | 1 | 2 | 5 | 3 | 4 |
| | 2 | 2 | 3 | 5 | 2 | 4 |
| | 3 | 1 | 3 | 5 | 2 | 4 |
| 5 | 1 | 4 | 1 | 5 | 2 | 3 |
| | 2 | 5 | 3 | 1 | 4 | 2 |
| | 3 | 2 | 5 | 4 | 3 | 1 |
| 6 | 1 | 2 | 4 | 5 | 1 | 3 |
| | 2 | 1 | 3 | 5 | 2 | 4 |
| | 3 | 2 | 3 | 4 | 1 | 5 |

| Technician | A M | A R | B M | B R | C M | C R | D M | D R | E M | E R | Median range |
|---|---|---|---|---|---|---|---|---|---|---|---|
| 1 | 2 | 1 | 3 | 1 | 4 | 1 | 1 | 1 | 5 | 2 | 1 |
| 2 | 1 | 0 | 3 | 1 | 5 | 1 | 2 | 1 | 4 | 1 | 1 |
| 3 | 1 | 1 | 4 | 1 | 5 | 0 | 2 | 1 | 3 | 1 | 1 |
| 4 | 1 | 1 | 3 | 1 | 5 | 0 | 2 | 1 | 4 | 0 | 1 |
| 5 | 4 | 3 | 3 | 4 | 4 | 4 | 3 | 2 | 2 | 2 | 3 |
| 6 | 2 | 1 | 3 | 1 | 5 | 1 | 1 | 1 | 4 | 2 | 1 |
| | | | | | | | Overall repeatability = | | | | 1.0 (median) |
| | 3 | | 1 | | 1 | | 2 | | 3 | | |
| | Reproducibility (range of median ranks) Overall reproducibility = 2 (median) | | | | | | | | | | |
| | 1 | | 1 | | 1 | | 1 | | 2 | | |
| | Reproducibility (eliminating technician 5) Overall reproducibility = 1 | | | | | | | | | | |

M = median of ranks R = range of ranks

precision of the measurement process used to generate continuous data without requiring additional measurements. An example of using control charts to study the precision of a measurement process is included in Sec. 7.2.

Another way to evaluate the precision of a measurement method is to analyze the same sample or material over a period of time. For example, a sample from the same production material could be analyzed

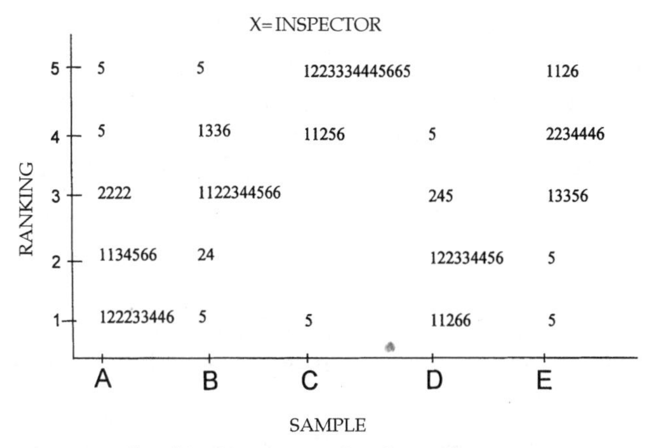


**Figure B.7** Graphical summary of ranking data.

each day that the measurement system is used. The individual results could be analyzed on an $X$ chart. The moving range of the daily results could be used to estimate the precision of the test method. This method can potentially underestimate the precision because of the familiarity with the sample being measured. Repeated analysis of a standard could also be used to evaluate precision if the sources of variation for the standard are shown to be similar to the sources of variation of production material.

How precise does a measurement process have to be in order to be useful? The answer depends on how it will be used. Sometimes measurement systems that are precise enough for inspection situations do not have adequate precision to allow for learning about other sources of variation in a process. Often for inspection, a requirement for precision of a measurement process is expressed as a ratio of the standard deviation of the measurement process (repeatability or reproducibility) and the width of the applicable specifications for the quality characteristic of interest. For example, the General Motors *Statistical Process Control Manual* (1984) states that gage repeatability and reproducibility should be less than 10 percent of the tolerance to be generally acceptable. Ford Motor Company (1986) requires the capability ($\pm 3\sigma$ spread) of a measurement process to be equal to or less than 10 percent of the tolerance for the characteristic being evaluated.

For use in process improvement, the measurement variation should be expressed as a percentage of the total process variation. McNeese and Klein describe a capable measurement process as one that is useful for monitoring and improving a process over time.

### Evaluating bias of a measurement process

To study the bias of a measurement process, some type of standard or reference method must be used. In some industries, it is common to lack standards for some important characteristics. In these cases, the stability of the bias over time is more important than the magnitude of the bias. Bias can be evaluated in two ways:

1. A material with a sufficiently well-established value or an accepted value is measured.
2. A method which is considered the standard (the reference method) is used to measure the same material that is analyzed by the measurement system of interest.

Standard control charts can be modified to study the bias and accuracy of a process for measurement. Figure B.8 shows a schematic of the modification. Each time that a subgroup is analyzed, some type of standard is also analyzed. Alternatively, one of the process samples could be analyzed by the reference method.

The individual measurements of the production samples are used to construct the $\overline{X}$ and $R$ charts while the measurement of the standard is used to construct an individual chart. The moving range of the repeated measurements of the standard is used to establish control limits. If the chart for the standard is in statistical control, the difference between the centerline of the individual chart and the accepted standard value is the bias of the measurement process. The stability of the individual chart measures the consistency of the bias.

Figure B.9 shows an example of a set of control charts modified in this manner. A sample of material is taken from the process every hour and analyzed for protein content. The results are grouped into subgroups every 3 h, and a reference material (certified at 16.0 percent protein) is analyzed. The stability of the $X$ chart for the analysis of the standard provides good evidence that the special causes in the process are not due to the measurement process. The average of the analyses of the standard material (centerline = 15.96) indicates that the analytical method has a very small bias.

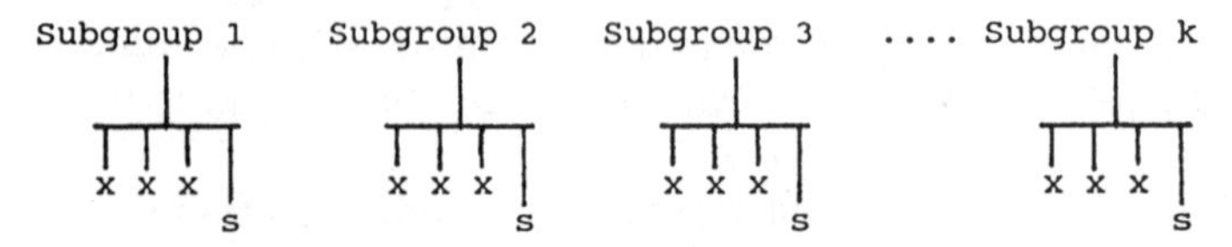


**Figure B.8** Modified $\overline{X}$ and $R$ chart to evaluate bias of measurement.

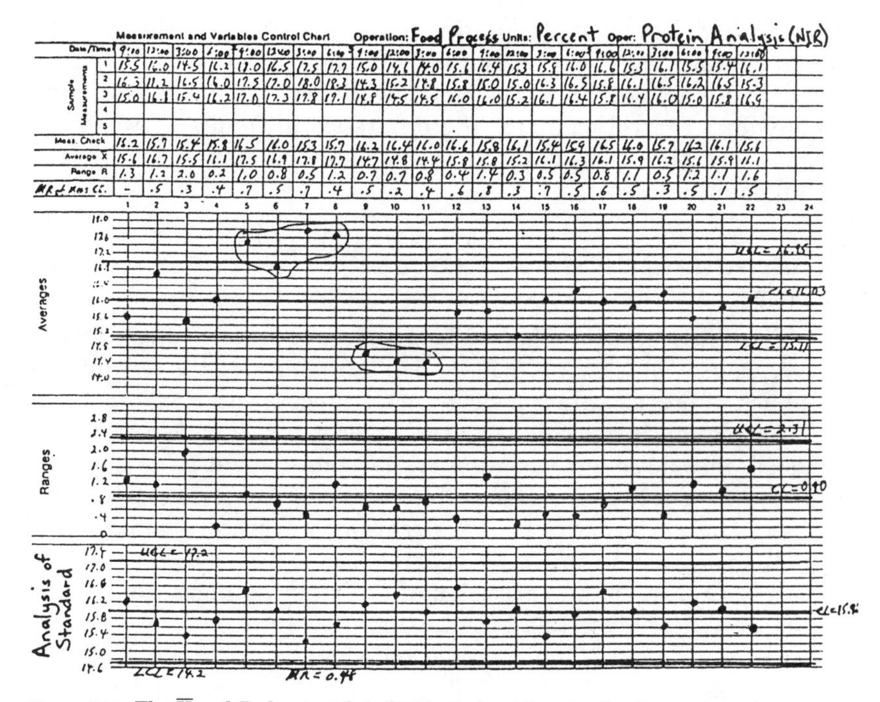


**Figure B.9** The $\overline{X}$ and $R$ chart with individual chart for standard.

The most important aspect of bias in a measurement process is stability. The example in Fig. B.9 (the chart at the bottom of the figure) shows a method which is stable and whose average is near the target of 16 percent. If the chart did not indicate stability, it would not be useful to draw conclusions about the bias of the method.

If multiple measures of a standard are available within a given time period, an $\overline{X}$ and R chart can be developed to evaluate the bias. The $\overline{X}$ chart, based on averages of subgroups, would describe the bias of the method.

Bias for classification, count, or rank data can also be evaluated by using standards or reference methods. For classification data, the bias would be expressed as the percentage of correct classifications. For example, for the precision study summarized in Table B.1, the classifications of the interns could be compared to the determinations of an experienced doctor. The percentage of classifications that disagreed with the doctor's determination would be the measure of bias. The bias would indicate a tendency of the measurement to favor one of the two classifications.

The bias of a method yielding count data would be stated as the average difference in counts, using the measurement method and a reference method. For example, routine procedures to count the inventory in an operating production system could be evaluated for bias once per month by doing a thorough count while the plant is shut down. An individual control chart for the difference in counts between the two methods could then be used to evaluate the bias of the routine measurement process.

### Effect of precision and bias of a measurement process

How will an imprecise or a biased measurement process affect the use of the measurements? What can be done to minimize the adverse effects? Lack of precision in a measurement process will show up as increased variation in measured materials or samples. When tests are conducted, measurement variation or bias can make it difficult to assess the impact of changes. If $\overline{X}$ and $R$ charts are being used to control a process, the variation from the measurement process will become part of the variation in the range chart.

One way to increase the precision of the reported measurements is by doing repeat analyses, and then using the average of the analyses as the "measurement." The expected reduction in variation of the reported averages is a function of the square root of the number of repeated analyses. Table B.3 summarizes this relationship. The measurement process must be repeated 4 times ($n = 4$) to reduce the measurement variation by one-half.

The effect of variation in a measurement process can become much greater if control charts are not used in making decisions to change or adjust the process for which the measurements are being made. What if the process is making satisfactory product but, because of variation in the measurement, the reported sample result is outside specifications? If the process is adjusted in this case without consideration of whether the process is in control, the overall variation could be increased even more.

The effect of bias in a measurement process will show up differently from the effect of precision. Bias will affect the location of the measurement data, not the variation. If a process is being studied using $\overline{X}$ and $R$ charts, a consistent bias will not affect the range chart, but it will affect the values on the $\overline{X}$ chart. The averages will be consistently high or low by the amount of bias in the measurement method. This will not affect the performance of the control chart, but may affect the interpretation of the data.

Good calibration practices can minimize the impact of bias of a measurement process. Control charts (such as in Fig. B.5) should always be used to make decisions on when to make calibration adjustments.

The bottom line is that the measurement process must be useful for

**TABLE B.3 Reduction in Variation from Repeated Measurements**

| Number of replications $n$ | Expected variation from measurement process: Standard deviation of average of $n$ measurements ($\sigma_m$ is standard deviation of individual measurement) |
|---|---|
| 1 | $\sigma_m = \sigma_m$ |
| 2 | $\sigma_m/\sqrt{2} = 0.71\sigma_m$ |
| 3 | $\sigma_m/\sqrt{3} = 0.58\sigma_m$ |
| 4 | $\sigma_m/\sqrt{4} = 0.50\sigma_m$ |
| 9 | $\sigma_m/\sqrt{9} = 0.33\sigma_m$ |
| Standard deviation of average of $n$ measurements $= \sigma_{\text{avg}} = \sigma_m/\sqrt{n}$ | |

the intended application. The following definition of a *capable measurement system* is based on McNeese and Klein (1991–1992):

> A measurement process that is stable with respect to both the average and variation, whose average equals the reference value, and that is responsible for less than 10 percent of the total process variation.

## B.5 Using Planned Experimentation to Improve a Measurement Process

When a measurement system has an unacceptable level of precision (repeatability or reproducibility), the methods of planned experimentation can be used to

- Identify major sources of variation in the measurement process
- Determine the effect of various factors in the measurement process
- Evaluate alternative measurement methods
- Develop a measurement process that is not affected by minor departures in the method

The case study in the next section includes an application of planned experimentation methods to evaluate the most important source of variation in the measurement system. Some aspects of the experiment are discussed in Chap. 7 (Sec. 7.3) on evaluating sources of variation.

## B.6 Case Study: Evaluating a Measurement Process

A series of studies were done to evaluate the measurement process used to determine the gap width in a machining process. Process documentation for the measurement process specified an unbiased test method with repeat measurements within 0.2 mm. The measurement process involved two key steps—a setup of the part to be measured and the gaging of the part by using calipers. The studies follow the flow diagram for monitoring and improving a measurement process, presented in Fig. B.4.

**Develop control charts for the measurement process.** Once a measurement process is selected for study, the first step is to use the analysis of reference materials and product samples to develop control charts for the measurement process. This can be done in a number of ways, as presented in Sec. B.4. In this study, two charts were developed to evaluate this measurement system. The first chart is used to evaluate the bias of the measurement process. The individual chart in Fig. B.10 was developed using a reference standard.

The second control chart is for the precision of the measurement process. Repeat measurements of parts selected from the process were made by submitting each part to the entire measurement process twice. A second operator conducted the second measurement about 1 hour after the initial test. Figure B.11 shows the completed $R$ chart for the measurement process.

**Stability.** Study of the control charts in Figs. B.10 and B.11 is used to establish the stability of the measurement process. The individual chart in Fig. B.10 is based on measurement of a reference standard (a device purchased from a calibration laboratory which simulates a production part) with an accepted value of 4.00 mm for gap width. The chart was prepared after the reference standard was measured about once per hour over a 3-day period (20 measurements). Sometime during each hour, the reference part was "set up" on the fixture and measured one time using the calipers. The moving range between consecutive measurements was used to establish control limits for the individual measurements. The completed chart indicated that the measurement process was stable (no special causes) relative to bias.

The range chart for the repeated measurements of the production parts in Fig. B.11 indicated that the measurement process was stable relative to precision. The entire measurement process was repeated each time (setup and measurement with calipers) a measurement was taken by each of the operators.

Since both the control chart for bias and the chart for precision indicate that the measurement process is stable, the next step is to evaluate the acceptability of the measurement process. Information on the control charts established to study the stability of the measurement process can be used to determine the acceptability of the bias and precision.

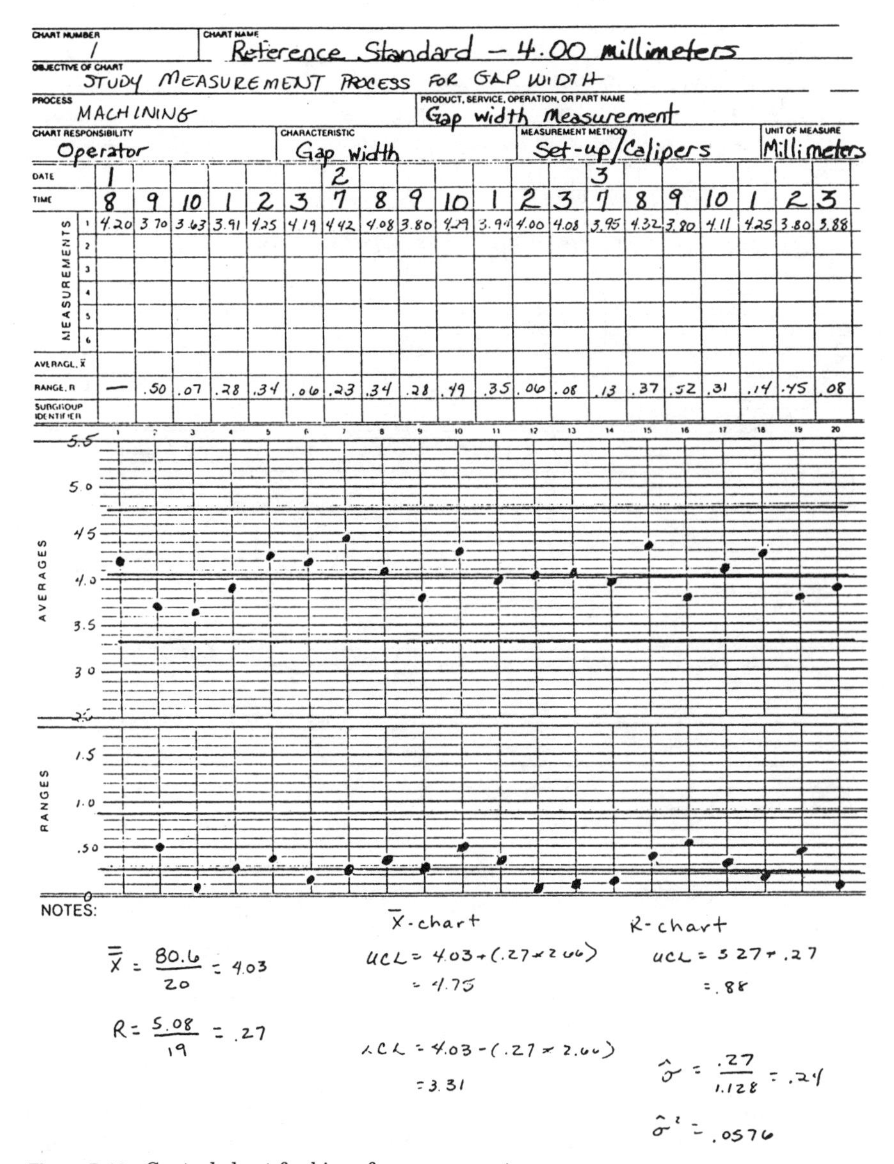

CHART NUMBER: 1

CHART NAME: Reference Standard – 4.00 millimeters

OBJECTIVE OF CHART: STUDY MEASUREMENT PROCESS FOR GAP WIDTH

PROCESS: MACHINING

PRODUCT, SERVICE, OPERATION, OR PART NAME: Gap width Measurement

CHART RESPONSIBILITY: Operator

CHARACTERISTIC: Gap width

MEASUREMENT METHOD: Set-up/Calipers

UNIT OF MEASURE: Millimeters

| | 1 | 2 | 3 | 4 | 5 | 6 | 7 | 8 | 9 | 10 | 11 | 12 | 13 | 14 | 15 | 16 | 17 | 18 | 19 | 20 |
|---|---|---|---|---|---|---|---|---|---|---|---|---|---|---|---|---|---|---|---|---|
| DATE | 1 | | | | | | 2 | | | | | | | 3 | | | | | | |
| TIME | 8 | 9 | 10 | 1 | 2 | 3 | 7 | 8 | 9 | 10 | 1 | 2 | 3 | 7 | 8 | 9 | 10 | 1 | 2 | 3 |
| MEASUREMENTS 1 | 4.20 | 3.70 | 3.63 | 3.91 | 4.25 | 4.19 | 4.42 | 4.08 | 3.80 | 4.29 | 3.94 | 4.00 | 4.08 | 3.95 | 4.32 | 3.80 | 4.11 | 4.25 | 3.80 | 3.88 |
| 2 | | | | | | | | | | | | | | | | | | | | |
| 3 | | | | | | | | | | | | | | | | | | | | |
| 4 | | | | | | | | | | | | | | | | | | | | |
| 5 | | | | | | | | | | | | | | | | | | | | |
| 6 | | | | | | | | | | | | | | | | | | | | |
| AVERAGE, $\bar{X}$ | | | | | | | | | | | | | | | | | | | | |
| RANGE, R | — | .50 | .07 | .28 | .34 | .06 | .23 | .34 | .28 | .49 | .35 | .06 | .08 | .13 | .37 | .52 | .31 | .14 | .45 | .08 |
| SUBGROUP IDENTIFIER | | | | | | | | | | | | | | | | | | | | |



NOTES:

$\bar{\bar{X}} = \frac{80.6}{20} = 4.03$

$R = \frac{5.08}{19} = .27$

$\bar{X}$-chart

$UCL = 4.03 + (.27 \times 2.66) = 4.75$

$LCL = 4.03 - (.27 \times 2.66) = 3.31$

R-chart

$UCL = 3.27 \times .27 = .88$

$\hat{\sigma} = \frac{.27}{1.128} = .24$

$\hat{\sigma}^2 = .0576$

**Figure B.10** Control chart for bias of measurement process.

**Evaluation of bias.** To evaluate bias, the accepted value of the reference standard (4.00 mm) was compared to the average of repeated analyses of the reference standard. The centerline of the individual chart in Fig. B.10 provided this average (4.03 mm). The bias of the measurement method is the average minus the accepted value (4.03−4.00 = +0.03 mm). This amount of bias was considered small by the engineer responsible for the machining process. The bias is also

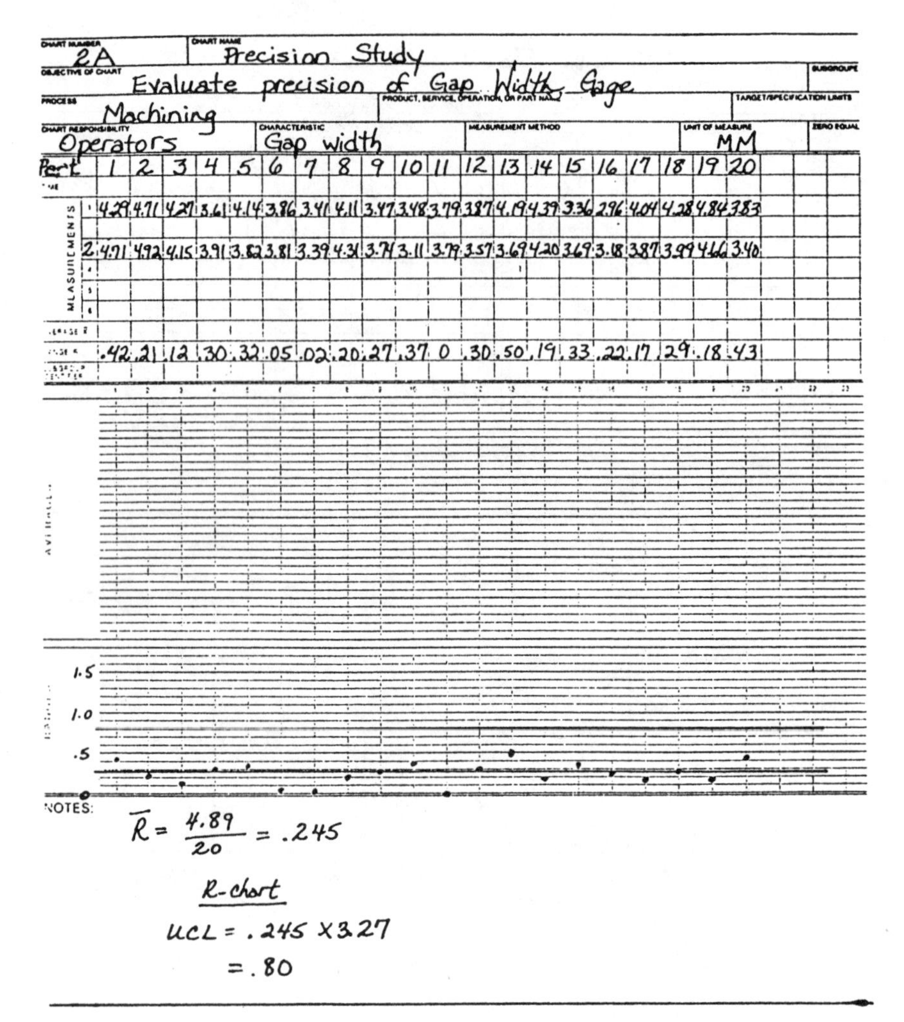
CHART NUMBER: 2A
CHART NAME: Precision Study
OBJECTIVE OF CHART: Evaluate precision of Gap Width Gage
PROCESS: Machining
CHART RESPONSIBILITY: Operators
CHARACTERISTIC: Gap width
UNIT OF MEASURE: MM

| Part | 1 | 2 | 3 | 4 | 5 | 6 | 7 | 8 | 9 | 10 | 11 | 12 | 13 | 14 | 15 | 16 | 17 | 18 | 19 | 20 |
|---|---|---|---|---|---|---|---|---|---|---|---|---|---|---|---|---|---|---|---|---|
| Measurement 1 | 4.29 | 4.71 | 4.27 | 3.61 | 4.14 | 3.86 | 3.41 | 4.11 | 3.47 | 3.48 | 3.79 | 3.87 | 4.19 | 4.39 | 3.36 | 2.96 | 4.04 | 4.28 | 4.84 | 3.83 |
| Measurement 2 | 4.71 | 4.92 | 4.15 | 3.91 | 3.82 | 3.81 | 3.39 | 4.31 | 3.74 | 3.11 | 3.79 | 3.57 | 3.69 | 4.20 | 3.69 | 3.18 | 3.87 | 3.99 | 4.66 | 3.40 |
| Range R | .42 | .21 | .12 | .30 | .32 | .05 | .02 | .20 | .27 | .37 | 0 | .30 | .50 | .19 | .33 | .22 | .17 | .29 | .18 | .43 |

Range chart scale: 1.5, 1.0, .5

NOTES:

$$\bar{R} = \frac{4.89}{20} = .245$$

R-chart

$$UCL = .245 \times 3.27$$
$$= .80$$

Figure B.11 Control chart for precision of measurement process.

small relative to the control limits for the individual measurements of the standard (3.31 to 4.75 mm). The engineer concluded that the current procedures for calibration were adequate for controlling the bias and that the measurement system could be considered unbiased.

**Evaluation of precision: Repeatability and reproducibility.** The study of precision of a test method includes consideration of the sources of variation (operators, time, equipment, etc.) which are included in the determination of precision. Repeatability represents the variation of a test method under the best conditions (minimum sources of variation), while reproducibility refers to variation with additional sources of variation present. The control chart for precision in Fig. B.11 includes repeatability of the test method, as well as variation due to setup, operator, time, and other variables that could change over a 1-hour period. The average range for this chart was 0.245 mm.

Since this range was larger than expected, a quick study was done to isolate the repeatability of the measurement instrument. A setup was performed on a production part, and then one of the operators measured the part, using the caliper 20 times over a 1-hour period. The control chart in Fig. B.12 shows the results of this study.

The variation is considerably less (average range = 0.045) than that of the initial precision study (Fig. B.11) where other measurement variables were not held constant. The information from the control charts in Figs. B.11 and B.12 can be combined to summarize the repeatability and reproducibility of the measurement method. Table B.4 shows these results. The magnitude of the reproducibility was not considered acceptable, so a small study was planned to understand the important source of variation.

**Planned experiment for the measurement process.** An experiment was designed to evaluate the sources of variation in the measurement process. This experiment is described in Chap. 7. The analysis of data from the experiment revealed that

- 74 percent of the variation is attributable to part differences.
- 18 percent is attributable to the measurement setup procedure.
- 8 percent is due to the measurement procedure (calipers).

Reductions in variation in the process could come from improving the setup procedure in the measurement process and then concentrating on factors that cause the part-to-part variation.

**Improvements to the measurement process.** A meeting was held to consider ways to improve the reproducibility of the measurement process. One operator pointed out that when the planned experiment was done, all the measurements were rounded to the nearest 0.05 mm in an attempt to get more consistent readings. This practice of rounding had originated at a time when the calipers did not have the current scale sensitivity. The new calipers being used would allow data to be recorded to 0.01 mm rather than rounding to the nearest 0.05 mm. Although this change affected only the repeatability of the measurement process, it was decided to clarify the procedures and always record the data to 0.01 mm.

The process engineer reported on a new type of setup fixture that she had seen being used in another plant that employed tension springs to locate the part in the fixture. This would take much of the

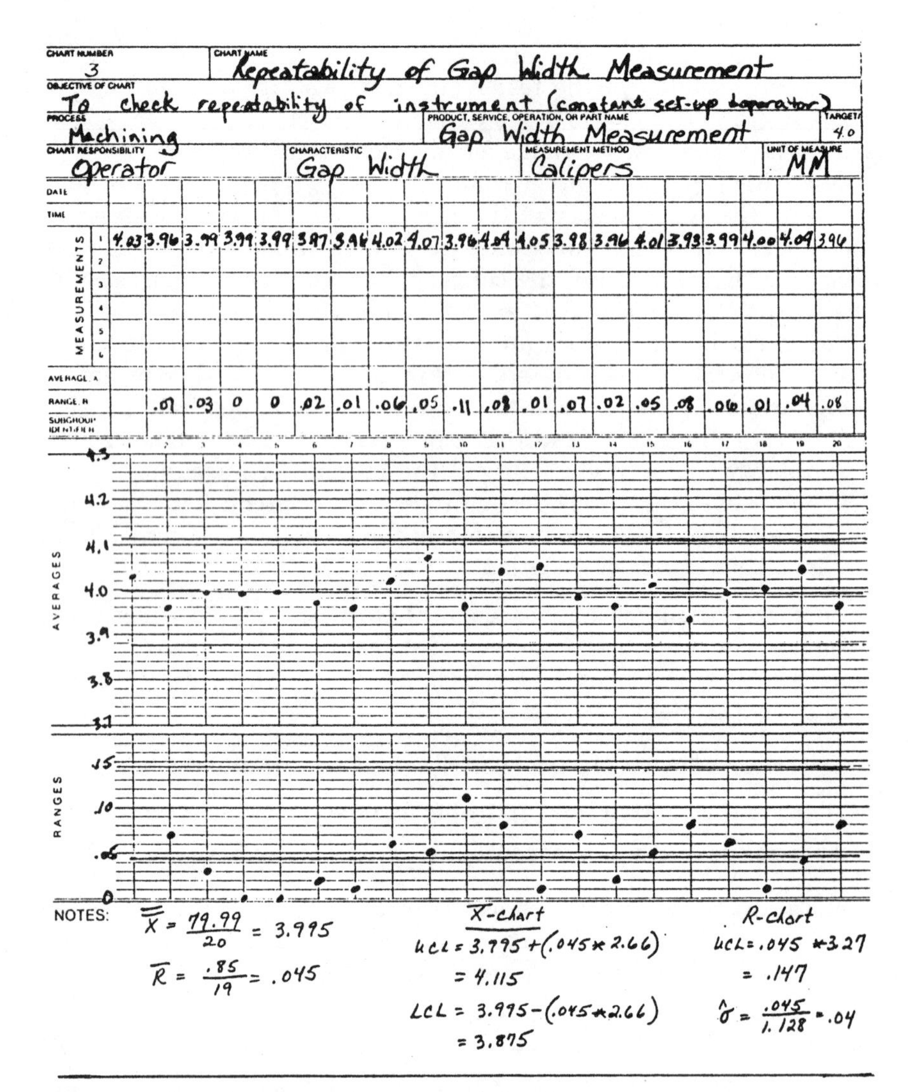


**Figure B.12** Control chart to evaluate repeatability of measurement process.

**TABLE B.4 Repeatability and Reproducibility of the Measurement Method**

| | Average range $\bar{R}_m$ | $\sigma_m$ | $\pm 3\sigma_m$ |
|---|---|---|---|
| First study—reproducibility (Fig. B.11) | 0.245 | 0.217 | 0.65 |
| Second study—repeatability (Fig. B.12) | 0.045 | 0.040 | 0.12 |

$\sigma_m$ = standard deviation of measurement ($\sigma_m = \bar{R}_m/1.128$, the $d_2$ factor)

manual clamping out of the setup step. It was decided to obtain one of these fixtures and conduct some tests.

After the installing of the new fixture and training of the operators in its use, the initial precision study was repeated. Ten parts selected from the process were submitted to the entire measurement process twice. As before, a second operator conducted the second measurement about 1 hour after the initial test. The control limits from the initial $R$ chart (Fig. B.11) were extended, and the ten new data points plotted on the same chart. Figure B.13 shows the extended chart. All the ranges were less than the centerline, so it was concluded that the changes to the measurement process had resulted in significant improvements.

**Updating control charts for the measurement process.** After confirmation that the changes in the measurement process had resulted in improvement, the initial control charts for bias and precision were updated. Since the repeatability of the caliper measurement was similar on both the reference standard and the production parts, the reference standard chart could be used to document repeatability.

To develop the new charts, a part was selected once per line from the

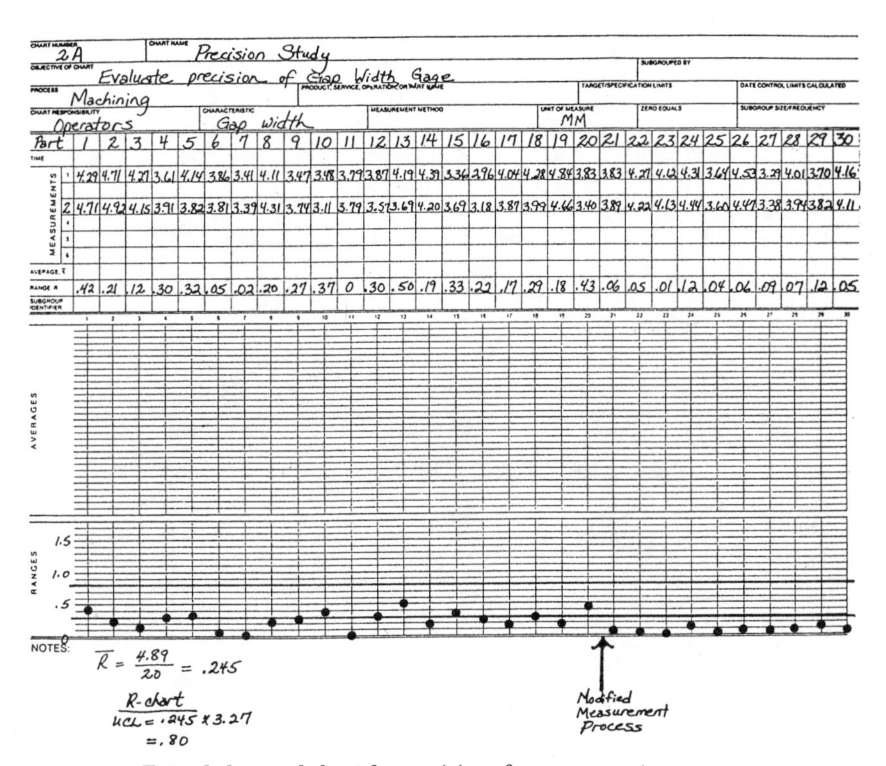
CHART NUMBER: 2A
CHART NAME: Precision Study
OBJECTIVE OF CHART: Evaluate precision of Gap Width Gage
SUBGROUPED BY:
PROCESS: Machining
PRODUCT, SERVICE, OPERATION OR PART NAME:
TARGET/SPECIFICATION LIMITS:
DATE CONTROL LIMITS CALCULATED:
CHART RESPONSIBILITY: Operators
CHARACTERISTIC: Gap width
MEASUREMENT METHOD:
UNIT OF MEASURE: MM
ZERO EQUALS:
SUBGROUP SIZE/FREQUENCY:

| Part | 1 | 2 | 3 | 4 | 5 | 6 | 7 | 8 | 9 | 10 | 11 | 12 | 13 | 14 | 15 |
|---|---|---|---|---|---|---|---|---|---|---|---|---|---|---|---|
| TIME | | | | | | | | | | | | | | | |
| Measurement 1 | 4.29 | 4.71 | 4.27 | 3.61 | 4.14 | 3.86 | 3.41 | 4.11 | 3.47 | 3.48 | 3.79 | 3.87 | 4.19 | 4.39 | 3.36 |
| Measurement 2 | 4.71 | 4.92 | 4.15 | 3.91 | 3.82 | 3.81 | 3.39 | 4.31 | 3.74 | 3.11 | 3.79 | 3.57 | 3.69 | 4.20 | 3.69 |
| AVERAGE, $\bar{X}$ | | | | | | | | | | | | | | | |
| RANGE, R | .42 | .21 | .12 | .30 | .32 | .05 | .02 | .20 | .27 | .37 | 0 | .30 | .50 | .19 | .33 |

| Part | 16 | 17 | 18 | 19 | 20 | 21 | 22 | 23 | 24 | 25 | 26 | 27 | 28 | 29 | 30 |
|---|---|---|---|---|---|---|---|---|---|---|---|---|---|---|---|
| TIME | | | | | | | | | | | | | | | |
| Measurement 1 | 2.96 | 4.04 | 4.28 | 4.84 | 3.83 | 3.83 | 4.27 | 4.62 | 4.31 | 3.64 | 4.53 | 3.29 | 4.01 | 3.70 | 4.16 |
| Measurement 2 | 3.18 | 3.87 | 3.99 | 4.66 | 3.40 | 3.89 | 4.22 | 4.13 | 4.44 | 3.60 | 4.47 | 3.38 | 3.94 | 3.82 | 4.11 |
| AVERAGE, $\bar{X}$ | | | | | | | | | | | | | | | |
| RANGE, R | .22 | .17 | .29 | .18 | .43 | .06 | .05 | .01 | .12 | .04 | .06 | .09 | .07 | .12 | .05 |



NOTES: $\bar{R} = \frac{4.89}{20} = .245$

R-chart
UCL = .245 × 3.27
= .80

**Figure B.13** Extended control chart for precision of measurement process.

production process. The first operator performed the setup of the part and made the measurement. The operator then removed the part, performed the setup on the reference standard, and measured the reference standard twice. Later in the hour, a second operator performed another setup of the selected part and made and recorded a measurement. Thus, the requested measurements of the reference standard were used to evaluate repeatability, and the repeated measurements of the production parts were used to evaluate reproducibility.

Figure B.14 shows the control charts developed after 2 days. All three charts were stable. The bias was evaluated by using the centerline of the average chart for the reference standard. The bias was −0.005 mm (3.995−4.000), which was considered negligible.

The repeatability and reproducibility were evaluated by using the two range charts. Table B.5 shows these results. The repeatability was about the same as that of the original study, but the reproducibility was decreased by greater than a factor of 3. The reproducibility was now considered acceptable for use. Based on the part-to-part variation from the planned experiment, the measurement variation now represents less than 2 percent of the total variation in the

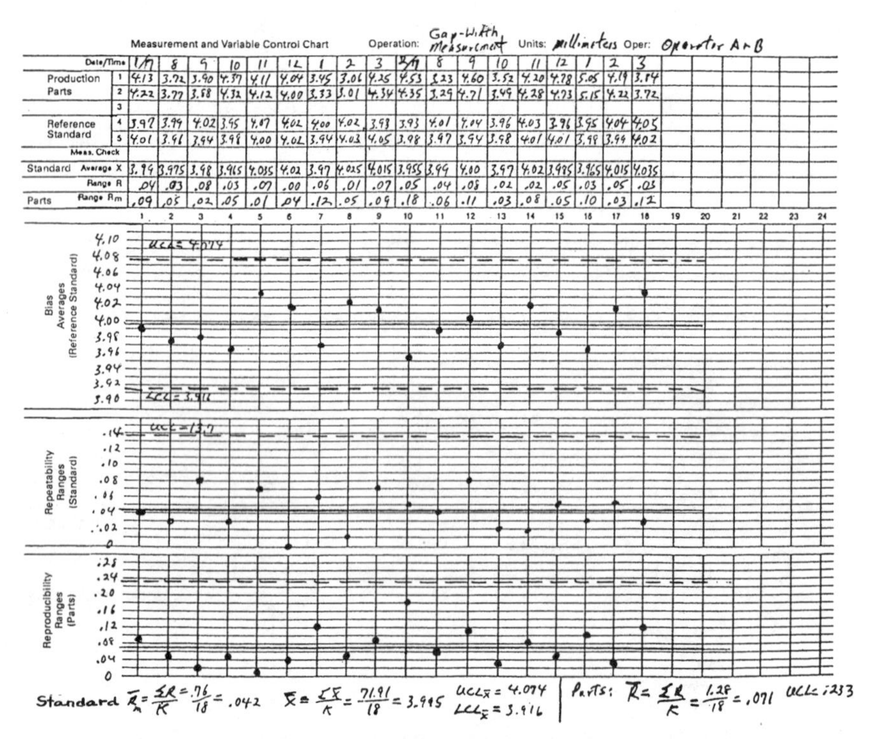


**Figure B.14** Control charts for improved measurement process.

**TABLE B.5 Repeatability and Reproducibility of Improved Measurement Process**

| | Prior to improvement $\sigma_m$ | After improvement $\sigma_m$ | After improvement $\pm 3\sigma_m$ |
|---|---|---|---|
| Reproducibility | 0.217 | 0.063 | $\pm$0.19 |
| Repeatability | 0.040 | 0.037 | $\pm$0.11 |

process. Future improvement efforts can now focus on the production process variation.

**Monitoring the measurement process.** The control charts in Fig. B.14 will be used for ongoing monitoring of the measurement process. The frequency of evaluation of a standard and production part will be reduced from hourly to twice per day. If the measurement process remains stable for a month, this frequency will be further reduced to once per day.

## B.7 Summary

The study and improvement of processes for measurement are important aspects of the improvement process. The measurement process is the window through which we view our processes of production and service (and our products). If that window does not provide a predictable, consistent view of the process, intelligent decisions about actions to be taken on the process cannot be made. The quality of an individual measurement cannot be directly assessed; the assessment has to be made of the measurement process.

Some important concepts about measurement were presented in this appendix:

1. There are many different quality characteristics of a measurement process. The most common are precision, bias, and accuracy. An accurate test method has both a useful level of precision and bias. Establishing the stability of the measurement process through the use of control charts must be done before the level of these characteristics can be evaluated.
2. The study of a measurement process will differ depending on the type of data generated by the process: continuous, rank, count, or classification.
3. Study of the precision of a measurement requires repeating the measurement process on the same materials (or similar materials when testing is destructive). The determination of precision will be affected by sources of variation that are allowed to vary when the measurements are repeated.
4. The concept of calibration, standards, and reference methods is important to the study of the bias of a measurement process.

5. The effect of variation of the current measurement process on the output of the production or service process being measured can be minimized by using good calibration practices (to control bias) and by repeating measurements and reporting the average (to increase precision).

6. Improving a measurement process is no different from improving a process for production or service. Planned experimentation can be used to identify the important sources of variation in the measurement process.

### Definitions of terms used with measurement

The definitions of the characteristics used to describe the measurement process vary greatly from application to application. Operational definitions should be stated whenever these terms are used. The following definitions have been used in a number of measurement references:

**Sensitivity** A measurement of precision that does not depend on the scale that is used; useful for comparing alternative methods. Sensitivity evaluates the ability of a measurement method to detect small changes in the property being measured (Mandel and Stiehler, 1954).

**Limit of detection** The lowest concentration level that can be determined to be statistically different from background noise or a "blank" material (American Chemical Society, 1983).

**Robustness (or ruggedness)** The degree to which a measurement system is immune to modest (and inevitable) departures from the procedures and controls in the methods (Youden, 1967).

**Linearity** The degree to which the calibration of a measurement system can be accomplished with a straight line. A nonlinear system requires more complex calibration procedures (Hunter, 1981).

**Reliability** Ability of a measurement system to perform under stated conditions for a stated period of time. (*Note:* In the social sciences, reliability is often used to describe the concept of precision.)

**Validity** The extent to which a measure reflects only the desired attribute of interest (used in the social sciences).

**Cost** The resources required to use a measurement system to conduct a measurement. This characteristic is important in comparing alternative measurement systems (Mandel and Stiehler, 1954).

**Speed** Time required to complete a measurement (also referred to as *turnaround time*).

**Stability** Tendency of the measurement system to remain in statistical control. A stable system is affected only by common causes of variation, while an unstable system is affected by both common and special causes of variation.

**Simplicity** Ease of use of the measurement system.

**Calibration** The comparison of a set of measurements to an accepted or standard value for the purpose of detecting or correcting any bias in the measurement process or for establishing a relationship between the measurements and the standard or accepted values.

**Reference standard** A material or measurement method that has an accepted value of a unit of measurement. The reference standard can be either a reference material or a reference method.

**Reference material** A material or substance with properties sufficiently well established to be used for the calibration or the assessment of a measurement method, or for assigning values to materials.

**Reference method** A measurement process that has been studied and has an acceptable level of accuracy to give accepted values of a unit of measure.

**Check standard** A stable, well-characterized, in-house standard remeasured periodically to ensure that the measurement process remains in a state of statistical control. Check standards are also called QC standards, QA standards, QC samples, monitors, etc. The difference between a check standard and a reference material is the level of certification and traceability to some accepted standard (such as a reference standard).

## References

American Chemical Society, Committee on Environmental Improvement (1983): "Principles of Environmental Analysis," *Analytical Chemistry,* vol. 55, pp. 2210–2218, December.

American National Standards Institute (1987a): *Definitions, Symbols, Formulas, and Tables for Control Charts,* ANSI/ASQC A1-1987, American Society for Quality Control, Milwaukee, Wisc.

American National Standards Institute (1987b): *American National Standard for*

*Calibration Systems,* ANSI/ASQC M1-1987, American Society for Quality Control, Milwaukee, Wisc.
American Society for Testing and Materials (1980): "Use of the Terms Precision and Accuracy as Applied to Measurement of a Property of a Material," E177-71. ASTM Committee E11 on Statistical Methods, West Conshohocken, Pa.
Bishop, Lane, W. J. Hill, and W. S. Lindsay (1987): "Don't Be Fooled by the Measurement System," *Quality Progress,* pp. 35–38, December.
Deming, W. E. (1986): *Out of the Crisis,* chap. 9, "Operational Definitions," MIT Center for Advanced Engineering Study, Cambridge, Mass.
Eisenhart, C. (1963): "Realistic Evaluation of the Precision and Accuracy of Instrument Calibration Systems," *Journal of Research of the National Bureau of Standards,* vol. 67c, no. 2, pp. 161–187.
Ford Motor Company (1986): *Q-101 Quality System Standard,* Quality Office, Detroit, Mich.
General Motors Corporation (1984): *Statistical Process Control Manual,* Detroit, Mich., p. 3–10.
Grubbs, F. E. (1973): "Errors of Measurement, Precision, Accuracy and the Statistical Comparison of Measuring Instruments," *Technometrics,* vol. 15, no. 1, pp. 53–66, February.
Holmes, D. S., and A. E. Mergen (1991–1992): "A Discussion of the Unit of Calibration Required for a Gauge," *Quality Engineering,* vol. 4, no. 1, pp. 1–7.
Hunter, J. S. (1981): "Calibration and Straight Line: Current Statistical Practices," *Journal of the Association of Official Analytical Chemists,* vol. 64, no. 3.
Mandel, J. (1971): "Repeatability and Reproducibility," *Material Research and Standards,* vol. 11, no. 8, pp. 8–16.
Mandel, J., and R. D. Stiehler (1954): "Sensitivity—A Criterion for the Comparison of Methods of Test," *Journal of Research of the National Bureau of Standards,* vol. 53, no. 3, pp. 155–159, September.
McNeese, W. H., and R. A. Klein (1991–1992): "Measurement Systems, Sampling, and Process Capability," *Quality Engineering,* vol. 4, no. 1, pp. 21–39.
Propst, A. L. (1989–1990): "Verification of a Shop Floor Measurement System," *Quality Engineering,* vol. 2, no. 1, pp. 1–12.
Schumacher, R. B. F. (1983): "Measurement Assurance through Control Charts," *ASQC Technical Conference Transactions,* ASQC, Milwaukee, Wisc., pp. 401–409.
Shewhart, W. A. (1931): *Economic Control of Quality of Manufactured Product,* American Society for Quality Control, Milwaukee, Wisc. (reprinted).
Shewhart, W. A. (1939): *Statistical Method from the Viewpoint of Quality Control,* W. E. Deming, ed., The Graduate School. U.S. Department of Agriculture, Washington, D.C.
Snee, R. D., L. B. Hare, and J. R. Trout (1985): *Experiments in Industry, Design, Analysis, and Interpretation of Results,* American Society for Quality Control, Milwaukee, Wisc.
Taylor, J. K. (1986): "Measurement and Calibration," *Chemtech,* pp. 756–763, December.
Taylor, J. K. (1981): "Quality Assurance of Chemical Measurements," *Analytical Chemistry,* vol. 53, no. 14, pp. 1588A–1596A, December.
Troxell, J. R. (1992): "Variance Components in Measurement Assurance Studies," *1992—ASQC Quality Congress Transactions—Nashville,* American Society of Quality Control, Milwaukee, Wisc., pp. 412–418.
Wernimont, G. (1946): "Use of Control Charts in the Analytical Laboratory," *Industrial and Engineering Chemistry,* vol. 18, no. 10, pp. 587–592.
Wheeler, D. J. (1992): "Problems with Gauge R&R Studies," *1992—ASQC Quality Congress Transactions—Nashville,* American Society of Quality Control, Milwaukee, Wisc., pp. 179–185.
Wheeler, D. J., and David S. Chambers (1986): *Understanding Statistical Process Control,* Statistical Process Controls, Inc., Knoxville, Tenn.
Wheeler, D. J., and R. W. Lyday (1989): *Evaluating the Measurement Process,* 2d ed., Statistical Process Controls, Inc., Knoxville, Tenn.
Youden, W. J. (1967): *Statistical Techniques for Collaborative Tests,* Association of Analytical Chemists, Washington, D.C.

Appendix

# C

# Forms

## C.1 Worksheet for Documenting a PDSA Cycle

Figure C.1 is a worksheet for the PDSA cycle that has been used extensively in the application of the model for improvement. The description of the model was developed in Sec. 1.3.

## C.2 Form for Documentation of a Planned Experiment

Figure C.2 is a form for documenting a planned experiment. The form was introduced in Sec. 3.5.

## C.3 Tables of Random Numbers

Randomization is the objective assignment of factor level combinations to experimental units. Tables of random numbers have been developed for use in making this random assignment. This section contains two different types of random number tables: a general random number table and a table of random permutations. When random permutations are used, each number of a series is selected without replacement until all numbers in the series have been selected.

Table C.1 contains random numbers from 00 to 99. Multiple columns can be combined to obtain larger numbers if needed. For example, the first digit from the second column can be combined with the first column to obtain random numbers from 000 to 999.

Tables C.2 and C.3 contain random permutations. The numbers 1 through 16 are permuted in Table C.2; each column contains three permutations. The numbers 1 through 8 are permuted in Table C.3; each column contains five random permutations.

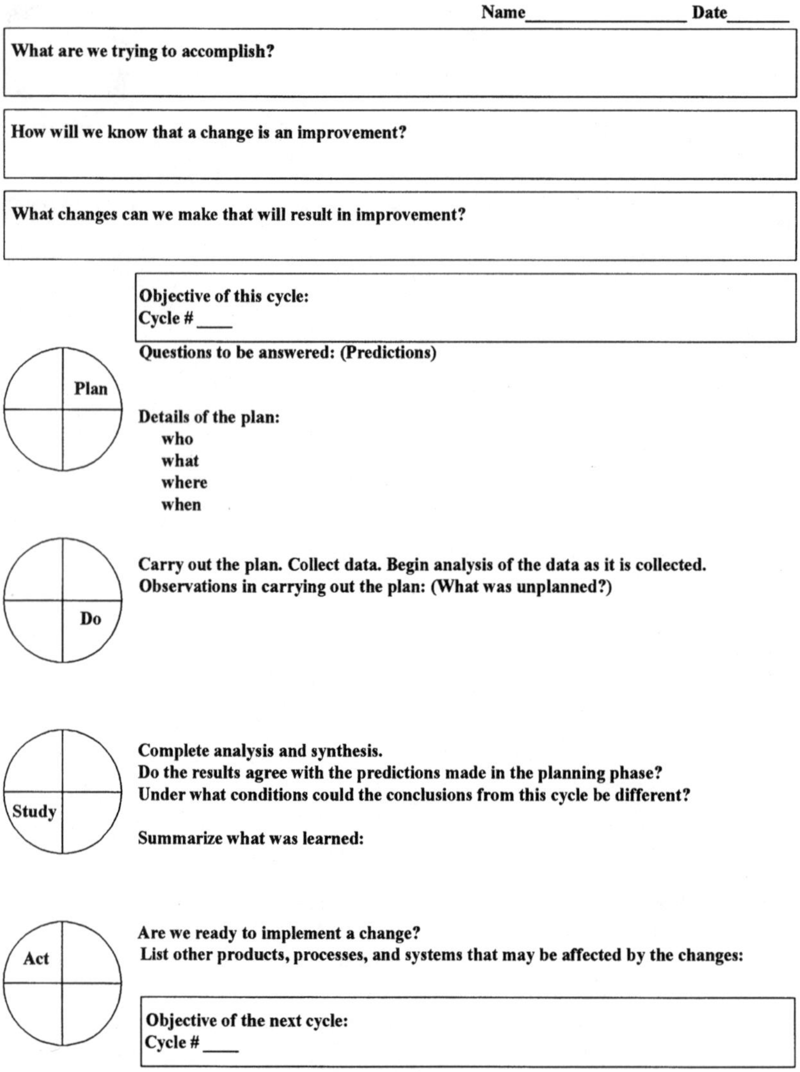

Name_______________ Date______

What are we trying to accomplish?

How will we know that a change is an improvement?

What changes can we make that will result in improvement?

Objective of this cycle:
Cycle # ____

Questions to be answered: (Predictions)

Plan

Details of the plan:
who
what
where
when

Do

Carry out the plan. Collect data. Begin analysis of the data as it is collected.
Observations in carrying out the plan: (What was unplanned?)

Study

Complete analysis and synthesis.
Do the results agree with the predictions made in the planning phase?
Under what conditions could the conclusions from this cycle be different?

Summarize what was learned:

Act

Are we ready to implement a change?
List other products, processes, and systems that may be affected by the changes:

Objective of the next cycle:
Cycle # ____

**Figure C.1** Worksheet for the PDSA cycle.

There are a number of ways to select numbers from the tables. An informal procedure would be to drop a pencil on the page of interest and start selecting numbers from the point of the pencil. A more formal approach would be to randomly select a row and column number from Table C.2 and use these numbers to designate the starting point of the permutation. One could also start in one corner of the table and mark off the numbers or permutations as they are used.

1. **Objective:**

2. **Background information:**

3. **Experimental variables:**

| A. | Response variables | Measurement technique |
|---|---|---|
| 1. | | |
| 2. | | |
| 3. | | |

| B. | Factors under study | Levels |
|---|---|---|
| 1. | | |
| 2. | | |
| 3. | | |
| 4. | | |
| 5. | | |
| 6. | | |
| 7. | | |

| C. | Background variables | Method of control |
|---|---|---|
| 1. | | |
| 2. | | |
| 3. | | |

4. **Replication:**

5. **Methods of randomization:**

6. **Design Matrix:** (attach copy)

7. **Data collection forms:** (attach copies)

8. **Planned methods of statistical analysis:**

9. **Estimated cost, schedule, and other resource considerations:**

**Figure C.2** Form for documentation of a planned experiment.

**TABLE C.1 Random Numbers**

| | 1 | 2 | 3 | 4 | 5 | 6 | 7 | 8 | 9 | 10 | 11 | 12 | 13 | 14 | 15 | 16 | 17 | 18 | 19 | 20 |
|---|---|---|---|---|---|---|---|---|---|---|---|---|---|---|---|---|---|---|---|---|
| **1** | 09 | 40 | 31 | 49 | 80 | 12 | 45 | 78 | 42 | 07 | 65 | 81 | 25 | 95 | 58 | 23 | 93 | 62 | 57 | 20 |
| **2** | 58 | 98 | 25 | 75 | 13 | 53 | 43 | 18 | 26 | 58 | 09 | 54 | 53 | 22 | 76 | 67 | 30 | 12 | 09 | 50 |
| **3** | 11 | 23 | 37 | 28 | 19 | 74 | 47 | 58 | 81 | 11 | 97 | 11 | 64 | 98 | 47 | 84 | 22 | 33 | 03 | 35 |
| **4** | 72 | 02 | 97 | 35 | 03 | 33 | 22 | 84 | 06 | 53 | 56 | 58 | 41 | 17 | 92 | 89 | 79 | 76 | 02 | 12 |
| **5** | 15 | 27 | 99 | 71 | 42 | 18 | 80 | 76 | 64 | 35 | 54 | 84 | 01 | 69 | 15 | 57 | 31 | 05 | 42 | 76 |
| **6** | 80 | 01 | 44 | 18 | 33 | 58 | 85 | 93 | 39 | 72 | 55 | 12 | 93 | 40 | 78 | 84 | 99 | 41 | 75 | 09 |
| **7** | 33 | 49 | 64 | 04 | 33 | 17 | 56 | 52 | 80 | 39 | 63 | 98 | 69 | 83 | 60 | 96 | 16 | 01 | 06 | 35 |
| **8** | 45 | 33 | 54 | 88 | 44 | 52 | 79 | 41 | 11 | 43 | 02 | 16 | 04 | 21 | 27 | 78 | 87 | 22 | 09 | 75 |
| **9** | 06 | 63 | 02 | 80 | 67 | 17 | 76 | 40 | 16 | 38 | 47 | 51 | 04 | 64 | 32 | 43 | 26 | 42 | 09 | 05 |
| **10** | 23 | 32 | 09 | 37 | 73 | 26 | 99 | 45 | 06 | 67 | 22 | 65 | 79 | 04 | 82 | 31 | 54 | 26 | 39 | 42 |
| **11** | 26 | 78 | 51 | 43 | 82 | 85 | 95 | 22 | 87 | 36 | 20 | 79 | 55 | 70 | 11 | 09 | 70 | 11 | 70 | 98 |
| **12** | 92 | 79 | 45 | 85 | 62 | 10 | 77 | 45 | 40 | 12 | 37 | 31 | 12 | 53 | 99 | 94 | 79 | 35 | 37 | 72 |
| **13** | 14 | 76 | 60 | 56 | 39 | 71 | 24 | 05 | 21 | 23 | 69 | 33 | 95 | 94 | 30 | 01 | 21 | 28 | 20 | 39 |
| **14** | 61 | 32 | 61 | 34 | 33 | 45 | 32 | 65 | 33 | 87 | 25 | 09 | 46 | 21 | 96 | 29 | 83 | 20 | 71 | 61 |
| **15** | 47 | 53 | 35 | 97 | 69 | 05 | 01 | 71 | 98 | 75 | 81 | 66 | 59 | 92 | 19 | 07 | 03 | 74 | 94 | 58 |
| **16** | 94 | 54 | 30 | 03 | 66 | 54 | 58 | 28 | 13 | 04 | 65 | 79 | 76 | 51 | 29 | 14 | 36 | 65 | 99 | 48 |
| **17** | 82 | 87 | 13 | 22 | 35 | 58 | 11 | 14 | 33 | 11 | 74 | 84 | 83 | 09 | 66 | 88 | 83 | 67 | 47 | 84 |
| **18** | 36 | 67 | 26 | 13 | 15 | 96 | 08 | 00 | 46 | 10 | 33 | 36 | 72 | 23 | 10 | 05 | 56 | 82 | 06 | 22 |
| **19** | 97 | 65 | 95 | 60 | 61 | 28 | 66 | 43 | 75 | 66 | 39 | 22 | 30 | 96 | 82 | 72 | 34 | 03 | 93 | 37 |
| **20** | 37 | 14 | 98 | 95 | 57 | 75 | 99 | 03 | 30 | 00 | 47 | 89 | 30 | 74 | 72 | 78 | 82 | 06 | 11 | 93 |
| **21** | 74 | 91 | 65 | 50 | 67 | 44 | 71 | 81 | 43 | 39 | 19 | 36 | 40 | 71 | 24 | 05 | 36 | 42 | 64 | 06 |
| **22** | 50 | 04 | 94 | 15 | 85 | 04 | 68 | 19 | 60 | 64 | 84 | 68 | 61 | 60 | 83 | 43 | 00 | 81 | 42 | 32 |
| **23** | 92 | 89 | 70 | 59 | 85 | 33 | 67 | 62 | 66 | 50 | 17 | 85 | 42 | 68 | 81 | 12 | 42 | 23 | 32 | 67 |
| **24** | 53 | 66 | 53 | 52 | 16 | 44 | 33 | 42 | 10 | 79 | 14 | 80 | 67 | 76 | 98 | 88 | 75 | 85 | 00 | 98 |
| **25** | 94 | 62 | 73 | 33 | 36 | 87 | 39 | 96 | 78 | 05 | 20 | 90 | 02 | 31 | 67 | 80 | 28 | 16 | 91 | 75 |
| **26** | 68 | 57 | 06 | 61 | 56 | 04 | 84 | 05 | 43 | 45 | 74 | 74 | 75 | 66 | 33 | 87 | 47 | 31 | 72 | 81 |
| **27** | 06 | 84 | 59 | 23 | 81 | 29 | 05 | 48 | 37 | 23 | 45 | 71 | 68 | 33 | 49 | 68 | 74 | 09 | 83 | 93 |
| **28** | 85 | 79 | 41 | 68 | 14 | 79 | 11 | 28 | 89 | 05 | 52 | 31 | 49 | 68 | 45 | 06 | 64 | 39 | 12 | 91 |
| **29** | 16 | 39 | 54 | 23 | 35 | 94 | 86 | 85 | 64 | 32 | 22 | 01 | 90 | 81 | 02 | 73 | 39 | 39 | 75 | 83 |
| **30** | 90 | 03 | 87 | 19 | 44 | 10 | 79 | 38 | 89 | 16 | 90 | 65 | 36 | 77 | 50 | 19 | 87 | 33 | 81 | 21 |
| **31** | 18 | 68 | 85 | 69 | 43 | 94 | 89 | 88 | 45 | 61 | 52 | 75 | 23 | 68 | 43 | 82 | 70 | 19 | 94 | 10 |
| **32** | 78 | 92 | 03 | 53 | 60 | 27 | 38 | 11 | 92 | 17 | 90 | 28 | 93 | 55 | 88 | 85 | 42 | 09 | 35 | 34 |
| **33** | 59 | 54 | 68 | 32 | 48 | 10 | 95 | 67 | 39 | 22 | 59 | 88 | 30 | 94 | 01 | 92 | 56 | 47 | 38 | 29 |
| **34** | 09 | 79 | 59 | 43 | 18 | 43 | 76 | 68 | 49 | 33 | 98 | 05 | 32 | 19 | 64 | 02 | 36 | 76 | 65 | 06 |
| **35** | 54 | 37 | 70 | 85 | 29 | 42 | 78 | 70 | 59 | 00 | 06 | 43 | 56 | 68 | 41 | 38 | 74 | 77 | 85 | 90 |
| **36** | 14 | 92 | 32 | 82 | 02 | 15 | 56 | 65 | 78 | 32 | 45 | 61 | 92 | 40 | 54 | 29 | 76 | 50 | 33 | 65 |
| **37** | 12 | 67 | 52 | 09 | 54 | 74 | 44 | 95 | 06 | 83 | 21 | 99 | 29 | 61 | 02 | 98 | 59 | 45 | 72 | 19 |
| **38** | 03 | 61 | 29 | 07 | 74 | 38 | 76 | 38 | 29 | 65 | 43 | 02 | 18 | 75 | 19 | 45 | 69 | 30 | 27 | 81 |
| **39** | 89 | 90 | 89 | 65 | 63 | 99 | 97 | 65 | 55 | 91 | 25 | 25 | 65 | 22 | 01 | 25 | 42 | 31 | 69 | 43 |
| **40** | 44 | 46 | 80 | 83 | 08 | 27 | 66 | 78 | 03 | 56 | 82 | 42 | 45 | 13 | 03 | 67 | 75 | 09 | 49 | 26 |
| **41** | 95 | 60 | 34 | 69 | 60 | 77 | 02 | 97 | 22 | 82 | 38 | 35 | 96 | 74 | 57 | 16 | 47 | 98 | 94 | 38 |
| **42** | 62 | 51 | 42 | 73 | 96 | 38 | 71 | 11 | 43 | 19 | 69 | 79 | 75 | 39 | 17 | 66 | 26 | 22 | 17 | 99 |
| **43** | 24 | 43 | 17 | 93 | 11 | 81 | 58 | 47 | 73 | 19 | 28 | 37 | 50 | 09 | 12 | 39 | 30 | 67 | 76 | 73 |
| **44** | 22 | 05 | 09 | 58 | 26 | 18 | 43 | 53 | 13 | 75 | 25 | 76 | 42 | 92 | 05 | 31 | 57 | 15 | 69 | 51 |
| **45** | 21 | 52 | 96 | 85 | 32 | 49 | 16 | 44 | 79 | 39 | 07 | 43 | 87 | 24 | 34 | 59 | 38 | 72 | 18 | 30 |

**TABLE C.2 Random Permutations of the Numbers 1 to 16**

| | | | | | | | | | | | | | | | |
|---|---|---|---|---|---|---|---|---|---|---|---|---|---|---|---|
| 9 | 4 | 6 | 9 | 8 | 2 | 6 | 12 | 4 | 3 | 12 | 9 | 16 | 13 | 5 | 10 |
| 8 | 11 | 13 | 1 | 14 | 1 | 2 | 5 | 2 | 12 | 5 | 2 | 15 | 7 | 4 | 4 |
| 11 | 15 | 16 | 2 | 5 | 8 | 7 | 13 | 3 | 8 | 13 | 4 | 13 | 1 | 11 | 3 |
| 2 | 5 | 15 | 8 | 13 | 11 | 13 | 15 | 7 | 15 | 15 | 16 | 12 | 16 | 3 | 8 |
| 15 | 9 | 7 | 3 | 9 | 4 | 11 | 3 | 6 | 10 | 3 | 5 | 10 | 9 | 1 | 14 |
| 10 | 1 | 3 | 16 | 12 | 15 | 8 | 2 | 16 | 2 | 2 | 13 | 11 | 11 | 8 | 6 |
| 3 | 13 | 10 | 14 | 15 | 12 | 1 | 10 | 9 | 6 | 10 | 6 | 2 | 2 | 12 | 1 |
| 5 | 7 | 11 | 13 | 6 | 13 | 5 | 11 | 8 | 4 | 11 | 10 | 1 | 12 | 2 | 12 |
| 6 | 2 | 4 | 11 | 2 | 3 | 3 | 4 | 13 | 11 | 4 | 11 | 9 | 6 | 10 | 16 |
| 13 | 10 | 2 | 12 | 4 | 16 | 16 | 9 | 14 | 1 | 9 | 3 | 3 | 10 | 15 | 5 |
| 16 | 14 | 12 | 7 | 16 | 6 | 12 | 1 | 5 | 9 | 1 | 1 | 4 | 15 | 13 | 7 |
| 12 | 6 | 5 | 6 | 3 | 7 | 10 | 7 | 10 | 5 | 7 | 12 | 14 | 3 | 14 | 11 |
| 14 | 3 | 8 | 4 | 10 | 10 | 9 | 16 | 1 | 14 | 16 | 8 | 5 | 4 | 9 | 9 |
| 1 | 16 | 9 | 10 | 11 | 14 | 4 | 14 | 15 | 7 | 14 | 7 | 8 | 5 | 16 | 13 |
| 7 | 8 | 1 | 5 | 1 | 9 | 15 | 6 | 12 | 13 | 6 | 14 | 7 | 8 | 7 | 15 |
| 4 | 12 | 14 | 15 | 7 | 5 | 14 | 8 | 11 | 16 | 8 | 15 | 6 | 14 | 6 | 2 |

| | | | | | | | | | | | | | | | |
|---|---|---|---|---|---|---|---|---|---|---|---|---|---|---|---|
| 16 | 5 | 10 | 16 | 2 | 8 | 5 | 14 | 3 | 6 | 9 | 4 | 14 | 13 | 2 | 2 |
| 13 | 4 | 4 | 13 | 16 | 3 | 1 | 9 | 9 | 16 | 12 | 11 | 15 | 15 | 3 | 15 |
| 3 | 11 | 3 | 3 | 11 | 7 | 9 | 1 | 12 | 13 | 3 | 3 | 4 | 3 | 4 | 8 |
| 9 | 3 | 8 | 9 | 15 | 11 | 10 | 5 | 11 | 12 | 14 | 15 | 7 | 4 | 10 | 16 |
| 5 | 1 | 14 | 5 | 5 | 14 | 13 | 10 | 7 | 15 | 7 | 5 | 5 | 5 | 16 | 4 |
| 10 | 8 | 6 | 10 | 6 | 16 | 14 | 15 | 5 | 11 | 10 | 8 | 10 | 9 | 7 | 9 |
| 6 | 12 | 1 | 6 | 4 | 4 | 8 | 16 | 13 | 14 | 1 | 9 | 11 | 1 | 9 | 13 |
| 1 | 2 | 12 | 1 | 10 | 13 | 12 | 3 | 10 | 8 | 2 | 7 | 13 | 11 | 1 | 5 |
| 2 | 10 | 16 | 2 | 1 | 1 | 15 | 2 | 2 | 10 | 5 | 6 | 12 | 2 | 12 | 14 |
| 15 | 15 | 5 | 15 | 12 | 12 | 2 | 13 | 14 | 2 | 4 | 1 | 9 | 16 | 11 | 10 |
| 8 | 13 | 7 | 8 | 13 | 5 | 7 | 8 | 6 | 4 | 11 | 12 | 1 | 12 | 13 | 6 |
| 4 | 14 | 11 | 4 | 9 | 6 | 4 | 4 | 1 | 9 | 8 | 13 | 6 | 7 | 5 | 3 |
| 11 | 9 | 9 | 11 | 3 | 15 | 16 | 11 | 4 | 1 | 6 | 14 | 3 | 14 | 14 | 7 |
| 14 | 16 | 13 | 14 | 8 | 2 | 11 | 7 | 16 | 7 | 13 | 10 | 16 | 10 | 6 | 11 |
| 12 | 7 | 15 | 12 | 14 | 10 | 6 | 12 | 15 | 5 | 16 | 16 | 2 | 6 | 8 | 1 |
| 7 | 6 | 2 | 7 | 7 | 9 | 3 | 6 | 8 | 3 | 15 | 2 | 8 | 8 | 15 | 12 |

| | | | | | | | | | | | | | | | |
|---|---|---|---|---|---|---|---|---|---|---|---|---|---|---|---|
| 4 | 3 | 7 | 1 | 14 | 6 | 7 | 1 | 13 | 9 | 5 | 6 | 1 | 8 | 13 | 7 |
| 8 | 6 | 2 | 10 | 1 | 7 | 5 | 4 | 15 | 11 | 14 | 15 | 16 | 13 | 16 | 13 |
| 16 | 9 | 14 | 14 | 13 | 2 | 12 | 7 | 9 | 14 | 12 | 10 | 7 | 2 | 4 | 4 |
| 7 | 1 | 5 | 11 | 9 | 5 | 4 | 13 | 5 | 16 | 8 | 16 | 11 | 3 | 6 | 2 |
| 11 | 7 | 10 | 8 | 8 | 9 | 1 | 11 | 16 | 6 | 2 | 7 | 9 | 12 | 8 | 9 |
| 1 | 8 | 8 | 4 | 6 | 10 | 9 | 2 | 6 | 3 | 6 | 12 | 3 | 15 | 3 | 12 |
| 2 | 15 | 1 | 12 | 16 | 1 | 14 | 15 | 2 | 8 | 9 | 9 | 14 | 5 | 2 | 14 |
| 9 | 5 | 4 | 16 | 4 | 12 | 13 | 8 | 7 | 13 | 7 | 11 | 5 | 16 | 7 | 3 |
| 14 | 11 | 11 | 3 | 2 | 15 | 15 | 6 | 10 | 12 | 11 | 5 | 10 | 10 | 10 | 11 |
| 15 | 10 | 15 | 7 | 5 | 11 | 8 | 14 | 14 | 7 | 10 | 14 | 15 | 11 | 12 | 16 |
| 6 | 12 | 3 | 15 | 11 | 4 | 16 | 5 | 12 | 1 | 4 | 2 | 12 | 14 | 14 | 10 |
| 10 | 2 | 12 | 13 | 15 | 8 | 10 | 3 | 11 | 2 | 3 | 3 | 13 | 4 | 9 | 8 |
| 3 | 4 | 16 | 2 | 3 | 14 | 3 | 12 | 8 | 5 | 1 | 1 | 8 | 7 | 1 | 6 |
| 12 | 14 | 13 | 9 | 7 | 16 | 2 | 10 | 1 | 15 | 16 | 8 | 6 | 9 | 15 | 15 |
| 5 | 16 | 6 | 5 | 10 | 3 | 11 | 9 | 4 | 4 | 13 | 4 | 2 | 6 | 11 | 5 |
| 13 | 13 | 9 | 6 | 12 | 13 | 6 | 16 | 3 | 10 | 15 | 13 | 4 | 1 | 5 | 1 |

**TABLE C.3 Random Permutations of the Numbers 1 to 8**

| | | | | | | | | | | | | | | | | | | | |
|---|---|---|---|---|---|---|---|---|---|---|---|---|---|---|---|---|---|---|---|
| 2 | 3 | 1 | 3 | 3 | 6 | 7 | 4 | 3 | 2 | 1 | 1 | 7 | 7 | 5 | 4 | 3 | 3 | 8 | 5 |
| 5 | 5 | 8 | 1 | 8 | 2 | 8 | 6 | 6 | 4 | 8 | 6 | 2 | 4 | 4 | 5 | 7 | 8 | 7 | 6 |
| 7 | 2 | 6 | 7 | 4 | 3 | 6 | 3 | 7 | 6 | 2 | 3 | 8 | 3 | 1 | 7 | 5 | 7 | 3 | 2 |
| 1 | 7 | 5 | 8 | 7 | 8 | 3 | 7 | 8 | 8 | 4 | 4 | 4 | 6 | 7 | 1 | 6 | 4 | 2 | 1 |
| 8 | 4 | 2 | 2 | 6 | 4 | 1 | 8 | 1 | 7 | 5 | 7 | 3 | 1 | 6 | 6 | 1 | 2 | 4 | 7 |
| 4 | 1 | 4 | 5 | 5 | 7 | 5 | 2 | 4 | 5 | 3 | 2 | 1 | 2 | 8 | 2 | 4 | 5 | 6 | 4 |
| 6 | 8 | 7 | 4 | 1 | 1 | 4 | 5 | 5 | 3 | 7 | 8 | 5 | 8 | 3 | 8 | 2 | 6 | 5 | 8 |
| 3 | 6 | 3 | 6 | 2 | 5 | 2 | 1 | 2 | 1 | 6 | 5 | 6 | 5 | 2 | 3 | 8 | 1 | 1 | 3 |
| 5 | 4 | 5 | 4 | 5 | 2 | 5 | 6 | 8 | 2 | 8 | 8 | 8 | 6 | 4 | 6 | 6 | 3 | 8 | 6 |
| 4 | 3 | 8 | 5 | 4 | 6 | 3 | 5 | 3 | 4 | 5 | 6 | 6 | 7 | 6 | 1 | 5 | 7 | 7 | 2 |
| 8 | 7 | 2 | 2 | 3 | 1 | 1 | 4 | 7 | 1 | 1 | 2 | 7 | 1 | 2 | 4 | 4 | 5 | 3 | 3 |
| 6 | 8 | 4 | 1 | 8 | 8 | 2 | 7 | 5 | 7 | 6 | 1 | 4 | 4 | 7 | 5 | 2 | 4 | 1 | 7 |
| 2 | 5 | 3 | 6 | 7 | 3 | 8 | 3 | 1 | 8 | 7 | 4 | 1 | 5 | 1 | 2 | 8 | 8 | 2 | 5 |
| 1 | 2 | 7 | 8 | 6 | 7 | 4 | 1 | 2 | 6 | 3 | 7 | 2 | 8 | 5 | 7 | 1 | 2 | 5 | 8 |
| 3 | 1 | 6 | 3 | 1 | 5 | 6 | 8 | 6 | 5 | 4 | 5 | 3 | 2 | 3 | 8 | 3 | 6 | 6 | 1 |
| 7 | 6 | 1 | 7 | 2 | 4 | 7 | 2 | 4 | 3 | 2 | 3 | 5 | 3 | 8 | 3 | 7 | 1 | 4 | 4 |
| 8 | 6 | 6 | 4 | 4 | 8 | 3 | 4 | 2 | 4 | 3 | 5 | 4 | 8 | 6 | 7 | 6 | 1 | 8 | 2 |
| 3 | 3 | 2 | 6 | 6 | 2 | 8 | 2 | 6 | 3 | 8 | 7 | 8 | 1 | 1 | 8 | 5 | 6 | 7 | 1 |
| 6 | 7 | 3 | 1 | 1 | 4 | 6 | 5 | 1 | 6 | 7 | 6 | 7 | 5 | 4 | 6 | 7 | 5 | 5 | 6 |
| 5 | 8 | 8 | 3 | 7 | 5 | 5 | 8 | 7 | 7 | 1 | 3 | 6 | 4 | 5 | 5 | 3 | 4 | 2 | 8 |
| 7 | 1 | 4 | 5 | 8 | 1 | 2 | 3 | 5 | 8 | 6 | 2 | 3 | 2 | 2 | 1 | 8 | 2 | 4 | 5 |
| 2 | 2 | 7 | 2 | 5 | 6 | 1 | 6 | 3 | 2 | 2 | 8 | 5 | 6 | 7 | 2 | 4 | 7 | 3 | 3 |
| 1 | 5 | 5 | 7 | 3 | 7 | 7 | 7 | 4 | 5 | 5 | 1 | 2 | 3 | 3 | 3 | 2 | 3 | 1 | 7 |
| 4 | 4 | 1 | 8 | 2 | 3 | 4 | 1 | 8 | 1 | 4 | 4 | 1 | 7 | 8 | 4 | 1 | 8 | 6 | 4 |
| 7 | 2 | 1 | 8 | 8 | 7 | 8 | 3 | 1 | 5 | 8 | 2 | 1 | 6 | 7 | 8 | 5 | 1 | 4 | 5 |
| 6 | 5 | 4 | 5 | 4 | 6 | 6 | 2 | 8 | 6 | 6 | 8 | 4 | 5 | 3 | 7 | 3 | 8 | 3 | 8 |
| 4 | 8 | 8 | 2 | 3 | 5 | 3 | 5 | 6 | 8 | 7 | 6 | 2 | 7 | 1 | 4 | 1 | 4 | 2 | 1 |
| 2 | 1 | 2 | 4 | 6 | 3 | 5 | 7 | 3 | 3 | 3 | 3 | 8 | 8 | 4 | 3 | 6 | 2 | 8 | 6 |
| 8 | 7 | 7 | 3 | 7 | 1 | 2 | 6 | 4 | 1 | 2 | 1 | 3 | 1 | 6 | 5 | 8 | 3 | 6 | 3 |
| 5 | 3 | 6 | 6 | 1 | 2 | 7 | 4 | 2 | 7 | 1 | 4 | 6 | 3 | 2 | 2 | 4 | 6 | 7 | 2 |
| 3 | 6 | 3 | 1 | 5 | 8 | 1 | 8 | 5 | 4 | 4 | 5 | 7 | 4 | 8 | 1 | 2 | 5 | 1 | 4 |
| 1 | 4 | 5 | 7 | 2 | 4 | 4 | 1 | 7 | 2 | 5 | 7 | 5 | 2 | 5 | 6 | 7 | 7 | 5 | 7 |
| 3 | 8 | 5 | 5 | 2 | 2 | 6 | 3 | 1 | 6 | 6 | 6 | 3 | 2 | 6 | 1 | 6 | 1 | 3 | 7 |
| 1 | 4 | 8 | 7 | 6 | 6 | 2 | 7 | 3 | 5 | 2 | 5 | 5 | 1 | 7 | 3 | 5 | 7 | 4 | 8 |
| 6 | 5 | 4 | 1 | 1 | 7 | 5 | 6 | 8 | 8 | 5 | 3 | 6 | 7 | 3 | 6 | 8 | 2 | 8 | 1 |
| 4 | 1 | 1 | 6 | 4 | 5 | 1 | 1 | 5 | 7 | 8 | 1 | 7 | 4 | 4 | 8 | 3 | 5 | 5 | 3 |
| 5 | 3 | 3 | 3 | 5 | 3 | 4 | 2 | 4 | 3 | 4 | 2 | 2 | 8 | 5 | 7 | 7 | 6 | 2 | 2 |
| 8 | 2 | 2 | 2 | 8 | 4 | 8 | 4 | 6 | 1 | 1 | 8 | 1 | 5 | 2 | 5 | 1 | 4 | 6 | 4 |
| 2 | 7 | 6 | 8 | 7 | 1 | 7 | 5 | 2 | 2 | 7 | 7 | 4 | 3 | 8 | 4 | 4 | 3 | 7 | 6 |
| 7 | 6 | 7 | 4 | 3 | 8 | 3 | 8 | 7 | 4 | 3 | 4 | 8 | 6 | 1 | 2 | 2 | 8 | 1 | 5 |

## C.4 Design Matrices for Low Current Knowledge

**TABLE C.4 One-half of a $2^3$ Factorial Design**

(Any two factors form a full factorial pattern)

| | | $2^{3-1}$ Design matrix | | | |
|---|---|---|---|---|---|
| | | 1 | 2 | 3 | |
| Test | Run order | 23 | 13 | 12 | Response |
| 1 | | − | − | + | |
| 2 | | + | − | − | |
| 3 | | − | + | − | |
| 4 | | + | + | + | |
| Divisor = 2 | | | | | |
| Effect | | | | | |

**TABLE C.5 One-Sixteenth of a $2^7$ Factorial Design**

(Any two factors form a full factorial pattern)

| | | $2^{7-4}$ Design matrix | | | | | | | |
|---|---|---|---|---|---|---|---|---|---|
| | | 1 | 2 | 3 | 4 | 5 | 6 | 7 | |
| | | 24 | 14 | 15 | 12 | 13 | 23 | 34 | |
| | Run | 35 | 36 | 26 | 56 | 46 | 45 | 25 | |
| Test | order | 67 | 57 | 47 | 37 | 27 | 17 | 16 | Response |
| 1 | | − | − | − | + | + | + | − | |
| 2 | | + | − | − | − | − | + | + | |
| 3 | | − | + | − | − | + | − | + | |
| 4 | | + | + | − | + | − | − | − | |
| 5 | | − | − | + | + | − | − | + | |
| 6 | | + | − | + | − | + | − | − | |
| 7 | | − | + | + | − | − | + | − | |
| 8 | | + | + | + | + | + | + | + | |
| Divisor = 4 | | | | | | | | | |
| Effect | | | | | | | | | |

**TABLE C.6 Design Matrix for a $2^{15-11}$ Pattern**

(Any two factors form a full factorial pattern)

| Test | Run order | 1 | 2 | 3 | 4 | 5 | 6 | 7 | 8 | 9 | 10 | 11 | 12 | 13 | 14 | 15 |
|---|---|---|---|---|---|---|---|---|---|---|---|---|---|---|---|---|
| | | ................................ | | | | | Two-factor interactions | | | | | ................................ | | | | |
| 1 | | − | − | − | − | + | + | + | + | + | + | − | − | − | − | + |
| 2 | | + | − | − | − | − | − | − | + | + | + | + | + | + | − | − |
| 3 | | − | + | − | − | − | + | + | − | − | + | + | + | − | + | − |
| 4 | | + | + | − | − | + | − | − | − | − | + | − | − | + | + | + |
| 5 | | − | − | + | − | + | − | + | − | + | − | + | − | + | + | − |
| 6 | | + | − | + | − | − | + | − | − | + | − | − | + | − | + | + |
| 7 | | − | + | + | − | − | − | + | + | − | − | − | + | + | − | + |
| 8 | | + | + | + | − | + | + | − | + | − | − | + | − | − | − | − |
| 9 | | − | − | − | + | + | + | − | + | − | − | − | + | + | + | − |
| 10 | | + | − | − | + | − | − | + | + | − | − | + | − | − | + | + |
| 11 | | − | + | − | + | − | + | − | − | + | − | + | − | + | − | + |
| 12 | | + | + | − | + | + | − | + | − | + | − | − | + | − | − | − |
| 13 | | − | − | + | + | + | − | − | − | − | + | + | + | − | − | + |
| 14 | | + | − | + | + | − | + | + | − | − | + | − | − | + | − | − |
| 15 | | − | + | + | + | − | − | − | + | + | + | − | − | − | + | − |
| 16 | | + | + | + | + | + | + | + | + | + | + | + | + | + | + | + |
| Divisor = 8 | | | | | | | | | | | | | | | | |
| Effect | | | | | | | | | | | | | | | | |

## C.5 Design Matrices for Moderate Current Knowledge

**TABLE C.7 Some Balanced Incomplete Block Designs**

| | | | | Block number | | | | | | | | | |
|---|---|---|---|---|---|---|---|---|---|---|---|---|---|
| *t* | *b* | *k* | *r* | 1 | 2 | 3 | 4 | 5 | 6 | 7 | 8 | 9 | 10 |
| 3 | 2 | 3 | 2 | A | A | B | | | | | | | |
| | | | | B | C | C | | | | | | | |
| 4 | 2 | 6 | 3 | A | A | A | B | B | C | | | | |
| | | | | B | C | D | C | D | D | | | | |
| 4 | 3 | 4 | 3 | A | A | A | B | | | | | | |
| | | | | B | B | C | C | | | | | | |
| | | | | C | D | D | D | | | | | | |
| 5 | 2 | 10 | 4 | A | A | A | A | B | B | B | C | C | D |
| | | | | B | C | D | E | C | D | E | D | E | E |
| 5 | 3 | 10 | 6 | A | A | A | A | A | A | B | B | B | C |
| | | | | B | B | B | C | C | D | C | C | D | D |
| | | | | C | D | E | D | E | E | D | E | E | E |
| 5 | 4 | 5 | 4 | A | A | A | A | B | | | | | |
| | | | | B | B | C | B | C | | | | | |
| | | | | C | C | D | D | D | | | | | |
| | | | | D | E | E | E | E | | | | | |
| 6 | 3 | 10 | 5 | A | A | A | A | A | B | B | B | C | D |
| | | | | B | B | C | C | D | C | C | D | E | E |
| | | | | E | F | D | F | E | D | E | F | F | F |
| 6 | 4 | 15 | 10 | A | A | A | A | A | A | A | A | | |
| | | | | B | B | B | B | B | B | C | C | | |
| | | | | C | C | C | D | D | E | D | D | | |
| | | | | D | E | F | E | F | F | E | F | | |
| | | | | A | A | B | B | B | B | C | | | |
| | | | | C | D | C | C | C | D | D | | | |
| | | | | E | E | D | D | E | E | E | | | |
| | | | | F | F | E | F | F | F | F | | | |
| 6 | 5 | 6 | 5 | A | A | A | A | A | B | | | | |
| | | | | B | B | B | B | C | C | | | | |
| | | | | C | C | C | D | D | D | | | | |
| | | | | D | D | E | E | E | E | | | | |
| | | | | E | F | F | F | F | F | | | | |

$t$ = Number of factor levels or combinations
$b$ = Block size (number of experimental units per block)
$k$ = Number of blocks required for balanced design
$r$ = Number of replications of each factor level required
A, B, C, D, E, and F represent factor levels or factor combinations.

**TABLE C.8 One-Half of a $2^4$ Factorial Design**

(Any three factors form a full factorial pattern)

| Test | Run order | $2^{4-1}$ Design matrix | | | | | | | Response |
|---|---|---|---|---|---|---|---|---|---|
| | | 1 | 2 | 3 | 4 | 14<br>23 | 24<br>13 | 34<br>12 | |
| 1 | | − | − | − | − | + | + | + | |
| 2 | | + | − | − | + | + | − | − | |
| 3 | | − | + | − | + | − | + | − | |
| 4 | | + | + | − | − | − | − | + | |
| 5 | | − | − | + | + | − | − | + | |
| 6 | | + | − | + | − | − | + | − | |
| 7 | | − | + | + | − | + | − | − | |
| 8 | | + | + | + | + | + | + | + | |
| Divisor = 4 | | | | | | | | | |
| Effect | | | | | | | | | |

**TABLE C.9 One-Half of a $2^5$ Factorial Design**

(Any four factors form a full factorial pattern)

| Test | Run order | $2^{5-1}$ Design matrix | | | | | | | | | | | | | | | Response |
|---|---|---|---|---|---|---|---|---|---|---|---|---|---|---|---|---|---|
| | | 1 | 2 | 3 | 4 | 5 | 12 | 13 | 14 | 15 | 23 | 24 | 25 | 34 | 35 | 45 | |
| 1 | | − | − | − | − | + | + | + | + | − | + | + | − | + | − | − | |
| 2 | | + | − | − | − | − | − | − | − | − | + | + | + | + | + | + | |
| 3 | | − | + | − | − | − | − | + | + | + | − | − | − | + | + | + | |
| 4 | | + | + | − | − | + | + | − | − | + | − | − | + | + | − | − | |
| 5 | | − | − | + | − | − | + | − | + | + | − | + | + | − | − | + | |
| 6 | | + | − | + | − | + | − | + | − | + | − | + | − | − | + | − | |
| 7 | | − | + | + | − | + | − | − | + | − | + | − | + | − | + | − | |
| 8 | | + | + | + | − | − | + | + | − | − | + | − | − | − | − | + | |
| 9 | | − | − | − | + | − | + | + | − | + | + | − | + | − | + | − | |
| 10 | | + | − | − | + | + | − | − | + | + | + | − | − | − | − | + | |
| 11 | | − | + | − | + | + | − | + | − | − | − | + | + | − | − | + | |
| 12 | | + | + | − | + | − | + | − | + | − | − | + | − | − | + | − | |
| 13 | | − | − | + | + | + | + | − | − | − | − | − | − | + | + | + | |
| 14 | | + | − | + | + | − | − | + | + | − | − | − | + | + | − | − | |
| 15 | | − | + | + | + | − | − | − | − | + | + | + | − | + | − | − | |
| 16 | | + | + | + | + | + | + | + | + | + | + | + | + | + | + | + | |
| Divisor = 8 | | | | | | | | | | | | | | | | | |
| Effect | | | | | | | | | | | | | | | | | |

**TABLE C.10 One-Sixteenth of a $2^8$ Factorial Design**

(Any three factors form a full factorial pattern)

| | | $2^{8-4}$ Design matrix | | | | | | | | | | | | | | | |
|---|---|---|---|---|---|---|---|---|---|---|---|---|---|---|---|---|---|
| Test | Run order | 1 | 2 | 3 | 4 | 5 | 6 | 7 | 8 | 12<br>37<br>56<br>48 | 13<br>27<br>46<br>58 | 14<br>36<br>57<br>28 | 15<br>26<br>47<br>38 | 16<br>25<br>34<br>78 | 17<br>23<br>45<br>68 | 24<br>35<br>67<br>18 | Response |
| 1 | | − | − | − | + | + | + | − | + | + | + | − | − | − | + | − | |
| 2 | | + | − | − | − | − | + | + | + | − | − | − | − | + | + | + | |
| 3 | | − | + | − | − | + | − | + | + | − | + | + | − | + | − | − | |
| 4 | | + | + | − | + | − | − | − | + | + | − | + | − | − | − | + | |
| 5 | | − | − | + | + | − | − | + | + | + | − | − | + | + | − | − | |
| 6 | | + | − | + | − | + | − | − | + | − | + | − | + | − | − | + | |
| 7 | | − | + | + | − | − | + | − | + | − | − | + | + | − | + | − | |
| 8 | | + | + | + | + | + | + | + | + | + | + | + | + | + | + | + | |
| 9 | | + | + | + | − | − | − | + | − | + | + | − | − | − | + | − | |
| 10 | | − | + | + | + | + | − | − | − | − | − | − | − | + | + | + | |
| 11 | | + | − | + | + | − | + | − | − | − | + | + | − | + | − | − | |
| 12 | | − | − | + | − | + | + | + | − | + | − | + | − | − | − | + | |
| 13 | | + | + | − | − | + | + | − | − | + | − | − | + | + | − | − | |
| 14 | | − | + | − | + | − | + | + | − | − | + | − | + | − | − | + | |
| 15 | | + | − | − | + | + | − | + | − | − | − | + | + | − | + | − | |
| 16 | | − | − | − | − | − | − | − | − | + | + | + | + | + | + | + | |
| Divisor = 8 | | | | | | | | | | | | | | | | | |
| Effect | | | | | | | | | | | | | | | | | |

**TABLE C.11 Design Matrix for a $2^{16-11}$ Pattern**
(Any three factors form a full factorial design)

| Test | Run order | 1 | 2 | 3 | 4 | 5 | 6 | 7 | 8 | 9 | 10 | 11 | 12 | 13 | 14 | 15 | 16 | Two-factor interactions | | | | | | | | | | | | | | |
|---|---|---|---|---|---|---|---|---|---|---|---|---|---|---|---|---|---|---|---|---|---|---|---|---|---|---|---|---|---|---|---|---|
| 1 | | − | − | − | − | + | + | + | + | + | + | − | − | − | − | + | + | − | − | − | − | + | + | + | + | + | + | − | − | − | − | + |
| 2 | | + | − | − | − | − | − | − | + | + | + | + | + | + | − | − | + | + | − | − | − | − | − | − | + | + | + | + | + | + | − | − |
| 3 | | − | + | − | − | − | + | + | − | − | + | + | + | − | + | − | + | − | + | − | − | − | + | + | − | − | + | + | + | − | + | − |
| 4 | | + | + | − | − | + | − | − | − | − | + | − | − | + | + | + | + | + | + | − | − | + | − | − | − | − | + | − | − | + | + | + |
| 5 | | − | − | + | − | + | − | + | − | + | − | + | − | + | + | − | + | − | − | + | − | + | − | + | − | + | − | + | − | + | + | − |
| 6 | | + | − | + | − | − | + | − | − | + | − | − | + | − | + | + | + | + | − | + | − | − | + | − | − | + | − | − | + | − | + | + |
| 7 | | − | + | + | − | − | − | + | + | − | − | − | + | + | − | + | + | − | + | + | − | − | − | + | + | − | − | − | + | + | − | + |
| 8 | | + | + | + | − | + | + | − | + | − | − | + | − | − | − | − | + | + | + | + | − | + | + | − | + | − | − | + | − | − | − | − |
| 9 | | − | − | − | + | + | + | − | + | − | − | − | + | + | + | − | + | − | − | − | + | + | + | − | + | − | − | − | + | + | + | − |
| 10 | | + | − | − | + | − | − | + | + | − | − | + | − | − | + | + | + | + | − | − | + | − | − | + | + | − | − | + | − | − | + | + |
| 11 | | − | + | − | + | − | + | − | − | + | − | + | − | + | − | + | + | − | + | − | + | − | + | − | − | + | − | + | − | + | − | + |
| 12 | | + | + | − | + | + | − | + | − | + | − | − | + | − | − | − | + | + | + | − | + | + | − | + | − | + | − | − | + | − | − | − |
| 13 | | − | − | + | + | + | − | − | − | − | + | + | + | − | − | + | + | − | − | + | + | + | − | − | − | − | + | + | + | − | − | + |
| 14 | | + | − | + | + | − | + | + | − | − | + | − | − | + | − | − | + | + | − | + | + | − | + | + | − | − | + | − | − | + | − | − |
| 15 | | − | + | + | + | − | − | − | + | + | + | − | − | − | + | − | + | − | + | + | + | − | − | − | + | + | + | − | − | − | + | − |
| 16 | | + | + | + | + | + | + | + | + | + | + | + | + | + | + | + | + | + | + | + | + | + | + | + | + | + | + | + | + | + | + | + |
| 17 | | + | + | + | + | − | − | − | − | − | − | + | + | + | + | − | − | − | − | − | − | + | + | + | + | + | + | − | − | − | − | + |
| 18 | | − | + | + | + | + | + | + | − | − | − | − | − | − | + | + | − | + | − | − | − | − | − | − | + | + | + | + | + | + | − | − |
| 19 | | + | − | + | + | + | − | − | + | + | − | − | − | + | − | + | − | − | + | − | − | − | + | + | − | − | + | + | + | − | + | − |
| 20 | | − | − | + | + | − | + | + | + | + | − | + | + | − | − | − | − | + | + | − | − | + | − | − | − | − | + | − | − | + | + | + |
| 21 | | + | + | − | + | − | + | − | + | − | + | − | + | − | − | + | − | − | − | + | − | + | − | + | − | + | − | + | − | + | + | − |
| 22 | | − | + | − | + | + | − | + | + | − | + | + | − | + | − | − | − | + | − | + | − | − | + | − | − | + | − | − | + | − | + | + |
| 23 | | + | − | − | + | + | + | − | − | + | + | + | − | − | + | − | − | − | + | + | − | − | − | + | + | − | − | − | + | + | − | + |
| 24 | | − | − | − | + | − | − | + | − | + | + | − | + | + | + | + | − | + | + | + | − | + | + | − | + | − | − | + | − | − | − | − |
| 25 | | + | + | + | − | − | − | + | − | + | + | + | − | − | − | + | − | − | − | − | + | + | + | − | + | − | − | − | + | + | + | − |
| 26 | | − | + | + | − | + | + | − | − | + | + | − | + | + | − | − | − | + | − | − | + | − | − | + | + | − | − | + | − | − | + | + |
| 27 | | + | − | + | − | + | − | + | + | − | + | − | + | − | + | − | − | − | + | − | + | − | + | − | − | + | − | + | − | + | − | + |
| 28 | | − | − | + | − | − | + | − | + | − | + | + | − | + | + | + | − | + | + | − | + | + | − | + | − | + | − | − | + | − | − | − |
| 29 | | + | + | − | − | − | + | + | + | + | − | − | − | + | + | − | − | − | − | + | + | + | − | − | − | − | + | + | + | − | − | + |
| 30 | | − | + | − | − | + | − | − | + | + | − | + | + | − | − | + | − | + | − | + | + | − | + | + | − | − | + | − | − | + | − | − |
| 31 | | + | − | − | − | + | + | + | − | − | − | + | + | + | − | + | − | − | + | + | + | − | − | − | + | + | + | − | − | − | + | − |
| 32 | | − | − | − | − | − | − | − | − | − | − | − | − | − | − | − | − | + | + | + | + | + | + | + | + | + | + | + | + | + | + | + |

Divisor = 16
Effect

**TABLE C.12 A $2^3$ Factorial Design in Two Blocks**

(Any three factors form a full factorial pattern)

| Test | Run order | Design matrix 1 | 2 | 3 | $B$ | $1B$ 23 | $2B$ 13 | $3B$ 12 | Response |
|---|---|---|---|---|---|---|---|---|---|
| Block 1 | | | | | | | | | |
| 1 | | − | − | + | + | − | − | + | |
| 2 | | + | − | − | + | + | − | − | |
| 3 | | − | + | − | + | − | + | − | |
| 4 | | + | + | + | + | + | + | + | |
| Block 2 | | | | | | | | | |
| 5 | | + | + | − | − | − | − | + | |
| 6 | | − | + | + | − | + | − | − | |
| 7 | | + | − | + | − | − | + | − | |
| 8 | | − | − | − | − | + | + | + | |
| Divisor = 4 | | | | | | | | | |
| Effect | | | | | | | | | |

$B$ = Blocking variable.

**TABLE C.13 A $2^4$ Design in Two Blocks of 8**

(Any four factors form a full factorial pattern)

| Test | Run order | 1 | 2 | 3 | 4 | $B$ | 12 | 13 | 14 | 1$B$ | 23 | 24 | 2$B$ | 34 | 3$B$ | 4$B$ | Response |
|---|---|---|---|---|---|---|---|---|---|---|---|---|---|---|---|---|---|
| Block 1 | | | | | | | | | | | | | | | | | |
| 1 | | − | − | − | − | + | + | + | + | − | + | + | − | + | − | − | |
| 2 | | + | + | − | − | + | + | − | − | + | − | − | + | + | − | − | |
| 3 | | + | − | + | − | + | − | + | − | + | − | + | − | − | + | − | |
| 4 | | − | + | + | − | + | − | − | + | − | + | − | + | − | + | − | |
| 5 | | + | − | − | + | + | − | − | + | + | + | − | − | − | − | + | |
| 6 | | − | + | − | + | + | − | + | − | − | − | + | + | − | − | + | |
| 7 | | − | − | + | + | + | + | − | − | − | − | − | − | + | + | + | |
| 8 | | + | + | + | + | + | + | + | + | + | + | + | + | + | + | + | |
| Block 2 | | | | | | | | | | | | | | | | | |
| 9 | | − | + | − | − | − | − | + | + | + | − | − | − | + | + | + | |
| 10 | | − | − | + | − | − | + | − | + | + | − | + | + | − | − | + | |
| 11 | | + | + | + | − | − | + | + | − | − | + | − | − | − | − | + | |
| 12 | | − | − | − | + | − | + | + | − | + | + | − | + | − | + | − | |
| 13 | | + | + | − | + | − | + | − | + | − | − | + | − | − | + | − | |
| 14 | | + | − | + | + | − | − | + | + | − | − | − | + | + | − | − | |
| 15 | | − | + | + | + | − | − | − | − | + | + | + | − | + | − | − | |
| 16 | | + | − | − | − | − | − | − | − | − | + | + | + | + | + | + | |
| Divisor = 8 | | | | | | | | | | | | | | | | | |
| Effect | | | | | | | | | | | | | | | | | |

$B$ = Blocking variable.

**TABLE C.14 Design Matrix for a $2^{15-10}$ Pattern in Two Blocks**
(Any three factors form a full factorial pattern)

| Test | Run order | 1 | 2 | 3 | 4 | 5 | 6 | 7 | 8 | 9 | 10 | 11 | 12 | 13 | 14 | 15 | *B* | ..........Two-factor interactions.......... | | | | | | | | | | | | | | |
|---|---|---|---|---|---|---|---|---|---|---|---|---|---|---|---|---|---|---|---|---|---|---|---|---|---|---|---|---|---|---|---|---|
| Block 1 | | | | | | | | | | | | | | | | | | | | | | | | | | | | | | | | |
| 1 | | − | − | − | − | + | + | + | + | + | + | − | − | − | − | + | + | − | − | − | − | + | + | + | + | + | + | − | − | − | − | + |
| 2 | | + | − | − | − | − | − | − | + | + | + | + | + | + | − | − | + | + | − | − | − | − | − | − | + | + | + | + | + | + | − | − |
| 3 | | − | + | − | − | − | + | + | − | − | + | + | + | − | + | − | + | − | + | − | − | − | + | + | − | − | + | + | + | − | + | − |
| 4 | | + | + | − | − | + | − | − | − | − | + | − | − | + | + | + | + | + | + | − | − | + | − | − | − | − | + | − | − | + | + | + |
| 5 | | − | − | + | − | + | − | + | − | + | − | + | − | + | + | − | + | − | − | + | − | + | − | + | − | + | − | + | − | + | + | − |
| 6 | | + | − | + | − | − | + | − | − | + | − | − | + | − | + | + | + | + | − | + | − | − | + | − | − | + | − | − | + | − | + | + |
| 7 | | − | + | + | − | − | − | + | + | − | − | − | + | + | − | + | + | − | + | + | − | − | − | + | + | − | − | − | + | + | − | + |
| 8 | | + | + | + | − | + | + | − | + | − | − | + | − | − | − | − | + | + | + | + | − | + | + | − | + | − | − | + | − | − | − | − |
| 9 | | − | − | − | + | + | + | − | + | − | − | − | + | + | + | − | + | − | − | − | + | + | + | − | + | − | − | − | + | + | + | − |
| 10 | | + | − | − | + | − | − | + | + | − | − | + | − | − | + | + | + | + | − | − | + | − | − | + | + | − | − | + | − | − | + | + |
| 11 | | − | + | − | + | − | + | − | − | + | − | + | − | + | − | + | + | − | + | − | + | − | + | − | − | + | − | + | − | + | − | + |
| 12 | | + | + | − | + | + | − | + | − | + | − | − | + | − | − | − | + | + | + | − | + | + | − | + | − | + | − | − | + | − | − | − |
| 13 | | − | − | + | + | + | − | − | − | − | + | + | + | − | − | + | + | − | − | + | + | + | − | − | − | − | + | + | + | − | − | + |
| 14 | | + | − | + | + | − | + | + | − | − | + | − | − | + | − | − | + | + | − | + | + | − | + | + | − | − | + | − | − | + | − | − |
| 15 | | − | + | + | + | − | − | − | + | + | + | − | − | − | + | − | + | − | + | + | + | − | − | − | + | + | + | − | − | − | + | − |
| 16 | | + | + | + | + | + | + | + | + | + | + | + | + | + | + | + | + | + | + | + | + | + | + | + | + | + | + | + | + | + | + | + |

**TABLE C.14 Design Matrix for a $2^{15-10}$ Pattern in Two Blocks (*Continued*)**

(Any three factors form a full factorial pattern)

| Test | Run order | 1 | 2 | 3 | 4 | 5 | 6 | 7 | 8 | 9 | 10 | 11 | 12 | 13 | 14 | 15 | *B* | ..........Two-factor interactions.......... | | | | | | | | | | | | | | |
|---|---|---|---|---|---|---|---|---|---|---|---|---|---|---|---|---|---|---|---|---|---|---|---|---|---|---|---|---|---|---|---|---|
| Block 2 | | | | | | | | | | | | | | | | | | | | | | | | | | | | | | | | |
| 17 | | + | + | + | + | − | − | − | − | − | − | + | + | + | + | − | − | − | − | − | − | + | + | + | + | + | + | − | − | − | − | + |
| 18 | | − | + | + | + | + | + | + | − | − | − | − | − | − | + | + | − | + | − | − | − | − | − | − | + | + | + | + | + | + | − | − |
| 19 | | + | − | + | + | + | − | − | + | + | − | − | − | + | − | + | − | − | + | − | − | − | + | + | − | − | + | + | + | − | + | − |
| 20 | | − | − | + | + | − | + | + | + | + | − | + | + | − | − | − | − | + | + | − | − | + | − | − | − | − | + | − | − | + | + | + |
| 21 | | + | + | − | + | − | + | − | + | − | + | − | + | − | − | + | − | − | − | + | − | + | − | + | − | + | − | + | − | + | + | − |
| 22 | | − | + | − | + | + | − | + | + | − | + | + | − | + | − | − | − | + | − | + | − | − | + | − | − | + | − | − | + | − | + | + |
| 23 | | + | − | − | + | + | + | − | − | + | + | + | − | − | + | − | − | − | + | + | − | − | − | + | + | − | − | − | + | + | − | + |
| 24 | | − | − | − | + | − | − | + | − | + | + | − | + | + | + | + | − | + | + | + | − | + | + | − | + | − | − | + | − | − | − | − |
| 25 | | + | + | + | − | − | − | + | − | + | + | + | − | − | − | + | − | − | − | − | + | + | + | − | + | − | − | − | + | + | + | − |
| 26 | | − | + | + | − | + | + | − | − | + | + | − | + | + | − | − | − | + | − | − | + | − | − | + | + | − | − | + | − | − | + | + |
| 27 | | + | − | + | − | + | − | + | + | − | + | − | + | − | + | − | − | − | + | − | + | − | + | − | − | + | − | + | − | + | − | + |
| 28 | | − | − | + | − | − | + | − | + | − | + | + | − | + | + | + | − | + | + | − | + | + | − | + | − | + | − | − | + | − | − | − |
| 29 | | + | + | − | − | − | + | + | + | + | − | − | − | + | + | − | − | − | − | + | + | + | − | − | − | − | + | + | + | − | − | + |
| 30 | | − | + | − | − | + | − | − | + | + | − | + | + | − | + | + | − | + | − | + | + | − | + | + | − | − | + | − | − | + | − | − |
| 31 | | + | − | − | − | + | + | + | − | − | − | + | + | + | − | + | − | − | + | + | + | − | − | − | + | + | + | − | − | − | + | − |
| 32 | | − | − | − | − | − | − | − | − | − | − | − | − | − | − | − | − | + | + | + | + | + | + | + | + | + | + | + | + | + | + | + |
| Divisor = 16 | | | | | | | | | | | | | | | | | | | | | | | | | | | | | | | | |
| Effect | | | | | | | | | | | | | | | | | | | | | | | | | | | | | | | | |

*B* = Blocking variable.

**TABLE C.15 Design Matrix for a $2^{7-3}$ Pattern in Two Blocks**

(Any three factors form a full factorial pattern)

| Test | Run order | 1 | 2 | 3 | 4 | 5 | 6 | 7 | *B* | 12 37 56 4*B* | 13 27 46 5*B* | 14 36 57 2*B* | 15 26 47 3*B* | 16 25 34 7*B* | 17 23 45 6*B* | 24 35 67 1*B* | Response |
|---|---|---|---|---|---|---|---|---|---|---|---|---|---|---|---|---|---|
| Block 1 | | | | | | | | | | | | | | | | | |
| 1 | | − | − | − | + | + | + | − | + | + | + | − | − | − | + | − | |
| 2 | | + | − | − | − | − | + | + | + | − | − | − | − | + | + | + | |
| 3 | | − | + | − | − | + | − | + | + | − | + | + | − | + | − | − | |
| 4 | | + | + | − | + | − | − | − | + | + | − | + | − | − | − | + | |
| 5 | | − | − | + | + | − | − | + | + | + | − | − | + | + | − | − | |
| 6 | | + | − | + | − | + | − | − | + | − | + | − | + | − | − | + | |
| 7 | | − | + | + | − | − | + | − | + | − | − | + | + | − | + | − | |
| 8 | | + | + | + | + | + | + | + | + | + | + | + | + | + | + | + | |
| Block 2 | | | | | | | | | | | | | | | | | |
| 9 | | + | + | + | − | − | − | + | − | + | + | − | − | − | + | − | |
| 10 | | − | + | + | + | + | − | − | − | − | − | − | − | + | + | + | |
| 11 | | + | − | + | + | − | + | − | − | − | + | + | − | + | − | − | |
| 12 | | − | − | + | − | + | + | + | − | + | − | + | − | − | − | + | |
| 13 | | + | + | − | − | + | + | − | − | + | − | − | + | + | − | − | |
| 14 | | − | + | − | + | − | + | + | − | − | + | − | + | − | − | + | |
| 15 | | + | − | − | + | + | − | + | − | − | − | + | + | − | + | − | |
| 16 | | − | − | − | − | − | − | − | − | + | + | + | + | + | + | + | |
| Divisor = 8 | | | | | | | | | | | | | | | | | |
| Effect | | | | | | | | | | | | | | | | | |

*B* = Blocking variable.

**TABLE C.16 Four-Factor Design ($3^{4-1}$)**

| Code | Interpretation |
|---|---|
| – | Low level of factor |
| 0 | Middle level of factor |
| + | High level of factor |

| Test | Run order | *A* | *B* | *C* | *D* | Response |
|---|---|---|---|---|---|---|
| 1 | | – | – | – | – | |
| 2 | | + | + | 0 | – | |
| 3 | | 0 | 0 | + | – | |
| 4 | | 0 | + | – | – | |
| 5 | | – | 0 | 0 | – | |
| 6 | | + | – | + | – | |
| 7 | | + | 0 | – | – | |
| 8 | | 0 | – | 0 | – | |
| 9 | | – | + | + | – | |
| 10 | | + | + | – | 0 | |
| 11 | | 0 | 0 | 0 | 0 | |
| 12 | | – | – | + | 0 | |
| 13 | | – | 0 | – | 0 | |
| 14 | | + | – | 0 | 0 | |
| 15 | | 0 | + | + | 0 | |
| 16 | | 0 | – | – | 0 | |
| 17 | | – | + | 0 | 0 | |
| 18 | | + | 0 | + | 0 | |
| 19 | | 0 | 0 | – | + | |
| 20 | | – | – | 0 | + | |
| 21 | | + | + | + | + | |
| 22 | | + | – | – | + | |
| 23 | | 0 | + | 0 | + | |
| 24 | | – | 0 | + | + | |
| 25 | | – | + | – | + | |
| 26 | | + | 0 | 0 | + | |
| 27 | | 0 | – | + | + | |

*Note:* Most of the information on two-factor interactions is clear. The following effects have some confounding: *AB* with *CD*, *AC* with *BD*, *AD* with *BC*. If the interaction effects with factor *A* can be considered negligible, then the remaining interactions are clear.

**TABLE C.17 Design Matrices for Three-Level Factorials**

| Code | Interpretation |
|---|---|
| − | Low level of factor |
| 0 | Middle level of factor |
| + | High level of factor |

Five-factor design ($3^{5-1}$)

| Test | *A* | *B* | *C* | *D* | *E* | Test | *A* | *B* | *C* | *D* | *E* | Test | *A* | *B* | *C* | *D* | *E* |
|---|---|---|---|---|---|---|---|---|---|---|---|---|---|---|---|---|---|
| 1 | − | − | − | − | − | 28 | 0 | 0 | − | − | 0 | 55 | + | + | − | − | + |
| 2 | + | − | 0 | − | − | 29 | − | 0 | 0 | − | 0 | 56 | 0 | + | 0 | − | + |
| 3 | 0 | − | + | − | − | 30 | + | 0 | + | − | 0 | 57 | − | + | + | − | + |
| 4 | 0 | + | − | − | − | 31 | + | − | − | − | 0 | 58 | − | 0 | − | − | + |
| 5 | − | + | 0 | − | − | 32 | 0 | − | 0 | − | 0 | 59 | + | 0 | 0 | − | + |
| 6 | + | + | + | − | − | 33 | − | − | + | − | 0 | 60 | 0 | 0 | + | − | + |
| 7 | 0 | 0 | 0 | − | − | 34 | + | + | 0 | − | 0 | 61 | − | − | 0 | − | + |
| 8 | − | 0 | + | − | − | 35 | 0 | + | + | − | 0 | 62 | + | − | + | − | + |
| 9 | + | 0 | − | − | − | 36 | − | + | − | − | 0 | 63 | 0 | − | − | − | + |
| 10 | 0 | + | + | 0 | − | 37 | + | − | + | 0 | 0 | 64 | − | 0 | + | 0 | + |
| 11 | − | + | − | 0 | − | 38 | 0 | − | − | 0 | 0 | 65 | + | 0 | − | 0 | + |
| 12 | + | + | 0 | 0 | − | 39 | − | − | 0 | 0 | 0 | 66 | 0 | 0 | 0 | 0 | + |
| 13 | + | 0 | + | 0 | − | 40 | − | + | + | 0 | 0 | 67 | 0 | − | + | 0 | + |
| 14 | − | 0 | − | 0 | − | 41 | + | + | − | 0 | 0 | 68 | − | − | − | 0 | + |
| 15 | 0 | 0 | 0 | 0 | − | 42 | 0 | + | 0 | 0 | 0 | 69 | + | − | 0 | 0 | + |
| 16 | − | − | − | 0 | − | 43 | − | 0 | − | 0 | 0 | 70 | 0 | + | − | 0 | + |
| 17 | 0 | − | 0 | 0 | − | 44 | + | 0 | 0 | 0 | 0 | 71 | − | + | 0 | 0 | + |
| 18 | + | − | + | 0 | − | 45 | 0 | 0 | + | 0 | 0 | 72 | + | + | + | 0 | + |
| 19 | + | 0 | 0 | + | − | 46 | − | + | 0 | + | 0 | 73 | 0 | − | 0 | + | + |
| 20 | 0 | 0 | + | + | − | 47 | + | + | + | + | 0 | 74 | − | − | + | + | + |
| 21 | − | 0 | − | + | − | 48 | 0 | + | − | + | 0 | 75 | + | − | − | + | + |
| 22 | − | − | 0 | + | − | 49 | 0 | 0 | 0 | + | 0 | 76 | + | + | 0 | + | + |
| 23 | + | − | + | + | − | 50 | − | 0 | + | + | 0 | 77 | 0 | + | + | + | + |
| 24 | 0 | − | − | + | − | 51 | + | 0 | − | + | 0 | 78 | − | + | − | + | + |
| 25 | − | + | + | + | − | 52 | 0 | − | + | + | 0 | 79 | + | 0 | + | + | + |
| 26 | + | + | − | + | − | 53 | − | − | − | + | 0 | 80 | 0 | 0 | − | + | + |
| 27 | 0 | + | 0 | + | − | 54 | + | − | 0 | + | 0 | 81 | − | 0 | 0 | + | + |

*Note:* All two-factor interactions are clear in this design.

## C.6 Design Matrices for High Current Knowledge

**TABLE C.18** $2^2$ **Factorial Design**

| Test | Run order | 1 | 2 | 12 | Response |
|---|---|---|---|---|---|
| 1 | | − | − | + | |
| 2 | | + | − | − | |
| 3 | | − | + | − | |
| 4 | | + | + | + | |
| Divisor = 2 | | | | | |
| Effect | | | | | |

**TABLE C.19** $2^3$ **Factorial Design**

| | | Design matrix | | | | | | | |
|---|---|---|---|---|---|---|---|---|---|
| Test | Run order | 1 | 2 | 3 | 12 | 13 | 23 | 123 | Response |
| 1 | | − | − | − | + | + | + | − | |
| 2 | | + | − | − | − | − | + | + | |
| 3 | | − | + | − | − | + | − | + | |
| 4 | | + | + | − | + | − | − | − | |
| 5 | | − | − | + | + | − | − | + | |
| 6 | | + | − | + | − | + | − | − | |
| 7 | | − | + | + | − | − | + | − | |
| 8 | | + | + | + | + | + | + | + | |
| Divisor = 4 | | | | | | | | | |
| Effect | | | | | | | | | |

**TABLE C.20** **Design Matrix for a** $2^4$ **Factorial Pattern**

| Test | Run order | 1 | 2 | 3 | 4 | 12 | 13 | 14 | 23 | 24 | 34 | 123 | 124 | 134 | 234 | 1234 | Response |
|---|---|---|---|---|---|---|---|---|---|---|---|---|---|---|---|---|---|
| 1 | | − | − | − | − | + | + | + | + | + | + | − | − | − | − | + | |
| 2 | | + | − | − | − | − | − | − | + | + | + | + | + | + | − | − | |
| 3 | | − | + | − | − | − | + | + | − | − | + | + | + | − | + | − | |
| 4 | | + | + | − | − | + | − | − | − | − | + | − | − | + | + | + | |
| 5 | | − | − | + | − | + | − | + | − | + | − | + | − | + | + | − | |
| 6 | | + | − | + | − | − | + | − | − | + | − | − | + | − | + | + | |
| 7 | | − | + | + | − | − | − | + | + | − | − | − | + | + | − | + | |
| 8 | | + | + | + | − | + | + | − | + | − | − | + | − | − | − | − | |
| 9 | | − | − | − | + | + | + | − | + | − | − | − | + | + | + | − | |
| 10 | | + | − | − | + | − | − | + | + | − | − | + | − | − | + | + | |
| 11 | | − | + | − | + | − | + | − | − | + | − | + | − | + | − | + | |
| 12 | | + | + | − | + | + | − | + | − | + | − | − | + | − | − | − | |
| 13 | | − | − | + | + | + | − | − | − | − | + | + | + | − | − | + | |
| 14 | | + | − | + | + | − | + | + | − | − | + | − | − | + | − | − | |
| 15 | | − | + | + | + | − | − | − | + | + | + | − | − | − | + | − | |
| 16 | | + | + | + | + | + | + | + | + | + | + | + | + | + | + | + | |
| Divisor = 8 | | | | | | | | | | | | | | | | | |
| Effect | | | | | | | | | | | | | | | | | |

**TABLE C.21 Two-Factor Design ($3^2$)**

| Test | *A* | *B* |
|---|---|---|
| 1 | − | − |
| 2 | 0 | − |
| 3 | + | − |
| 4 | − | 0 |
| 5 | 0 | 0 |
| 6 | + | 0 |
| 7 | − | + |
| 8 | 0 | + |
| 9 | + | + |

**TABLE C.22 Three-Factor Design ($3^3$)**

| Test | *A* | *B* | *C* |
|---|---|---|---|
| 1 | − | − | − |
| 2 | 0 | − | − |
| 3 | + | − | − |
| 4 | − | 0 | − |
| 5 | 0 | 0 | − |
| 6 | + | 0 | − |
| 7 | − | + | − |
| 8 | 0 | + | − |
| 9 | + | + | − |
| 10 | − | − | 0 |
| 11 | 0 | − | 0 |
| 12 | + | − | 0 |
| 13 | − | 0 | 0 |
| 14 | 0 | 0 | 0 |
| 15 | + | 0 | 0 |
| 16 | − | + | 0 |
| 17 | 0 | + | 0 |
| 18 | + | + | 0 |
| 19 | − | − | + |
| 20 | 0 | − | + |
| 21 | + | − | + |
| 22 | − | 0 | + |
| 23 | 0 | 0 | + |
| 24 | + | 0 | + |
| 25 | − | + | + |
| 26 | 0 | + | + |
| 27 | + | + | + |

| Code | Interpretation |
|---|---|
| − | Low level of factor |
| 0 | Middle level of factor |
| + | High level of factor |

# Glossary

**accuracy** Degree of variation in individual measurements from the accepted standard value.

**analytic study** A study in which action will be taken on a cause-and-effect system to improve performance of a product or a process in the future.

**attribute data (classification data or count data)** For classification data, the quality characteristic recorded in one of two classes; for count data, the number of incidences of a particular type recorded.

**background variable** A variable that potentially can affect a response variable in an experiment but is not of interest as a factor; sometimes called a *noise variable* or *blocking variable.*

**bias** Amount of deviation of the average of individual measurements from the accepted standard value.

**blocks** Groups of experimental units treated in a similar way in an experimental design; usually defined by background variables.

**capability of a process** A prediction of the individual outcomes or measurements of a quality characteristic from a stable process.

**cause system** A particular combination of causes of variation that affects a quality characteristic.

**cause-and-effect diagram** A diagram that organizes potential causes into general categories, such as methods, materials, machines, and people, and illustrates the common relationships with quality characteristics.

**chunk variable** A variable developed by forming blocks of a certain combination of background variables.

**common causes** Those causes that are inherent in the process over time, affect everyone working in the process, and affect all outcomes of the process.

**composite design** A design for evaluating nonlinear factor effects that is constructed by adding selected factor combinations to two-level factorial designs.

**confirmatory study** A study used to increase the degree of belief that the results hold from "best settings" of factors based on previous experiments.

**confounded effects** The average effect of a factor, or a differential effect between factors (interactions), combined indistinguishably with the effects of other factor(s), block factor(s), or interaction(s).

**conjoint analysis** A marketing application of a planned experiment used to test different features for a new product on potential customers.

**control chart** A statistical tool used to distinguish variation in a process due to common causes and variation due to special causes. Common types are the $C$ chart, $U$ chart, $NP$ chart, $P$ chart, $X$ chart, $\overline{X}$ and $R$ chart, and $\overline{X}$ and $S$ chart.

**control factors** Factors that can be assigned at specific levels, set by those designing the product or process (not directly changed by the customer).

**current knowledge** A priori knowledge essential to planning any change.

**customer** The person or group that receives or uses the outcome of a process.

**data** Documentation of an observation or a measurement.

**degree of belief** The extent to which an experimenter has drawn conclusions from an analytic study; cannot be quantified.

**design matrix** A simple listing of the combinations of factors to be run in a study.

**dot diagram** A plot, using a number line centered at zero, of the estimated effects from a factorial study. Effects clustered near zero on the diagram cannot be distinguished from variation due to nuisance variables and should not be considered important.

**dot-frequency diagram** A graph used to analyze data from a nested experimental pattern. The graph is designed to visually partition the variation in the data among the nested factors.

**effect** The change in the response variable that occurs when a factor or background variable is changed from one level to another. The effect must be further described in the context in which it is used (a linear effect, an interaction effect, etc.).

**enumerative study** A study in which action will be taken on the universe.

**evolutionary operation (EVOP)** A strategy to run experiments on a process in operation. The EVOP studies were designed to be run by operators on a full-scale manufacturing process while the process continued to produce output of satisfactory quality.

**experiment** An analytic study to provide a basis for action.

**experimental pattern** The arrangement of factor levels and experimental units in the design.

**experimental unit** The smallest division of material in an experiment such that any two units may receive different combinations of factors. Examples of experimental units are a part, a batch, one pound of material, an individual person, or a 10-$ft^2$ plot of ground.

**external noise factors** Noise factors relating to the environment in which the product is used or distributed.

**factor** A variable that is deliberately varied or changed in a controlled manner in an experiment to observe its effect on the response variable; sometimes called an *independent variable* or *causal variable.*

**flowchart** A display of the various stages in a process, using different types of symbols, to demonstrate the flow of product or service over time.

**fractional factorial design** An experimental pattern in which only a particular fraction of the factor combinations required for the complete factorial experiment is selected to be run.

**frame** A list of identifiable, tangible units, some or all of which belong to the universe, and any number of which may be selected and studied.

**improvement cycle** An adaptation of the scientific method (consisting of four steps—plan, do, study, act) used to increase a team's knowledge about the product, process, or system and to provide a systematic way of accomplishing change.

**incomplete block design** An experimental pattern in which the number of experimental units in a block is less than the number of combinations of factors and levels.

**interaction** A situation in which the effect that a factor has on the response may depend on the levels of some of the other factors.

**internal noise factors** Noise factors relating to product deterioration with age or use.

**judgment samples** A selection of experimental units and conditions that is based on a judgment other than random selection.

**level** A given value or specific setting of a quantitative factor, or a specific option of a qualitative factor, that is included in the experiment. The levels of a factor selected for study in the experiment may be fixed at certain values of interest, or they may be chosen from many possible values.

**mixture design** A special type of study in which the factors are ingredients that are mixed together. The response variables are thought to depend on the relative proportions of the components of the mixture rather than on the absolute concentrations.

**Model for Improvement** An improvement framework made up of two components. The first component includes three questions: (1) What are we trying to accomplish? (2) How will we know that a change is an improvement? (3) What changes can we make that will result in improvement. The second component consists of the PDSA cycle. Application of the model allows individuals and groups to gain and apply knowledge to the improvement of product, process, or system.

**needs** Objectives defined by the producer that will add value to a society. Customers are defined, and then products and services are designed to satisfy the customers.

**nested or hierarchal design** An experiment to examine the effect of two or more factors in which the same level of a factor cannot be used with all levels of other factors.

**noise factors** Factors that can potentially affect the quality characteristic but cannot be controlled at the design phase.

**nonlinear effect** A change in a response variable that is not linearly related to the corresponding change in a factor.

**nuisance variable** An unknown background variable that can affect a response variable in an experiment, sometimes called a *lurking variable* or *extraneous variable.*

**observational study** A study used to observe the impact of a change by comparing the results before and after the change.

**operational definition** A definition that gives communicable meaning to a concept by specifying how the concept is applied within a particular set of circumstances.

**paired-comparison experiments** A single factor at two levels and one background variable.

**planned experimentation** A collection of methods and a strategy to make a change to a product, process, or system and to observe the effect of that change on one or more quality characteristics with the purpose of helping experimenters gain the most information with the resources available.

**planned grouping** Arrangement of experimental units in blocks.

**precision of a measurement process** Degree of variation in individual measurements of the same item.

**prediction** A declaration of the value or state of some characteristics of a product, process, or system in the future.

**process** A set of causes and conditions that repeatedly come together to transform inputs to outcomes.

**process improvement** The continuous endeavor to learn about the cause system in a process and to use this knowledge to change the process to reduce variation and complexity and to improve customer satisfaction.

**QFD relation diagram** A matrix for relating the quality characteristics to factors that should be addressed in the design of a new product or process. This matrix helps define the current knowledge for a product and is analogous to the cause-and-effect diagram for current knowledge of a process.

**quality characteristic** A trait, preferably measurable, of an input or outcome of a process or a measure of performance of a process used to define quality.

**quality characteristics diagram** A matrix used to define quality by relating needs to quality characteristics.

**random sample** A sample of items selected using a random number table or similar device such that the selection of any particular unit depends purely on chance.

**randomization** The objective assignment of combinations of factor levels to experimental units.

**randomized block design** An experimental pattern in which the size of the block equals the number of combinations of factor and levels (either a single background variable or a chunk variable).

**replication** Repetition of experiments, experimental units, measurements, treatments, etc. as part of the planned experiment.

**response plot** A plot illustrating the relationship between the response and important factors in an experiment.

**response variable** A variable observed or measured in an experiment, sometimes called a *dependent variable*. The response variable is the outcome of an experiment and is often a quality characteristic or a measure of performance of the process.

**run chart** Plot of data in time order.

**screening design** A set of fractional factorial designs used by the experimenter with a low level of knowledge to screen out unimportant factors.

**sequential experimentation** Sequential building of knowledge using the improvement cycle (PDSA) with prediction as the aim.

**special causes** Causes that are not in the process all the time or do not affect everyone, but arise because of specific circumstances.

**stable process** A process in which variation in outcomes arises only from common causes.

**subgrouping** Organizing (classifying, stratifying, grouping, etc.) data from the process in a way that is likely to give the greatest chance for the data in each subgroup to be alike and the greatest chance for data in other subgroups to be different. The aim of rational subgrouping is to include only common causes of variation within a subgroup, with all special causes of variation occurring between subgroups.

**supplier** The person or group that provides an input to the process.

**Taguchi loss function** A quadratic function that relates loss of customer satisfaction with distance from a target of some quality characteristic.

**unit-to-unit noise factors** Variations in the manufacturing process.

**universe** The entire group (e.g., people, material, invoices) possessing certain properties of interest.

**unstable process** A process in which variation is a result of both common and special causes.

**variable data** Measured numerical values of the quality characteristic: a dimension, physical attribute, cost, or time.

**variance components** Estimates of the variation due to each factor in a nested study.

# Index

## ABOUT THE AUTHORS

RONALD D. MOEN, THOMAS W. NOLAN, and LLOYD P. PROVOST have worked as a team in the field of planned experimentation for more than 25 years. Their involvement in planned experiments began in the early 1970s at the U.S. Department of Agriculture, continued as they worked with managers, engineers, and scientists engaged in improving quality, and culminated in their founding of Associates in Process Improvement, a consulting group that assists organizations in making quality a key business strategy.